Epistemology versus Ontology

LOGIC, EPISTEMOLOGY, AND THE UNITY OF SCIENCE

VOLUME 27

Logic, Epistemology, and the Unity of Science aims to reconsider the question of the unity of science in light of recent developments in logic. At present, no single logical, semantical or methodological framework dominates the philosophy of science. However, the editors of this series believe that formal techniques like, for example, independence friendly logic, dialogical logics, multimodal logics, game theoretic semantics and linear logics, have the potential to cast new light on basic issues in the discussion of the unity of science.

This series provides a venue where philosophers and logicians can apply specific technical insights to fundamental philosophical problems. While the series is open to a wide variety of perspectives, including the study and analysis of argumentation and the critical discussion of the relationship between logic and the philosophy of science, the aim is to provide an integrated picture of the scientific enterprise in all its diversity.

For further volumes:
http://www.springer.com/series/6936

Peter Dybjer • Sten Lindström • Erik Palmgren
Göran Sundholm
Editors

Epistemology versus Ontology

Essays on the Philosophy and Foundations of Mathematics in Honour of Per Martin-Löf

Springer

Editors
Peter Dybjer
Department of Computer Science
and Engineering
Chalmers University of Technology
Göteborg
Sweden

Sten Lindström
Department of Historical, Philosophical
and Religious Studies
Umeå University
Umeå
Sweden

Erik Palmgren
Department of Mathematics
Stockholm University
Stockholm
Sweden

Göran Sundholm
Philosophical Institute
Leiden University
Netherlands

ISBN 978-94-007-4434-9 ISBN 978-94-007-4435-6 (eBook)
DOI 10.1007/978-94-007-4435-6
Springer Dordrecht Heidelberg New York London

Library of Congress Control Number: 2012942911

Printed on acid-free paper

Springer is part of Springer Science+Business Media (www.springer.com)

Contents

Introduction

1 Background

The present anthology has its origin in an international conference that was arranged at the *Swedish Collegium for Advanced Studies* (SCAS) in Uppsala, May 5–8, 2009, "Philosophy and Foundations of Mathematics: Epistemological and Ontological Aspects". The conference was dedicated to Per Martin-Löf on the occasion of his retirement.

The aim of the conference was to bring together philosophers, mathematicians, and logicians to penetrate both current and historically important problems in the philosophy and foundations of mathematics. Starting with the pioneering work of Dag Prawitz and Per Martin-Löf in the 1960s, Swedish logicians and philosophers have made important contributions to the foundations and philosophy of mathematics. In philosophy, one has been concerned with the opposition between constructivism and classical mathematics and the different ontological and epistemological views that are reflected in this opposition. Swedish logicians have made significant contributions to the foundations of mathematics, for example, in proof theory, proof-theoretic semantics, and constructive type theory. These contributions have had a strong impact on computer science, for example, through *Martin-Löf's intuitionistic type theory* (*MLTT*), particularly in programming languages and proof assistants.

A suggested basis for the discussions during the conference was current *foundational frameworks* for mathematics. These frameworks give rise to – and some of them purport to solve – important epistemological and ontological problems about mathematics. The dominant, or "mainstream", foundational framework for current mathematics is based on classical logic and set theory with the axiom of choice. Within this framework, a mathematical proof is considered rigorous if it can be formalized in Zermelo-Fraenkel set theory (ZFC), at least in principle. This framework is, however, laden with philosophical difficulties. Set-theoretic Platonism involves a transfinite hierarchy of infinite sets and is associated with serious epistemological problems. Moreover, the encoding of mathematical entities as iterative sets is unnatural and arbitrary and is not in accordance with standard mathematical practice. Two alternative foundational programmes that are actively pursued today are (1) *predicativistic constructivism* and (2) *category-theoretic foundations*:

1. Predicativistic constructivism can be based on MLTT, Aczel-Myhill's constructive set theory, or similar systems. The practice of the Bishop school of constructive mathematics fits well into this framework. Associated philosophical foundations are meaning theories in the tradition of Wittgenstein, Dummett, Prawitz, and Martin-Löf. What is the relationship between proof-theoretical semantics in the tradition of Gentzen, Prawitz, and Martin-Löf, on the one hand, and Wittgensteinian or other accounts of meaning-as-use, on the other hand? What can proof-theoretical analysis tell us about the scope and limits of constructive and (generalized) predicative mathematics? To what extent is it possible to reduce classical mathematical frameworks to constructive ones? Such reductions often reveal computational content of classical existence proofs. Is computational content enough to solve the epistemological questions?
2. Category-theoretic foundations are closely related to a structural vision of mathematics, where the unity of mathematics is obtained through a systematic use of abstraction, enabling transfer of results between seemingly unrelated fields. This has its historical roots in the works of Dedekind and Hilbert. In certain areas of mathematical practice, structuralism is strongly manifested through the extensive use of category theory. Structuralism can also be taken as an ontological view. What is the relationship between structuralism as a way of practicing mathematics and the philosophical point of view that goes under the same name? And what is the relationship between category theory and structuralism as a philosophy of mathematics?

Similarities between type theory and category theory have been observed and exploited since the early 1970s in the form of topos theory. Developments of various forms of "algebraic" set theory (Lawvere, Joyal, Moerdijk) suggest that mathematics can be fruitfully based on category theory. Philosophical critics of category theory as a foundation of mathematics include Feferman and Kreisel. Hellman and Shapiro have proposed alternative foundational frameworks for structuralist mathematics. Such systems may, for instance, be based on modal concepts and may be eliminatory with respect to higher-order mathematical objects. It is of interest to discuss structuralist views in the philosophy of mathematics in connection with developments in topos theory and category theory.

The conference was successful in bringing together a number of internationally renowned mathematicians and philosophers around common concerns. Most of the papers in this collection originate from the Uppsala Conference, but a few additional papers of relevance to the issues discussed there have been solicited especially for this volume.

2 Martin-Löf: Pioneer and Land Clearer

"Per Martin-Löf, incomparable défricheur", Jean-Yves Girard writes in the dedication in the beginning of his contribution to this volume. It is as a pioneer and "land clearer" that we know Martin-Löf – one of the principal clarifiers of the syntax and semantics of constructive mathematics.

Martin-Löf's work began already in the 1960s with an important contribution to the foundations of algorithmic randomness, while studying with Andrei Kolmogorov, forerunner in foundations of both probability and constructivism. With his Ph.D. thesis, "Notes on Constructive Mathematics" 1968, Martin-Löf's long exploration of the foundations of constructive mathematics started. We refer to Göran Sundholm's article for a comprehensive account of the development of this journey and of Martin-Löf's philosophical outlook. We also refer to Dag Prawitz' article that discusses Martin-Löf's philosophy in a more critical vein.

Martin-Löf's work ranges over several fields: not only constructive mathematics, mathematical logic, and philosophy but also statistics and, not least, the foundations of computer science. His influential paper "Constructive Mathematics and Computer Programming" from 1979 explains why "constructive mathematics and computer programming are the same" and why "intuitionistic type theory is a programming language". Shortly after that paper was written, computer scientists in Göteborg and at Cornell University started putting Martin-Löf's ideas into practice. Among other things, several "proof editors" or "proof assistants" based on intuitionistic type theory were developed. These are computer systems, which can be used both for formalizing constructive mathematics and for developing programs satisfying given specifications.

Intuitionistic type theory as presented in "Constructive Mathematics and Computer Programming" had an extensional equality, but did not satisfy the normalization and decidability properties of his original proposals. Martin-Löf considered this unsatisfactory, and in 1986 the theory was again revised. By changing the rules for equality, he made it "intensional" and recovered the normalization and decidability properties. Moreover, he separated an underlying "theory of types", a lambda calculus with dependent types providing a "logical framework", from a "theory of sets". The decidability of the judgements was exploited by many of the proof assistants for the theory, which were soon to be implemented. Intuitionistic type theory had now found its final form, although it has later been extended and modified in many ways by computer scientists to make it more practical. Examples of such languages based on intuitionistic type theory are ALF, Agda, and Epigram, and their impredicative cousins, such as Coq. These are all functional languages with dependent types, which incorporate a number of useful programming language features including general methods for inductive definitions and pattern matching, and modules systems.

During the 1980s, Martin-Löf also dedicated much of his time to the theory of choice sequences (infinite streams in the terminology of computer science). He developed his own approach to domain theory in 1983–1984 based on Dana Scott's neighbourhood systems and information systems. This domain theory models an extension of intuitionistic type theory with partial computations and infinite streams. He also investigated nonstandard analysis and developed in "Mathematics of Infinity" a nonstandard type theory based on the identification of propositions and types. The idea of "formal neighbourhoods" in Martin-Löf's approach to domain theory was also a first step towards the formal topology, which he began developing together with Giovanni Sambin.

3 Contributions to This Volume

We have divided this book into two parts. In the first part, we have collected papers on a more philosophical and nontechnical nature.

3.1 Part I: Philosophy of Logic and Mathematics

Many of the papers in the first part of the book concern various versions of constructivism. Mark van Atten investigates the roots of Brouwerian intuitionism in Kant's philosophy of mathematics. According to Kant, a mathematical entity exists only if it is in principle constructible in human intuition, which by its very nature is finite. On Kant's view, $\sqrt{2}$ exists as a geometrical magnitude, but not as a number. Brouwer, on the other hand, identified the irrational number $\sqrt{2}$ with a potentially infinite sequence of rational numbers. Van Atten discusses the systematic reasons why in Kant's philosophy this identification is impossible.

A special restrictive form of constructivism going back to Leopold Kronecker (1823–1891) is *finitism*, according to which (1) the natural numbers are taken as primitive, (2) all other mathematical objects ought to be constructed by finitary means from the natural numbers, and (3) all statements about numbers ought to be decidable algorithmically in finitely many steps. In his paper, W. W. Tait discusses Skolem's Primitive Recursive Arithmetic (PRA), which can be viewed as a formal realization of Kronecker's finitist programme for arithmetic. PRA does not contain bound variables, induction is over quantifier free expressions only, and definition of functions by primitive recursion is freely allowed. Tait discusses the historical roots of PRA, its relationship to the requirement of Hilbert's programme that metamathematical proofs of consistency be finitary, and its relation to Kant's philosophy of mathematics.

Gödel's two incompleteness theorems obviously had a profound influence on Hilbert's finitist consistency programme for classical mathematics. Wilfried Sieg's contribution aims at a nuanced and deepened understanding of how Gödel's results affected a transformation in proof theory between 1930 and 1934. The starting point is Gödel's announcement of a restricted form of his first incompleteness theorem in

Königsberg on 7 September 1930 and the endpoint is the first consistency proof for full arithmetic that Gentzen completed in December 1934. Sieg argues that Hilbert played a significant role in the development between these points. In his last publication "Beweis des tertium non datur" from 1931, Hilbert responds to Gödel's second incompleteness theorem – without mentioning Gödel (!) – and presented novel directions and concrete problems that needed to be addressed. Gentzen did resolve the problems in surprising new ways, but according to Sieg fully in the spirit of Hilbert's view that true contentual thinking consists in operations on proofs. The main point is that there is genuine continuity between Hilbert's "old" proof theory and the "new" proof theory initiated by Gentzen.

Dag Prawitz discusses the question whether – from an intuitionist point of view – logic is in essence epistemological or ontological. To be more specific: Are the concepts of truth and proof epistemological or ontological? Prawitz is especially concerned with examining Martin-Löf's views on this matter. Mathematical intuitionists are usually taken to view both proof and truth as epistemic concepts. That proof is an epistemic concept seems to be fairly uncontroversial, and since intuitionists define the truth of a proposition as the existence of a proof of it, it seems to follow that truth is also an epistemic notion. In his paper, Prawitz argues that Martin-Löf has changed his mind on this issue. Originally he held to the standard intuitionist view that both proof and truth are epistemic concepts. But this does not seem to be Martin-Löf's present opinion. Martin-Löf makes a distinction between two senses of proof, one ontological and one epistemological. Proofs in the ontological sense he calls *proof-objects*, and proofs in the epistemological sense he calls *demonstrations*. What intuitionists refer to as "proofs" in their explanations of meaning and truth for propositions should properly be understood as proof-objects, not demonstrations. But then it seems to follow that both truth and proof in intuitionism are ontological concepts. Prawitz gives a critical examination of what he takes to be Martin-Löf's reasons for adopting this view.

Two papers in this collection concern Wittgenstein's philosophy of mathematics. Sören Stenlund discusses Wittgenstein's philosophy of mathematics in the "middle period" (roughly 1929–1936). Stenlund is concentrating on the change in Wittgenstein's thinking that takes place mainly in the beginning of the 1930s. By examining certain crucial features in this change, he tries to show that Wittgenstein received decisive impulses and ideas from new developments in mathematics and natural science at the time. Hertz, Hilbert, and Einstein are important sources of inspiration. Stenlund argues that these ideas affected not only Wittgenstein's thinking about mathematics but also his thinking about language and the nature of philosophy in general.

Juliet Floyd discusses a series of remarks that Wittgenstein wrote on July 30, 1947. It begins with the remark "Turing's 'machines': these machines are humans who calculate. And one might express what he says also in the form of games". Immediately after his remark about "Turing's machines", Wittgenstein formulates what he calls a "variant" of Cantor's diagonal proof. Wittgenstein's argument has the form of a language game that involves a rule, which is circular and cannot be executed. Floyd presents and assesses Wittgenstein's variant, claiming that it is a distinctive form of proof, and an elaboration rather than a rejection of Turing or Cantor.

In Sambin's article, we find a discussion on how "real" and "ideal" are treated differently in relation to constructive and classical mathematics. He shows that the communication between the two is much stronger in the constructive stance and exemplifies with constructive topology and his and Maietti's so-called Minimalist Foundations.

Jan Smith's paper is a contribution to evolutionary epistemology. The question discussed here is whether our ability to reason logically and develop mathematics can be explained in evolutionary terms. Smith also poses the question, "Given that evolution explains why there is mathematics, can it single out any of the views on the foundations of mathematics as the correct one?" His answer is negative, "Although Formalism, Platonism and Intuitionism have very different explanations of mathematics, it seems to me to be possible for a devotee of any of them to argue for an evolutionary origin".

3.2 *Part II: Foundations*

We now turn to the contributions to the second part of the book.

The first constructive versions of Zermelo-Fraenkel Set Theory arose in the work of Harvey Friedman and John Myhill. Myhill introduced a theory called Constructive Set Theory intended to be able to formalize Bishop-style constructive mathematics. It was based on intuitionistic logic and avoided impredicative construction principles such as power sets and full separation. A convincing argument for it being truly constructive and predicative (in the generalized sense) was given by Peter Aczel, who constructed a model of the theory inside Martin-Löf's Intuitionistic Type Theory (MLTT). Based on this model, he developed the axioms further to what is now commonly called Constructive ZF (CZF), or Aczel-Myhill Set Theory, and taken to be one of the standard systems for formalizing constructive mathematics. Aczel's model, viewing sets as (infinitary) trees and their equality governed by bisimilarity, has proved very fruitful and flexible. It gives a standard method for comparing type theories and set theories. Rathjen's contribution "Constructive Zermelo-Fraenkel Set Theory, Power Sets and the Calculus of Constructions" deals with the extension of CZF with power sets and its relation to a type theory akin to Coquand and Huet's Calculus of Constructions, as well as an extension of Kripke-Platek set theory.

Väänänen argues in the contribution "Second Order Logic, Set Theory and Foundations of Mathematics" that the difference between classical second-order logic and set theory is, contrary to widespread belief, illusory when it comes to questions of categoricity.

The view that Category Theory gives a good realization of structuralist philosophy of mathematics has been well defended by several authors such as Colin McLarty and Steve Awodey. A first case demonstrating this was Lawvere's

Elementary Theory of the Category of Sets (ETCS) from 1964. In the contribution by Palmgren, a constructive version of ETCS is investigated. It is related to Bishop's informal set theory that was presented by Bishop in 1967. Bishop was essentially taking a type-theoretic view of sets, namely that a set is a type together with a defined equivalence relation, that is, in modern parlance a *setoid.* The setoids of a type theory behaves largely as sets from a category-theoretic point of view, depending of course on axioms on the underlying type theory. A perhaps surprising feature of MLTT is that the *axiom of choice is a theorem* of the theory. It is a triviality when understood that it refers to choices over types and that the existential quantifier is interpreted by the Σ-type construction. However, to make this version of choice valid for setoids, one needs a uniform method of introducing minimal equivalence relations on types. This is what the identity types of MLTT do. These constructions do more than was probably intended: the identity types of standard (intensional) MLTT introduce a natural groupoid structure on types, as was discovered by Hofmann and Streicher in 1994. A groupoid is the category where all arrows are invertible. It can be considered as a common generalization of a group and an equivalence relation and figures prominently in Homotopy Theory, as groupoids of paths in space. Around 2005–2006 Steve Awodey and Vladimir Voevodsky discovered, independently, a deeper and surprising relation between homotopy theory and MLTT. Awodey gives in his contribution "Type Theory and Homotopy Theory" an overview of this active field of research.

There are at least two major classes of models for constructive systems. One may be called the realizability class, which includes the Brouwer-Heyting-Kolmogorov interpretation, and which often give computational interpretation to the system. Another may loosely be called the forcing class of models, which include Kripke- and Beth-semantics and more generally sheaf models. Models of this class are often devoid of direct computational sense, but are instead capable of expressing epistemic states. However, Coquand and Jaber show in their contribution that certain forcing models of type theory can indeed be given natural computational interpretations.

Girard started his programme Geometry of Interaction to give interpretations of dynamical aspects of logic, such as Gentzen's cut-elimination, in terms of operator algebras. In his contribution "Normativity in logic", he considers the application of this method to an example of computational complexity.

The meaning explanations of MLTT are discussed in Dybjer's contribution to this volume and are considered from a program testing point of view. A type-theoretic judgement is interpreted as a conjecture about program correctness that can be tested by computing the output of the program for all inputs that are possible to generate. Furthermore, testing of higher-order functions becomes an interactive process akin to game semantics.

Logical semantics of natural languages was pioneered by Frege and further developed in detail by Montague using an extension of classical simple type theory, Montague's intensional logic. Aarne Ranta started, as a student of Martin-Löf, to investigate what MLTT with its rich structure of contexts and dependent types could say more about natural languages. Notable successes include the treatment of anaphora and discourses. Ranta has since then developed a type theory, the Grammatical Framework (GF), adapted to describe languages and to make machine translations between multiple languages. In his contribution, he gives an introduction to machine translation and GF.

The Lorenzen-Prawitz inversion principle for natural deduction gives a way of deriving the elimination rules from given introduction rules. This is an important principle used as well in Martin-Löf type theory. In Setzer's contribution to this volume, he considers an inversion principle that goes in the opposite direction, generating introduction rules from elimination rules. For the so-called coinductively defined types – a typical example is a potentially infinite stream of data – this is shown to be the natural approach. Meaning explanations are given for this new class of types.

Acknowledgments

We thank the Swedish Collegium for Advanced Study (SCAS) in Uppsala and the Centre for Interdisciplinary Mathematics at Uppsala University for invaluable organizational support in connection with the Uppsala Conference held in May 2009 dedicated to Per Martin-Löf. We are especially grateful to the director of SCAS, Professor Björn Wittrock, for supporting the project throughout and to the people at SCAS for all their efforts in connection with the conference. We are grateful to Professor Dag Prawitz (Stockholm University), Professor Sören Stenlund (Uppsala University), and Professor Viggo Stoltenberg-Hansen (Uppsala University) for scientific advice and support. We also wish to thank Anna Eriksson-Trenter, Anton Hedin, Tore Hållander, Kristina Lindgren, and Olov Wilander for all their help with organizational matters. We gratefully acknowledge the financial support from the Swedish Research Council, The Royal Swedish Academy of Sciences, The Royal Swedish Academy of Letters, History and Antiquities, and the Departments of Mathematics and the Departments of Philosophy in Stockholm and Uppsala.

Göteborg	Peter Dybjer
Umeå	Sten Lindström
Stockholm	Erik Palmgren
Leiden	Göran Sundholm

On the Philosophical Work of Per Martin-Löf

Göran Sundholm

Per Martin-Löf began his work on logic and foundational issues in 1966 with a definition of the notion of a random sequence that has become classic. It was continued with the doctoral dissertation *Notes on Constructive Mathematics* that was written in 1968. Martin-Löf spent the academic year 1968–1969 in the United States, first at Chicago where, in December 1968, W. A. Howard gave him a copy of the handwritten manuscript *The formulae-as-types notion of construction* (that was subsequently published (1980) in the Curry Festschrift). In it an isomorphism à la Curry is established between axiomatic Hilbert-style systems for predicate logic and arithmetic and matching calculi in combinatory logic. The correspondence struck Martin-Löf as being of profound significance and he was determined to understand it fully. His first contribution to this end, as pointed out by Howard in a note added to the published paper, was to carry over the idea from Howard's Hilbert-style calculi to the framework of Gentzen's Natural Deduction that was known to him from Dag Prawitz's (1965) dissertation. In Martin-Löf's formulation, proof-theoretic reductions of natural-deduction derivations correspond to conversions of terms in an enriched lambda calculus. The now customary "Curry-Howard" isomorphism between natural-deduction derivations and the terms of a matching lambda-calculus (rather than the combinatory logic that was used by Howard) was then written up in a paper on *Infinite terms and a system of natural deduction* that was circulated in March 1969. Armed with the insight that natural-deduction derivations and lambda-calculus terms are essentially equivalent, Martin-Löf began a search for the optimal way of proving normalization results for systems of natural deduction. His investigation of iterated generalized inductive definitions, which was completed by March 1970, carried over the computability method of W. W. Tait (1967) to proofs of normalization for natural-deduction derivations, and the paper was presented at the *Second Scandinavian Logic Symposium* at Oslo in June 1970. Kreisel (1975, p. 177, footnote 6) reports that when travelling – with Prawitz and Martin-Löf – by train to Oslo, he mentioned J.-Y. Girard's very recent, but still unpublished, work on how to extend Gödel's *Dialectica* interpretation to second-order arithmetic ("Analysis") and gave Girard's manuscript to Martin-Löf, who already at the Oslo meeting convinced himself that Girard's insight would carry over into a normalization theorem also for second-order logic, giving a procedure of normalization, rather than completeness of the cut-free sequent-calculus rules for second-order logic (as in Prawitz's (1967) elegant demonstration of Takeuti's conjecture). After the Oslo meeting, Girard, Martin-Löf, and Prawitz all gave normalization proofs for second-order logic that were published in its *Proceedings* (Fenstad (1971)), and Girard (1971, p. 64) confirms that "les remarques de M. Martin-Löf sur la possibilité d'une démonstration de normalisation de l'Analyse ... on été déterminantes pour la suite de ce travail". Martin-Löf quickly extended the computability approach also to the intuitionistic simple theory of types (1970c). A characterization of the provable well-orderings of the theory of species also belongs to this period (1970d) of great creativity. Finally, in the autumn of 1970, the first system of Intuitionistic Type Theory was designed and presented in a seminar lecture at Stockholm as a deliberate attempt to extend the computability approach that had so amply demonstrated its worth at the Oslo conference to even stronger systems. Its main building blocks were Gentzen's natural-deduction style of formalization, with proof-theoretic reduction steps from Prawitz (1965), Gödel's *Dialectica* system T (1958), the Curry-Howard isomorphism, and Tait's (1967) computability method for normalization proofs. The written presentation, in a preprint dated February, revised October (1971a), was submitted for publication to *Acta Mathematica*. Martin-Löf's invited lecture at the Bucharest fourth LMPS conference in 1971b was also devoted to his novel intuitionistic type theory. However, both submissions were withdrawn from publication pursuant to Girard's discovery that the extreme impredicativity of the system allowed for the derivation of a version of Burali-Forti's paradox. Martin-Löf's revised version of his Intuitionistic Type Theory from 1972 was finally published in the proceedings of the conference at Venice that was held in 1995 to commemorate *Twenty-five Years of Constructive Type Theory*, and the proceedings of the 1973 Bristol Logic Colloquium contain the first published presentation of a – predicative – version of the type theory.

The task that faced a mathematical logician from Frege's days until 1930 was a challenging one: to graft the Fregean notion of a formal system onto the Aristotelian conception of demonstrative science. That is, one should design a sizeable formal system with clearly delimited axioms and rules of inference. The system should be adequate for the needs of analysis after the then novel fashion of Weierstraß and Dedekind. In particular, it should admit of classical logic (as well as impredicative quantification). The axioms and (primitive) rules of inference should be rendered immediately evident from the meaning explanations for the (primitive signs of the) formal language of the formal system in question. Frege, Whitehead-Russell, Lesniewski, and the early logical works of Curry, Church, and Quine, are good examples here; as is well known, the early logicist attempts were not successful, owing to the use of axioms, such as Reducibility, that were not rendered evident by the relevant meaning-explanations. Carnap's contribution to the famous conference on the philosophy and foundations of mathematics at Königsberg in November 1930 can be seen as the last stand of Logicism. Thereafter, it seemed clear that Logicism was no longer a live option. This meant that the foundations of mathematics were faced with the two horns of a dilemma: either we retain classical logic in a mathematical object-language, but give up hope for meaning explanations, or we insist on retaining a contentual language with meaning explanations, but have to jettison classical logic. Hilbert in his programme chose the first option, whereas the second one was preferred by Brouwer and elaborated by Heyting in his articles on the formalization and interpretation of intuitionist logic.

Martin-Löf's first logical writings belong to the metamathematical paradigm. The formal languages and systems dealt with are principally objects of metamathematical study, and in that spirit normalization *theorems* are established for the early versions of type theory and other systems. However, in 1974, stimulated by reading Wittgenstein, as well as by listening to Michael Dummett's lecture *The Philosophical Basis of Intuitionistic Logic* at the Bristol Logic Colloquium 1973, Martin-Löf turned to the theory of meaning, and thereby brought type theory within the contentual approach in logic. Peter Hancock's words in the lectures that marked the meaning-theoretical turn (1975b, p. 13) speak with quiet confidence and could be used unchanged also today:

> Especially in its later stages, Martin-Löf's work has been developed with the principal aim that it should admit a detailed and coherent semantics. We are not concerned here, except by implication, to subject other languages to a destructive criticism. We are ignorant of a comparable project that has been carried through for a different language. So we do not have to say at this stage why our account is to be preferred to another. You will just have to make up your mind about that if and when there *is* another account.

He has devoted himself ever since to the realization of the second horn of the above dilemma in the foundations of mathematics with a project that is comparable to that of Frege's *Grundgesetze*, but without classical logic and impredicativity: design a full-scale formal language, with explicit meaning-explanations, that is adequate for the needs of mathematical analysis in the style of Errett Bishop's (1967) revolutionary constructive presentation. In the first two papers after his meaning-theoretical turn, Martin-Löf made an experimental attempt to view the natural-deduction *elimination* rules as basic, or meaning giving. A trace of this approach can be found in the sole reference (1991, p. 280) to Martin-Löf in Michael Dummett's William James lectures (that were given at Harvard in 1976, shortly after Martin-Löf's course at All Souls in Michaelmas Term 1975, and was later published as Dummett (1975)). The paper (1975a), written jointly with Peter Hancock, on primitive recursive arithmetic, has attracted a measure of attention also as a contribution to the proper interpretation of mathematics in Wittgenstein's *Tractatus*, whereas the lecture notes (1975b) gave meaning explanations for the language of type theory. The experiment with the elimination rules as basic was soon abandoned, though, and the lecture notes left uncompleted, after it became clear that the approach could not be carried through. In Martin-Löf's (1979) lecture at LMPS 6 in Hannover, the pattern of meaning explanations based on the introduction-rule constructors that yield canonical proof-objects is introduced. These explanations have essentially remained constant ever since, particularly in the Padova lecture notes from 1980 by Giovanni Sambin that were published by Bibliopolis. (The only notable change from the expositions in 1979 and 1980 lies in Martin-Löf's current use of *intensional* identity rather than the previous extensional one.) In this mature form, Martin-Löf's constructive type theory, from the point of view of the foundations of mathematics, constitutes an impressive, mathematically precise rendering of the BHK meaning-explanations that were first given by Arend Heyting in 1930.

Around this time, that is, the late 1970s, Martin-Löf also began a study of the philosophy of Edmund Husserl and from then on the phenomenological perspective has been a rich source of inspiration. In conversation Martin-Löf has indicated that he regards his syntactic-semantic method of logical exploration as a version of phenomenology. To this time belongs also a second period of experimentation, in which Martin-Löf attempted to avoid the use of type-theoretical abstractions, and instead explored the alternative of working within predicate logic, eschewing the use of proof-objects and the form of judgement a: proof(A), working instead with the predicate logic form A true as principal form of judgement. This work culminated in a set of influential lectures that were given in 1983 at Siena, on which basis a compact course was published, and where considerable systematic effort was spent on relating the work to issues in the philosophy of logic. With the creation of the higher type structure, Martin-Löf returned to a type-theoretical formulation of his ideas. The full language of type theory, now using both sets and types, has been in place since 1986, when it was first presented in a lecture at Edinburgh. A lecture series given at Florence in 1987 gave meaning explanations also for the higher type theory. It was further extended in 1992 with a calculus of explicit substitutions in order to make the treatment fully formal. In 1993, Martin-Löf gave a semester-long series of lectures on the *Philosophical Aspects of Type Theory* at Leiden, in which he presented the meaning explanations in full detail and dealt with a number of topics from the philosophy of logic and language.

Martin-Löf's mature philosophical outlook is characterized by three main tenets that make it unique among contemporary philosophical positions within the foundations of mathematics. First, and most significantly, constructive type theory is an *interpreted formal language*. The importance of this cannot be stressed firmly enough. Today, as a rule, the metamathematical "expressions" employed in mathematical logic are mere objects of study, but do not express. On the contrary, they are objects that may serve as referents of real expressions. In constructive type theory, on the other hand, the expressions used are real expressions that carry meaning. In a nutshell, the language is endowed with meaning by turning the proof-theoretic reductions into steps of meaning explanation. Just like the formulae of Frege's ideography, or of the language of *Principia Mathematica*, the type-theoretic formulae are actually intended to say something. They do not essentially serve as objects of metamathematical study. This, of course, in no way precludes Martin-Löf's contentual system from being studied metamathematically.

Secondly, Martin-Löf has restored the notion of *judgement* to pride of place in logic. In the metamathematical tradition, the same well-formed formulae serve in different roles: wff's are built up from wff's that have been generated earlier, for instance, when φ and χ are wff's, then so is $(\varphi \supset \chi)$. Furthermore, a wff φ may be a derived theorem, that is, the end-formula of a closed derivation (with no open assumptions). From Martin-Löf's contentual point of view, a proposition A is a set of (canonical) proof-objects. Such propositions serve as building blocks for more complex propositions. The contentual role of a derivational end-formula, however, is not propositional; here we do not have just an occurrence of the proposition A, but an *assertion* that

proposition A is true.

Martin-Löf's propositions are given via *proof conditions*, that is, to each proposition A there is associated a type proof(A) that is explained in terms of how canonical proof-objects for A may be formed and when two such proof-objects are the same. Truth of propositions is explained by the "truth-maker" analysis

proposition A is true = proof(A) exists,

where the existence in question is the constructive Brouwer-Weyl notion that is explained in terms of possession of an instance. It should be noted that these are proofs of *propositions*. Previously in the history of logic, all proving (better: all *demonstration*) took place at the level of assertions, and not at the level of propositions (that are traditionally seen as unasserted contents of theorems). Such proofs of propositions were first considered in Heyting's seminal writings from the early 1930s. Martin-Löf's crucial notion of a judgement is explained in terms of an *assertion condition* that lays down what one has to know in order to have the right to make the judgement in question. Thus for instance, in order to have the right to make the judgement that the proposition A&B is true, one has to have exhibited a proof-object c that either is of, or evaluates to, the form <a, b>, where a is a proof-object for the proposition A and b is a proof-object for the proposition B. This reintroduction of judgements into logic, with the concomitant distinction between judgements and propositions, also leads to a crucial distinction between (epistemic) *demonstrations* of judgements, and *proofs*, in the sense of proof-objects, of propositions. This distinction between propositions and judgements, and the ensuing distinction between proofs of propositions and demonstrations of judgements, also brings about a corresponding distinction between (epistemic) *inference* from premise to conclusion judgements and relations of *consequence* between antecedent and consequent propositions.

Finally, the emphasis on judgement and the acquaintance with the works of Husserl has led Martin-Löf to an uncommon epistemological perspective. In contemporary analytical philosophy, epistemology takes a very "factual", almost ontological stance. Knowledge is invariably seen from a third person perspective as an ontological state that obtains in the world and makes true propositions such that *agent A knows proposition p*. Here the main concern is not what it is to know something oneself, but rather what it is for *someone else* to know something. Thus, on the linguistic level, one is concerned with the meaning of the locution "A knows p" rather than with "I know p". Martin-Löf's approach to meaning, on the other hand, is squarely *first person*. To him every assertion by means of an utterance of a declarative sentence contains an *implicit* first-person knowledge claim. Martin-Löf explains a declarative by means of an "assertion condition" that lays down *what one has to know* in order to have the right to make the assertion by means of an utterance of the declarative in question. Accordingly, the *legitimacy* of the counter-question *How do you know that?* serves as a criterion by which assertoric uses of declarative sentences can be recognized.

The effect of this can be seen, for instance, in Moore's paradox concerning an assertion by means of:

> "It is raining, but I do not believe it."

As is well known, the use of the present tense and the first person is essential here. No paradox results in the imperfect: "It was raining, but I did not believe it." Similarly, non-assertoric occurrences of the crucial sentence are not problematic: "If it is raining, but I do not believe it, then I will get wet when I go outside." Also any third-person assertion by me, of "It is raining, but X does not believe it," where X is placeholder for the name of a person, is not paradoxical, even though we may choose X = Göran Sundholm here. Only the first person poses problems, owing to the implicit first-person "I know" that is contained in every assertion and that contradicts the second clause of the Moorean assertion. Martin-Löf's insistence on this kind of first-person knowledge comes out time and again in his philosophy, for instance, in the quotation above from 1975b, but perhaps most clearly in his pertinent request to the reader in the first full contentual exposition of constructive type theory (1979, p. 166):

> For each of the rules of inference, the reader is asked to try to make the conclusion evident on the presupposition that he knows the premises. This does not mean that further verbal explanations are of no help in bringing about an understanding of the rules, only that this is not the place for such detailed explanations. But there are also certain limits to what verbal explanations can do when it comes to justifying axioms and rules of inference. In the end, everybody must understand for himself.

In a series of published philosophical lectures from 1985 to 2004, Martin-Löf has explored central notions within the philosophy of logic, such as judgement, evidence, rightness, and knowledge, as well as in the philosophy of mathematics, for instance concerning the Axiom of Choice, and spelled out consequences for his views on metaphysics and epistemology. He has, however, been a frequent invited speaker at conferences and the list of topics covered in unpublished material is long: it comprises Frege's distinction between *Sinn* and *Bedeutung*, intensionality of objects, Tarski's truth definition and the notion of a model for type theory, predicativity in mathematics, propositions versus contents, categories, and general methodology in the philosophy of logic. In recent years, since his Gödel lecture *The two layers of logic* at the annual meeting of the Association for Symbolic Logic at Montréal in 2006, Martin-Löf's philosophical work has been directed to the question, Logic, epistemological or ontological? that also gave the title for his lecture at the Uppsala meeting.

Works by Per Martin-Löf

Martin-Löf has often been reticent in publishing his papers and sometimes there is a considerable delay between the distribution of the preprint version and its ultimate publication, if any. I therefore cite his works after the year of distribution rather than of publication.

Martin-Löf, Per. 1966. The definition of random sequences. *Information and Control* 9(6): 602–619.

Martin-Löf, Per. 1968. *Notes on constructive mathematics* (doctoral dissertation). Stockholm: Almqvist & Wiksell, 1970.

Martin-Löf, Per. 1969. Infinite terms and a system of natural deduction. *Compositio Mathematica* 24 (1972): 93–103.

Martin-Löf, Per. 1970a. Hauptsatz for the intuitionistic theory of iterated inductive definitions. In *Proceedings of the 2nd Scandinavian logic symposium*, ed. J.E. Fenstad, 179–216. Amsterdam: North-Holland Publishing Company, 1971.

Martin-Löf, Per. 1970b. Hauptsatz for the intuitionistic theory of species. In *Proceedings of the 2nd Scandinavian logic symposium*, ed. J.E. Fenstad, 217–233. Amsterdam: North-Holland Publishing Company, 1971.

Martin-Löf, Per. 1970c. Hauptsatz for intuitionistic simple type theory. In *Logic, methodology and philosophy of science IV, Proceedings of the fourth international congress for logic, methodology and philosophy of science, Bucharest, 1971*, ed. P. Suppes, L. Henkin and A. Joja, 279–290. Amsterdam: North-Holland Publishing Company, 1973.

Martin-Löf, Per. 1970d. A construction of the provable wellorderings of the theory of species. In *Logic, meaning and computation*, ed. A. Anthony Anderson and M. Zeleny, 343–352. Dodrecht: Kluwer, 1997.

Martin-Löf, Per. 1971a. An intuitionistic theory of types, unpublished preprint.

Martin-Löf, Per. 1971b. On the strength of intuitionistic reasoning, preprint of a contribution to the Symposium on Perspectives in the Philosophy of Mathematics, *LMPS* IV, Bucharest, September 1971.

Martin-Löf, Per. 1972. An intuitionistic theory of types. In *Twenty-five years of constructive type theory*, ed. G. Sambin and Jan J. Smith, 127–172. Oxford: Clarendon Press, 1998.

Martin-Löf, Per. 1973. An intuitionistic theory of types: Predicative part. In *Logic colloquium'73*, ed. H.E. Rose and J. Shepherdson, 73–118. Amsterdam: North-Holland Publishing Company, 1975.

Martin-Löf, Per. 1975a. Syntax and semantics of the language of primitive recursive functions (written jointly with Peter Hancock), unpublished preprint.

Martin-Löf, Per. 1975b. Syntax and semantics of mathematical language (Notes by Peter Hancock on a lecture series given at All Souls College, Oxford, Michaelmas term, 1975), unpublished preprint.

Martin-Löf, Per. 1979. Constructive mathematics and computer programming. In *Logic, methodology and philosophy of science VI, Proceedings of the 1979 international congress at Hannover, Germany*, ed. L.J. Cohen, J. Los, H. Pfeiffer and K.-P. Podewski, 153–175. Amsterdam: North-Holland Publishing Company, 1982.

Martin-Löf, Per. 1980. Intuitionistic type theory: Notes by Giovanni Sambin of a series of lectures given in Padua, June 1980. Napoli: Bibliopolis 1984.

Martin-Löf, Per. 1983. On the meanings of the logical constants and the justifications of the logical laws. *Nordic Journal of Philosophical Logic* 1(1): 11–60, 1996 ("Siena lectures").

Martin-Löf, Per. 1985. Truth of a proposition, evidence of a judgement, validity of a proof. *Synthese* 73: 407–420.

Martin-Löf, Per. 1988. Mathematics of infinity. In *Colog 88*, ed. P. Martin-Lof and G. Mints, 146–197. Berlin: Springer, 1990.

Martin-Löf, Per. 1991. A path from logic to metaphysics. In *Atti del Congresso Nuovi Problemi della Logica e della Filosofia della Scienza, Viareggio, 8–13 gennaio, 1990*, ed. G. Corsi and G. Sambin, 141–149. Bologna: CLUEB, 1991.

Martin-Löf, Per. 1992. Analytic and synthetic judgements in type theory. In *Kant and contemporary epistemology*, ed. P. Parrini, 87–99. Dordrecht: Kluwer, 1994.

Martin-Löf, Per. 1994. Verificationism then and now. In *The foundational debate: Complexity and constructivity in mathematics and physics*, ed. W. DePauli-Schimanovich et al, 187–196. Dordrecht: Kluwer, 1995.

Martin-Löf, Per. 1995. Truth and knowability: On the principles C and K of Michael Dummett. In *Truth in mathematics*, ed. H.G. Dales and G. Oliveri, 105–114. Oxford: Clarendon Press, 1998.

Martin-Löf, Per. 2004. 100 years of Zermelo's Axiom of Choice: What was the problem with it? In *Logicism, intuitionism, and formalism – What has become of them?*, ed. S. Lindström, E. Palmgren, K. Segerberg and V. Stoltenberg-Hansen, 209–219. Dordrecht: Springer, 2009.

Martin-Löf, Per. 2008. The Hilbert-Brouwer controversy resolved? In *One hundred years of intuitionism (1907–2007). The Cerisy conference*. Publications of the Henri Poincaré Archives. Science around 1900, ed. M. van Atten, P. Boldini, M. Bourdeau and G Heinzmann, 243–256. Basel: Birkhäuser, 2008.

Works by Others

Bishop, Errett. 1967. *Foundations of constructive analysis*. New York: McGraw-Hill.

Dummett, Michael. 1975. The philosophical basis of intuitionistic logic. In *Logic colloquium'73*, ed. H.E. Rose and J. Shepherdson, 5–40. Amsterdam: North-Holland Publishing Company.

Dummett, Michael. 1991. *The logical basis of metaphysics*. London: Duckworth.

Fenstad, J.E. (ed.). 1971. *Proceedings of the second Scandinavian logic symposium*. Amsterdam: North-Holland Publishing Company.

Girard, J.-Y. 1971. Une extension de l'interprétation de Gödel à l'analyse, et son application à l'élimination des coupures dans l'analyse et la théorie des types. In *Proceedings of the 2nd Scandinavian logic symposium*, ed. J.E. Fenstad, 63–92. Amsterdam: North-Holland Publishing Company, 1971.

Gödel, Kurt. 1958. Über eine bisher noch nocht benützte Erweiterung des finiten Standpunktes. *Dilalectica* 12: 280–287.

Howard, W.A. 1980. The formulae-as-types notion of construction. In *To H. B. Curry: Essays on combinatory logic, lambda calculus and formalism*, ed. J.P. Seldin and J.R. Hindley, 479–490. London: Academic.

Kreisel, Georg. 1975. Observations on a recent generalization of completeness theorems due to Schütte. In *ISILC Proof Theory Symposion, Dedicated to Kurt Schütte on the occasion of his 65th birthday. Proceedings of the International Summer Institute and Logic Colloquium, Kiel 1974. Springer Lecture Notes in Mathematics*, vol. 500, ed. J. Diller and G.H. Müller, 164–181. Berlin: Springer.

Prawitz, Dag. 1965. *Natural deduction* (doctoral dissertation). Stockholm: Almqvist & Wiksell. (Second edition at Dover Publishing Company, 2006).

Prawitz, Dag. 1967. Completeness and Hauptsatz for second order logic. *Theoria* 33: 246–258.

Prawitz, Dag. 1971. Ideas and results in proof theory. In *Proceedings of the 2nd Scandinavian logic symposium*, ed. J.E. Fenstad, 235–307. Amsterdam: North-Holland Publishing Company, 1971.

Tait, W. W. 1967. Intensional interpretations of functionals of finite type. *Journal of Symbolic Logic* 32: 198–212.

Notes on the Contributors

Steve Awodey is professor of philosophy at Carnegie Mellon University. His research is focused on Logic and Category Theory, as well as the History and Philosophy of Logic. In addition to numerous research articles, he is the author of a textbook, *Category Theory* (Oxford Logic Guides 52). His most recent work is centered on the recently discovered connection between constructive logic and homotopy theory.

Thierry Coquand is professor of computing science at the University of Gothenburg. His main research areas are constructive mathematics and intuitionistic type theory.

Peter Dybjer is professor of computing science at Chalmers University of Technology. His research is in the junction of computer science and logic with a focus on intuitionistic type theory, program verification, and programming language semantics.

Juliet Floyd is professor of philosophy at Boston University, working primarily on the interplay between logic, mathematics, and philosophy in late nineteenth and early-twentieth-century philosophy, as well as on Kant, aesthetics, and eighteenth-century philosophy. She has written articles on Kant, Frege, Russell, Wittgenstein, Quine, Rawls, and Gödel and edited (with Sanford Shieh) *Future Pasts: The Analytic Tradition in Twentieth Century Philosophy* (Oxford University Press, 2001; Oxford Scholarship Online, 2004). Her current research examines Wittgenstein's reactions to Turing in the mid-1930s.

Jean-Yves Girard is a logician working in proof theory. His contributions include a proof of strong normalization in a system of second-order logic called system F; the invention of linear logic; the geometry of interaction; and ludics. He is senior researcher (directeur de recherches) at the Centre National de la Recherche Scientifique (CNRS) affiliated with the Institute of Mathematics at Luminy.

Guilhem Jaber is a Ph.D. student at the Ecole des Mines de Nantes, working on forcing and realizability, and their application to proofs of program correctness.

Sten Lindström is professor of theoretical philosophy at Umeå University. His main research areas are philosophical logic, philosophy of language, and philosophy of mathematics.

Erik Palmgren is professor of mathematical logic at Stockholm University. His research revolves around constructive mathematics and its foundations, type theory, and categorical logic.

Dag Prawitz is professor emeritus of theoretical philosophy at Stockholm University. His main research has been in proof theory and theory of meaning.

Aarne Ranta is professor of computer science at the University of Gothenburg. He was a Ph.D. student of Per Martin-Löf in Stockholm in 1987–1990. Ranta's work is focused on the analysis of natural languages in type theory and on computer programs that apply the analysis in machine translation and human–computer interaction. Most of this work is carried out in GF, Grammatical Framework, which is a grammar formalism based on type theory.

Michael Rathjen is professor of pure mathematics at the University of Leeds, England. His main research area is mathematical logic, especially proof theory, infinitary logics, type theory, non-classical set theories, and theories based on intuitionistic logic.

Giovanni Sambin is professor of mathematical logic at the University of Padua. He wrote up the lecture notes that until this day is the only textbook by Per Martin-Löf on his constructive type theory. He put type theory into action, giving life to formal topology (i.e., topology as developed over type theory), a field which he has always worked on since then. He has always cultivated an active interest in the foundations of mathematics.

Anton Setzer is a reader at the Department of Computer Science at Swansea University. He has a Diploma (German M.Sc.) and Ph.D. in mathematics from the University of Munich (Germany), and a docentship in Mathematical Logic from Uppsala University (Sweden). He held research assistantships in Munich, Uppsala (Sweden), Gothenburg (Sweden), Leeds (England), and had longer visits at Hiroshima and Kobe (Japan), the Mittag-Leffler Institute (Stockholm), and the Newton Institute (Cambridge). He is specialized in proof theory, especially of extensions of Martin-Löf Type Theory, and on theoretical basis of the interactive theorem prover and dependently typed programming language Agda.

Jan Smith obtained his Ph.D. in mathematics from the University of Gothenburg in1978, supervised by Per Martin-Löf. His research has since then mainly been on type theory as a logical framework for programming and he is now professor in computing science at Chalmers University of Technology. Over the years, Jan Smith has remained in close contact with Per Martin-Löf on research questions related to type theory.

Wilfried Sieg is the Patrick Suppes Professor of Philosophy and Mathematical Logic at Carnegie Mellon University. He works in proof theory, philosophy and history of modern mathematics, computation theory and, relatedly, in the foundations of cognitive science. Most relevant to his paper in this volume is his editorial work concerning Kurt Gödel, David Hilbert, and Paul Bernays. As to Hilbert, he is editing (with William Ewald) Hilbert's unpublished notes for lectures on arithmetic and logic from the 1890s to the early 1930s.

Sören Stenlund is professor emeritus of philosophy at Uppsala University. He is the author of *Language and Philosophical Problems* (Routledge 1990) and of the article *Different senses of finitude: an inquiry into Hilbert's finitism* (Synthese). He has published several other books and articles on various themes in the philosophies of language, logic, and mathematics. Problems concerning the nature and history of philosophy are other themes dealt with in Stenlund's publications, some of which are available only in Swedish.

Göran Sundholm studied at Lund, Uppsala, and Oxford and has held the Chair of Logic at Leiden University since 1987, serving as Visiting Professor also at the universities of Utrecht, Siena, Stockholm, Nancy, and Paris. He has written extensively on the philosophy of mathematical constructivism, applying the perspective of Martin-Löf's Type Theory, as well as on the history of modern logic, and is a frequent speaker on such topics at international conferences.

William Tait is a professor emeritus of philosophy at the University of Chicago. He has worked in logic, principally in proof theory, and in the foundations of mathematics and its history.

Jouko Väänänen is a professor of mathematics at the Department of Mathematics and Statistics of the University of Helsinki. He is also affiliated to the Institute of Logic, Language and Computation (ILLC) of the University of Amsterdam. He specializes in mathematical logic, set theory, model theory, and foundations of mathematics. Recently he has focused on the logic of concepts of dependence and independence, and on second-order logic. His "Dependence Logic" appeared in 2007 and "Models and Games" in 2011, both published by Cambridge University Press. He is currently writing a monograph on second-order logic.

Mark van Atten is senior researcher (directeur de recherches) at the Centre National de la Recherche Scientifique (CNRS), affiliated with the Institut d'Histoire et de Philosophie des Sciences et des Techniques (IHPST), Paris. His main interests being phenomenology and the foundations of mathematics, he has published on Brouwer, Gödel, Husserl, and Leibniz.

Part I
Philosophy of Logic and Mathematics

Chapter 1
Kant and Real Numbers

Mark van Atten

dedicated to Per Martin-Löf

1.1 Introduction

Consider the following three concepts:

1. The square root of 2
2. The diagonal of a square with sides of length 1
3. The infinite sequence of rational numbers

$$1, 1.4, 1.41, \ldots$$

   given by a rule that ensures that the square of the successive rationals converges to 2.

Nowadays we say that under each of these concepts falls an object. The length of the diagonal of the square is given by $\sqrt{2}$ but this root exists independently from geometry. $\sqrt{2}$ is a real number, and this real number can, if one wishes to do so, be identified with the infinite sequence determined by the third concept (or, alternatively, an equivalence class of such sequences).

But Kant stopped short, like a horse in front of a fence, of introducing real numbers by identifying them with infinite sequences. Indeed, he viewed the relations between the three concepts above differently. The main source that documents Kant's view on real numbers is his reply of Autumn 1790 to a letter from

M. van Atten (✉)
IHPST, 13 rue du Four, 75006 Paris, France
e-mail: Mark.vanAtten@Univ-Paris1.fr

P. Dybjer et al. (eds.), *Epistemology versus Ontology*, Logic, Epistemology, and the Unity of Science 27, DOI 10.1007/978-94-007-4435-6_1,

August Rehberg.[1] This view was, as such, perfectly traditional in Kant's days,[2] but it is interesting to see that they readily fit his newly proposed foundations of mathematics:[3]

1. The concept of the magnitude $\sqrt{2}$ is not empty, because it can be instantiated geometrically:

   > That a middle proportional magnitude can now be found between one that equals 1 and another that equals 2, and is therefore not an empty concept (without an object), geometry shows with the diagonal of the square. [AAXI:208][4]

2. But $\sqrt{2}$ cannot be instantiated numerically, because genuine numbers are composed out of units and hence rational:

   > But the pure schema of magnitude (quantitatis), as a concept of the understanding, is number, a representation which comprises the successive addition of homogeneous units. [A142/B182, also referred to by Rehberg][5]

---

[1]Rehberg's letter: AAXI:205–206; Kant's reply: AAXI:207–210. Rehberg did not reply in turn, but much later published excerpts from Kant's letter to him, together with his dissatisfied comments on it, in the first volume of his *Saemtliche Schriften* of 1828 (Rehberg 1828, pp. 52–60). With the publication of Kant's Nachlass, also two drafts for his reply to Rehberg became known [AAXIV:53–55, 55–59]. For an amusing description of Rehberg, see Jachmann's letter to Kant of October 14, 1790, AAXI:215–227, in particular p. 225. It is Jachmann's letter that tells us that Rehberg's letter came to Kant via Nicolovius. For detailed information on Rehberg's life and work, see Beiser (2008).

[2]However, Stevin had already argued in 1585 that $\sqrt{8}$ is a number because it is part of 8, which is a number: 'La partie est de la mesme matiere qu'est son entier; Racine de 8 est partie de son quarré 8: Doncques $\sqrt{8}$ est de la mesme matiere qu'est 8: Mais la matiere de 8 est nombre; Doncques la matiere de $\sqrt{8}$ est nombre: Et par consequent $\sqrt{8}$, est nombre.' Of course, Stevin did not go on to provide an arithmetization of real numbers (Stevin 1585, p. 30).

[3]Kant did not publish this view in his lifetime, and it seems it first appeared in print in Rehberg's later comments on their exchange (Rehberg 1828, pp. 52–60). However, three remarks to the same effect were published within a framework close to Kant's in Solomon Maimon's book on Kant's philosophy, *Versuch über die Transscendentalphilosophie* of Autumn 1789, the year before Kant's exchange with Rehberg (Maimon 1790; the title page states 1790, but see its editor's remark in footnote 1 on p. II of the edition used here). The remarks in question appear on p. 374, 229/374, and 374, respectively. There seems to be no evidence as to whether Kant had seen Maimon's remarks before writing to Rehberg (or later). (Warda's list (Warda 1922) and the more comprehensive database 'Kants Lektüre' (http://web.uni-marburg.de/kant/webseitn/ka_lektu.htm) suggest that Kant did not own Maimon's book. But that does not show that he did not see it at some point.) Note that Rehberg, in his later comments (Rehberg 1828), does not mention Maimon's book either. We will come back to the exchange between Rehberg and Kant from a systematical point of view in Sect. 1.3.

[4]'Daß nun die mittlere Proportionalgröße zwischen einer die = 1 und einer anderen welche = 2 gefunden werden könne, mithin jene kein leerer Begrif (ohne Object) sey, zeigt die Geometrie an der Diagonale des Qvadrats.'

[5]'Das reine Schema der Größe aber (quantitatis), als eines Begriffs des Verstandes, ist die Zahl, welche eine Vorstellung ist, die die sukzessive Addition von Einem zu Einem (gleichartigen) zusammenbefaßt.'

> So the only question is why for this quantum [$\surd 2$] no number can be found that represents the quantity (the ratio to unity) clearly and completely in a concept. ... That, however, the understanding, which arbitrarily makes for itself the concept of $\surd 2$, cannot also bring forth the complete number concept, namely its rational ratio to unity, ... [AAXI:208][6]

3. We may have a rule to generate a potentially infinite sequence of rationals that will approximate an irrational 'number' such as $\surd 2$:

> that for every number one should be able to find a square root, if necessary one that is itself no number, but only the rule to approximate it as closely as one wishes, ... [AAXI:210][7]

But for Kant $\surd 2$ and a sequence of rational approximations to it are two different things. This becomes clear when in his reply to Rehberg he writes of such a sequence as

> a sequence of fractions that, because it can never be completed, although it can be brought as near to completion as one wishes, expresses the root (but only in an irrational way) [AAXI:209][8]

Indeed, this incompletability served to characterize 'irrational' in one of Kant's *Reflexionen* of the same period:

> Concepts of irrational ratios are those, that cannot be exhausted by any approximation. [AAXVIII:716, 1790–1795?][9]

If Kant had thought that the square root could be identified with the potentially infinite sequence, he could not have said that the latter is an incomplete expression of the former.[10] Thus, although Kant at times speaks of 'irrational numbers',[11] he

[6] 'Es ist also nur die Frage warum für dieses Qvantum [$\surd 2$] keine Zahl gefunden werden könne welche die Qvantität (ihr Verhaltnis zur Einheit) deutlich und vollständig im Begriffe vorstellt. ... Daß aber der Verstand, der sich willkürlich den Begrif von $\surd\surd 2$ macht, nicht auch den vollständigen Zahlbegrif, nämlich durch das rationale Verhaltnis derselben zur Einheit hervorbringen könne, ... '

[7] 'daß sich zu jeder Zahl eine Qvadratwurzel finden lassen müsse, allenfalls eine solche, die selbst keine Zahl, sondern nur die Regel der Annäherung derselben, wie weit man es verlangt, ... '

[8] 'eine ... Reihe von Brüchen ..., die, weil sie nie vollendet seyn kan, obgleich sich der Vollendung so nahe bringen läßt als man will, die Wurzel (aber nur auf irrationale Art) ausdrückt'.

[9] 'Begriffe irrationaler Verhaltnisse sind solche, die durch keine Annäherung erschopft werden können.'

[10] Note that, in the same sense, an infinite decimal expansion such as 0.333 ... would also be 'an irrational way' to express a magnitude, but in that case there is also the rational way of expressing it as a complete object, i.e. the fraction $1/3$. Hegel called the expression by infinite and hence incompletable means of something that can also be expressed finitely and hence completely a case of 'bad infinity' ('schlechte Unendlichkeit') (his example being the infinite decimal expansion 0.285714... and the fraction $2/7$) (Hegel 1979, pp. 287–289). I thank Pirmin Stekeler-Weithofer for bringing this to my attention.

[11] A480/B508; AAXI:209; AAXIV:57; AAXVII:718.

also made it clear that this is a façon de parler, and that we actually only have rules for generating approximations.

In a foundational setting, to introduce real numbers as infinite sequences, one has to do two things:

1. Give a foundational account of infinite sequences as objects;
2. Explain in what sense such sequences can be considered to be numbers.

So when Kant rejects the identification, that can be on account of his concept of number, or on account of his foundational ideas about infinite sequences. Indeed, in his writings and correspondence, one finds objections of both kinds.

Kant's conception, mentioned above, that a number is composed out of given units, and that accordingly only whole numbers and rationals are numbers in the proper sense, goes back to the Greek.[12] Note that Kant never adopted the more general concept of number as proportion, as, for example, Newton had;[13] for the question addressed in the present paper, it would have made little difference if he had.[14]

Today, on the other hand, one would defend the claim that certain infinite sequences can be said to be numbers by referring to the algebraic concept of a field extension of the rationals. That is the result of a development starting with Abel and Galois and in which Hankel's *Theorie der complexen Zahlensysteme* of 1867 was of particular importance.[15] Note that this algebraic concept is abstract enough not to depend on the particular way real numbers are implemented. Indeed, while Hankel thus extended the traditional concept of number, he still held that the *existence* of irrational numbers is shown only geometrically (Hankel 1867, p. 59);[16] his work

[12]For example, both Euclid (Elements, Book VII, def. 2) and Diophantus (Arithmetic, Book I, Introduction) define numbers as multitudes of units; while Euclid did not accept rational numbers, Diophantus did, in the sense that, as Klein explains it, 'by a fraction Diophantus meant nothing but a number of fractional parts' (Klein 1968, p. 137).

[13]'By number we understand not a multitude of units, but rather the abstract ratio of any one quantity to another of the same kind taken as unit. Numbers are of three sorts; integers, fractions, and surds: an integer is what the unit measures, the fraction what a submultiple part of the unit measures, and a surd is that with which the unit is incommensurable.' ('Per numerum non tam multitudinem unitatum quam abstractam quantitatis cujusvis ad aliam ejusdem generis quantitatem quae pro unitate habetur rationem intelligimus. Estque triplex; integer, fractus & surdus: Integer quem unitas metitur, fractus quem unitatis pars submultiplex metitur, & surdus cui unitas est incommensurabilis.') Newton, *Arithmetica universalis, sive de compositione et resolutione arithmetica liber* (Cambridge, 1707) as quoted and translated in Petri and Schappacher (2007, p. 344).

[14]Eudoxus' theory of proportions was, however, of great importance to Kant's views on the relations between arithmetic, geometry, and algebra. For an extensive treatment of that topic, see Sutherland (2006).

[15]I thank Carl Posy for drawing my attention to this.

[16](Tennant 2010, Why arithmetize the reals? why not geometrize them? Unpublished typescript), addresses the following question: 'Who was the first major foundationalist thinker explicitly to reject (on the basis of reasons or argument, however inconclusive) recourse to geometric concepts or intuitions or principles or understanding, in the attempt to provide a satisfactory foundation for real analysis?', and argues that it was Bolzano. I am grateful to Tennant for sharing his typescript with me.

preceeded the arithmetization of real numbers as infinite sequences of rationals by Cantor in 1872 (Cantor 1872).[17] Cantor comfortably counted these sequences among the 'numerical quantities' ('Zahlengrösse'),[18] and emphasized that on this conception, the real number is not an object that is distinct from the sequence, as limits in the original sense were.

Interestingly, Charles Méray, who 3 years before Cantor published the same mathematical ideas (Méray 1869), and thus holds priority,[19] had still preferred to reserve the term 'nombres' for whole numbers and rationals (Méray 1869, p. 284),[20] and considered incommensurable numbers to be 'fictions' (Méray 1872, p. 4). The infinite sequences he called 'variables progressives convergentes' instead. Thus, Méray in effect held a middle position between Kant's and Cantor's. The difference between Méray and Cantor is of course by no means merely terminological.[21]

But even if Kant would have had occasion to consider extending the number concept the way Hankel would later suggest,[22] he would, as we saw, still have seen reason to object to the identification of $\sqrt{2}$ and an appropriate potentially infinite sequence, because of the essential incompleteness of the latter. Now, compare this with Brouwer's position. Like Kant, Brouwer based his foundations of mathematics on an a priori intuition of time.[23] Yet, Brouwer accepts the modern concept of number and moreover does identify real numbers and certain potentially infinite sequences:[24]

---

[17]Cantor's idea was first published, with credit, by Heine (1872, p. 173).

[18]Kant also used that term (e.g. AAXI:208), but, as we will see, for him it did not refer to infinite sequences.

[19]As far as I know, Méray (1835–1911) and Cantor (1845–1918) have never been in contact; in particular, both as subject and as object Méray is completely absent from Cantor's known, rich correspondence with the French (Décaillot 2008). Méray states his priority claim on p. XXIII of the 'préface' to Méray (1894).

[20]I do not know yet whether a reaction of Méray on Hankel's work is known.

[21]For a detailed history of the arithmetization of real numbers, see Boniface (2002) and Petri and Schappacher (2007).

[22]One is reminded of the footnote (there, concerning the term 'analytic') in section 5 of the *Prolegomena*, which begins: 'It is impossible to prevent that, as knowledge advances further and further, certain expressions that have already become classical, dating from the infancy of science, should subsequently be found insufficient and badly fitting' ('Es ist unmöglich zu verhüten, daß, wenn die Erkenntniß nach und nach weiter fortrückt, nicht gewisse schon classisch gewordne Ausdrücke, die noch von dem Kindheitsalter der Wissenschaft her sind, in der Folge sollten unzureichend und übel anpassend gefunden werden') [AAIV:276n.].

[23]Indeed, in his inaugural lecture 'Intuitionisme en formalisme' of 1912, Brouwer presented his position as fundamentally Kantian (Brouwer 1913, p. 85). That general qualification is absent from his later work; in the light of the considerations in the present paper, that seems, conceptually if not historically as well, to be no coincidence.

[24]That Brouwer here describes a sequence of nested intervals, and not of rationals, is not essential to the question at hand.

> We call such an indefinitely proceedable sequence of nested ... intervals a point P or a real number P. We must stress that for us the sequence ... *itself* is the point P. ... For us, a point and hence also the points of a set, are always unfinished. (Brouwer 1992, p. 69, original emphasis)[25]

In what follows, I will leave aside the fact that Brouwer here also includes sequences that are constructed not according to a rule but by free choices. Even Brouwer's lawlike sequences, such as one for $\sqrt{2}$, would not themselves be real numbers for Kant. On the other hand, for Brouwer there is nothing *irrational* about the expression of a magnitude by an incompletable sequence.

Even before we ask whether or not a potentially infinite sequence is a rational way to express any kind of number, we can ask the ontological question whether the concept of such a sequence has constructible instances at all. There are three theses of Kant's that, taken together, at first sight seem to lead to a positive answer, along lines very similar to Brouwer:

1. We have an a priori intuition of time;
2. Time is given to us as infinite;
3. 'Time is in itself a sequence (and the formal condition of all sequences)' [A411/B438].[26]

Couldn't a potentially infinite sequence then be accepted as an object by labelling moments in the sequence of time with its elements? We will see, however, that Kant's understanding of these three theses is such that the answer to this question is negative.

Lisa Shabel has observed that Kant 'doesn't claim that the rule for the approximation of an irrational magnitude constitutes a "construction" of any kind' (Shabel 1998, p. 597n.12); I took that to mean, among other things, that Kant does not claim that to generate a potentially infinite sequence of rationals according to an appropriate rule is to effect the construction of the mathematical concept of an irrational magnitude.[27] Here I will defend the stronger thesis that for Kant, it would have been impossible to make that particular claim, as in his system we have no means to construct the concept of a potentially infinite sequence. A fortiori, Kant could not have arithmetized irrational quantities by infinite sequences.

Our considerations must begin with a review of what, for Kant, determines whether a mathematical concept can be constructed or not.

---

[25] 'Ein derartige unbegrenzte Folge ineinander geschachtelter ... Intervalle nennen wir einen Punkt P order eine reelle Zahl P. Wir betonen, dass bei uns die Folge ... *selbst* der Punkt P ist ... Bei uns sind ein Punkt und daher auch die Punkte einer Menge immer etwas Werdendes.'

[26] 'Die Zeit ist an sich selbst eine Reihe (und die formale Bedingung aller Reihen).' In Kemp-Smith's translation, I have replaced 'series' by 'sequence'.

[27] In an email, Lisa Shabel has confirmed to me that this indeed is included in her observation; I thank her for this clarification.

1.2 Mathematics Within Subjective Limits

Kant takes the existence of mathematical knowledge as a given [B20]. He considers philosophy and mathematics two different enterprises that cannot and should not change one another.[28] But they are closely related: mathematics can serve as an instrument in philosophy, and one of the tasks of philosophy is to give an answer to the transcendental question how mathematical knowledge (such as we indeed have it) is possible.[29] In his lectures on logic, Kant emphasizes that the content of mathematics as a science is not influenced by the answer to the transcendental question. In the explanation what makes mathematical knowledge possible we cannot find motivations for revisions of mathematics.[30]

The starting point for Kant's transcendental clarification of mathematics is his dictum 'Thoughts without content are empty, intuitions without concepts are blind' [A51/B75]. Mathematical knowledge can only be had if a concept and an intuition of an object are brought together. An intuition is necessary to show that a concept is related to an object, in other words, that a concept has objective reality [A155/B194].[31] Even instances of analytic judgements count as

[28]Various places in *Kritik der reinen Vernunft*, the *Prolegomena*; also AAXXIII:201.

[29]'For the possibility of mathematics must itself be demonstrated in transcendental philosophy." ('Denn sogar die Möglichkeit der Mathematik muß in der Transscendentalphilosophie gezeigt werden.') [A733/B761]; also A149/B188–189.

[30]'Whatever the fact of the matter may be [on the relation between logic and the science of knowledge], this much is agreed: in any case logic remains, within its domain, unchanged, as far as the essential is concerned; and the transcendental question whether the logical propositions are capable and in need of a derivation from a higher, absolute principle, can have as little influence on logic itself and on the validity and evidence of its laws as the transcendental task, How are are synthetic judgements a priori possible in mathematics?, can have on pure mathematics regarding its scientific content. Like the mathematician as mathematician, so the logician as logician can calmly and safely continue his course of explaining and proving, without having to worry about the question of the transcendental philosopher and the philosopher of science, which lies outside his sphere: How is pure mathematics or pure logic as a science possible?' ('Welche Bewandtniß es nun aber auch immer hiermit haben möge, so viel ist ausgemacht: in jedem Fall bleibt die Logik im Innern ihres Bezirkes, was das Wesentliche betrifft, unverändert; und die transscendentale Frage: ob die logischen Sätze noch einer Ableitung aus einem höhern, absoluten Princip fähig und bedürftig sind, kann auf sie selbst und die Gültigkeit und Evidenz ihrer Gesetze so wenig Einfluß haben, als auf die reine Mathematik in Ansehung ihres wissenschaftlichen Gehalts die transscendentale Aufgabe hat: Wie sind synthetische Urtheile a priori in der Mathematik möglich? So wie der Mathematiker als Mathematiker, so kann auch der Logiker als Logiker innerhalb des Bezirks seiner Wissenschaft beim Erklären und Beweisen seinen Gang ruhig und sicher fortgehen, ohne sich um die außer seiner Sphäre liegende transscendentale Frage des Transscendental-Philosophen und Wissenschaftslehrers bekümmern zu dürfen: Wie reine Mathematik oder reine Logik als Wissenschaft mäglich sei?') [AAIX:008].

[31]For phenomenologists, it is of interest that this is how Kant defines the notion of 'evidence':

> When objective certainty is intuitive, it is called 'evidence' ('Wenn die obiective Gewisheit anschauend ist, so heisst sie evidentz.') [AAXVI:375 (1769? 1770?)]

mathematical knowledge only to the extent that they have been combined with appropriate intuitions:

> Some few fundamental propositions, presupposed by the geometrician, are, indeed, really analytic, and rest on the principle of contradiction; ... And even these propositions, though they are valid according to pure concepts, are only admitted in mathematics because they can be exhibited in intuition. [B16–17][32]

According to Kant, the only kind of intuition humans have is sensuous intuition [A51/B75]. This means that we can only have intuitions of objects that are given to us either in sense perception or in the imagination. Kant denies that humans can have intuition of (what we would call) abstract objects; we do not have intellectual intuition. On the other hand, he acknowledges that we do have purely mathematical knowledge. Kant is able to combine those two views by pointing out that a mathematical concept can be combined with a *sensuous* intuition, namely if the concept is exemplified or instantiated in it.[33] In particular, then, on Kant's conception mathematics is not about sui generis mathematical objects, but about possible empirical instantiations of mathematical concepts.[34]

For example, the concept of the number 5 is instantiated in an image of 5 dots. Moreover, Kant says, when we think of a number (be it small or large) we are not so much thinking of such an image, as of a rule for producing images

> Mathematical certainty is also called evidence, as intuitive knowledge is clearer than discursive knowledge. ('Die mathematische Gewißheit heißt auch Evidenz, weil ein intuitives Erkenntniß klärer ist als ein discursives.') [AAIX:70]
>
> Concepts a priori (in discursive knowledge) can never be a source of intuitive certainty, i.e. evidence, however much the judgement may otherwise be apodictically certain. ('Aus Begriffen a priori (im diskursiven Erkenntnisse) kann aber niemals anschauende Gewißheit, d.i. Evidenz entspringen, so sehr auch sonst das Urteil apodiktisch gewiß sein mag.') [A734/B762]

But it is not a term that Kant actually uses often.

[32] 'Einige wenige Grundsätze, welche die Geometer voraussetzen, sind zwar wirklich analytisch und beruhen auf del Satze des Widerspruchs; ... Und doch auch diese selbst, ob sie gleich nach bloßen Begriffe gelten, werden in der Mathematik nur darum zugelassen, weil sie in der Anschauung können dargestellet werden.'

[33] 'mathematica per constructionem conceptus secundum intuitionem sensitivam' [AAXVII:425 (1769? 1773–1775?)]; and various other places.

[34] 'mathematics ... the object of that science is to be found nowhere except in possible experience' ('die Mathematik, [die] ihren Gegenstand nirgend anders, als in der möglichen Erfahrung hat') [A314/B371n.]; 'Consequently, the pure concepts of understanding, even when they are applied to a priori intuitions, as in mathematics, yield knowledge only in so far as these intuitions—and therefore indirectly by their means the pure concepts also—can be applied to empirical intuitions' ('Folglich verschaffen die reinen Verstandesbegriffe, selbst wenn sie auf Anschauungen a priori (wie in der Mathematik) angewandt werden, nur so fern Erkenntniß, als diese, mithin auch die Verstandesbegriffe vermittelst ihrer auf empirische Anschauungen angewandt werden können') [B147]; '[the] mathematician ... who likewise deals only with possible objects of the outer senses' ('[der] Mathematiker ... der es auch blos mit möglichen Gegenständen äußerer Sinne zu thun hat') [AAXX:418] (1790).

showing that number of objects [A140/B179]. The rule prescribes a series of acts in which an appropriate image will be brought about. Now, the number 5 will be equally well instantiated in an image of 5 dots, strokes, or yet another kind of object. By not stipulating that we use any of these in particular, but merely requiring that we be able to consider the things we are adding as in some sense homogeneous, the rule assumes a generality that accounts for the possibility of obtaining general knowledge through the acts of producing what is, after all, a particular image (see also what Kant says on triangles at A713–714/B741–742). It is here that the inner sense of time comes in. Kant holds that all that we need to be able successively to add units into one image is the inner sense of time [A142–143/B182].[35] A rule for producing images that instantiate a number concept need therefore not appeal to more than that inner sense. Because of this sufficiency, Kant can say that the foundation of arithmetic (tacitly, as a variety of human knowledge—see below) is the a priori intuition of time. In the series of acts prescribed by a rule, Kant says in his particular idiom, we 'construct the concept' [A713/B741]. Such a construction may be actually carried out (resulting in, e.g. an actual image of 5 dots) or, alternatively, be conceived of as an in some appropriate sense ideal possibility (an ideally possible image of 1,000 clearly distinguishable and surveyable dots) [A140/B179]. What matters to Kant is not actual construction but ideal constructibility (see also Kant's reply to Eberhard in this matter: AAVIII:210–212, and the footnote on 191–192). This invites of course a discussion what 'in principle' amounts to; for Kant, the idealization involved is constrained by what he takes to be the essential properties of the human mind.[36]

This view on numbers allowed Kant to accept as humanly constructible mathematical concepts not only the natural numbers but the rational numbers, too, by taking, to arrive at a particular rational number, whatever part of 1 is appropriate for unit.[37] The concept of such a fractional unit is given intuitive content geometrically, by assigning length 1 to a given line segment and then constructing geometrically the required part of that segment (for example by the method of Euclid book VI, Proposition 9).

But for Kant, to irrational numbers correspond no humanly constructible concepts. As mentioned, Kant held on to the Greek conception of number, which he could readily ground by his particular transcendental account of our mathematical knowledge:

> The concept of magnitude in general can never be explained except by saying that it is that determination of a thing whereby we are enabled to think how many times a unit is posited

---

[35]Hence, as Kant emphasizes in reflection 6314 [AAXVIII:616 (1790–1791)], for the representation of a number both time and space are necessary, as an image has a spatial character. See also 4629 [AAXVII:614] from between 1771 and 1775.

[36]In the *Kritik der Urteilskraft* [AAV:254], Kant distinguishes between 'comprehensive' and 'progressive apprehension' ('comprehensive' and 'progressive Auffassung'), but to my mind in both cases what is aimed for is one (ideal) image; here I disagree with von Wolff-Metternich (1995, pp. 57–60).

[37]Kant does this at, e.g. AAXIV:057 (draft to Rehberg) and AAXI:208 (letter to Rehberg).

in it. But this how-many-times is based on successive repetition, and therefore on time and the synthesis of the homogeneous in time. [A242/B300][38]

In the following elaboration of Kant's transcendental account, I will refer to a number of passages in his *Reflexionen*. Although the *Reflexionen* in general can certainly not be granted the same status as Kant's published work, the specific passages used below, which are all from 1769 or later, present a coherent view, which in turn is coherent with that presented in the first *Kritik*.

For Kant, to obtain an image out of a manifold of elements requires a synthesis of the imagination, which necessarily occurs in time. But, as a particularity of the human mind, in a finite time span, we can generate a manifold of only finitely many elements:

> Progression. The infinity of the sequence as such is possible, but not the infinity of the aggregate. The former is an infinite possibility (of additions), the latter an infinite (actual) comprehension. [AAXVII:414, around 1769–1771][39]

and, more generally,

> What is only given by composition, is for that reason always finite, even though composition can go on infinitely. [AAXVIII:378 no. 5897 around 1780–1789?][40]

As a consequence, Kant cannot accept any actually infinite totalities as objects of human mathematical knowledge. In particular, it would not be open to Kant to accept irrational numbers (and, more generally, real numbers) as actually infinite sums of rational numbers. But he also says that the ground of the impossibility of infinite composition lies not in the mathematical concept of infinity, but in the limits to the capacities of the human mind. Kant does not exclude that minds of a different type can grasp an infinite aggregate as a whole:

> When a magnitude is given as a thing in itself, the whole precedes its composition, and in that case I cannot conclude from the fact that this putting together can never be finished and hence its quantitas can never be completely known, that such a thing, to the extent that it is an infinite quantum, is impossible. It is only impossible for us to know it completely according to our way of measuring magnitudes, because it is not measurable. From that it does not follow that a different understanding could not know the quantum as such completely without measuring. Similar for division. [AAXVIII:242–243, no. 5591 (1778–1789)][41]

---

[38] 'Den Begriff der Größe überhaupt kann niemand erklären, als etwa so: daß sie die Bestimmung eines Dinges sei, dadurch, wie vielmal Eines in ihm gesetzt ist, gedacht werden kann. Allein dieses Wievielmal gründet sich auf die sukzessive Wiederholung, mithin auf die Zeit und die Synthesis (des Gleichartigen) in derselben.'

[39] 'Progression. Die Unendlichkeit der Reihe als solche ist möglich, aber nicht die Unendlichkeit des Aggregats. Jenes ist eine unendliche Möglichkeit (der Hinzuthuungen), dieses eine unendliche (wirkliche) Zusammennehmung.'

[40] 'Was nur durch die composition gegeben wird, ist darum immer endlich, obgleich die composition ins Unendliche geht.'

[41] 'Wenn eine Größe als ein Ding an sich selbst gegeben ist, so geht das Ganze vor der composition voraus, und da kann ich darum, daß diese zusammensetzung niemals vollendet werden und also

(Exactly the same point was already made in the Inaugural Dissertation of 1770 [AAII:388 note**].)

Indeed, Kant says explicitly that our impossibility to grasp an infinite magnitude as a whole, an impossibility which follows from the dependence of our grasp of magnitudes on time, is not objective, but subjective:

> In the infinite, the difficulty is to reconcile the totality with the impossibility of a complete synthesis. Therefore the difficulty is subjective. On the other hand, the potential infinite (infinity of potential coordination) is very well understandable, but without totality. [AAXVII:452 no. 4195. 1769–1770?][42]

and

> How the conflict with subjective conditions or their presupposition mirrors the truth of the objective conditions and forces itself upon [unterschiebe] the latter. For example, a mathematical infinite is possible, as it does not conflict with the rules of the intellect [der Einsicht]; it is impossible, as it conflicts with the conditions of comprehension. [AAXVIII:135 1776–1778?][43]

Kant's answer to the question what these subjective limits are for us is 'that which can be represented a priori in intuition, that is, space and time and change in time'. [AAXVII:701 (around 1775–1777)][44] I take it, then, that Kant's remark in the quotation above from A242/B300 that the explanation of the notion of magnitude must depend on the notion of successive repetition and hence on time is limited to the specific context of human mathematical cognition, and that the same also holds for his statement at A142–143/B182 that 'Number is therefore simply the unity of the synthesis of the manifold of a homogeneous intuition in general, a unity due to my generating time itself in the apprehension of the intuition'.[45] I will now turn to the question what Kant's conception of mathematics within the subjective limits proper to us means for his view on real numbers.[46]

---

die quantitas derselben niemals ganz erkannt werden kann, nicht schließen, daß ein solches qua unendliche quantum unmöglich sey. Es ist uns nur unmoglich, nach unserer Art großen zu messen es gantz zu erkennen, weil es unermeßlich ist. Daraus folgt nicht, daß nicht ein anderer Verstand ohne Messen das quantum als ein solches Ganz erkennen könne. Ebenso mit der Teilung.' Also e.g. AAXVIII:379 no. 5903.

[42]'Im Unendlichen ist die Schwierigkeit, die totalitaet mit der unmöglichkeit einer synthesis completae zu vereinbaren. folglich ist die Schwierigkeit subiectiv. Dagegen ist das potentialiter infinitum (infinitum coordinationis potentialis) sehr wohl begreiflich, aber ohne totalitaet.'

[43]'Wie der Wiederstreit der subiectiven Bedingungen oder ihre Voraussetzung die Wahrheit der obiectiven nachahme und unterschiebe. e.g. Ein Mathematisch unendliches ist möglich, weil es den regeln der Einsicht nicht wiederstreitet; es ist unmöglich, weil es den Bedingungen der comprehension wiederstreitet.'

[44]'Welches sind die Grenzen der mathematischen Erkenntnis? Das, was a priori in der Anschauung kann vorgestellt werden, also Raum und Zeit und Veranderung in der Zeit.'

[45]'Also ist die Zahl nichts anderes, als die Einheit der Synthesis des Mannigfaltigen einer gleichartigen Anschauung überhaupt, dadurch, daß ich die Zeit selbst in der Apprehension der Anschauung erzeuge.'

[46]Maimon, in his *Versuch über die Transscendentalphilosophie*, also emphasizes the dependency on subjective conditions. Describing the division of a line segment into parts, he writes:

1.3 Kant's Discussion with Rehberg

Rehberg's primary concern when writing to Kant in 1790 was not the ontological status of real numbers, but the issue whether the intuition of time is really a condition of the possibility of mathematics for us. For our present purpose, the main interest of Rehberg's letter lies in two specific questions that are raised in it:

1. What are the conditions of the possibility of knowing that $\sqrt{2}$ is irrational? Rehberg disputes Kant's claim at A149/B188–189 in the *Kritik der reinen Vernunft* that 'mathematical principles ... are derived solely from intuition, but not from the pure concept of understanding'.[47] While Rehberg agrees for the case of geometry, he disagrees for the case of arithmetic and algebra, and claims that in those domains the a priori intuitions of time and space are not necessary to obtain knowledge, but only the concepts themselves [AAXI:205–206]. In his later comments, Rehberg calls the corresponding kind of intuition 'pure intellectual intuition' ('reine Anschauung des Verstandes') (Rehberg 1828, p. 57).
2. 'Why is our understanding, which produces numbers spontaneously, unable to think numbers corresponding to $\sqrt{2}$?' [AAXI:206][48] It is not clear what Rehberg exactly means by 'thinking a number',[49] but the very fact that Rehberg, who must

---

> In case the parts are infinite [in number], then this division is, for a finite being, impossible, not, however, in itself. (Maimon 1790, p. 375) ('Sind also die Theile unendlich, so ist diese Theilung, in Beziehung auf ein endliches Wesen, unmöglich, nicht aber an sich.')

And, on infinite numbers:

> An absolute understanding, on the other hand, thinks the concept of an infinite number without invoking a temporal sequence, all at once. Therefore, that which for the understanding [i.e. the human understanding] is, in accordance with its limitations, a mere idea, is, with respect to its absolute existence, a true object. (Maimon 1790, p. 228) ('Bei einem absoluten Verstande hingegen, wird der Begrif einer unendlichen Zahl, ohne Zeitfolge, auf einmal, gedacht. Daher ist das was der Verstand [i.e. the human understanding] seiner Einschränkung nach, als bloße Idee betrachtet, seiner absoluten Existenz nach ein reelles Objekt.')

It seems, then, that Maimon explicitly leaves open the possibility that infinite minds could admit into arithmetic not only whole and rational numbers, but also real numbers, as actually infinite sums of fractions. The human mind, however, cannot do this.

[47] '[D]ie mathematischen Grundsätze ... [sind] nur aus der Anschauung, aber nicht aus dem reinen Verstandesbegriffe gezogen'.

[48] 'Warum kann er [i.e. der Verstand], der Zahlen willkührlich hervorbringt keine $\sqrt{\sqrt{2}}$ Zahlen denken?' From Rehberg's letter and his later elaboration of his view (Rehberg 1828, p. 56), it is clear that by 'willkürlich' he does not mean 'subject to no condition at all'. While he claims, against Kant, that it is a spontaneity that is unconstrained by the forms of time and space, he also thinks it is subject to constraints of a different kind (see footnote 50 below), and takes the impossibility, as he sees it, to think $\sqrt{2}$ in numbers as a proof of that fact.

[49] Longuenesse claims that Rehberg means by it 'thinking in multiples or fractions of the unit, that is, in rational numbers' (Longuenesse 1998, p. 262n.38). (Also Dietrich reads him that way (Dietrich 1916, p. 118).) I do not find evidence for this in Rehberg's letter or his later comments. In

have known the infinite series, raises this question, suggests that generating an infinite series is not an example.[50]

In his reply to Rehberg, Kant argues that time is, after all, involved in our coming to know the irrationality of $\sqrt{2}$, as follows.[51] From the mere concept of a given natural number it cannot be seen whether its square root is rational or irrational. To determine this, Kant appeals to the following theorem: if the square root of a natural number n is not itself a natural number, then it is not a rational number either [AAXI:209].[52] We can only find out whether n is the square of a natural number by testing. The test proceeds by constructing the natural numbers from 1 onward until the square is equal to or greater than n.[53] But constructing numbers involves the intuition of time. And, although Kant does not remark on this in the letter, more

effect, on that reading Rehberg is asking why the understanding cannot think an irrational number as a rational one. I read Rehberg differently; see the next footnote. (Of course, when Rehberg writes, 'Es heißt zwar p. 182 der Critik, daß die Zahl eine successive Addition sey' [AAXI:205], this formulation is neutral as to whether he agrees.)

[50]Rehberg's own suggestion for an answer is that the ground of this impossibility lies in the transcendental faculty of the imagination and its connection to the understanding [AAXI:206], which he thinks has a property that limits our capacity of generating numbers in such a way that thinking (a quantum) in numbers for us is limited to 'discretely generated magnitudes' ('discretive erzeugten Größen') (Rehberg 1828, pp. 57, 59); see also Parsons (1984, p. 111). In his letter he qualifies the nature of this faculty as 'transcending all human capacities of investigation' ('alles menschliche Untersuchungsvermögen übersteigend') [AAXI:206], but nevertheless goes on to suggest the possibility of a 'transcendental system of algebra' ('transscendentales System der Algebra'), which would serve to determine a priori, on the basis of principles, which equations we can solve and how. In one of the drafts for his reply, Kant says he can answer Rehberg's question 'without having to look into the first grounds of the possibility of a science of numbers' ('ohne auf die ersten Gründe der Moglichkeit einer Zahlwissenschaft zurüksehen zu dürfen') [AAXIV:55–56], but it is interesting that, decades earlier, he himself in a note had remarked: 'Philosophical insight into geometrical and arithmetical problems would be excellent. It would open the way to an art of discovery. But it is very difficult.' ('Ein philosophisch Erkentniß der geometrischen und Arithmetischen Aufgaben würde vortreflich seyn. sie würde den Weg zur Erfindungskunst bahnen. aber sie ist sehr schweer.') [AAXVI:55 (1752–W.S. 1755/56)]

[51]Given Kant's remarks quoted at the end of Sect. 1.2, I disagree with Friedman's claim that for Kant, 'the fact of the irrationality of $\sqrt{2}$, which is presumably a fact of *pure* arithmetic, is itself based on successive enumeration and hence on time' (Friedman 1992, p. 116, original emphasis). What depends on time is rather the possibility for humans to come to know that fact. See also Kant's letter to Schultz of November 25, 1788 [AAX:556–557] and Parsons' comments on it (Parsons 1984, pp. 116–117).

[52]This is known as Theaetetus' Theorem, although Plato's dialogue to which it owes its name gives no proof; for the ancient history of the theorem and its proofs, see Mazur (2007). Kant (who does not call the theorem by that name) may well have seen it, with a proof, in Sect. 137 of Johann Segner's *Anfangsgründe der Arithmetik* (von Segner 1764) to which he refers, in a different context, at B15. The method to extract the square root of larger numbers that Kant refers to at AAXI:209 corresponds to the method given by Segner in Sect. 136. (The same material is also present in Michael Stifel's *Arithmetica Integra* (Stifel 1544) of which Kant owned a copy (Warda 1922, p. 40).)

[53]Note that the procedure to extract roots in effect starts with the same test.

generally, for him any algebraic means of establishing the irrationality of $\sqrt{2}$ could be said to depend on the a priori intuition of time, as for him it is characteristic of an algebraic proof as such that it 'exhibits all the procedure through which magnitude is generated and altered in accordance with certain rules in intuition' [A717/B745].[54] 'Step by step', as Friedman comments on that statement (Friedman 1992, p. 120n.42).

It is in reply to Rehberg's second question, why the understanding cannot think $\sqrt{2}$ in numbers, that Kant rejects, as we have seen, the identification of that square root with a certain potentialy infinite sequence, because of the essential incompletability of the latter. In the following, I attempt to reconstruct the ground on which for Kant this incompletability is objectionable.

1.4 Infinite Sequences as Concepts and as Objects

One difference between a potentially infinite sequence and an image is that the parts of an image all exist simultaneously, whereas the parts (elements) of a potentially infinite sequence do not.[55] In an image, the elements of the sequence that are not yet there obviously cannot be shown. Moreover, the fact *that* there are further elements yet to come, which is part of the concept of potentially infinite sequence, cannot itself be intuitively represented in the image.[56] This is because for Kant, there is nothing to the sequence that can be given intuitively, and hence synthesized, but the elements constructed so far themselves.[57] (When we write 0.333 . . . , we understand

[54]'so stellet sie alle Behandlung, die durch die Größe erzeugt und verändert wird, nach gewissen allgemeinen Regeln in der Anschauung dar'.

[55]Compare AAXVII:397 no. 4046 (1769? 1771?): 'The omnitudo collectiva in One or totality rests on the positione simultanea. From the multitudine distributiva I can conclude to the unitatem collectivam, but not from the omnitudine, because the progression is infinite and not complete.' ('Die omnitudo collectiva in Einem oder totalitaet beruhet auf der positione simultanea. Aus der multitudine distributiva kan ich auf die unitatem collectivam schließen, aber nicht aus der omnitudine, weil die Progression unendlich ist und nicht complet.'); also AAXVII:700 (around 1775–1777): 'The infinite of continuation or of collection. The infinitely small of composition or decomposition. Where the former is the condition, the latter does not occur." ('Unendlich der Fortsetzung oder der Zusammennehmung. unendlich klein der composition oder decomposition. Wo das erstere die Bedingung ist, findet das letztere nicht statt.')

[56]Note that ideal, adequate givenness of a potentially infinite sequence does not consist in its being given as an actually infinite sequence (for that would contradict the essence of the object qua potentially infinite), but in the givenness of the whole finite initial segment generated so far, however large the number of its elements may be, together with the open horizon that indicates the ever present possibility to construct further elements of the sequence. The absence of such further elements from an intuition of the sequence at a given moment does not make it an inadequate intuition, because they do not yet even exist. In contrast, the reason why our intuition of a physical object at a given moment is necessarily inadequate is precisely that, as a matter of three-dimensional geometry, any concrete view of it hides parts that do at that moment exist.

[57]The order relation is represented by the relation between left and right, but that already requires an act of the understanding: do we take the order in a sequence to be from left to right, or from right to left?

what the three dots stand for, but the concept they instantiate is not that of infinity but of the number 3.) The understanding gives form to our sensuous intuitions by combining them in certain ways, but these forms are not themselves given to us in their own kind of intuition. In Husserl there is categorial intuition, but not in Kant.[58] Rather, Kant characterizes the human understanding as one

> whose whole power consists in thought, consists, that is, in the act whereby it brings the synthesis of a manifold, given to it from elsewhere in intuition, to the unity of apperception—a faculty, therefore, which by itself knows nothing whatsoever, but merely combines and arranges the material of knowledge, that is, the intuition, which must be given to it by the object. [B145][59]

We can therefore represent a potentially infinite sequence as a concept, and indeed use the concept to construct ever longer finite sequences, but we can never wholly instantiate that concept itself in an intuition. Hence, for Kant the concept of such a sequence is not mathematically constructible.

Note that the impossibility of a potential infinite sequence as a constructible mathematical concept has its ground in the requirement of an image rather than in a property of our capacity of synthesis. For it is the requirement of an image that imposes a condition of completeness (i.e. the simultaneous presence of all its parts).[60] This is also why for Kant it is irrational to try to arrive at a representation

---

[58]'It is true that in Kant's thought the categorial (logical) functions play a great role; but he never arrives at the fundamental extension of the concepts of perception and intuition over the categorial realm' ('In Kants Denken spielen zwar die kategorialen (logischen) Functionen eine große Rolle; aber er gelangt nicht zu der fundamentalen Erweiterung der Begriffe Wahrnehmung und Anschauung über das kategoriale Gebiet.') (Husserl 1984, p. 732).

[59]'dessen ganzes Vermögen im Denken besteht, d.i. in der Handlung, die Synthesis des Mannigfaltigen, welches ihm anderweitig in der Anschauung gegeben worden, zur Einheit der Apperception zu bringen, der also für sich gar nichts erkennt, sondern nur den Stoff zum Erkenntniß, die Anschauung, die ihm durchs Object gegeben werden muß, verbindet und ordnet'. See also A51/B75; B138–139; A147/B186; B302–303n.; A289/B345; *Prolegomena* sections 22, 39 and 57.

[60]According to Kant, in pure mathematics all questions have a definite answer (or else the senselessness of the question can be demonstrated), and the same holds for transcendental philosophy and pure ethics [A476/B504ff.]; see for discussion Posy (1984, pp. 127–128). The general reason Kant gives for this is that in these purely rational sciences, 'the answer must issue from the same sources from which the question proceeds' ('die Antwort aus denselben Quellen entspringen muß, daraus die Frage entspringt') [A476/B504]. It seems to me that, when the details of this answer are spelled out for the case of pure mathematics, the condition of completeness that is imposed by Kant's requirement of an image must enter into the explanation. For intuitionistic mathematics is equally wholly concerned with spontaneous constructions in a priori intuition—where Kant speaks of questions raised by pure reason as concerned with its 'inner constitution' (innere Einrichtung) [A695/B723], Brouwer calls mathematics 'inner architecture' (Brouwer 1949, p. 1249). But in intuitionism, the most we can justify in general is the weaker claim that there are no unanswerable questions, as $\neg\neg(p \vee \neg p)$ is demonstrable while $p \vee \neg p$ is not. For example, consider a potentially infinite lawless sequence of natural numbers α (which, as follows from the considerations in the present paper, for Kant would not be a mathematically constructible concept). We cannot, in general, show that $\exists n(\alpha(n) = 0) \vee \neg\exists n(\alpha(n) = 0)$, due to the open-endedness of such a sequence. We can show $\exists n(\alpha(n) = 0)$ as soon as we have indeed chosen 0 in the sequence, but we are never obliged to make that choice. On the other hand, we can at

of a quantum by generating a potentially infinite sequence. On the other hand, Kant acknowledges that in principle the acts of synthesis can always be continued:

> The infinity of synthesis in a sequence [is], as in a progression, only potential. [AAXVIII:277][61]

and, to repeat an earlier quotation,

> What is only given by composition, is for that reason always finite, even though composition can go on infinitely.[62]

Of course, we may create an image of a finite sequence to construct the concept of an *initial segment* of a potentially infinite sequence, but the potentially infinite sequence is not thereby itself given in intuition. The difference is both philosophically and mathematically important: the collection of all finite sequences is denumerable, the collection of all potentially infinite sequences is not.

We also can associate to the concept of the potentially infinite sequence a schema, as a method to construct in intuition ever longer initial segments. Indeed, in his letter to Rehberg, Kant says that $\sqrt{2}$ 'is actually no number, but only a determination of magnitude by means of a rule of enumeration', and he seems to hold this for (irrational) real numbers more generally.[63] But as Kant emphasizes, in a different text and for a different reason, a schema is not itself an image [A142/B181]. So Kant is not only saying that $\sqrt{2}$ is no number (in his sense), but that it is no proper object. Note also that the rule is, in one sense, given in intuition when written down. But that is not the sense needed here: the written rule is a finite object, whereas what is under discussion here is the intuition of a potentially infinite sequence.[64]

Kant's acknowledgement that composition can go on infinitely certainly involves a knowledge that time is in some sense infinite, as all composition takes place in, and hence presupposes, time. What, then, of the suggestion (above, p. 8) that Kant's theses of time as infinite, given, and sequential, could provide a basis for the construction of infinite sequences?

---

any time show $\neg\neg(\exists n(\alpha(n) = 0) \vee \neg\exists n(\alpha(n) = 0))$ (which also shows that the original question is not senseless). Intuitionism, however, accepts Kant's claim for questions that ask whether a given construction of finite character is possible in a given finite system; e.g. Brouwer (1949, p. 1245).

[61] 'Die Unendlichkeit der Synthesis in einer Reihe [ist] wie im progressu blos potential.'

[62] XVIII:378 no. 5897 around 1780–1789?: 'Was nur durch die composition gegeben wird, ist darum immer endlich, obgleich die composition ins Unendliche geht.'

[63] 'eine Irrationalzahl ... ist ... wirklich keine Zahl, sondern nur eine Großenbestimmung durch eine Regel des Zählens' [AAXIV:57] Compare in one of the drafts: 'a square root, if necessary one that is itself no number, but only the rule to approximate it as closely as one wishes' ('eine Qvadratwurzel ..., allenfalls eine solche, die selbst keine Zahl, sondern nur die Regel der Annäherung zu derselben, wie weit man es verlangt') [AAXI:210].

[64] Compare on this point also Wittgenstein: '"We know the infinity from the description." Well, then only this description exists and nothing else.' ('"Wir kennen die Unendlichkeit aus der Beschreibung." Nun, dann gibt es eben nur diese Beschreibung und nichts sonst.') (Wittgenstein 1964, p. 155).

Kant says that time, in its original representation, is not a concept, but is *given* to us, and as unlimited at that:

> 5. The infinitude of time signifies nothing more than that every determinate magnitude of time is possible only through limitations of one single time that underlies it. The original representation, time, must therefore be given as unlimited. But when an object is so given that its parts, and every quantity of it, can be determinately represented only through limitation, the whole representation cannot be given through concepts, since they contain only partial representations; on the contrary, such concepts must themselves rest on immediate intuition. [B47–48][65]

But Kant denies that we can represent time itself in a mode of intuition proper to it, and repeatedly says that time itself cannot be perceived, e.g.:[66]

> For time is not viewed as that wherein experience immediately determines position for every existence. Such determination is impossible, inasmuch as absolute[67] time is not an object of perception with which appearances could be confronted. [A215/B262][68]

According to Kant, we can represent time as an object only indirectly, by analogy [A33/B50], 'under the image of a line, in so far as we draw it' ('unter dem Bilde einer Linie, so fern wir sie ziehen') [B156].[69] As soon as we conceptualize time, that is, come to think of it as an object to which concepts apply, then it has to be represented by a construction in space.[70] Indeed, for Kant the intuitiveness of our representation of time is concluded to from the possibility to represent it spatially, and we derive all properties of time not from a direct representation of it, but from

[65]'(5) Die Unendlichkeit der Zeit bedeutet nichts weiter, als daß alle bestimmte Größe der Zeit nur durch Einschränkungen einer einigen zum Grunde liegenden Zeit mäglich sei. Daher muß die ursprüngliche Vorstellung Zeit als uneingeschränkt gegeben sein. Wovon aber die Teile selbst, und jede Größe eines Gegenstandes, nur durch Einschränkung bestimmt vorgestellt werden können, da muß die ganze Vorstellung nicht durch Begriffe gegeben sein (denn die enthalten nur Teilvorstellungen), sondern es muß ihnen unmittelbare Anschauung zum Grunde liegen.'

[66]Here also, Brouwer and Husserl disagree with Kant; e.g. Brouwer (1907, pp. 104–105), claims that the one-dimensional temporal intuitive continuum is given as an object without requiring the givenness of any other object; for Husserl, see Husserl (1928), in particular pp. 436–437 and 471–473.

[67][By 'absolute', I take Kant here to mean 'not in relation to any objects whose appearances are temporally determined', in analogy to his explanation of the term 'absolute space' in the note at A429/B457.]

[68]'die Zeit wird nicht als dasjenige angesehen, worin die Erfahrung unmittelbar jedem Dasein seine Stelle bestimmte, welches unmöglich ist, weil die absolute Zeit kein Gegenstand der Wahrnehmung ist'. Also A32–33/B49; A37/B54; B219; B225; B226; B233; B257.

[69]Following Böhme (1974, p. 272), I take it that Kant is referring not to time as such but to time in this relation to space and movement when he writes that 'The pure image ... of all objects of the senses in general is time' ('Das reine Bild ... aller Gegenstände der Sinne aber überhaupt, die Zeit') [A142/B181–182].

[70]AAXIV:55 (1790): 'But without space, time itself would not be represented as a magnitude and this concept would have no object at all.' ('Aber ohne Raum würde Zeit selbst nicht als Größe vorgestellt werden und überhaupt diese Begriff keine Gegenstand haben.')

the line [A33/B50] (except that the reference to the act of drawing is essential for the representation of succession [B154–155]).

A consequence for Kant's view is that the intrinsic possibilities and limitations of spatial representation also condition our representation of time as an object. In the Transcendental Aesthetic, Kant argues that space is given to us as infinite [B39–40]. An elucidation is given in a later manuscript, 'Über Kästners Abhandlungen' of 1790.[71] Kant there distinguishes between mathematical infinity and metaphysical infinity. It is the latter that according to Kant is an 'actual (but only metaphysically real) infinity'.[72] It is actual because it is present in all of our experiences, and Kant therefore says this infinity is *given* to us. It is also metaphysical, because by that qualification Kant means that it pertains to the subjective forms of our sensibility. (The more usual notion of metaphysics Kant refers to as 'dogmatic metaphysics'.) At the same time, Kant repeats the point he had made, in somewhat different words, in the Transcendental Aesthetic [B39] that actual, metaphysical space cannot be brought under a concept that we would be capable of constructing. In fact, metaphysical, actual infinity is the precondition for the potential infinity of our mathematical constructions. It is the former that guarantees the presence of indeterminate space in which mathematicians construct determinate parts.[73]

As any such constructed determinate part will be finite, we can represent in a determinate way only finite segments of time in spatial intuition. When we represent such a finite segment of time by a finite line, the part of time that is yet to come, the future, is represented in an indeterminate way by the part of metaphysical, given space into which we have not yet extended the line but can do so if we wish.[74] But as according to Kant metaphysical space as such is unconceptualizable, the finite line we have drawn and metaphysical, given space do not together make up an image in which the concept of a potentially infinite segment of time is constructed. Metaphysical space is not an image or part thereof, but a condition of possibility for images (see also footnote 69). This means that we cannot represent time in intuition

[71] AAXX:410–423, in particular 417ff. Written for, and indeed used by, Johannes Schultz; see the latter's 'Rezension von Johann August Eberhard, Philosophisches Magazin' (Schultz 1790), and Kant's letters to Schultz of Summer 1790: AAXI:183; AAXI:184; AAXI:200; AAXI:200–201.

[72] That concise phrase occurs in a longer passage that Kant deleted; but the content of the passage agrees with the main text (in particular pp. 420–421). The sentence containing this phrase runs: 'For that one can extend a line into the infinite, or surfaces as far as one wishes, this potential infinity, which is the only one that the mathematician needs to base his determinations of space on, presupposes that actual (but only metaphysically real) infinity and is possible only under this presupposition.' ('Denn daß man eine Linie ins Unendliche fortziehen oder Ebenen so weit man will aus einander rücken kan diese potentiale Unendlichkeit welche der Mathematiker allein seinen Raumesbestimmungen zum Grunde zu legen nöthig hat setzt jene actuelle (aber nur metaphysisch wirkliche) Unendlichkeit voraus und ist nur unter dieser Voraussetzung möglich.') [AAXX:418].

[73] As the Transcendental Aesthetic is concerned with metaphysical infinity, not mathematical infinity, it gives necessary, but not sufficient conditions for mathematical cognition. These need to be completed by the Axioms of Intuition. See for a detailed discussion of this point Sutherland (2005).

[74] See on this point Michel (2003, p. 112).

as a potentially infinite object. It follows that, although there is for Kant a specific sense in which time is given to us and is given to us as unlimited, this does not provide us with a basis for the construction of the concept of a potentially infinite sequence.

1.5 Concluding Remark

The above arguments are general: for Kant the concept of no potentially infinite sequence whatsoever can be constructed by us, be it in a mathematical context or not. An incompletable process, even when fully specified, can never result in one, finished image.[75] In the case of the natural numbers, this means that Kant's position allows him to construct every one of them, one after the other, but not the potentially infinite sequence of them. It also means that Kant's position does not allow him to identify real numbers with potentially infinite sequences. (Likewise, any other explicit construction of a real number as an object out of infinitely many elements, such as a Dedekind cut, is impossible.) This changes when one recognizes what Husserl called 'categorial intuition', and accepts that the flow of time, together with its structuring moments of retentions and protentions, is given in an intuition proper to it; for this opens the possibility of applying categorial intuition to the flow of time, and then on that basis construct potentially infinite sequences as objects in intuition, as Brouwer did. That leads to a far richer mathematics.[76]

Acknowledgements Earlier versions were presented at CUNY, New York, November 6, 2008; at REHSEIS, Paris, January 16, 2009; at the Oskar Becker Tagung, Bad Neuenahr/Ahrweiler, February 6, 2009; at IHPST, Paris, March 23, 2009; at Philosophy and Foundations of Mathematics: Epistemological and Ontological Aspects (a conference dedicated to Per Martin-Löf on the occasion of his retirement), Uppsala, May 7, 2009; at the meeting of the Société des études kantiennes en langue française, Lyon, September 8, 2009; at the joint philosophy-mathematics seminar (CEPERC/UFRAM) in Marseille, March 10, 2010; and at the logical-philosophical seminar at Charles University, Prague, March 28, 2011. I thank the audiences for their questions and comments, and also Carl Posy, Ofra Rechter, Lisa Shabel, Pirmin Stekeler-Weithofer, Neil Tennant, Robert Tragesser, and an anonymous referee.

[75]In this sense, for Kant potentially infinite sequences would seem to be even more problematic than actually infinite ones; the latter might still be representable in an image by other minds than ours.

[76]This is not to suggest that Husserl actually influenced Brouwer; rather, in my view, the ideas that Brouwer independently developed are best understood in the framework that Husserl provides. See van Atten (2007) for a phenomenological analysis of Brouwer's choice sequences.

References

Kant's works are referred to as follows:

A *Kritik der reinen Vernunft*. First edition. Hartknoch, Riga, 1781. Edition used: W. Weischedel (ed.), Suhrkamp, Frankfurt, 1974.

AA *Gesammelte Schriften*. Vols. I–XXIX. Hrsg. von der Königlich-Preussischen Akademie der Wissenschaften zu Berlin, 1902–.

B *Kritik de reinen Vernunft*. Second edition. Hartknoch, Riga, 1787. Edition used: W. Weischedel (ed.), Suhrkamp, Frankfurt, 1974.

English translations of AA are my own; those of A and B are taken from N. Kemp Smith's translation *Immanuel Kant's Critique of Pure Reason*, St. Martin's Press, New York, 1965.

van Atten, M. 2007. *Brouwer meets Husserl. On the phenomenology of choice sequences*. Dordrecht: Springer.

Beiser, F. 2008. August Wilhelm Rehberg. In *The Stanford encyclopedia of philosophy*, ed. E. Zalta, CSLI. http://plato.stanford.edu/archives/win2008/entries/august-rehberg/.

Böhme, G. 1974. *Zeit und Zahl. Studien zur Zeittheorie bei Platon, Aristoteles, Leibniz und Kant*. Frankfurt: Klostermann.

Boniface, J. 2002. *Les constructions des nombres réels dans le mouvement d'arithmétisation de l'analyse*. Paris: Ellipses.

Brouwer, L.E.J. 1907. *Over de grondslagen der wiskunde*. PhD thesis, Universiteit van Amsterdam.

Brouwer, L.E.J. 1913. Intuitionism and formalism. *Bulletin of the American Mathematical Society* 20:81–96.

Brouwer, L.E.J. 1949. Consciousness, philosophy and mathematics. *Proceedings of the 10th International Congress of Philosophy, Amsterdam 1948*, vol. 3, 1235–1249.

Brouwer, L.E.J. 1992. *Intuitionismus*. Mannheim: Bibliographisches Institut, Wissenschaftsverlag.

Cantor, G. 1872. Über die Ausdehnung eines Satzes aus der Theorie der trigonometrischen Reihen. *Mathematische Annalen* 5:123–132.

Dietrich, A.J. 1916. *Kants Begriff des Ganzen in seiner Raum-Zeitlehre und das Verhältnis zu Leibniz*. Halle: Niemeyer.

Décaillot, A.-M. 2008. *Cantor et la France. Correspondance du mathématicien allemand avec les Français à la fin du XIXe siècle*. Paris: Éditions Kimé.

Friedman, M. 1992. *Kant and the exact sciences*. Cambridge, MA: Harvard University Press.

Hankel, H. 1867. *Vorlesungen über die complexen Zahlen und ihre Functionen. I. Teil. Theorie der complexen Zahlensysteme*. Leipzig: Voss.

Hegel, G.F.W. 1979. *Wissenschaft der Logik, 1. Teil, 1. Band: Die Lehre vom Sein*, volume 5 of *Werke*. Frankfurt: Suhrkamp.

Heine, E. 1872. Die Elemente der Functionenlehre. *Journal für die reine und angewandte Mathematik* 74:172–188.

Husserl, E. 1928. Vorlesungen zur Phänomenologie des inneren Zeitbewußtseins. *Jahrbuch für Philosophie und phänomenologische Forschung* IX:VIII–X, 367–498.

Husserl, E. 1984. *Logische Untersuchungen. Zweiter Band, 2. Teil*, volume XIX/2 of Husserliana, ed. U. Panzer. Den Haag: Martinus Nijhoff.

Klein, J. 1968. *Greek mathematical thought and the origin of algebra*. Cambridge, MA: MIT Press.

Longuenesse, B. 1998. *Kant and the capacity to judge*. Princeton: Princeton University Press.

Maimon, S. 1790. *Versuch über die Transscendentalphilosophie*. Berlin: Voß und Sohn. Digital edition by Andreas Berger, Tübingen, October 2003 (Version A 1.1ß). http://tiss.zdv.uni-tuebingen.de/webroot/fp/fpsfr01_W0304/dokumente/Maimon-VTP-Normal-3.pdf. Page references are to the original.

Mazur, B. 2007. How did Theatetus prove his theorem? In *The envisoned life: Essays in honor of Eva Brann*, ed. P. Kalkavage and E. Salem, 227–250. Philadelphia: Paul Dry Books.

Méray, C. 1869. Remarques sur la nature des quantités définies par la condition de servir de limites à des variables données. *Revue des Sociétés Savantes* 4:280–289.
Méray, C. 1872. *Nouveau précis d'analyse infinitésimale*. Paris: Savy.
Méray, C. 1894. *Leçons nouvelles sur l'analyse infinitésimale et ses applications géométriques. Première partie. Principes généraux*. Paris: Gauthiers-Villars et fils.
Michel, K. 2003. *Untersuchungen zur Zeitkonzeption in Kants Kritik der reinen Vernunft*. Berlin: De Gruyter.
Parsons, C. 1984. Arithmetic and the categories. *Topoi*, 3(2):109–122.
Petri, B., and Schappacher, N. 2007. On arithmetization. In *The shaping of arithmetic: After C.F. Gauss's Disquisitiones Arithmeticae*, ed. N. Schappacher, C. Goldstein and J. Schwermer, 343–374. Berlin: Springer.
Posy, C. 1984. Kant's mathematical realism. *The Monist* 67(1):115–134.
Rehberg, A. 1828. *Saemtliche Schriften*, vol. I. Hannover: Hahn.
Schultz, J. 1790. Rezension von Johann August Eberhard, Philosophisches Magazin. *Jenaer Litteraturzeitung*: 281–284.
von Segner, J.A. 1764. *Anfangsgründe der Arithmetik*. Halle: Renger. Translation and revision by J.W. Segner of his father's *Elementa arithmeticae geometriae* (2nd ed. 1756).
Shabel, L. 1998. Kant on the symbolic construction of mathematical concepts. *Studies in the History and Philosophy of Science* 29(4):589–621.
Stevin, S. 1585. *L'Arithmétique*. Leyde: Plantin.
Stifel, M. 1544. *Arithmetica Integra*. Nürnberg: Petreius.
Sutherland, D. 2005. The point of Kant's axioms of intuition. *Pacific Philosophical Quarterly* 86: 135–159.
Sutherland, D. 2006. Kant on arithmetic, algebra, and the theory of proportions. *Journal of the History of Philosophy* 44(4):533–558.
Warda, A. 1922. *Kants Bücher*. Berlin: Breslauer.
Wittgenstein, L. 1964. *Philosophische Bemerkungen*. Oxford: Blackwell.
von Wolff-Metternich, B.-S. 1995. *Die Überwindung des mathematischen Erkenntnisideals. Kants Grenzbestimmung von Mathematik und Philosophie*. Berlin: De Gruyter.

Chapter 2
Wittgenstein's Diagonal Argument: A Variation on Cantor and Turing[1]

Juliet Floyd

2.1 Introduction

On 30 July 1947 Wittgenstein began writing what I call in what follows his "1947 remark"[2]:

> Turing's 'machines'. These machines are humans who calculate. And one might express what he says also in the form of games. And the interesting games would be such as brought one via certain rules to nonsensical instructions. I am thinking of games like the "racing game".[3] One has received the order "Go on in the same way" when this makes no sense,

[1]Thanks are due to Per Martin-Löf and the organizers of the Swedish Collegium for Advanced Studies (SCAS) conference in his honor in Uppsala, May 2009. The audience, especially the editors of the present volume, created a stimulating occasion without which this essay would not have been written. Helpful remarks were given to me there by Göran Sundholm, Sören Stenlund, Anders Öberg, Wilfried Sieg, Kim Solin, Simo Säätelä, and Gisela Bengtsson. My understanding of the significance of Wittgenstein's Diagonal Argument was enhanced during my stay as a fellow 2009–2010 at the Lichtenberg-Kolleg, Georg August Universität Göttingen, especially in conversations with Felix Mühlhölzer and Akihiro Kanamori. Wolfgang Kienzler offered helpful comments before and during my presentation of some of these ideas at the Collegium Philosophicum, Friedrich Schiller Universität, Jena, April 2010. The final draft was much improved in light of comments provided by Sten Lindström, Sören Stenlund and William Tait.

[2]This part of the remark is printed as §1096 of Wittgenstein et al. (1980), hereafter abbreviated RPP I. See footnote 21 below for the manuscript contexts.

[3]I have not been able to identify with certainty what this game is. I presume that Wittgenstein is thinking of a board game in which cards are drawn, or knobs turned so as to move pieces in a simulated horse race. See below for specifics.

J. Floyd (✉)
Department of Philosophy, Boston University, Boston, USA
e-mail: jfloyd@bu.edu

P. Dybjer et al. (eds.), *Epistemology versus Ontology*, Logic, Epistemology, and the Unity of Science 27, DOI 10.1007/978-94-007-4435-6_2,

say because one has got into a circle. For that order makes sense only in certain positions. (Watson.[4])

The most sustained interpretation of this remark was offered some time ago by Stewart Shanker, who argued (1987, 1998) that its primary focus is philosophy of mind, and specifically the behaviorism embedded within the cognitivist revolution that Turing spawned. Shanker maintains that Wittgenstein is committed to denying Church's thesis, viz., that all (humanly) computable functions are Turing computable. In what follows I shall leave aside Church's thesis: too many issues about it arise for me to profitably canvas the associated problems here, and Shanker is quite clear that he is reconstructing the implications of Wittgenstein's remark and not its specific, local, content. Nor shall I contest the idea – forwarded not only by Shanker, but also by Kripke and Wright (among many others) – that there are fundamental criticisms of functionalism, reductionism, and computationalism about the mind that may be drawn out of Wittgenstein's later thought.[5] Shanker is surely right to have stressed the broad context of Wittgenstein's 1947 remark, which is a lengthy exploration of psychological concepts. And Wittgenstein did investigate the sense in which any model of computation such as Turing's could be said to give us a description of how humans (or human brains or all possible computing machines) actually work, when calculating. Turing offers, not a definition of "state of mind", but what Wittgenstein thought of as a "language game", a simplified model or snapshot of a portion of human activity in language, an object of comparison forwarded for a specific analytic purpose.

Turing sent Wittgenstein an offprint of his famous (1937a) paper "On Computable Numbers, With an Application to the *Entscheidungsproblem*".[6] It contains terminology of "processes", "motions" "findings" "verdicts", and so on. This talk had the potential for conflating an analysis of Hilbert's *Entscheidungsproblem* and the purely logical notion of possibility encoded in a formal system with a description of human computation. As Shanker argues, such conflations without due attention to the idealizations involved were of concern to Wittgenstein. However, as I am confident Shanker would allow, there are other issues at stake in Wittgenstein's remark than philosophy of mind or Church's thesis. Turing could not have given a negative resolution of the *Entscheidungsproblem* in his paper if his proof had turned on a specific thesis in philosophy of mind. Thus it is of importance to stress that in his 1947 remark Wittgenstein was directing his attention, not only to psychological concepts, but to problems in the foundations of logic and mathematics, and to one problem in particular that had long occupied him, viz., the *Entscheidungsproblem*.

In the above quoted 1947 remark Wittgenstein is indeed alluding to Turing's famous (1937a) paper. He discussed its contents and then recent undecidability results with (Alister) Watson in the summer of 1937, when Turing returned to

[4]Alister Watson discussed the Cantor diagonal argument with Turing in 1935 and introduced Wittgenstein to Turing. The three had a discussion of incompleteness results in the summer of 1937 that led to Watson (1938). See Hodges (1983), pp. 109, 136 and footnote 7 below.

[5]Kripke (1982), Wright (2001), Chapter 7. See also Gefwert (1998).

[6]See Hodges (1983), p. 136. Cf. Turing (1937c).

Cambridge between years at Princeton.[7] Since Wittgenstein had given an early formulation of the problem of a decision procedure for all of logic,[8] it is likely that Turing's (negative) resolution of the *Entscheidungsproblem* was of special interest to him. These discussions preceded and, I believe, significantly stimulated and shaped Wittgenstein's focused work on the foundations of mathematics in the period 1940–1944, especially his preoccupation with the idea that mathematics might be conceived to be wholly *experimental* in nature: an idea he associated with Turing. Moreover, so far as we know Wittgenstein never read Turing's "Computing Machinery and Intelligence" Turing (1950), the paper that injected the AI program, and Church's thesis, into philosophy of mind.[9] Instead, in 1947 Wittgenstein was recalling discussions he had had with Watson and Turing in 1937–1939 concerning problems in the foundations of mathematics.

In general, therefore, I agree with Sieg's interpretation of Turing's model in relation to Wittgenstein's 1947 remark. Sieg cites it while arguing, both that Turing was not the naive mechanist he is often taken to be, and also that Wittgenstein picked up on a feature of Turing's analysis that was indeed crucial for resolving the *Entscheidungsproblem*.[10] What was wanted to resolve Hilbert's famous problem was an analysis of the notion of a "definite method" in the relevant sense: a "mechanical procedure" that can be carried out by human beings, i.e., computers, with only limited cognitive steps (recognizing a symbolic configuration, seeing that one of finitely many rules applies, shifting attention stepwise to a new symbolic configuration, and so on).[11] An analysis like Turing's that could connect the notion with (certain limited aspects of possible) *human* cognitive activity was, then, precisely what was wanted. The human aspect enters at one pivotal point, when Turing claims that a human computer can recognize only a bounded number of different discrete configurations "at a glance", or "immediately".[12] Sieg's conceptual analysis explains what makes Turing's analysis of computability more vivid, more pertinent and (to use Gödel's word) more epistemologically satisfying than Church's or

---

[7]Hodges (1983), p. 135; cf. Floyd (2001).

[8]In a letter to Russell of later November or early December 1913; see R. 23 in McGuinness (2008) or in Wittgenstein (2004). For a discussion of the history and the philosophical issues see Dreben and Floyd (1991).

[9]Malcolm queried by letter (3 November 1950, now lost) whether Wittgenstein had read "Computing Machinery and Intelligence", asking whether the whole thing was a "leg pull". Wittgenstein answered (1 December 1950) that "I haven't read it but I imagine it's no leg-pull". (Wittgenstein (2004), McGuinness (2008), p. 469).

[10]Sieg (1994), p. 91; Sieg (2008), p. 529.

[11]The *Entscheidungsproblem* asks, e.g., for an algorithm that will take as input a description of a formal language and a mathematical statement in the language and determine whether or not the statement is provable in the system (or: whether or not a first-order formula of the predicate calculus is or is not valid) in a finite number of steps. Turing 1937a offered a proof that there is no such algorithm, as had, albeit with a different proof, the earlier Church (1936).

[12]As Turing writes (1937a, p. 231), "the justification lies in the fact that the human memory is necessarily limited"; cf. §9 of the paper.

Gödel's extensionally equivalent demarcations of the class of recursive functions, though without subscribing to Gödel's and Church's own accounts of that epistemic advantage.[13]

It is often held (e.g., by Gödel[14]) that Turing's analogy with a human computer, drawing on the assumption that a (human) computer scans and works with only a finite number of symbols and/or states, involves strong metaphysical, epistemological and/or psychological assumptions that he intended to use to *justify* his analysis. From the perspective adopted here, this is not so. Turing's model only makes explicit certain characteristic features earmarking the concept that is being analyzed in the specific, Hilbertian context (that of a recognizeable *step within* a computation or a formal system, a "definite procedure" in the relevant sense). It is not a thesis in philosophy of mind or mathematics, but instead an assumption taken up in a spirit analogous to Wittgenstein's idea that a proof must be perspicuous (*Übersichtlich, Übersehbar*), i.e., something that a human being can take in, reproduce, write down, communicate, verify, and/or articulate *in some systematic way or other.*[15]

If we look carefully at the context of Wittgenstein's 1947 remark, we see that it is Turing's *argumentation* as such that he is considering, Turing's *use* of an abstract model of human activity to make a diagonal argument, and not any issue concerning the explanation or psychological description of human mental activity as such. This may be seen, not only by emphasizing, as Sieg does, that Turing's analysis requires no such general description, but also by noticing that immediately after this 1947 remark Wittgenstein frames a novel "variant" of Cantor's diagonal argument.

The purpose of this essay is to set forth what I shall hereafter call *Wittgenstein's Diagonal Argument*. Showing that it *is* a distinctive argument, that it is a *variant* of Cantor's and Turing's arguments, and that it *can* be used to make a proof are my primary aims here. Full analysis of the 1947 remarks' significance within the context of Wittgenstein's philosophy awaits another occasion, though in the final section I shall broach several interpretive issues.

As a contribution to the occasion of this volume, I dedicate my observations to Per Martin-Löf. He is a unique mathematician and philosopher in having used proof-theoretic semantics to frame a rigorous analysis of the notions of judgment and proposition at work in logic, and in his influential constructive type theory.[16] I like to think he would especially appreciate the kind of "variant" of the Cantor proof that Wittgenstein sketches.

---

[13]See Sieg (2006a, b). Compare Gandy (1988). On Gödel's attitude, see footnote 28 below.

[14]See the note Gödel added to his "Some remarks on the undecidability results" (1972a), in Gödel (1990), p. 304, and Webb (1990). Gödel (somewhat unfairly) accuses Turing of a "philosophical error" in failing to admit that "*mind, in its use, is not static, but constantly developing*", as if the appropriateness of Turing's analysis turns on denying that mental states might form a continuous series.

[15]Wittgenstein's notion of *perspicuousness* has received much attention. Two works which argue, as I would, that it does not involve a restrictive epistemological thesis or reductive anthropologism are Marion (2011) and Mühlhölzer (2010).

[16]See, e.g., Martin-Löf (1984, 1996).

In presenting Wittgenstein's Diagonal Argument I proceed as follows. First (Sect. 2.2.1), I briefly rehearse the Halting Problem, informed by a well-known application of diagonal argumentation. While that argument itself does not, strictly speaking, appear in Turing's (1937a) paper, a closely related one does, at the beginning of its §8 (Sect. 2.2.2). However, Turing frames another, rather different argument immediately afterward, an argument that appeals to the notion of computation by machine in a more concrete way, through the construction of what I shall call a *Pointerless Machine* (Sect. 2.2.3). Next (3) I present Wittgenstein's Diagonal Argument, arguing that it derives from his reading of Turing's §8. And then (4) I present a "positive" version of Russell's paradox that is analogous to Wittgenstein's and Turing's arguments and which raises interesting questions of its own. Finally (5), I shall canvas a few of the philosophical and historical issues raised by these proofs.

2.2 Three Diagonal Arguments

2.2.1 The Halting Problem

Though it does not, strictly speaking, occur in Turing (1937a), the so-called "Halting Problem" is an accessible and well-known example of diagonal argumentation with which we shall begin.[17]

The totality of Turing machines in one variable can be enumerated. In his (1937a) Turing presented his machine model in terms of "skeleton tables" and associated with each particular machine a unique "description number" (**D.N.**), thus Gödelizing; nowadays it is usual to construe a Turing machine as a set of quadruples. In the modern construal, a Turing machine t has as its input-output behavior a partial function $f: N \rightarrow N$ as ollows: t is presented with an initial configuration that codes a natural number j according to a specified protocol, and t then proceeds through its instructions. In the event that t goes into a specified halt state with a configuration that codes a natural number k according to protocol, then $f(j) = k$ and f is said to *converge at* j, written "$f(j)\downarrow$". Otherwise, f is said to *diverge at* j, written "$f(j)\uparrow$". In general, f is partial because of the latter possibility.

Enumerating Turing machines as t_i, we have corresponding partial functions f_i : $N \rightarrow N$, and a partial function $g : N \rightarrow N$ is said to be *computable* if it is an f_i. The set of Turing machines is thus definable and enumerable, but represents the set of *partial* computable functions. Because of this, it is not possible to diagonalize out

[17]Turing's argument in 1937a in §8 is not formulated as a halting problem; this was done later, probably by Martin Davis in a lecture of 1952. For further details on historical priority, see *http://en.wikipedia.org/wiki/Halting_problem#History_of_the_halting_problem* and Copeland (2004), p. 40 n 61.

of the list of computable functions, as it is from a list of, e.g., real numbers in binary representation (as in Cantor's 1891 argument). In other words, the altered diagonal sequence, though it may be defined as a function, is not a computable function in the Turing sense.

The last idea is what is to be proved. (Once the equivalence to formal systems is made explicit, this result yields Turing's negative resolution of the *Entscheidungsproblem.*)

To fix ideas, consider a binary array, conceived as indicating *via* "↑" that Turing machine t_i diverges on input j, and *via* "↓" that it converges on input j. Each t_i computes a partial function $f_i : N \to N$ on the natural numbers, construed as a binary sequence.

t_1	↑	↑	↓	↓	↑	...
t_2	↓	↓	↑	↑	↓	...
t_3	↓	↓	↓	↑	↑	...
t_4	↑	↑	↑	↑	↓	...
t_5	↓	↑	↓	↑	↓	...
...						

Cantor's method of diagonal argument applies as follows. As Turing showed in §6 of his (1937a), there is a universal Turing machine UT_1. It corresponds to a partial function $f(i, j)$ of two variables, yielding the output for t_i on input j, thereby simulating the input-output behavior of every t_i on the list. Now we construct D, the Diagonal Machine, with corresponding one-variable function which on input i computes UT_1 (i, i). D is well-defined, and corresponds to a well-defined (computable, partial) function.

We suppose now that we can define a "Contrary" Turing machine C that reverses the input-output behavior of D as follows: C, with the initial configuration coding j, first proceeds through the computation of D(j) and then follows this rule:

$$(*)\quad \text{If } \mathrm{D}(j)\downarrow, \text{then } \mathrm{C}(j) = \uparrow;$$
$$\text{If } \mathrm{D}\ (j) \uparrow, \text{then } \mathrm{C}\ (j) = 1$$

In other words, if D(j) converges then proceed to instructions that never halt, and if D(j) diverges, then output the code for 1 and enter the halting state.

But there is a contradiction with assuming that this rule can be followed, or implemented by a machine that is somewhere on the list of Turing machines. Why? If C were a Turing machine, it would be t_k for some k. Then consider t_k on input k. By rule (*), *if* t_k converges on k, then it diverges on k; but if it diverges on k, then it converges on k. So t_k converges on k if and only if it diverges on k. This contradiction indicates that our supposition was false.

Rule (*) assumes Halting Knowledge, i.e., that machine C can reach a conclusion about the behavior of D on any input j, and follow rule (*). But to have such knowledge requires going through all the (possibly) infinitely many steps of the D machine. And that is not itself a procedure that we can express by a rule for a one-variable Turing machine. In other words Halting Knowledge is not Turing computable.

Classical philosophical issues about negation in infinite contexts – the worry about what it means to treat a completed totality of steps as just another step – emerge. Turing himself acknowledged as much. In (1937b) he published some corrections to his (1937a) paper. The first fixed a flaw in a definition pointed out by Bernays, thereby narrowing a reduction class he had framed for the Decision Problem. The second, also stimulated by Bernays, made his analysis more general, showing that his definition of "computable number" serves independently of a choice of logic. Turing wrote to Bernays (22 May 1937) that when he wrote the original paper of (1937a), "I was treating 'computable' too much as one might treat 'algebraic', with wholesale use of the principle of excluded middle. Even if this sounds harmless, it would be as well to have it otherwise" (1937d). In his (1937b) correction he modified the means by which computable numbers are associated with computable sequences, citing Brouwer's notion of an overlapping choice sequence, as Bernays suggested he do.[18] This avoids what Turing calls a "disagreeable situation" arising in his initial arguments: although the law of the excluded middle may be invoked to show that a Turing machine *exists* that will compute a function (e.g., the Euler constant), we may not have the means to *describe* any such machine (Turing 1937b, p. 546). The price of Turing's generalization is that real numbers no longer receive unique representations by means of sequences of figures. The payoff is that his definition's applicability no longer depends upon invoking the law of the excluded middle in infinite contexts. The loss, he explains, "is of little theoretical importance, since the [description numbers of Turing machines] are not unique in any case" and the "totality of computable numbers [remains] unaltered" (Turing 1937b, p. 546). In other words, his characterization of the computable numbers is robust with respect to its representation by this or that formal system, this or that choice of logic, or any specific analysis of what a real number really *is*. Today we would say that the class of computable numbers is *absolute* with respect to its representation in this or that formal system.[19] And this too is connected with

[18]Cf. Bernays to Turing 24 September 1937 (Turing 1937d). The corrections using Brouwer's notion of an overlapping sequence are explained in Petzold (2008), pp. 310ff. Petzold conjectures that conversations with Church at Princeton (or with Weyl) may have stimulated Turing's interest in recasting his proof, though he suspects that "Turing's work and his conclusions are so unusual that ... he wasn't working within *anyone's* prescribed philosophical view of mathematics" (2008, p. 308). I agree. But in terms of possible influences on Turing, Bernays should be mentioned, and Wittgenstein should be added to the mix. The idea of expressing a rule as a table-cum-calculating device read off by a human being was prevalent in Wittgenstein's philosophy from the beginning, forming part of the distinctive flavor in the air of Cambridge in the early 1930s, and discussed explicitly in his Wittgenstein (1980).

[19]Gödel, concerned with his own notion of general recursiveness when formulating the absoluteness property (in 1936) later noted the importance of this notion in connection with the independence of Turing's analysis from any particular choice of formalism. He remarked that with

the anthropomorphic quality of his model. For it is not part of the ordinary activity of a human computer, or the general concept of a person working *within* a formal system of the kind involved, to take a stance on the law of the excluded middle.

2.2.2 *Turing's First Argument*

Turing's (1937a) definitions are as follows. A ***circle-free machine*** is one that, placed in a particular initial configuration, prints an infinite sequence of 0's and 1's (blank spaces and other symbols are regarded by Turing as aids to memory, analogous to scratch paper; only these scratch symbols are ever erased). A ***circular machine*** fails to do this, never writing down more than a finite number of 0s and 1s. (Unlike a contemporary Turing Machine, then, for Turing the *satisfactory* machines print out infinite sequences of 0's and 1's, whereas the *unsatisfactory* ones "get stuck" (see footnote 26).) A ***computable number*** is a real number differing by an integer from a number computed by a circle-free machine (i.e., its decimal (binary) expansion will, in the non-integer part, coincide with an infinite series of 0's and 1's printed by some circle-free machine); this is a real number whose decimal (binary) expression is said to be ***calculable by finite means***. A ***computable sequence*** is one that can be represented (computed) by a circle-free machine.

The First Argument begins §8. Turing draws a distinction between the application of Cantor's original diagonal argument and the version of it he will apply in his paper:

> It may be thought that arguments which prove that the real numbers are not enumerable would also prove that the computable numbers and sequences cannot be enumerable. [n. Cf. Hobson, *Theory of functions of a real variable* (2nd ed., 1921), 87, 88]. It might, for instance, be thought that the limit of a sequence of computable numbers must be computable. This is clearly only true if the sequence of computable numbers is defined by some rule.
>
> Or we might apply the diagonal process. "If the computable sequences are enumerable, let α_n be the n-th computable sequence, and let $\phi_n(m)$ be the m-th figure in α_n. Let β be the sequence with $1-\phi_n(n)$ as its n-th figure. Since β is computable, there exists a number K such that $1-\phi_n(n) = \phi_K(n)$ all n. Putting $n = K$, we have $1 = 2\phi_K(K)$, *i.e.* 1 is even. This is impossible. The computable sequences are therefore not enumerable".

The argument Turing offers in quotation marks purports to show that the computable numbers are not enumerable in just the same way as the real numbers are not, according to Cantor's original diagonal argument. (We should notice that

---

Turing's analysis of computability "one has for the first time succeeded in giving an absolute definition of an interesting epistemological notion, i.e., one not depending on the formalism chosen" (Gödel here means a formal system of the relevant (recursively axiomatizeable, finitary language) kind). See Gödel's 1946 "Remarks before the Princeton bicentennial conference on problems in mathematics", in Gödel (1990), pp. 150–153; Compare his Postscriptum to his 1936a essay "On the Length of Proofs", Ibid., p. 399. See footnote 28, and Sieg (2006a, b), especially pp. 472ff.

its structure is reminiscent of the Contrary Machine, framed in the Halting Problem above, which switches one kind of binary digit to another, "negating" all the steps along the diagonal.) However, Turing responds:

> The fallacy in this argument lies in the assumption that β is computable. It would be true if we could enumerate the computable sequences by finite means [JF: i.e., by means of a circle-free machine], but the problem of enumerating computable sequences is equivalent to the problem of finding out whether a given number is the D.N of a circle-free machine, and we have no general process for doing this in a finite number of steps. In fact, by applying the diagonal process argument correctly, we can show that there cannot be any such general process.

This "correct" application of the diagonal argument is, globally, a *semantic* one in the computer scientist's sense: it deals with sequences (e.g. β) and the nature of their possible characterizations. The "fallacy" in thinking that Cantor's diagonal argument *can* apply to show that the computable numbers are not enumerable (i.e., in the original, Cantorian sense of enumerable as "countable") is that we will, as it turns out, be able to reject the claim that the sequence β is computable. So there is no diagonalizing out. The assumption that α_n, the enumeration of computable sequences, is enumerable *by finite means* is false. Turing's First Argument rejects that claim (much as in the Halting Argument above) by producing the contradiction he describes: it follows from treating the problem of enumerating all the computable sequences by finite means (i.e., by a circle-free machine) as "equivalent" to the problem of finding a general process for determining whether a given arbitrary number is or is not the description number of a circle-free machine. This, Turing writes – initially without argument – we cannot carry out in every case in a finite number of steps.

However, Turing immediately writes that this First Argument, "though perfectly sound", has a "disadvantage", namely, it may nevertheless "leave the reader with a feeling that 'there must be something wrong'". Turing has remained so far little more than intuitive about our inability to construct a circle-free machine that will determine whether or not a number is the description number of a circle-free machine, and he has not actually shown how to reduce the original problem to that one. At best he has leaned on the idea that an infinite tape cannot be gone through in a finite number of steps. While this is fine so far as it goes, Turing asks for something else, something more rigorous.

2.2.3 *The Argument from the Pointerless Machine*

Turing immediately offers a second argument, one which, as he says, "gives a certain insight into the significance of the idea "circle-free"". I shall call it the *Argument from the Pointerless Machine* to indicate a connection with Wittgenstein's idea of logic as comprised, at least in part, of tautologies, i.e., apparently sensical sentences which are, upon further reflection, *sinnlos*, directionless, like two vectors which when added yield nothing but a directionless point with "zero" directional information.[20] Since Turing's is the first in print ever to *construct* a machine model to argue over computability in principle, it is of great historic importance, and so worth rehearsing in its own right. More importantly for my purposes here, *it* is the argument that Wittgenstein's 1947 diagonal argument phrased in terms of games.

Turing's second argument is intended to isolate more perspicuously the difficulty indicated in his First Argument. It works by considering how to define a machine $\mathcal{H}$, using an enumeration of all Turing machines, to directly compute a certain sequence, β', whose digits are drawn from the $\phi_n(n)$ along the diagonal sequence issuing from the enumeration of all computable sequences α_n. Recall from 1.2 above that α_n is the nth computable sequence in the enumeration of computable sequences (i.e., those sequences computable by a circle-free machine); $\phi_n(m)$ is the mth figure in α_n. β, used in the First Argument, is the "contrary" sequence consisting of a series of 0's and 1's issuing from a switch of 0 to 1 and vice versa along the diagonal sequence, $\phi_n(n)$. By contrast β' is the sequence whose nth figure is the output of the *nth* circle-free machine on input n: it corresponds to $\phi_n(n)$, which we may think of as the *positive* diagonal sequence. Its construction will make clear how it is the way in which one conceives of the enumeration of α_n (by finite means or not by finite means) that matters.

The Turing machines may be enumerated, for each has a "standard" description number k. Now suppose that there is a definite process for deciding whether an arbitrary number is that of a circle-free machine, i.e., that there is a machine $\mathcal{D}$ which, given the standard description number k of an arbitrary Turing machine $\mathcal{M}$, will test to see whether k is the number of a circular machine or not. If $\mathcal{M}$ is circular, $\mathcal{D}$ outputs on input k "u" (for "unsatisfactory"), and if $\mathcal{M}$ is circle-free, $\mathcal{D}$ outputs on k "s" (for "satisfactory"). $\mathcal{D}$ enumerates α_n by finite means. Combining $\mathcal{D}$ with the universal machine $\mathcal{U}$, we may construct a machine $\mathcal{H}$. $\mathcal{H}$ is designed to compute the sequence β'. But it turns out to be (what I call) a *Pointerless Machine*, as we may see from its characterization.

$\mathcal{H}$ proceeds as follows to compute β'. Its motion is divided into sections. In the first $N-1$ sections the integers $1, 2, \ldots N-1$ have been tested by $\mathcal{D}$. A certain number of these, say $R(N-1)$, have been marked "s", i.e., are description numbers of circle-free machines. In the Nth section the machine $\mathcal{D}$ tests the number N. If N is satisfactory, then $R(N) = 1+R(N-1)$ and the first $R(N)$ figures of the sequence whose description number is N are calculated. $\mathcal{H}$ writes down the $R(N)$th figure of this sequence. This figure will be a figure of β', for it is the output on n of the nth circle-free Turing machine in the enumeration of α_n by finite means that $\mathcal{D}$ is assumed to provide. Otherwise, if N is not satisfactory, then $R(N) = R(N-1)$ and the machine goes on to the $(N + 1)$th section of its motion.

$\mathcal{H}$ is circle-free, by the assumption that $\mathcal{D}$ exists. Now let K be the D.N. of $\mathcal{H}$. What does $\mathcal{H}$ do on input K? Since K is the description number of $\mathcal{H}$, and $\mathcal{H}$ is circle-free, the verdict delivered by $\mathcal{D}$ cannot be "u". But the verdict also cannot be

[20]Compare the discussion in Dreben and Floyd (1991).

"s". For if it were, $\mathcal{H}$ would write down as the *Kth* digit of β' the Kth digit of the sequence computed by the Kth circle-free machine in α_n, namely by $\mathcal{H}$ itself. But the instruction for $\mathcal{H}$ on input K would be "calculate the first R(K) $= R(K-1)+1$ figures computed by the machine with description number K (that is, $\mathcal{H}$) and write down the $R(K)$th". The computation of the first $R(K)-1$ figures would be carried out without trouble. But the instructions for calculating the $R(K)$th figure would amount to "calculate the first $R(K)$ figures computed by $\mathcal{H}$ and write down the $R(K)$th". This digit "would never be found", as Turing says. For at the Kth step, it would be "circular", contrary to the verdict "s" and the original assumption that $\mathcal{D}$ exists ((1937a), p. 247). For its instructions at the Kth step amount to the "circular" order "do what you do".

The First Argument and Turing's Argument from the Pointerless Machine are constructive arguments in the classical sense: neither invokes the law of the excluded middle to reason about infinite objects. Moreover, as Turing's (1937b) correction showed, each may be set forth without presuming that standard machine descriptions are associated uniquely with real numbers, i.e., without presupposing the application of the law of excluded middle here either. Finally, both are, like the Halting argument, computability arguments: applications of the diagonal process in the context of Turing Machines.

But the Argument from the Pointerless Machine is more concrete than either the First Argument or the Halting Argument. And it is distinctive in not asking us to build the application of negation *into* the machine. The Pointerless Machine is one we construct, and then watch and trace out. The difficulty it points to is not that $\mathcal{H}$ gives rise to the possibility of constructing another contrary sequence which generates a contradiction. Instead, the argument is semantic in another way. The Pointerless Machine $\mathcal{H}$ gives rise to a command structure which is empty, tautologous, senseless. It produces, not a contradiction, but an empty circle, something like the order "Do what you are told to do". In the context at hand, this means that $\mathcal{H}$ cannot *do* anything. As Wittgenstein wrote in 1947, a command line "makes sense only in a certain positions".

2.3 Wittgenstein's Diagonal Argument

Immediately after his 1947 about Turing's "Machines" being "humans who calculate", Wittgenstein frames a diagonal argument of his own. This "expresses" Turing's argument "in the form of games", and should be counted as a part of that first remark.

> A variant of Cantor's diagonal proof:
> Let N=F (k, n) be the form of the law for the development of decimal fractions. N is the nth decimal place of the kth development. The diagonal law then is: N=F (n,n) = Def F′ (n).
> To prove that F′(n) cannot be one of the rules F (k, n).
> Assume it is the 100th. Then the formation rule of F′ (1) runs F (1, 1), of F′(2) F (2, 2) etc.

> But the rule for the formation of the 100th place of F′(n) will run F (100, 100); that is, it tells us only that the hundredth place is supposed to be equal to itself, and so for n = 100 it is *not* a rule.
> [I have namely always had the feeling that the Cantor proof did two things, while appearing to do only one.]
> The rule of the game runs "Do the same as..." – and in the special case it becomes "Do the same as you are doing".[21]

As we see, it is the Argument from the Pointerless Machine which Wittgenstein is translating into the vocabulary of language games in 1947. The reference to Turing and Watson is not extraneous. Moreover, the argument had a legacy. Wittgenstein was later credited by Kreisel with "a very neat way of putting the point" of Gödel's use of the diagonal argument to prove the incompleteness of arithmetic, in terms of the empty command, "Write what you write" (1950, p. 281n).[22]

Let us rehearse Wittgenstein's argument, to show that it constitutes a genuine proof. Wittgenstein begins by imagining a "form" of law for enumerating the "decimal fractions" (*Dezimalbrüchen*). We may presume that Wittgenstein has the rational numbers in mind, and in the case of the rational numbers, we know that such a law or rule (e.g., a listing) can exhaustively enumerate the totality. As Cantor showed, this is not true for the totality of real numbers. But the argumentation Wittgenstein sets forth applies whether the presentation of the list exhausts a set or not: all it assumes is that the presentation utilizes the expression of rules for the development of decimal fractions, a way of "developing" or writing them out that utilizes a countable mode of expression. Moreover, Wittgenstein's German speaks of decimal expansion development (*Entwicklung von Dezimalbrüchen*), and ordinarily in German this terminology (*Dezimalbruchentwicklung*) is taken to cover expansions of real numbers as well.[23] So Wittgenstein may well have had (a subset of) the real numbers, e.g., the computable real numbers, in mind as well. "Form" here assumes a space of *possible* representations: it means that we may imagine an enumeration in any way we like, and Wittgenstein does not restrict its presentation. He is articulating, in other words, a generalized *form* of diagonal argumentation. The argument is thus generally applicable, not only to decimal expansions, but to any purported listing or rule-governed expression of them; it does not rely on any particular notational device or preferred spatial arrangements of signs. In that sense, Wittgenstein's argument appeals to no picture, and it is not essentially

---

[21]Wittgenstein (1999), MS 135 p. 118; the square brackets indicate a passage later deleted when the remark made its way into Wittgenstein (1999), TS 229 §1764, published at RPP I §1097. (At *Zettel* §694 only this second remark concerning the proof is published, thereby separating it from the mention of Turing and Watson (Wittgenstein (1970), hereafter Z). The argument as written here occurs here with "F" replacing the original "ϕ", following the typescript.

[22]See also Stenius (1970) for another general approach to the antinomies distinguishing between contradictory rules (that cannot be followed) and contradictory concepts (e.g., "the round square") that is explicitly based on a reading of Wittgenstein (in this case, the *Tractatus*).

[23]On the German see http://de.wikipedia.org/wiki/Dezimalbruch and http://de.wikipedia.org/wiki/Dezimalsystem#Dezimalbruchentwicklung.

diagrammatical or representational, though it may be diagrammed (and of course, insofar as it is a *logical* argument, its logic may be represented formally).[24] Like Turing's arguments, it is free of a direct tie to any particular formalism. Unlike Turing's arguments, it explicitly invokes the notion of a language-game and applies to (and presupposes) an everyday conception of the notions of *rules* and the *humans who follow them*.[25] Every line in the diagonal presentation above is conceived as an instruction or command, analogous to an order given to a human being.

To fix ideas, let us imagine an enumeration of decimal fractions in the unit interval in binary decimal form. Now let $\mathrm{N} = \mathrm{F}(n, n) = \mathrm{Def}\ \mathrm{F}'(n)$, whose graph is given by the diagonal line in the picture below.

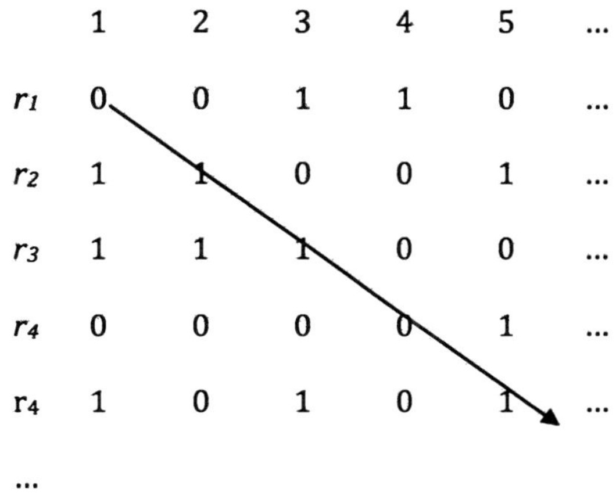

The rule for computing $\mathrm{F}'(n)$ is clear: go down the diagonal of this list, picking off the value of r_n on input n. This rule appears to be perfectly comprehensible and is in *that* sense well defined. But it is not determined, in the sense that at each and every step we know what to do with it. Why? Wittgenstein's "variant" of Cantor's Diagonal argument – that is, of Turing's Argument from the Pointerless Machine – is this.

Assume that the function F' is a development of one decimal fraction on the list, say, the 100th. The "rule for the formation" here, as Wittgenstein writes, "will run F(100, 100)." But this

---

[24]Recall that in his earlier 1938 remarks on the Cantor diagonal argument Wittgenstein was preoccupied with the idea that the proof might be thought to depend upon interpreting a particular kind of picture or diagram in a certain way. Wittgenstein (1978) Part II. There are many problematic parts of these remarks, and I hope to discuss them in another essay. For now I remark only that they are much earlier than the 1947 remarks I am discussing here, written down in the immediate wake of his summer 1937 discussions with Watson and Turing.

[25]Though Turing himself would write that "these [limitative] results, and some other results of mathematical logic, may be regarded as going some way towards a demonstration, within mathematics itself, of the inadequacy of 'reason' unsupported by common sense". Turing (1954), p. 23.

> ... tells us only that the hundredth place is supposed to be equal to itself, and so for $n = 100$ it is not a rule. The rule of the game runs "Do the same as..." – and in the special case it becomes "Do the same as you are doing". (RPP I §1097, quoted above).

We have here an order that, like Turing's $\mathcal{H}$ machine, "has got into a circle" (cf. RPP I §1096, quoted above).[26] If one imagines drawing a card in a board game that says "Do what this card tells you to do", or "Do what you are doing", I think we have a fair everyday representation of the kind of phenomenon upon which Wittgenstein draws.

Wittgenstein's form of circle is, unlike Turing's, explicitly expressed in terms of a tautology. And Turing's argument is distinctive, upon reflection, precisely in producing a tautology of a certain sort. In a sense, Wittgenstein is *literalizing* Turing's model, bringing it back down to the everyday, and drawing out the anthropomorphic, command-aspect of Turing's metaphors.

I have said that Wittgenstein presents a genuine proof in his 1947 remark, and I have been willing to regard it as a "variant" of Cantor's diagonal argumentation. But a qualification is in order. The argument cannot survive construal in terms of a purely extensional way of thinking, and that way of thinking is required for the context in which Cantor's argument is forwarded, a context in which infinite objects are reasoned about and with. What is shown in Wittgenstein's argument is that on the assumption, F′(100) cannot be computed. But not because of the task being infinite. Instead, we are given a rule, that, as Wittgenstein writes, "is *not* a rule" in the same sense. There is, extensionally speaking, something which *is* the value of F(100,100) in itself, and it is either 0 or 1. But if we ask *which* digit it is, we end up with the answer, "F(100,100)", which doesn't say one way or the other what it is, because that will depend upon the assumption that this sequence is the value of F′(100) at 100. The diagonal rule, in other words, cannot be applied at this step. And we have no other means of referring to the *it* that is either 0 or 1 by means of any other rule or articulation on the list that we can *follow*.

One outcome of both Turing's and Wittgenstein's proofs is that the extensional point of view is not or exclusive as a perspective in the foundations of mathematics. Wittgenstein's version of the Argument from the Pointerless Machine shows that the particular rule, F′(n), cannot be identified with any of the rules on the list, because it cannot be applied if we try to think of it as a particular member of the list. The

[26] Watson uses the metaphor that the machine "gets stuck" (Watson 1937, p. 445), but I have not found that metaphor either in Wittgenstein or Turing: it is rather ambiguous, and does not distinguish Turing's First Argument from that of the Pointerless Machine. Both Watson and Turing attended Wittgenstein's 1939 lectures at Cambridge; see (Wittgenstein 1989) where the metaphor of a contradiction "jamming" or "getting stuck" is criticized. I assume this is in response to a worry about the way of expressing things found in Watson 1937. He worries that the machine metaphor may bring out a perspective on logic that is either too psychologistic, or too experimental. He emphasizes, characteristically, that instead what matters if we face a contradiction is that we do not recognize any action to be the fulfillment of a particular order, we say, e.g., that it "makes no sense". As he writes in the 1947 remarks considered here, "an order only makes sense in certain positions". Recall Z §689: "Why is a contradiction to be more feared than a tautology"?

argument shows a "crossing of pictures" or concepts which yields something new. If one likes, it proves that there is a number which is not a number given on the list, for it shows how to construct a rule for a sequence of 0s and 1s which cannot be a rule on the list like the others. The argument would apply, moreover, in any context in which the rule-articulable ("computable") real numbers were asserted to be listed or enumerated in any way according to a rule – including, of course, any context in which, more controversially, one assumed that *only* rule-articulable real numbers *are* real numbers. But this particular assumption is not essential, either to Turing's or to Wittgenstein's arguments, which involve no such necessarily revisionary constructivist or finitistic implications or assumptions.

To recapitulate. Unlike the Halting Problem or the First Argument presented above, Wittgenstein's argument does not apply the law of the excluded middle, or any explicit contradiction or negation *by* the machine. It is not propositional, but in a sense purely conceptual or performative, turning on the idea of a coherently expressed command that turns out, upon reflection, to be empty, thereby generating a rule that we *see* cannot be applied in the same way as other rules are applied. There is of course no direct appeal to community-wide standards of agreement or any explicit stipulation used to drawn the conclusion, so, it is not a purely "conventional" argument, though we see that the order could not be followed by anyone. Oddly, because it turns on a tautology, its conclusion is "positive": it "constructs" a formulable rule that cannot be literally identified with any of the rule-commands on the list of rules supposed to be given. The diagonal then gives one a positive way of creating something new, i.e., a directive that cannot be sensibly followed.

Before commenting further on this version of the proof, I want to underscore that as I have construed it there is no *rejection* of the results of Turing or Cantor involved in accepting Wittgenstein's Diagonal Argument. To make this clear, I shall briefly rehearse an analogous argument.

2.4 The Positive Russell Paradox

Consider the binary array of 0's and 1's anew, but this time as a membership chart for an arbitrary set S.

$x_i \in x_j$?	1	2	3	4	...	
1	1	0	0	1	1	...
2	0	1	0	1	1	...
3	1	1	1	0	1	...
4	0	0	0	0	1	...
...						
						???

Let the array be a diagram of membership relations. At the point (i,j) if we see a "0", this indicates that $x_i \notin x_j$; if we see "1", it means $x_i \in x_j$.

Now let $S = \{x_i | x_i \in x_i\}$. This is the exact complement, so to speak, of the usual Russell set of all sets that are *not* members of themselves: I think of it as the *positive* Russell set. Whenever there is a "1" at a point (i, i) along the diagonal, this means that $x_i \in S$. In a certain sense, S "comes before" Russell's set, for there is no use of negation in its definition.

Is $S = x_j$ for some j? Well there is a difficulty here. For $x_j \in x_j$ iff $x_j \in S$. But $x_j \in S$ iff $x_j \in x_j$. So we are caught in a circle of the form "it is what it is". This cannot be implemented.

An apparently unproblematic way of thinking is applied here, but two different ways of thinking about S are involved. They are at first blush buried, just as in Russell's usual form of the paradox, but they are there, and they are separable, viz., there is the thinking of S as an object or element that is a member of other sets, and the thinking of S as a concept, or defining condition.

We have here what might be regarded, following Turing and Wittgenstein, as a kind of performative or empty rule. You are told to do something depending upon what the rule tells you to do, but you cannot do anything, because you get into a loop or tautological circle. This set membership question cannot be a question on the list which you can apply, because you cannot apply the set's defining condition at every point. (An analogous line of reasoning may be applied to, e.g., "autological" in the Grelling paradox. Without negation, one does not get a contradiction, but one may generate a question that may be sensibly answered with a either Yes or No question, i.e., with a question that is unanswerable *in that sense*.)

Is the Positive Russell argument "constructive"? In a sense Yes. It does not have to be seen to apply to actually infinite objects and name them directly, or invoke any axioms of set theory involving the infinite, though of course it might.[27] So, in this other sense, No. Its outcome is that there is an essential lack of uniformity marking the notion of a rule that can be applied. It involves no use of negation in the rule itself. So what is essentially constructive here is the implication: *If* you write the list as a totality, *then* you will be able to formulate a new rule. And *it* will yield a question one cannot answer without further ado, i.e., *that* rule will not be applicable in the same sense.

The Positive Russell argument refers to an extensional context, that of sets. But there is a creative, "positive" aspect of the argument that emerges, just as it does in Turing's and Wittgenstein's Pointerless Arguments. One must appreciate something or see something about what does *not* direct (any)one to do a particular thing, or assert the existence of a particular solution – rather than being forced to admit the existence of something. Cantor's diagonal argument is often presented as doing the latter, and not the former. But, as Turing and Wittgenstein's proofs make clear, Cantor's argumentation is actually furnishing the materials for more than one

[27] S is empty by the axiom of foundation. Quine worked with *Urelemente* of the form $x=\{x\}$, sets whose only members are themselves. (Quine (1937), Reprinted in Quine (1953, 1980)).

kind of argument. Such, I suggest, is Wittgenstein's point in writing in the above-quoted remark of 1947 that Cantor did two different things. This is not to deny that Wittgenstein's argument is insufficient for Cantor's wider purposes, just as Turing's is, and for the same reason. These later "variants" of Cantor's argument are proofs with and about rules, not proofs utilizing or applying to actually infinite totalities. Nevertheless, we can distinguish Cantor's argumentation from his proof and from its applications, and regard what Turing and Wittgenstein do as "variants" of what Cantor did.

2.5 Interpreting Wittgenstein

The "pointerless" proofs I have considered are down-to-earth in the way Wittgenstein and Turing liked: the "entanglement" in the idea of an exhaustive listing of rules is exhibited in the form of a recipe for a further rule, and the diagonal argument is conceived as a kind of process of conceptualization that generates a new kind of rule. The reasoning in both cases, is, moreover, presented in a way unentangled with any expression in a particular formalism. This does not mean that the arguments are unformalizeable, of course: certainly they apply, as Turing taught us, to formal systems of a certain kind. And a Turing Machine may well be conceived of as a formal system, its activities encodable in, e.g., a system of equations. But Turing's Machines, being framed in a way that is unentangled with a specific formal system, also offer an analysis of the very notion of a formal system itself. This allows them to make general sense of the range of application of the incompleteness theorems, just as Gödel noted.[28]

Turing's and Wittgenstein's arguments from pointerless commands *evidently* do an end run around arguments over the application of the law of the excluded middle in infinite contexts, as other diagonal arguments do not. In this sense, they make logic (the question of a choice of logic) disappear. But I hope that my reconstruction of Wittgenstein's Diagonal Argument will go some distance toward in responding to the feeling some readers have had, namely, that Wittgenstein takes Cantor's proof to have no deductive content at all. It has been held that Wittgenstein took Cantor to provide only a picture or piece of applied mathematics warning against needless efforts to write down all the real numbers.[29] And it is true that Turing's and Wittgenstein's arguments require us to conceive of functions as presented through a collection of commands, rules, directives, in an *intensional* fashion. But they leave

---

[28]In a note added in 1963 to a reprinting of his famous 1931 incompleteness paper, Gödel called Turing's analysis "a precise and unquestionably adequate definition of the general notion of formal system", allowing a "completely general version" of his theorems to be proved. See Gödel (1986), p. 195. On the subject of "formalism freeness" in relation to Gödel see Kennedy (unpublished). Compare footnote 19.

[29]Hodges (1998).

open in what sense this notion, or the notion of a rule, is meant (i.e., the digits of 0s and 1s are a mere *façon de parler* in the way I have presented the arguments here). A critique of the idea that the extensionalist attitude is the *only* legitimate attitude is implied, though, as I have argued, no refutation of extensionalism, Cantor's Diagonal Proof, or set theory follows.

Of course, Wittgenstein's remarks criticizing extensionalism as an exclusively correct point of view are well known. So are his suggestions to look upon mathematical statements as commands. However, though I shall not argue the point here, it seems to me that taking Wittgenstein's Diagonal Argument seriously, at its word, should call into question the idea that he is either dogmatic or skeptical about the notion of following a rule and the "intensional" point of view – unless one means that the notion of a rule and the following of a rule in general are something to be *uniformly* understood in terms of a special kind of fact or intuitive insight. Neither Wittgenstein nor Turing believed this. Wittgenstein's Diagonal Argument serves, instead, to call into question forms of constructivism that take the notion of rule-following as clear or uniform. (I hope to discuss elsewhere the interpretations of Fogelin,[30] Kripke and Wright in light of the diagonal arguments I have discussed here.) His "everyday" version of the Argument from the Pointerless Machine, even more than Turing's, shows that there is a way of carrying out Cantor's argumentation that involves and applies to an "everyday" appeal to our sense of our ordinary activities when we compute or follow rules. In this sense, it makes the argumentation intelligible. One might want to say that it is more deeply or broadly anthropomorphic and intensional than Turing's. But that would be misleading. There is no scale involved here.

Thus it seems to me that one of the most important things to learn from Wittgenstein's argument is that the very idea of a single "intensional" approach is not clear off the bat – any more than are the ideas that perception, understanding, and/or thought are intensional. Wittgenstein's "game" argumentation involves, not merely the notion of a rule, recipe, representation or feasible procedure, but some kind of understanding of *us*, that is, those who are reading through the proof: we must *see* that we can do nothing with the rule that is formulated. Not all rules are alike, and we have to sometimes *look and see* how to operate or use a rule before we see it aright.

This last point is what Wittgenstein stressed just before the 1947 remarks I have discussed in this paper. He wrote,

> That we *calculate* with some concepts and with other do not, merely shows how different in kind conceptual tools are (how little reason we have ever to assume uniformity here). (RPP I §1095; cf. Z §347)

One of the most important themes in Wittgenstein's later philosophy starts from just this point. The difficulty in the grammar of the verb "to see" (or: "to follow a rule") is not so much disagreement (over a particular step, or a way of talking about *all* the steps), but instead that we often can get what we call "agreement" much

---

[30]Fogelin (1987).

too quickly, too easily. And thus we may be much too quickly inclined to think that we understand what is signified by (what we conceive of as) "agreement" and "disagreement" (or "rule of computation"). Quietism is one thing, unclear apparent agreement is another. Apparent agreement may well hide and mask the very basis and nature of that agreement itself, and an agreement may well turn out to rest upon a misunderstanding of what we share. Just as we may get someone much too quickly to agree that "Yes, of course the shape and colors are part of what I see", we may get someone much too quickly to agree that "Yes, of course it is not possible to list all the real numbers" (cf. RPP I §1107). The difficulty is not, in such a case, to decide on general grounds whether to revise the principles of logic or not, or whether to resolve an argument by taking sides Yes or No, e.g., with Hilbert or Brouwer. The difficulty is to probe wherein agreement does and does not lie, by drawing conceptual boundaries in a new way and paying attention to the details of a proof. Wittgenstein's and Turing's arguments as I have presented them here are neither revisionary nor anti-revisionary in a global way. What they do is to shift our understanding of what such global positions do and do not offer us.

References

Church, A. 1936. An unsolvable problem of elementary number theory. American Journal of Mathematics 58: 345–363.

Copeland, B.J. (ed.). 2004. The essential Turing: The ideas that gave birth to the computer age. Oxford: Clarendon Press.

Dreben, B., and J. Floyd. 1991. Tautology: How not to use a word. Synthese 87(1): 23–50.

Floyd, J. 2001. Prose versus proof: Wittgenstein on Gödel, Tarski and Truth. Philosophia Mathematica 3(9): 901–928.

Fogelin, R.J. 1987. Wittgenstein. London/New York: Routledge & K. Paul.

Gandy, R.O. 1988. The confluence of ideas in 1936. In: The universal Turing machine: A half-century survey, ed. R. Herken, 55–112. New York: Oxford University Press.

Gefwert, C. 1998. Wittgenstein on mathematics, minds and mental machines. Burlington: Ashgate Publishing.

Gödel, K. 1986. Kurt Gödel collected works. Volume I: Publications 1929–1936. New York: Oxford University Press.

Gödel, K. 1990. Kurt Gödel collected works. Volume II: Publications 1938–1974. New York: Oxford University Press.

Hodges, A. 1983. Alan Turing the enigma of intelligence. New York: Touchstone.

Hodges, W. 1998. An editor recalls some hopeless papers. Bulletin of Symbolic Logic 4(1): 1–16.

Kennedy, J. (unpublished). Gödel's quest for decidability: The method of formal systems; The method of informal rigor.

Kreisel, G. 1950. Note on arithmetic models for consistent formulae of the predicate calculus. Fundamenta Mathematicae 37: 265–285.

Kripke, S.A. 1982. Wittgenstein on rules and private language: An elementary exposition. Cambridge: Harvard University Press.

Marion, M. 2011. Wittgenstein on the surveyability of proofs. In The Oxford handbook to Wittgenstein, ed. M. McGinn. New York/Oxford: Oxford University Press.

Martin-Löf, P. 1984. Intuitionistic type theory. Napoli: Bibliopolis.

Martin-Löf, P. 1996. On the meanings of the logical constants and the justifications of the logical laws. Nordic Journal of Philosophical Logic 1(1): 11–60.
McGuinness, B. (ed.). 2008. Wittgenstein in Cambridge: Letters and documents, 1911–1951. Malden/Oxford: Blackwell.
Mühlhölzer, F. 2010. Braucht die Mathematik eine Grundlegung? Ein Kommentar des Teils III von Wittgensteins Bemerkungen über die Grundlagen der Mathematik. Frankfurt am Main: Vittorio Klostermann.
Petzold, C. 2008. The annotated Turing: A guided tour through Alan Turing's historic paper on computability and the Turing machine. Indianapolis: Wiley Publishing, Inc.
Quine, W.V. 1937. New foundations for mathematical logic. American Mathematical Monthly 44: 70–80.
Quine, W.V. 1953, 1980. From a logical point of view. Cambridge: Harvard University Press.
Shanker, S.G. 1987. Wittgenstein versus Turing on the nature of Church's thesis. Notre Dame Journal of Formal Logic 28(4): 615–649.
Shanker, S.G. 1998. Wittgenstein's remarks on the foundations of AI. New York: Routledge.
Sieg, W. 1994. Mechanical procedures and mathematical experience. In Mathematics and mind, ed. A. George, 91–117. New York/Oxford: Oxford University Press.
Sieg, W. 2006a. Gödel on computability. Philosophia Mathematica 14(2): 189–207.
Sieg, W. 2006b. Step by recursive step: Church's analysis of effective calculability. In Church's thesis after 70 years, ed. A. Olszewski, J. Wolenski and R. Janusz, 456–485. Frankfurt/Paris/Ebikon/Lancaster/New Brunswick: Ontos Verlag.
Sieg, W. 2008. On computability. In Handbook of the philosophy of science: Philosophy of mathematics, ed. A. Irvine. Amsterdam: Elsevier BV.
Stenius, E. 1970. Semantic antinomies and the theory of well-formed rules. Theoria 35–36(36): 142–160.
Turing, A.M. 1937a. On computable numbers, with an application to the Entscheidungsproblem. Proceedings of the London Mathematical Society 2(42): 230–265.
Turing, A.M. 1937b. On computable numbers, with an application to the Entscheidungsproblem. A correction. Proceedings of the London Mathematical Society 43(Part 7, 2nd Series): 544–546.
Turing, A.M. 1937c. Letter to Ethel Sarah Turing. Cambridge, U.K.: King's College Archives, K/1/54, February 11, 1937.
Turing, A.M. 1937d. Correspondence with Paul Bernays. Zürich: Eidgenössische Technische Hochschule Zürich/Swiss Federal Institute of Technology Zürich, Bibliothek.
Turing, A.M. 1950. Computing machinery and intelligence. Mind 59: 433–460.
Turing, A.M. 1954. Solvable and unsolvable problems. Science News 31: 7–23.
Watson, A.G.D. 1938. Mathematics and its foundations. Mind 47(188): 440–451.
Webb, Judson C. 1990 Remark 3, introductory note to Gödel (1972a), in Kurt Gödel collected works. Volume II: Publications 1938–1974, eds. S. Feferman, et al., 281–304. New York: Oxford University Press.
Wittgenstein, L. 1970. Zettel [Z]. Berkeley: University of California Press.
Wittgenstein, L. 1978. Remarks on the foundations of mathematics. Cambridge: MIT Press.
Wittgenstein, L. 1980. Wittgenstein's lectures, Cambridge 1930–32, from the notes of John King and Desmond Lee [DL]. Oxford: Blackwell.
Wittgenstein, L. 1999. The published works of Ludwig Wittgenstein [CD-Rom], Charlottesville, VA/Oxford: Intelex Corporation. Oxford University Press.
Wittgenstein, L. 2004. Ludwig Wittgenstein: Briefwechsel [CD-Rom, Innsbrucker elektronische Ausgabe], ed. M. Seekircher, B. McGuinness, A. Unterkircher, A. Janik and W. Methlagl. Charlottesville, VA: Intelex Corporation.
Wittgenstein, L., and G.H.V. Wright et al. 1980. Remarks on the philosophy of psychology, vol. 1 [RPP I]. Chicago/Oxford: University of Chicago Press/B. Blackwell.
Wittgenstein, L. 1989. Wittgenstein's lectures on the foundations of mathematics: Cambridge, 1939, ed. C. Diamond. Chicago: University of Chicago Press.
Wright, C. 2001. Rails to infinity: Essays on themes from Wittgenstein's philosophical investigations. Cambridge: Harvard University Press.

Chapter 3
Truth and Proof in Intuitionism

Dag Prawitz

Is logic in its essence epistemological or ontological? This question was the starting point of Per Martin-Löf's lecture at the conference at which the contributions to this volume were presented. In this essay I shall limit myself to the more specific question whether the concepts of truth and proof are epistemological or ontological. That proof is an epistemic concept is of course normally not in doubt, whereas opinions differ concerning truth. Some hold that sentences are true in virtue of a reality given independently of us, while others hold that our linguistic expressions are about our experiences or possible experiences and that truth therefore should be understood in terms of what it is to experience or know something. According to the first standpoint, known as realism, truth may be called an ontological concept. The second standpoint, often labelled anti-realism, takes truth to be instead an epistemic notion. Since mathematical intuitionists explain the meaning of their sentences and what it is for them to be true in terms of what counts as proofs of them, intuitionism has commonly been seen as a clear-cut example of an anti-realistic view.

However, Martin-Löf gives a different account of intuitionism. Although he explains the meaning of propositions in terms of proofs, and defines the truth of a proposition as the existence of a proof, he takes truth to be an ontological concept, not explained in terms of any epistemic notions. If one asks how this is possible, the answer is that he takes even proof to be a non-epistemic concept. More precisely, Martin-Löf makes a distinction between two senses of proofs. One sense is an ontological one, and he calls proofs in that sense *proof-objects*. He maintains that what intuitionists have called proofs in their explanations of meaning really amounts to what he calls proof-objects. The other sense of proof is the usual epistemic one, and he calls proofs in that sense *demonstrations*.

This is his present view, which he explained in his lecture and presented in print in 1998 – his latest publication on these issues. It stands in sharp contrast to

D. Prawitz (✉)
Department of Philosophy, Stockholm University, Stockholm, Sweden
e-mail: dag.prawitz@philosophy.su.se

P. Dybjer et al. (eds.), *Epistemology versus Ontology*, Logic, Epistemology, and the Unity of Science 27, DOI 10.1007/978-94-007-4435-6_3,

several earlier works, where he developed an intuitionistically inspired philosophy of mathematics and logic that agreed with the view of intuitionism as an example of an anti-realist standpoint or, as he put it, "an idealistic philosophy in the knowledge theoretical sense" (Martin-Löf 1987, p. 414). In this essay, I shall confront these two positions of Martin-Löf's with each other. The aim is to understand possible reasons for the shift of viewpoints and to evaluate them. The discussion is critical, but I want gratefully to acknowledge the great source of inspiration that Martin-Löf's work has been to me.

As a background, I shall recall some philosophical points from early presentations of intuitionism and the kind of verificationism that was modelled on intuitionistic explanations of meaning. The aim is here to examine the historical development from the perspective of the central theme of this essay, the opposition between what we may call an epistemic and an ontic view of intuitionism.[1]

3.1 Early Intuitionistic Accounts of Propositions, Assertions, and Proof

The notion of proof did not play such a prominent role in the early intuitionistic explanations of meaning as one may be led to think in view of what is now generally known as the *Brouwer-Heyting-Kolmogorov-interpretation*, or for short, the *BHK-interpretation*,[2] of intuitionistic logic. In fact, the idea to explain the meaning of a sentence or proposition A by stating recursive conditions for something to be a proof was never formulated explicitly by any of the three persons who are referred to in the acronym BHK (as now understood). Saying this I do not want to deny that Arend Heyting might very well have accepted a proof-theoretic interpretation of propositions, if one had been proposed to him.

3.1.1 Heyting on Propositions and Assertions

When Heyting explained in the 1930s[3] what a mathematical *proposition* is according to intuitionism, he said alternately that it expresses a *problem*, an *expectation*,

---

[1]For a general account of the development of intuitionistic logic, see van Atten (2009). Sundholm (1983, 1993) discusses in detail early intuitionistic accounts of proofs, drawing special attention to the fact that proofs appear both as objects and as acts and processes. My picture may look a little different from the one of van Atten and Sundholm because of concentrating on another double role of proofs.

[2]A term coined by Troelstra (1977), although the letter K then stood for Kreisel; see further Sect. 3.1.3 concerning this interpretation. Later, Troelstra and van Dalen (1988) let K stand for Kolmogorov, which is now the usual reading of K in the acronym the BHK-interpretation.

[3]Heyting (1930, 1931, 1934).

or an *intention* of (finding) a construction that satisfies certain conditions (while Kolmogorov confined himself to the first of these three alternatives). The term "proof" appeared in the account of propositions only when Heyting (1934, p. 14) explained the meaning of an implication $a \supset b$ as "the intention of a construction which from any proof of a leads to a proof of b".

An *assertion* on the other hand – Heyting followed Frege in distinguishing between propositions and assertions – was explained as signifying the *realization* of the expressed intention.

What is patently clear in all these explanations is that Heyting on behalf of intuitionism espouses an anti-realist standpoint. For instance, he emphasizes that the intention expressed by a proposition "does not concern the existence of a state of affair imagined as independent of us, but an experience imagined as possible".[4] In the same spirit, he contrasts the "classical assertion" with its reference to "a fact of transcendental nature" to the intuitionistic assertion, which ought to be taken as a "confirmation of an empirical fact, to know the realization of the intention expressed by the proposition" (Heyting 1930, pp. 958–959). He continues: "This is then the *Brouwerian assertion* of [a proposition] p: *One knows how to prove* p".[5]

Heyting avoids the term truth and rejects the idea that intuitionism could replace "p is true" by "there exists a proof of p" understood in a realistic vein. Instead, he speaks of the realization of the intention expressed by the proposition, and makes clear that this realization is to be understood epistemically as the actual experience of the construction intended by the proposition, not as the existence of an ontological fact.

One should note that on almost any meaning-theoretical view, an assertion is a knowledge claim, which in mathematics is justified by a proof. Both from a classical and from an intuitionistic point of view, by making the assertion, you indicate knowledge, which in mathematics amounts to having conclusive grounds for your claim in the form of a proof. That proofs are brought in when accounting for assertions is thus not something that characterizes intuitionism – the point is *where* they are brought in. On a classical, realistic view, an assertion normally concerns an objective state of affair. What you claim to know when making an assertion is that this possible state of affair is an actual fact, and to be warranted this claim has to be based on a proof. What is new in the view that Heyting expounds is that proofs are involved already in what you claim, that is, in the very content of the claim, not only in what the claim is based on. Heyting emphasizes this point by repeating: "We note again that, in classical logic as well as in intuitionistic logic, the assertion of a proposition is not the same as a proposition but is the affirmation of a fact. In classical logic it is a transcendent fact; in intuitionistic logic, it is an empirical fact."[6]

[4]"Die Intention geht ... nicht auf einen als unabhängig von uns bestehend gedachten Sachverhalt, sondern auf ein als möglich gedachtes Erlebnis." (Heyting 1931, p. 113).

[5]"Voici donc l'*affirmation brouwerienne* de p: *On sait démontrer* p." (Heyting 1930, p. 959).

[6]Heyting (1930, p. 959).

To make Heyting's point clearer, we may follow Martin-Löf in making a tripartition, distinguishing between assertion, the content of an assertion, and the proposition expressed by an assertion. We can then say that a proposition is classically determined by its truth-condition and intuitionistically by an intended construction; that the content of an assertion is classically that the truth-condition is satisfied – in other words, that the proposition is true, which may be a transcendental fact – while intuitionistically the content is that one has founded the intended construction – in other words, that the intention is realized, which is an empirical fact; and, finally, that an assertion is classically as well intuitionistically an indication of your knowing the fact in question.

3.1.2 Heyting on Proofs

Although proof is not the central concept in Heyting's meaning explanations, the notion is anyway quite present in his accounts of intuitionism as we saw above. When explaining propositions, the use of the term proof was confined to the clarification of implication, but the term proof is closely connected with that of construction, as appears from Heyting saying: "A proof is a mathematical construction which can itself be treated mathematically".[7] To the extent that conversely he was willing to take constructions to be proofs, he could have said that a proposition is the intention of a proof, instead of the intention of a construction.

Heyting's main use of the notion of proof is however when accounting for assertions, and the important point of a proof in this context is made clear in Heyting's declaration: "A proof of a proposition consists in the realization of the construction required by the proposition."[8] Since "the assertion of a proposition means the realization of the intention [expressed by the proposition]",[9] it should be clear that proofs in this context are what allows us to make assertions, in other words, that proofs in this context have their usual epistemic function.

This epistemic function may become obscured by proofs being constructions. Immediately after having explained that the content of a Brouwerian assertion is that one knows how to prove the proposition in question, Heyting stressed that proofs are constructions even when appearing as realizations of intentions of propositions, saying: "The word 'prove' ought to be taken in the sense 'prove by construction'".[10] It is noteworthy that intuitionistic accounts normally do not speak of proofs to explain the knowledge indicated in an assertion, but only to explain the content. In accounts from a realistic point of view, one appeals to proofs to explain what it is to know the fact referred to in an assertion. To do the same in an intuitionistic account

[7]Heyting (1931, p. 248).

[8]Heyting (1934, p. 14).

[9]Heyting (1931, 247).

[10]Heyting (1930, p. 959).

would be to appeal to proofs to say what it is to know that one has a realization of the construction intended by the proposition. It would be to say that to know this is to have proved the fact that the construction obtained is really a proof, in other words, that the proof is a proof. But normally one considers it unnecessary, or even inappropriate, to prove that a proof is a proof. A classical proof gets its epistemic force because of the compelling nature of the inference steps by which it is formed, not by proving that it is a proof of the theorem in question. Similarly, that a construction is a realization of the intention expressed by a proposition is seen from how it was constructed, in other words, from the nature of the steps taken in the construction. In a survey written much later, Heyting (1958) explains: "... every mathematical theorem is the expression of a result of a successful construction. The proof of the theorem consists in this construction itself, and the steps of the proof are the same as the steps of the mathematical construction". This should make it clear that, although proofs are mathematical constructions referred to in the content of a mathematical theorem, they have nevertheless epistemic function by being what establishes the theorem.

There is in this way a great burden put on the concepts of construction and proof: on one hand, they are mathematical objects, but on the other hand, they are processes or actions; an ambiguity that Sundholm has especially drawn attention to (see footnote 1). Above I have also wanted to draw attention to an additional double role played by proofs, namely, that proofs play for Heyting a semantic role in explaining the content of an assertion and an epistemic role in being what establishes an assertion. Since Heyting's semantic explanations do not make systematic use of a concept of proof, this double role is not yet as conspicuous as it will later become within intuitionism. The epistemic character of the meaning explanations is however a salient feature of his approach, and it is on this that his anti-realism is based. When Heyting (1956, sec. 7.1.1) came back in his book *Intuitionism* to semantic issues, he chose to explain the logical connectives by "giving necessary and sufficient conditions under which a complex expression can be asserted" (ibid, p. 97). The epistemic character of his approach thus remained, and his adherence to anti-realism was likewise unfailing.

3.1.3 The BHK-Interpretation

The work by Georg Kreisel (1962) brought the concept of proof into focus within the foundation of intuitionism. His idea was to determine the sense of a sentence A by specifying a construction r_A that for any construction c decides whether c proves that A is true. r_A is defined by recursion over the build-up of sentences. This inspired A. S. Troelstra (1977, p. 977) to present what he called the *Brouwer-Heyting-Kreisel-interpretation* (abbreviated the BHK-interpretation) of intuitionistic logic, part of which was stated as follows:

(a) A proof of $A \wedge B$ consists in a proof of A and a proof of B.
(b) A proof of $A \vee B$ consists in specifying a proof of A or a proof of B.

(c) A proof of $A \rightarrow B$ consists of a construction c which transforms any proof of A into a proof of B (*together* with the insight that c has the property: d proves $A \Rightarrow cd$ proves B).
(d) $\perp$ is an unprovable statement.
(e) If the variable x ranges over a "basic" domain D, we can explain a proof of ∀xAx as a construction c which on application to any $d \in D$ yields a proof cd of Ad, together with the insight that c has this property.
(f) For x ranging over a basic domain D, a proof of ∃xAx is given as a pair c, d, where c is a proof of Ad, and $d \in D$.

The meaning of the logical constants is meant to be explained here in terms of the primitive concept of proof, or, more precisely, by explaining what counts as a proof of a compound sentence in terms of what counts as proofs of the constituents.

Troelstra's presentation is deliberately less formal than Kreisel's, but they are both aiming at explicating from an intuitionistic point of view what it is for something to be a proof in the usual sense of establishing that something holds. For Kreisel a proof of a compound sentence is formed by applying an operation to proofs of the constituents of the sentence. Troelstra does not make such operations explicit, but also in his case proofs of implications and universal quantifications are constructions. The question of the epistemic force of a construction discussed in the previous section (Sect. 3.1.2) comes therefore up again and is especially acute in the case of implication and universal quantification. The view is now that a construction c does not get the epistemic force of proving $A \rightarrow B$ by just having *in fact* the property of yielding a proof of B when applied to a proof of A. A similar remark applies to $\forall xAx$. Consequently, there is this additional requirement in clause (c) and (e) about the insight that the construction c has the required property. Troelstra is here following Kreisel,[11] who said in effect:

> A proof of $A \rightarrow B$ consists of a construction c that transforms any proof of A into a proof of B, together with a proof of the fact that c has this property.

In later publications, Troelstra dropped the demand for insights (cf. footnote 2), thereby following instead Heyting, who never seemed to have been bothered by the problem that Kreisel's additional requirements were supposed to solve, probably for reasons indicated in Sect. 3.1.2: he took it to be evident from the steps by which the constructions had been formed that they had the required property. But we shall soon have reasons to return to the problems raised here.

[11] This is of course a further reason for attributing the interpretation to Kreisel among others; compare footnote 2.

3.2 Dummett's Verificationism

Michael Dummett's work on the theory of meaning was inspired by intuitionism to a great extent, and is of interest to consider here for the light it shows on how intuitionism was perceived philosophically and for seeing how intuitionistic ideas were incorporated in a wider philosophical context.

Wittgenstein was another important inspiration for Dummett, who made the slogan "meaning is use", often ascribed to Wittgenstein, into a more precise thesis, among other things in the form: we learn the meaning of a sentence by learning under what conditions it is correct to use the sentence for making an assertion. The *correctness* of an act of assertion depends on the ground that the speaker has for her assertion, and is to be distinguished from the truth of what is asserted. Dummett argues that the meaning of a sentence must therefore be determined by what counts as a ground for asserting the sentence, an idea that is also supported by other arguments.

Since in mathematics proofs are what count as grounds for assertions, Dummett finds the intuitionistic explanation of the logical constants in terms of proofs to provide a prototype for a theory of meaning built on a notion of ground. To generalize this to ordinary language, he suggests that we speak instead of *verifications*. Some statements in natural language are verified by making certain observations, while others require both observation and inference. In mathematics verification is by inference alone, which is thus a limiting case, opposite to that of observational parlance where verification is by observation alone.

These are the main ideas of what Dummett calls *verificationism*. In short, it may be described as a project to generalize an intuitionistic meaning-theory based on a notion of proof to ordinary language by taking the meaning of a sentence to be determined by what counts as a verification of the sentence.

3.2.1 A Correction of the Intuitionistic Meaning-Theory

Already in his paper *Truth*, Dummett (1959) wanted "to transfer to ordinary statements what intuitionists say about mathematical sentences", then thinking of Heyting's (1956) explanations of the meaning of statements in terms of assertion conditions. Later Dummett (1976) takes instead proof as the central concept, and the same holds for his book *Elements of Intuitionism* (Dummett 1977). There (ibid, pp. 12–13) he attends to the question discussed in Sect. 3.1.3 how the clause for implication is to be formulated, suggesting that a proof $A \rightarrow B$ is "a construction of which we can recognize that applied to any proof of A, it yields a proof of B", and makes a similar suggestion for universal quantification. Thus, in contrast to Kreisel and Troelstra, he does not add such recognition as an extra element to the proof of an implication, but takes it to be a requirement that proofs are to satisfy.

Furthermore, he notes that the intuitionistic meaning-theory needs a modification by distinguishing between, on the hand, *direct* or *canonical* proofs and, on the other, hand *indirect* or *non-canonical* proofs.[12] If proofs are viewed as what establish assertions, it is not right to say, for instance, that a proof of a disjunction has to consist of a proof of one of the disjuncts (as in clause (b) of the BHK-interpretation in Sect. 3.1.3): even from an intuitionistic perspective, the assertion of a disjunction is equally justified if we have inferred it in any other cogent way, say by universal instantiation from a universal statement proved by induction, in which case we may not have proved any of the disjuncts. We must therefore correct the intuitionistic meaning explanations by saying, firstly, that what is specified there is what constitutes direct or canonical proofs of statements of various forms and, secondly, that besides these proofs there are non-direct or non-canonical proofs, which are defined as method for finding direct proofs, and which constitute equally good grounds for assertions.[13] Verificationism must make a similar distinction between direct and indirect verifications for statements outside of mathematics, but it is unclear how indirect verifications are to be defined exactly in that case, because to say that they are methods for finding direct verifications is not appropriate.

3.2.2 *Truth in Verificationism and the Knowability Principle*

When one lays down what counts as a direct verification of a sentence and defines an indirect verification as a method of finding a direct verification, one says in effect what it is for an assertion of the sentence to be correct or warranted, namely that the speaker possesses such a (direct or indirect) verification, but one does not really fix the meaning of the sentence or, more precisely, the content of an assertion made by the use of the sentence. Now, to ask about the content of an assertion is to ask what it is for the asserted sentence to be true; what the speaker is claiming when asserting a sentence is undoubtedly that the sentence is true. Dummett has been unsure about how truth should be conceived of in verificationism. He has sometimes suggested that "the content of an assertion is that the statement asserted has been, or is capable of being, [directly] verified" Dummett (1976, p. 117), but more often he has tended to agree with Heyting's narrower interpretation of the content, which,

---

[12]Dummett (1973, pp. 27–30) discusses in a general way (not especially linked with verificationism) direct and indirect means of verifying statements, and speaks in this connection also of "the canonical means whereby the truth of sentences of various forms is to be established". Dummett (1975) recognizes that we must distinguish between *canonical proofs* referred to in intuitionistic explanations of meaning and *demonstrations*, which are to be seen as methods for finding canonical proofs. The term *canonical form* of a proof was used similarly by Prawitz (1974). Concerning the needed modification, see further footnote 13.

[13]More precisely, the correction must take the form of a definition by a simultaneous recursion of the two notions canonical and non-canonical proof, because when defining what counts as a canonical proof of a compound sentence one cannot in general (e.g. in the case of implication and universal generalization) avoid referring to non-canonical proofs of the constituents.

when expressed in terms of truth, identifies the truth of a sentence with a proof of it having been found, or has opted for a slightly broader notion of truth that is still tensed.

As we see, there are many options for how truth is to be understood that may be thought to be compatible with intuitionistic meaning explanations. At one extreme, truth is identified with the actual existence of a proof, and at the other extreme, with the existence of a proof in an abstract, tenseless sense.

Dummett has tended to be close to the first extreme, usually without really committing himself to a definite position,[14] but one principle that he has supported, and to which I shall return later, is a *knowability principle*, stated in the form of what he calls the *principal K*:

If a statement is true, it must be in principle possible to know that it is true.

3.3 Martin-Löf's Type Theory

In the seventies and eighties, Martin-Löf (1975, 1982, 1984) worked out an intuitionistic type theory and provided it with an explicit meaning-theory (most fully in Martin-Löf 1984). It did not only allow the formalization of constructive mathematics, but also developed previous philosophical ideas further and introduced several new ones. I shall summarize some points that I take to be the most important ones with respect to the theme of this essay.

(1) *Proofs of propositions appear as objects that are dealt with in the theory.*

Within type theory one can make judgements of the form "a is a proof of a proposition A", abbreviated $a : A$, which may also be read "a is an object of type A".[15] As Martin-Löf (1975) remarked, if the proofs of A are objects, they must form a type in view of the doctrine of types. Heyting's declaration quoted in Sect. 3.1.2, "A proof is a mathematical construction which can itself be treated mathematically", is realized here in that way.

(2) *"A proposition is defined by laying down what counts as a proof of the proposition."*[16]

In view of Martin-Löf's later position concerning proofs mentioned in my introduction, it should be noted that there is yet no idea about proofs of propositions lacking epistemic significance. In the absence of any such indication, it must be taken for granted that the notion of proof should be taken in its usual epistemic sense, and thus that the point of (2) is that a proposition is determined by how it is established as true. This seems also clear from other wordings of principle (2), for instance,

---

[14]For an exchange between Dummett and me on this issue, see Dummett (1998) and Prawitz (1998). The issue is discussed again by Prawitz (2012).

[15]Martin-Löf referred here to an essay by Howard (1980), at that time not yet published.

[16]Martin-Löf (1984, p. 11).

"A proposition is defined by how we are allowed to prove it" (Martin-Löf 1975, p. 76). Martin-Löf is thus here following Heyting and the verificationism of Dummett in taking an epistemic approach to meaning.

Principle (2) is only a first approximation of how propositions are understood, which is stated more precisely by:

(3) *A proposition is defined by prescribing how a canonical proof of A is formed as well as how equal canonical proofs of A are formed.*

Similarly, any type is defined by prescribing how its canonical elements and how equal canonical elements of that type are formed. The ideas that we met in Sect. 3.2.1 thus appear here in a more general form. The next principle is something quite new.

(4) *"$a : A$" is the notation for a judgement, not for a proposition.*

Proofs of propositions do not appear as ingredients of other propositions. Hence, there is no proposition to the effect that a is a proof of A. Instead, the notation "a: A" stands for the judgement that a is a proof of A. Frege's distinction between propositions and judgements, which Heyting also made use of (Sect. 3.1.1), gets a wider significance in Martin-Löf's type theory. Like Frege's *Begriffsschrift*, type theory has notations for propositions as well as for judgements that ascribe truth to propositions; if A is a proposition, "A true" stands for such a judgement. In addition, there are judgements of the form $a : A$, from which the judgement A *true* may be inferred. In other words, the truth of a proposition is established by constructing an object a of type A, establishing that a is of type A. The import of principle (4) is that the latter is not accomplished by constructing yet another proof of the kind referred to in principles (1)–(3). Instead, we have:

(5) *If $a : A$ holds, it can be established by inspection of how a has been formed.*

This inspection also takes the form of a proof, but a different kind of proofs than the ones referred to in principles (1)–(3). The proofs in type theory are thus of two kinds: firstly, there are proofs of propositions, which are talked about and dealt with in the theory by way of judgements of the forms $a : A$ and $a = b : A$ (saying that a and b are equal proofs of the proposition A), and secondly, there are proofs used as means to establish such judgements. The proofs of the second kind are built up by inferences and are displayed in tree-formed natural deductions. "When confusion might occur", Martin-Löf (1984, p. 6) remarks, the term "construction" is reserved for proofs of the first kind. But as noted above, this is not to be taken to intimate that there is a difference in essence between proofs of the two kinds with respect to their epistemic status.

Nevertheless, there is meant to be another essential difference in the status between the two kinds of proofs, besides the obvious one that proofs of the first kind are mentioned in the theory while proofs of the second kind are used. At a later stage, Martin-Löf (1994) expressed the difference by saying:

(6) *Judgements of the form "$a : A$" or "$a = b : A$" are analytic while those of the form "A true" are synthetic.*

By this he wanted to convey the idea that to make a judgement of the form *A true* evident, it is not sufficient to subject it to conceptual analysis – instead, we have to go beyond what is contained entirely within the judgement and construct an object a of type A. But a judgement of one of the first two forms, if evident at all, is evident solely by virtue of the meanings of the terms that occur in it. In type theory, a proof of such a judgement is based more precisely on four kinds of inference rules: rules of type formation, introduction rules for a type that say how canonical objects of that type are formed, elimination rules for a type that show how we can define functions on that type, and equality rules that show how a function introduced by an elimination rule operates on the canonical elements of the type in question.[17] The meaning explanations that the type theory is provided with are meant to make all these rules immediately evident.

While the principles (1) and (2) are in agreement with what Kreisel (1962) had tried to do, except that the objects in Martin-Löf's theory are typed, the principles (5) and (6) agree rather with Heyting's assumption discussed in Sects. 3.1.2 and 3.1.3 that no proof is needed to establish the fact that a construction has the desired properties which makes it to a proof of a proposition, because this is evident already from the steps by which it is formed. Martin-Löf's type theory may be seen as substantiating Heyting's idea in this way and to settle the problems concerning how implication and universal generalization are to be interpreted. In this context, principle (4), that is, the idea not to treat a judgement of the form $a : A$ as something that is proved in the same way as propositions, plays an important role.

3.4 Martin-Löf's Siena Lectures and a Subsequent Paper

The meaning-theory that Martin-Löf developed in his work on type theory concerned the language of that theory, but he soon entered into work on a general theory of meaning,[18] first presented at three lectures at Siena in 1983 (Martin-Löf 1985). Here he also developed a meaning-theory for the language of predicate logic as usually formulated, in other words, not as a part of his type theory.

The distinction between *judgements* (or *assertions,* the two terms are later used more or less synonymously by Martin-Löf) and *propositions*, which was so important in type theory, is now dealt with in depth, both historically and

[17] As Martin-Löf (1984) remarks, the introduction and elimination rules correspond to the rules that Gentzen (1935) called by these names and the equality rules correspond to the reduction rules for normalizing a deduction given by Prawitz (1965).

[18] I am following the terminology of Dummett in distinguishing between a meaning-theory, which specifies the meaning of the expressions of a particular language, and a theory of meaning, which has the ambition to elucidate general concepts such as meaning and truth.

philosophically. There is now only one kind of proofs, namely proofs of judgements. Thus, there are no proofs of propositions in the Siena lectures, but instead there are *verifications of propositions*. Dummett's term verification is adopted for what determines the meaning of a proposition, and, like in Dummett, verification is taken to be at bottom a definitely epistemic term, standing for "the act, or process, of verifying something" (ibid, p. 34). This is exemplified by saying that a verification of the atomic proposition 'the sun is shining' or 'the temperature is 10°C' is the direct seeing of the shining sun or a direct thermometer reading, respectively.

The *truth* of a proposition, a concept which plays an essential role in explaining the notions of assertion and proof of an assertion, is equated with the *verifiability* of the proposition; in other words, a proposition A is true if and only if A can be verified.

A *proof* of a judgement saying that A is true is now explained as an act by which one gets to know that A is true, which is what is required to have the right to assert that A is true. However, proofs of judgements turn out to have sometimes the nature of objects. To know that A is true is to know that A can be verified, which is in turn to know how A can be verified, or, in other words, to know a method of verifying A. A proof of the assertion that A is true is therefore also said to be the same as a method of verifying A, in other words, a method for finding a verification of A.

When we come to logically compound propositions in the language of predicate logic, a verification of a proposition A is defined in terms of proofs of assertions of (the truth of) the constituents of A. For instance, a verification of a conjunction $A \& B$ consists of a proof that A is true and a proof that B is true, or in other words, of a method of verifying A and a method of verifying B. A verification of an implication $A \supset B$ is a *hypothetical proof* that B is true under the assumption that A is true. What this means is in turn explained by saying that if it is supplement by a proof that A is true, it becomes a categorical proof that B is true.

As we see, verifications of propositions and proofs of judgements are defined by a simultaneous recursion in terms of each other, and both are epistemic acts or are built up of methods for finding verifications. One may then ask what the point is of distinguishing between verifications and proofs, a question that is not taken up by Martin-Löf. It should be clear by how they are defined that the two notions as used in the Siena lectures do not correspond to the two kinds of proofs in type theory. Instead, the answer is presumably that it is to take care of the distinction between canonical or direct proofs and non-canonical or indirect proofs described in Sect. 3.2.1, a distinction that was made for proofs of propositions in type theory. Direct verifications or canonical proofs now appear as verifications of propositions – they are required to have a certain form, and it is in terms of them that the meaning of propositions is explained. Indirect verifications or non-canonical proofs appear now as proofs of judgements – they are methods for finding (direct) verifications, and to be in possession of them is sufficient to be right in asserting the truth of the propositions in question. It is to be recalled that when defining direct proofs, we cannot in general avoid referring to indirect proofs (see footnote 13).

In a slightly later publication, Martin-Löf (1987) adopts the terminology direct and indirect proof in the sense explained in Sect. 3.2.1. He replaces "verification of a proposition" in the Siena lectures by "direct proof of a proposition", and it is to be understood (although not explicitly said) that when defining the direct proofs of a compound proposition, the references in the Siena lectures to proofs of *assertions* of (the truth of) the constituents of the propositions are to be replaced by references to indirect proofs *of propositions*, that is, indirect proofs of the constituents of the compound proposition. The notion *proof of an assertion* is then free to be used for epistemic acts that make assertions evident, which was the first explanation of this notion in the Siena lecture, and there is no more any need to say as in the Siena lectures that these acts are at the same time methods for verifying propositions. This terminology obviously fits much better to the line of thinking developed in connection with type theory.

The notion of proof of a proposition is still understood as an epistemic notion. This is obvious from the remark that was made in this connection and was quoted in my introduction, namely, that what "makes intuitionism into an idealistic philosophy in the knowledge theoretical sense" (ibid, p. 414) is the fact that in intuitionism the concept of proof of a proposition is conceptual prior to the concept of truth of a proposition. The truth of a proposition has in this connection been identified with direct provability.

In the Siena lectures, truth is not only explained in terms of epistemic notions, it is also argued that to be true is equivalent with being knowable, or more precisely that the following principle holds:

A proposition A is true if and only if A can be proved to be true.

It has been made clear in this connection that provable is meant to be the same as knowable. The implication from the left to the right in this equivalence, which amounts to the same thing as Dummett's principal K (Sect. 3.2.2), is said to hold in virtue of the validity of the proof rule

$$\frac{A \textit{ is true}}{(A \textit{ is true}) \textit{ is provable}}$$

by which Martin-Löf means that one has the right to make the assertion occurring as conclusion if one has the right to make the assertion occurring as premiss. The rule is obviously valid in that sense: As already said, if one has the right to make the assertion *A is true*, one knows how to verify *A*, which is the same as to know a method of verifying *A* or, in other words, a proof of the assertion *A is true*. Knowing such a proof, one also knows that the assertion *A is true* is provable, and then one has the right to assert: *(A is true) is provable*.

3.5 The Epistemic Approach to Meaning and Truth Being Abandoned

In the 1990s, there is a shift in Martin-Löf's thinking as to the nature of truth, proofs, and propositions, and he takes the position that I briefly sketched in the introduction to the essay. It was presented in a paper (Martin-Löf 1998) that had the title "Truth and knowability" and the subtitle "On the principles C and K of Michael Dummett". Principal C says: "If a statement is true, there must be something in virtue of which it is true". Principal K is the one we met in Sect. 3.2.2. Martin-Löf endorses principle C but rejects principle K, and says that the purpose of his paper is to resolve "the perplexities surrounding the principle K", which "has left me with uneasiness". He is concerned with, as he puts it, to find a "corrected form" or "emended version of the principal K". As a basis for his criticism of the principle K, Martin-Löf makes a number of conceptual distinctions.

The two crucial new ideas are to separate concepts that are epistemic from those that are non-epistemic and to distinguish between two radically different senses of proofs, of which one is epistemic and one is non-epistemic. As already indicated in the introduction, the epistemic notion appears when we speak of proofs of judgements, and for it Martin-Löf nowadays uses the term *demonstration*, while the non-epistemic notion appears when we speak of proofs of propositions. For it he uses the term verification or keeps the word proof; later *proof-object* has been the term commonly used for proofs in this sense, and I shall stick to that terminology here. As proofs of propositions did earlier, proof-objects appear in two forms, *canonical* and *non-canonical*; a non-canonical proof-object being as before a method of finding a canonical proof-object.

Among the non-epistemic concepts, we find, besides proof-object, *proposition* and *truth of a proposition*. A *proposition* is defined by its proof condition, which states what a canonical proof-object of the proposition looks like. What is then a canonical proof-object? The only answer given here is that it is the kind of thing that determines what a proposition is. Given the concepts proposition and proof-object, the *truth* of a proposition A is defined as the existence of a canonical proof-object of A.

Although there is only this circular explanation of proposition and proof-object, there are examples that indicate what is intended. An empirical proposition like "my fountain pen is blue", if true, is true in virtue of something in the world, "namely, the blueness of my fountain pen" (ibid, p. 106). So this is the proof-object in this case: it defines the proposition and its existence *makes the proposition true* or *verifies it*. One may say that "to verify" is now taken in a quite different sense from how it was used in the Siena lectures: it is not any longer the seeing of the shining sun but the shining sun itself which verifies the proposition "The sun is shining". But one may also say that the etymological root of "to verify", that is, to make true, is kept, and that it is the view of what it is that makes a proposition true which has changed.

If one restricts oneself to logic and mathematics, it is not so easy to see what the shift of position consists in, because the clauses that define what counts as proof-objects in the different cases (not stated in the paper) have to be understood as being the same as those that previously defined the notion of proof of proposition. The paper "Truth and knowability" is not especially tied to Martin-Löf's type theory, but is a general discussion of the concepts in question, although it is reasonable to see it as a result of a renewed reflection on type theory, which formally remains as before; terminologically, to call the two kinds of proofs that occur in type theory demonstration and proof-object, respectively, seems of course quite apt. What has changed is the perspective in which the formal development is seen, and this change is most easily understood in a context that includes empirical cases as above. The proof-object that makes an empirical proposition A true is an object in the empirical world, that is, something in the world that is given objectively without reference to a perceiving subject, while previously in the Siena lectures what made such a proposition true was the possibility of an act of perception. Thus, it is clear that the previous epistemic approach to meaning has now been abandoned.

What is new in the case of a mathematical proposition A is not that it is said that the existence of a proof of A is what makes A true. This Martin-Löf sometimes said also previously in the context of type theory. But in lack of any explanations to the contrary, by default so to say, the proof was understood as something epistemic, and it was therefore more natural to say that what makes a proposition true is its provability, or the fact that "a proof can be given" (Martin-Löf 1984, p. 11). The new perspective is that a proof-object is just like any other mathematical object, that is, there is nothing particular epistemic about it.

But it is not only that. What unites the mathematical and the empirical case is that what makes a proposition true is the existence of a related proof-object in the world – the empirical world or the mathematical world depending on the nature of proposition. Adopting this new view of proof-objects, Martin-Löf definition of truth in terms of them does not any longer make "intuitionism into an idealistic philosophy in the knowledge theoretical sense". On the contrary, it brings his new position close to, or at least closer to, realism. Asked what the proof-objects are after their epistemic connections have been severed, Martin-Löf and Sundholm often answers that they are just *truth-makers*. In current presentations such as the one I referred to in the introduction, Martin-Löf calls proof-objects, propositions, and truth ontological concepts. He says himself that his notion of truth "is really a version of the correspondence notion of truth, truth as agreement with reality", and adds that the only novelty is that we use the word proof-object for "that thing in reality which has to be there in order for the proposition to be true" (Martin-Löf 1998, p. 112).

Martin-Löf combines intuitionism or constructivism, which he still adheres to, with such an ontological position, which abandons the epistemic, anti-realistic understanding of meaning and truth that intuitionism has been connected with. He arrives in this way at an interesting and quite original combination of ideas.

This finishes my summary of Martin-Löf's new position. Before entering into a discussion of this position, a few more words about what he says on Dummett's principal K. Dummett speaks of true statements and says of them that they must be in principle possible to know as true. Martin-Löf translates this into his terminology as saying that if a proposition is true, then it can be known to be true. If "known" is replaced by "proved", this is verbatim what he himself argued for in the Siena lectures. Clearly, he does not accept this principle any more, but he does not really refute it by an argument; he says only that the principle "seems somehow unlikely". Instead he argues (among other things by referring to Leibniz's principle of sufficient reason) that Dummett's principle K has to be given the following "corrected form":

if a judgement 'A is true' is correct, then the proposition A can be known to be true.

This is his emendation of principle K, but since, according to Martin-Löf's analysis, "correct" means the same as "knowable", this reduces to a triviality: if it is knowable that a proposition A is true, then A can be known to be true.

3.6 Reasons for the Shift

Martin-Löf does not confront the position of the paper "Truth and knowability" with the ideas he held previously. It is therefore difficult to know how the shift of standpoints is to be understood and what may remain of previous ideas. Does he mean that the main previous views were incoherent or seriously mistaken, and that thereby the same holds for the early philosophical accounts of intuitionism and verificationism? If so, one would like to spot the errors. Or is it rather that he has found another philosophical perspective on intuitionism, and finds reasons to prefer it on the whole? In the rest of this paper, I shall indicate some possible answers to these questions. I shall confine myself to three questions: Whether the ontological standpoint is compatible with intuitionism, reasons for rejecting the knowability principle, and arguments for and against ascribing epistemic significance to proof-objects.

3.6.1 Is the Ontological Standpoint Compatible with Intuitionism?

One may ask whether it is really possible to combine Martin-Löf's new ontological view of truth with an intuitionistic standpoint. Investigating different ways in which one may argue in favour of intuitionistic versus classical mathematics, Dummett (1975) argued that one may very well take a realistic view of mathematical objects but nevertheless be convinced that a meaning-theory has to be construed along the lines of verificationism, which gives rise to intuitionistic instead of classical logic. In this connection, he quotes what he calls Kreisel's dictum: "The point is not the existence of mathematical objects, but the objectivity of mathematical truth".[19] But Martin-Löf now takes an ontological view not only of mathematical objects but also of truth, defined in terms of proof-objects existing in the world.

[19]See for instance Dummett (1975, p. 19). It is not known where Kreisel should have said this, but the dictum illustrates well what Dummett has seemed to be a crucial point in a philosophy that results in intuitionistic instead of classical logic.

Dummett has generally argued that what gives substance to a realistic attitude towards the world is the acceptance of a principle of bivalence for what exists and holds in the world, and that by so doing we have to accept the classical laws of logic. Applying his argument to in particular an objective world of proof-objects, we may reason as follows. Either there is a proof-object of a proposition A or there is not, and in the second case there exists instead a proof-object of $\neg A$. Hence, there exists a proof-object of either A or $\neg A$ and therefore also of the disjunction $A \vee \neg A$. Thus, if truth is equated with the existence of a proof object, we have shown that $A \vee \neg A$ is true, for an arbitrary proposition A.

However, Martin-Löf's ontological position is not quite as realistic as it may appear from what I have said so far. Although he equates truth with the existence of canonical proof-objects, which are objects on par with other mathematical objects without epistemic import, he holds that there is a certain conceptual order in which epistemic notions such as justification of a judgement or assertion come before ontological concepts such as truth of a proposition. The reason for this is that truth is after all not simply equated with the existence of a proof-object. What Martin-Löf calls "the official explanation" of truth occurs in the context of explaining what it is that you must know in order to have the right to make a judgement of the form *the proposition A is true* (Martin-Löf 1998, p. 112), and that explanation is that you must know a proof-object of A. It is only when "know" is cancelled at both sides of the equation

to know that A is true is the same as *to know that there is a proof-object of A* (3.1)

that we get the equation

A is true is the same as *there is a proof-object of A*. (3.2)

After the cancellation of "know" in (3.1), the proof-object in (3.2) can be taken to be canonical, since a non-canonical proof-object is a method for finding a canonical one.

This "official explanation" blocks the above argument for asserting the truth of $A \vee \neg A$, even if one accepts its initial use of the law of excluded middle. To have the right to assert the truth of $A \vee \neg A$ for a particular proposition A, we have to know a proof object of $A \vee \neg A$, which has to be understood as the requirement of being in possession of either a canonical proof-object or of a method for finding one. But the only thing we know according to the argument is that there is such a proof-object, and this knowledge comes from knowing that there is a proof-object of one of the disjuncts, without even knowing of which disjunct there is a proof-object. Because of how a proof-object of $A \vee \neg A$ is defined, knowledge of such a proof-object, would imply that one knows how to find a canonical proof-object of either A or $\neg A$, and it is a fact that for some propositions A we do not know this. Therefore, one cannot claim that one knows a proof-object of $A \vee \neg A$ for an arbitrary proposition A, and hence we have no right to assert instances of the law of excluded middle in general.

All that Martin-Löf has said "officially" about truth of propositions is thus that to know that A is true is the same as to know that there is a proof-object of A, which is Eq. 3.1 above. It is misleading to say that one can just cancel "know" in this equation (an operation that Martin-Löf (1987) had already performed but at a stage when proofs were conceived of epistemically). To cancel "know" in a principle does not in general give one a new principle that follows from the first one. In particular, Eq. 3.2 is a principle that is independent of (3.1). As we have seen (Sect. 3.2.2), to say what it is to know that a proposition is true is not to say what truth is. In particular, the Eq. 3.2 is quite ambiguous as long as the notion of existence is not explicated. It is worth discussing further whether Martin-Löf's adherence to the correspondence principle of truth is really compatible with the reading that he wants to give of the Eq. 3.2, but I have to leave that issue here.

3.6.2 Reasons for Rejecting the Knowability Principle

As we have seen, Martin-Löf denounces Dummett's principle K, and must then also reject his analogous claim in the Siena lectures that all truths are provable, since to be provable implies to be knowable. One may think that principle K follows when truth is understood epistemically but loses its validity when construed ontologically. But I do not think that this is the right way of looking at it.

In this case, it is in my opinion easy to explain the shift of standpoints by pointing to a definite mistake in the argument of the Siena lectures, even when assuming the epistemic definition of truth given there. As explained, Martin-Löf argues for the implication

if a proposition A is true, then A can be proved to be true,

by making use of the proof rule to go to the conclusion *(A is true) is provable* from the premiss *A is true* (end of Sect. 3.4). The rule is said to be valid in the sense that if one has the right to make the assertion occurring as premiss, the same holds for the assertion occurring as conclusion. In that sense the rule is certainly valid for the reason stated in Sect. 3.4, but this does not mean that it can be used validly in an argument from assumptions. In particular, it does not allow us to infer the conclusion of the proof rule from the assumption that A is true, distinct from the assumption that we have proved that A is true. Thus, we cannot establish the implication by using the rule in such an argument from assumptions, and hence does not get a ground in that way for asserting the implication.

In the Siena lectures he claims that what this implication means is just that the proof rule in question is valid in the particular sense indicated, but this is hardly congruent with the rest of his text, and, furthermore, given that meaning there would be no reason to reject Dummett's principle K.

However, when the truth of a proposition A is equated as in the Siena lectures with its verifiability, there is an argument for the knowability principle that is both cogent and easy: If A is true, then (by definition) A can be verified, in other words, it is possible (in principle) that A is verified. Furthermore, when A is verified, it is also known that A is verified; this is obvious for the examples of atomic propositions that were given, and when one verifies a compound proposition, one proves that certain propositions are true, and hence, because of the nature of proofs as understood in the Siena lectures, one knows that one is doing that. Having established the implication *if A is verified, A is known to be verified*, it follows that if it is possible that A is verified, it is possible that A is known to be verified, and so known to be true. Hence (by transitivity) if A is true, it is possible to know that A is true.

When the truth of a proposition A is defined as the existence of a canonical proof of A, as in "Truth and knowability" (and two slightly earlier papers, Martin-Löf, 1994, 1995), this argument is blocked. From the mere existence of a canonical proof of a proposition A, kept distinct from the knowledge that there is such a proof, we cannot conclude that A can be proved, that is, that we can find a proof of A. We have no general method of finding a proof for a given proposition A assumed to have a proof (as one has in the case of proofs in formal systems, assuming Markov's principle), and hence there is no ground for saying that we can get in possession of a proof of every true proposition, and thereby get to know that the proposition is true.

It is the definition of truth as the existence of a proof, taken in a tenseless sense, not as actual existence here and now, which is responsible for this blocking of the argument for the knowability principle. The status of the concept of proof, whether it is understood epistemologically or ontologically, does not matter here. If we instead identify truth with provability, it is difficult to see why a knowability principle would not follow, provided that when you come up with a proof, you know that what you have is a proof (which assumption will be discussed in Sect. 3.6.4).[20]

3.6.3 Are the Proofs of the BHK-Interpretation Representations of Proof Acts?

Martin-Löf says in the Siena lectures: "a proof is, not an object, but an act". But later in the same lecture, calling attention to "the tendency of our language towards hypostatization", he also says that "a proof that a proposition A is true is the same as a method of verifying ... A". Even if proofs are primarily acts, there are objects that relate in various ways to these acts and are also called proofs or, at least, are considered to have epistemic import. Furthermore, proof acts can be noted down, and what we then have are linguistic objects that can be seen as records of the acts. They can occur on blackboards or be published in journals, and we normally refer to these objects as proofs. Such a record of an act can be used by another agent to repeat the act, and can be seen as an instruction (Sundholm 1993 talks of "blueprint"

[20]Of course, we have also to assume that provability is taken in a modal sense as the possibility of finding a proof. In "Truth and knowability", Martin-Löf seems to allow himself to speak of provability without intending any such sense, meaning only that there is a proof.

or recipe) for how to perform a proof act. It has clearly epistemic import, and we may call it a *representation of a proof (act)* when there is need to distinguish it form the act itself.

Therefore, it is not reasonable to hold generally that mathematical objects cannot have an epistemic character. Hence, the not disputed fact that the proofs (occurring in BHK-interpretation or the proof-objects in type theory) in terms of which intuitionism explains the meaning of propositions are objects and not acts cannot be the reason for saying that they lack any epistemic import. A question that remains to be discussed is therefore why they are now said to carry no epistemological import, a standpoint that is formulated expressively by Sundholm (1998, p. 194) saying: "a proof-object is a mathematical object like any other, say a function in a Banach space, or a complex contour-integral, whence, from an epistemological point of view, it is of no more forcing than such objects".

Proofs in Gentzen's intuitionistic system of natural deduction are representations of proof acts in the sense suggested above. To show that such a proof formulated within a first order language is sound with respect to the BHK-interpretation, one converts it into a term in a suitably extended lambda calculus where the proofs of the BHK-interpretation can be defined; Martin-Löf's type theory is well suited for this, and we then get a proof-object in that theory. Such a conversion is especially easy in the case of Gentzen's natural deductions: indeed, the lambda-term or the proof-object in type theory can be seen as just a linear notation for a Gentzen deduction. A term or proof-object obtained in this way thus represents a proof act.

Can similarly all proof-objects be seen as representations of proof acts? If so, it would be a counter-argument against Martin-Löf's and Sundholm's claim that they lack epistemic character. If we restrict the extended lambda calculus or type theory to operations that correspond to inference steps in intuitionistic natural deduction, the answer is of course yes. But we cannot make such a restriction, if the BHK-interpretation is to remain a reasonable interpretation of an intuitionistic language. When we define a canonical proof of $A \supset B$ as a construction of the form $\lambda x t(x)$ where $t(u)$ is a proof, not necessarily canonical, of B, provided that u is a proof of A, it is crucial that the method $t(x)$ for transforming proofs of A to proofs of B is not restricted to what can be obtained by means of a given set of operations; otherwise the method does not get the intended generality.

In other words, we cannot argue that proofs of the BHK-interpretation or proof-objects in general correspond to proofs generated by a fixed set-up of inference rules. The proofs of the BHK-interpretation are defined by recursion over the sentences or propositions that they are proofs of. In the clauses for implication and universal implication they have to refer to arbitrary effective methods or functions by which certain transformations can be executed. It is not clear that such a BHK-proof can be translated to the representation of a proof act that is built up as a chain of inferences in an epistemic sense. Proof acts are generated inductively by inferences that have epistemic force, while BHK-proofs are defined by recursion over the linguistic objects that they are proofs of. There may be no guarantee that the latter can be seen as representations of the former.

To the extent that the view of proofs defined by BHK-interpretation or proofs of propositions in type-theory as epistemic entities of some kind was built on the idea that they could be seen as representations of proofs, it must thus be seen as unfounded. This is a conceivable argument for the standpoint that the concept of proof-object is non-epistemic.

3.6.4 An Alternative Argument for the Epistemic Nature of Proof-Objects

The project of verificationism was to account for linguistic meaning in terms of what counts as grounds for assertions and to do that in the form of a meaning-theory built on the intuitionistic explanations of the logical constants. At least the second part of this project must of course be seen as seriously mistaken, if Martin-Löf is right in his claim that these explanations should not be understood as referring to epistemic matters but to proof-objects whose main function seems to be that they are truth-makers. But even if it is right that the proofs referred to in the intuitionistic explanations of meaning are not proofs or representations of proofs, it may be too quick to draw the conclusion that they are truth-makers without any epistemic impact.

That the proof-objects are truth-makers should not be denied. On almost any theory of meaning, one may say that a proposition is determined by the kind of things or phenomena that could make it true. These things can naturally be called truth-makers. In the Siena lectures, acts of verifications were what made propositions true and they could thus be called truth-makers. The truth-maker terminology is thus compatible both with a classical realist theory of truth and with an epistemic theory. In the first case, the truth-makers are facts given independently of us, while in the second case they are our epistemic acts. In both cases the truth-makers are entities or phenomena that have already other functions or properties than being truth-makers. The idea that something is *merely* a truth-maker does not seem a comfortable view. In particular, it would be surprising if it turned out that proof-objects really had no relation to proofs or to epistemic matters.

It is worth noticing that within mathematics there can be no principal obstacles to the possibility of coming into possession of an intuitionistic truth-maker by constructing it, in contrast to what may be the case for what counts as truth-makers from a classical realist point of view. As we have seen, there are two views concerning whether we have the right to assert that a proposition A is true when we have come into possession of a proof-object of A. One view is that it remains to prove that the object in question is a proof-object of A. To take this view is in a way to respect Kreisel's requirement that a proof of, for instance, an implication $A \rightarrow B$ should consist of not only a construction c that gives a proof of B when applied to a proof of A, but also of a proof that c satisfies this property (Sect. 3.1.3). Another view is that from the way in which a proof-object of A has been constructed, it should be evident that it is a proof-object of A. This was Heyting's view, and as I have understood type theory, it is also Martin-Löf's view (Sect. 3.3, principles 4–6). That a constructed proof-object of A is a proof-object of A is not a proposition which has to be proved. It is enough to know the meaning of the terms involved and to be aware of how the proof-object has been constructed to know that the construction is a proof-object of the proposition in question. To be in possession of a proof-object of A in this way is thus a sufficient ground to have the right to make the judgement that A is true, which seems also to be in agreement with what is said in "Truth and provability".

On this second view, a proof-object of a proposition A amounts to what is commonly called a *ground* for asserting that A is true, that is, what a speaker must be in possession of in order to be correct or right in making the assertion. I think it is right to say that a proof-object of a proposition should not be looked upon as a representation of a proof for the reason indicated in the previous section, but that it should be seen as a ground for asserting the proposition. This is not the place to develop this idea further, which I have done elsewhere (Prawitz 2009, 2011a, b). But if I am right in this, the verificationistic project to account for meaning in terms of grounds for assertions was after all quite right in seeing an intuitionistic meaning-theory as an appropriate start. It would also mean that a proof-object, unlike mathematical objects in general, has an epistemic import, and that it is misleading to say that truth is not an epistemic notion when truth is defined in terms of proof-objects.

Acknowledgements The present paper is a revised and enlarged version of the paper that I presented at the conference "Philosophy and Foundations of Mathematics: Epistemological and Ontological Aspects". Parts of it have been presented at seminars at the Universities of Gothenburg and Stockholm and I thank the participants for discussions. I am especially grateful for constructive comments from Peter Dybjer, who read the next to final version of the paper. I am also grateful to Per Martin-Löf for remarks which inspired me to make some further revisions at a late stage.

References

Dummett, Michael. 1959. Truth. *Proceedings of the Aristotelian Society* 54: 141–162.

Dummett, Michael. 1973. *The justification of deduction*. London: British Academy.

Dummett, Michael. 1975. The philosophical basis of intuitionistic logic. In *Logic colloquium '73*, ed. H.E. Rose et al., 5–40. Amsterdam: North-Holland Publishing Company.

Dummett, Michael. 1976. What is a theory of meaning? (II). In *Truth and meaning*, ed. G. Evans and J. McDowell. Oxford: Clarendon Press.

Dummett, Michael. 1977. *Elements of intuitionism*. Oxford: Clarendon Press.

Dummett, Michael. 1998. Truth from the constructive standpoint. *Theoria* 64: 122–138.

Gentzen, Gerhard. 1935. Untersuchungen über das logische Schliessen. *Mathematische Zeitschrift* 39: 176–210, 405–431.

Heyting, Arend. 1930. Sur la logique intuitionniste. *Académie Royale de Belgique, Bullletin de la Classe des Sciences* 16: 957–963.

Heyting, Arend. 1931. Die intuitionistische Grundlegung der Mathematik. *Erkenntnis* 2: 106–115.

Heyting, Arend. 1934. *Mathematische grundlagenforschung, Intuitionismus, Beweistheorie.* Berlin: Springer.
Heyting, Arend. 1956. *Intuitionism, an introduction.* Amsterdam: North-Holland Publishing Company.
Heyting, Arend. 1958. Intuitionism in mathematics. In *Philosophy in the mid-century*, ed. R. Klibansky, 101–115. Florence: La Nuova Italia.
Howard, William. 1980. The formula-as-types notion of construction. In *To H. B. Curry: Essays on combinatory logic, lambda calculus and formalism*, ed. J. Seldin et al., 479–490. London: Academic.
Kreisel, Georg. 1962. Foundations of intuitionistic logic. In *Logic, methodology and philosophy of science*, ed. E. Nagel et al., 198–212. Stanford: Stanford University Press.
Martin-Löf, Per. 1975. An intuitionistic theory of types: Predicative part. In *Logic colloquium '73*, ed. H.E. Rose et al., 73–118. Amsterdam: North-Holland Publishing Company.
Martin-Löf, Per. 1982. Constructive mathematics and computer programming. In *Logic, methodology, and philosophy of science VI*, ed. L.J. Cohen et al., 153–175. Amsterdam: North-Holland Publishing Company.
Martin-Löf, Per. 1984. *Intuitionistic type theory.* Napoli: Bibliopolis.
Martin-Löf, Per. 1985. On the meanings of the logical constants and the justifications of the logical laws. In *Atti degli Incontri di Logica Matematica*, vol. 2, 203–281. Siena: Scuola di Specializzazione in Logica Matematica, Dipartimento di Matematica, Università di Siena. Republished in *Nordic Journal of Philosophical Logic* 1: 11–60, 1996.
Martin-Löf, Per. 1987. Truth of a proposition, evidence of a judgement, validity of a proof. *Synthese* 73: 407–420.
Martin-Löf, Per. 1994. Analytic and synthetic judgements in type theory. In *Kant and contemporary epistemology*, ed. P. Parrini, 87–99. Dordrecht: Kluwer Academic Publishers.
Martin-Löf, Per. 1995. Verificationism then and now. In *The foundational debate: Complexity and constructivity in mathematics and physics*, ed. W. DePauli-Schimanovich et al., 187–196. Dordrecht: Kluwer Academic Publishers.
Martin-Löf, Per. 1998. Truth and knowability: On the principles C and K of Michael Dummett. In *Truth in mathematics*, ed. H.G. Dales and G. Oliveri, 105–114. Oxford: Clarendon Press.
Prawitz, Dag. 1965. *Natural deduction. A proof-theoretical study.* Stockholm: Almqvist & Wiksell. Republished by New York: Dover Publications, 2006.
Prawitz, Dag. 1974. On the idea of a general proof theory. *Synthese* 27: 63–77.
Prawitz, Dag. 1998. Comments on the papers. *Theoria* 64: 283–337.
Prawitz, Dag. 2009. Inference and knowledge. In *The logica yearbook 2008*, ed. M. Pelis, 175–192. London: College Publications, King's College London.
Prawitz, Dag. 2011a. The epistemic significance of valid inference. *Synthese*, forthcoming, but available on line: http://dx.doi.org/10.1007/s11229-011-9907-7.
Prawitz, Dag. 2011b. Proofs and perfect syllogisms, In *Logic and Knowledge*, ed. C. Cellucci et al., 385–402. Newcastle upon Tyne: Cambridge Scholars Publication.
Prawitz, Dag. 2012. Truth as an epistemic notion. *Topoi*, forthcoming.
Sundholm, Göran. 1983. Constructions, proofs, and the meaning of logical constants. *Journal of Philosophical Logic* 12: 151–172.
Sundholm, Göran. 1993. Questions of proof. *Manuscrito* (Campinas) 16: 47–70.
Sundholm, Göran. 1998. Proof as acts and proofs as objects. *Theoria* 54: 17–216.
Troelstra, Anne Sjerp. 1977. Aspects of constructive mathematics. In *Handbook of mathematical logic*, ed. J. Barwise, 973–1052. Amsterdam: North-Holland Publishing Company.
Troelstra, Anne Sjerp, and Dirk van Dalen. 1988. *Constructivism in mathematics*, vol. 2. Amsterdam: North-Holland Publishing Company.
van Atten, Mark. 2009. The development of intuitionistic logic. In *The Stanford encyclopedia of philosophy* (Summer 2009 Edition), ed. Edward N. Zalta. URL = <http://plato.stanford.edu/archives/sum2009/entries/intuitionistic-logic-development/>.

Chapter 4
Real and Ideal in Constructive Mathematics*

Giovanni Sambin

Certain periods of life look as if they will last for ever. Then suddenly one is struck by the realization that time has gone by and that the period in question has become part of the past. Over 30 years have passed since Per Martin-Löf first came to Padua to give a course on his type theory. After 5 minutes of his first lecture, I felt sure that it was going to change my life. And that is exactly what happened.

Martin-Löf's type theory was very close to what I had been seeking for several years. I had the privilege of learning it from Per in person and of enjoying extensive discussion with him on philosophy and the foundations of mathematics, virtually throughout the whole of the 1980s.

Besides offering to write the notes on his lectures, my way of expressing my gratitude was to develop type theory further, namely by putting it into practice as a foundation for mathematics. This is how formal topology, which originated as topology over type theory, started in Spring 1984.

Around 10 years later Milly Maietti made me aware of the fact that, starting with the first paper (Sambin 1987), I had been pursuing a foundational vision for mathematics which is somewhat different from Per's and which leads to a variant of Martin-Löf type theory. Later on I will attempt to explain why this variant is necessary, in my opinion, for actually *doing* mathematics, and even indeed for formalizing it.

---

*This paper is based on the transcription of my actual talk at the conference Philosophy and Foundations of Mathematics: Epistemological and Ontological Aspects, dedicated to Per Martin-Löf on the occasion of his retirement, Uppsala, May 5–8, 2009, except for the last section, which is also based on my talk at Leeds Symposium on Proof Theory and Constructivism, Leeds (UK), 3–16 July 2009 and on Sambin (2011). I am grateful to John L. Bell and Milly Maietti for useful discussions, to John also for amending my abuses of English and to the editors for their patience with me.

G. Sambin (✉)
Università di Padova, Padova, Italy
e-mail: sambin@math.unipd.it

P. Dybjer et al. (eds.), *Epistemology versus Ontology*, Logic, Epistemology, and the Unity of Science 27, DOI 10.1007/978-94-007-4435-6_4,

The 1990s meant for me the beginning of a new vision of the foundations of mathematics and more generally of abstract thought. Central to that vision is the idea that everything in mathematics should be seen as arising from a dynamic process, and that indeed mathematics can be identified with the dynamic process itself. In particular, everything should be regarded as evolving in time.

More recently, I have found a dynamic view to be very fruitful, both for logic and for the foundations of mathematics. I would emphasize that this is also the case from a technical point of view, as this view supplies unexpected even revelatory insights and new discoveries which can be seen as applications.

So, after more than 30 years, I believe I now understand what characterizes my vision which is apparently different from and independent of all other proposals of foundations known to me. It is not my intention simply to reformulate what I have understood of Per's views, but rather I will try to present my personal views in a way I believe Per will understand.

4.1 Explanations from Above and Explanations from Below

Time marches on, human beings, and their works, are transitory. We agree with Hilbert when he asks: where else can we find absolute certainty, safe and stable truth if not in mathematics, and its foundations in particular?

We would like the foundations of mathematics to be able to explain in a conclusive way why mathematics is so certain and so effective. However, it is very easy to be misled by our wishes or psychological needs, and to replace what we want to achieve with an implicit assumption which is even less trustworthy and indeed even more difficult to explain than what we wanted to explain to begin with. It is easy to become trapped in a circular argument or to assume something which is easily stated, but of a degree of abstraction *higher* than what we intended to explain.

This arises in connection with any act of faith, or whatever one calls it: invoking demons in attempting to explain contagious diseases is of little help in finding a cure. An act of faith blocks a deeper understanding of how things are. And, conversely, giving up all supernatural principles forces us to grasp the nature of things in its full depth.

I believe that everything can and should be explained in accordance with the correct epistemological order of priority, that is, starting from things which arise – either in time or conceptually – before and not after the thing you want to explain.

Natural evolution provides an excellent illustration of what I mean. There are two ways of explaining why human beings have a strong reproductive desire. There is an explanation from above, which says that we are born with that desire, in the sense that it is an instinct for reproduction endowed in us by a supernatural entity. And there is the explanation from below, which I prefer, namely through natural selection, according to which people with a weaker reproductive impulse failed to multiply and as a result died off long ago!

This kind of explanation is not as silly as it may seem at first. It is something very deep, I believe, paradigmatic of the kind of explanations I would like to find.

It is a well-known phenomenon in evolutionary theory that one is psychologically inclined to explain things by using intentions which are actually not present, that is, to subvert the order of priority: for example, the claim that giraffes have long necks because they wanted to reach higher fruit. But again in fact the proper explanation is exactly the opposite: individuals with shorter necks did not survive in a certain environment. Through natural selection, only those quadrupeds tall enough to reach a certain variety of fruit survived and eventually became the giraffe, while the others failed to survive.

The giraffe species emerges through a dynamic interaction between pre-existing quadrupeds and their environment. There was a time when giraffes did *not* exist, and as a consequence of a certain process in which reproduction is restricted to a certain group, the existence of a species becomes a matter of fact.

My claim is that exactly the same kind of explanation applies to mathematics. Why is it so certain and effective? The explanation from above, or in the wrong order of priority, is that it follows from objective truth, or something equivalent to this. But then we have to explain something even more difficult, namely what objective truth is. And this is more difficult than explaining why mathematics is certain and effective.

Every form of Platonism is of this kind: an explanation from above. It is an act of faith which my own life experience has rendered inaccessible to me. But this forces me to understand things in a different way, and sometimes with a different perspective.

I am more interested in the *process* of constructing a concept (or getting to know a platonic entity), which one can touch and see, than in the putative completion of this process, which one can only postulate. What I regard as the process of constructing a concept is, from a platonist point of view, the process of approximating a pre-existing concept through our knowledge.

The explanation given by crude formalism is even less satisfactory: in fact, not only does it postulate something as self-existing, but also it postulates that it is given to us through formal systems. But it says nothing about where formal systems come from, how they are born, why they should work, and so on. A psychological reason perhaps explains why formalism is so successful: when mathematics is reduced to a formal game, it's easy to become an addict of the game.

Also in this case, as for giraffes and for the reproductive instinct, the explanation from below is very easy. Mathematics is certain and effective because mathematics *is*, by definition, among our intellectual products, which we use to interact with the environment and in our fight for survival, exactly that which *we* consider most certain, reliable and effective.

So the question: why is mathematics effective? has a simple answer: because mathematics *is* what has been selected as effective! Effectiveness of mathematics is a result, and not a cause, it is a result of a dynamic historical process.

My aim is not to try to convince readers to share my views, which doubtless some will regard merely as subjective opinions. I will continue with these considerations for a while only to show how, starting from them, one can go beyond subjective opinions to obtain facts. So one can consider my views, at least at the beginning, as suggestive metaphors which help to understand better some facts which I will explain later.

An idiomatic question in Italian is: *viene prima l'uovo o la gallina?* Which came first: the chicken or the egg? This too is a common question and answering it is not easy.

In fact, one can say that life begins exactly when an apparently circular equation is solved. Life is the ability of something to reproduce itself, through DNA. DNA (perhaps more precisely RNA) lives, by its very nature, on two different levels of abstraction: at one level DNA is a material part of the organism, but at the same time it is also, at a higher level, the information necessary to construct the organism itself. So it is both matter and also information supported by matter. Life begins as the solution of a fixed-point equation between two different levels of abstraction:

$$\text{DNA} = \text{DNA(DNA)}$$

DNA must be equal to DNA (as information) applied to DNA (as matter).

It looks like a circular problem, but we know that nature has been able to solve it, even if we don't know the details of how it was solved. What is certain is that here we can already see a pattern: two different levels of abstraction, which are linked by something and which interact with each other.

I believe a similar scheme applies to almost everything, from the DNA to organs (like a liver, a brain, ...) or to a species (giraffes, human beings, ...) but also beyond that, to all the tools which human beings have produced in the fight for survival. Thus it applies in particular to all concepts.

There is an interesting similarity, here, in the sense that one can describe both an organ and an abstract concept in exactly the same way. It is a tool which has been developed to perform a certain function which is useful for survival. In the case of concepts, communication is very important, and in fact it is essential. If I tell somebody that every day at a certain time there is a big and tasty animal going to drink, I have employed lots of abstract concepts, and I have communicated to my fellow human being something very important for survival.

One must always recall that there are two directions: one is from concrete to abstract, but there is one also from abstract to concrete. Any abstraction without an environment to which it can be applied has no positive meaning.

Let us see this better in the case of words. My wife Silvia is attractive. When I say so, I mean something very concrete: an emotion which I can feel in my body or my brain when I look at her. In this sense, I may say that Silvia is attractive, or that Anna is attractive, or that other women are attractive.

Then I can abstract from these specific judgements and obtain the predicate of attractiveness, that is being attractive. One can say that I understand what the word attractiveness means if I am able to apply it, for instance to Silvia, and obtain, ideally, the same judgement I meant when I said that Silvia is attractive:

Attractiveness(Silvia) is true if and only if Silvia is attractive.

So there is again an equation to be solved, and it is solved by the interaction of two levels:

Attractiveness

abstraction ↑ ↓ *application*

Silvia is attractive,
Anna is attractive, ...

At a certain moment this ceases to be felt as a vicious circle, or as an impredicative definition, and becomes an equation which holds trivially. That is precisely when something new is born (in one's mind), namely the concept of attractiveness.

Note that *first* there is the emotion, so the examples of somebody being attractive, and only *later* the concept attractiveness. The order of priority is clear: first I feel that Silvia is attractive, then I can form the concept of attractiveness and, if it works properly, the property of attractiveness applies to Silvia if and only if Silvia is attractive, in its first sense.

I purposely chose an example in which the difference between concrete instances and abstract concept is well visible. My claim is that the same happens with all words, for instance apple. Apple is the solution of an equation

$$Apple(x) \text{ is true if and only if } x \text{ is an apple}$$

describing the interaction between some kind of fruit, which at first are big, sweet, easy to eat, and yet nameless and only later it will happen that I call apples, and the concept apple which I obtain by abstraction.

To know that one understands the concept, one must be able to apply it in the most economical way. By this I mean that if the concept of apple should apply to bottles of water, or even worse to chairs, it would be very ineffective. If I say to Peter that there is an apple in the other room, Peter understands that there something to which the concept of apple applies. But if in the room he can only find chairs, he would rightly become angry, with me or with himself, for this kind of misunderstanding. This is to underline that both directions are essential, abstraction *and* application.

4.2 The Dynamic Process in Logic and in Foundations

As I have hinted, my claim is that also all concepts of mathematics are of this kind. Natural numbers are an abstraction of counting, real numbers are an abstraction of measuring, etc. In general, each notion of mathematics comes by abstraction from some kind of reality (by which I mean also previous notions of a lower level of abstraction).

To have a correct foundation of mathematics means simply to be able to say, for any of its abstract concepts, from which kind of reality it comes. One has to clarify both ways, abstraction and application, since both directions are an essential ingredient of the dynamic process generating a concept, and hence they must be present in any wellfounded explanation of that concept.

Incidentally, a dynamic view seems to be the only way which allows one to hope that certain mistakes can be corrected. If a dynamic process were not *always* active, it would be difficult to explain how some universal and apparently stable views can be changed when they are mistakes in the scale of history. Examples are, in my opinion, the Zermelo-Fraenkel axiomatic set theory with Choice in mathematics or some modern forms of fascism in social life.

The same pattern we have seen in the generation of an abstract concept can be seen very well in logic, with a dynamic explanation of logical constants. Contrary to the first impressions, I claim that first comes the notion of judgement, such as the proposition A is true, and then that of A being a proposition. This is analogous to the fact that first comes "Silvia is attractive" and only later the abstract proposition x being attractive and its application to Silvia. The usual terminology could be misleading here, since it says that the proposition A is in the object language and that the judgement A true is in the meta-language; so we should say that first comes meta-language and then (object) language. My claim is confirmed by two arguments.

First, one can observe that students have some difficulty in grasping the difference between the proposition A and the judgement that A is true. They tend to attach every proposition with a propositional attitude, usually that of being true. This is a sign of the fact that the judgement A is true is more concrete than the pure proposition A. Also some professional mathematicians have difficulties in grasping implication as a proposition, namely $A \longrightarrow B$, while it is easy for them to understand the compound assertion that A true gives B true as a consequence, namely $A \vdash B$, which is equivalent to $A \longrightarrow B$ true.

However, the most compelling argument in my opinion is that starting from this idea one can explain the meaning of *all* logical constants in a very convincing way by what I call the *principle of reflection.* I will give just one example, that of disjunction; one can treat in the same way all logical constants, including the quantifiers (Sambin et al. 2000; Maietti and Sambin 2005).

For all propositions A, B, the proposition $A \vee B$ is the abstraction of all situations in which $A \vee B$ can be asserted true. And the truth of $A \vee B$ in a certain context is by definition governed by the equivalence:

$$\Gamma, A \vee B \vdash C \quad \text{if and only if} \quad \Gamma, A \vdash C \quad \text{and} \quad \Gamma, B \vdash C \tag{4.1}$$

for all lists $\Gamma \equiv C_1, \ldots, C_n$ and all C. Here we understand that $\Gamma, A \vee B \vdash C$ is a short way of expressing that (C_1 true and ... and C_n true and $A \vee B$ true) yields C true, and similarly for $\Gamma, A \vdash C$ and $\Gamma, B \vdash C$. So (4.1) defines under which conditions, or in which context, the assertion of the new compound proposition $A \vee B$ is expected to be equivalent to a compound assertion involving A true and B

true. One can look at it as an "equation" which defines $A \vee B$ implicitly by saying that it must be the "reflection" at object level of the meta-linguistic conjunction of two assertions. It is easy to "solve" it, that is, one can derive from it a wellfounded explanation of $A \vee B$ in terms of its inference rules and such that (4.1) is satisfied.

The direction from right to left in (4.1) has already the form of an acceptable inference rule. In fact, expressing as usual the conjunction of two premises by a blank space, it gives[1]:

$\vee$-formation

$$\frac{\Gamma, A \vdash C \quad \Gamma, B \vdash C}{\Gamma, A \vee B \vdash C}$$

We can see that it brings from a situation with A true and B true to one with the more complex $A \vee B$ true. The opposite direction is not as good, since it brings from a situation with $A \vee B$ true, which is what we want to define, to a simpler situation with A true or with B true:

$$\frac{\Gamma, A \vee B \vdash C}{\Gamma, A \vdash C} \qquad \frac{\Gamma, A \vee B \vdash C}{\Gamma, B \vdash C}$$

That is why I call it implicit reflection. However, it can be made explicit, that is, one can show that it is equivalent, as an inference rule, to one in which $A \vee B$ appears in the conclusion:

$\vee$-reflection

$$\frac{\Gamma \vdash A}{\Gamma \vdash A \vee B} \qquad \frac{\Gamma \vdash B}{\Gamma \vdash A \vee B}$$

In fact, from implicit reflection by considering Γ empty and $C \equiv A \vee B$, since $A \vee B \vdash A \vee B$ holds one obtains as a conclusion:

$\vee$-axiom

$$A \vdash A \vee B \qquad B \vdash A \vee B$$

Conversely, from $\Gamma, A \vee B \vdash C$ and the axiom $A \vdash A \vee B$ by cutting $A \vee B$ one has $\Gamma, A \vdash C$.

Now from the axiom and $\Gamma \vdash A$ by cut one has $\vee$-reflection. Conversely, by the substitution $\Gamma \equiv A$ one obtains the axiom.

In this way we have shown that $\vee$-reflection and implicit reflection have the same deductive power, although one is wellfounded and the other is not. Hence by choosing $\vee$-formation and $\vee$-reflection as the inference rules for disjunction we obtain an explanation of what it means to assert $A \vee B$ true which is wellfounded, since it uses only something which comes before, namely the assertions A true and B true.

The dynamic attitude I have been describing applies also to the foundations of mathematics. A natural consequence of what I have been saying so far is that

[1]The usual terms to denote inference rules, namely introduction and elimination, would be highly confusing in this context.

in this case too one should observe an interaction between two different levels of abstraction. Indeed, this is the central epistemological principle inspiring the approach to foundations which Milly Maietti and I introduced and called the *minimalist foundation* (Maietti and Sambin 2005). As we have seen in all our examples, the two levels of abstraction must be clearly linked:

set theories
extensional theory

abstraction ↑ ↓ *application*
= forget some information *= implementation*
= pass to quotients *= restore information*

intensional theory
type theories

One can pass from the intensional theory to the extensional theory by abstraction, which means forgetting some information. In practice, the intensional theory is a variant of Martin-Löf type theory (Martin-Löf 1984; Nordström et al. 1990). Then abstraction is simply closure under quotients. That is, we *forget* the information given by proof-terms, or elements of a set, and we identify some or all of them.

We also need a way back, that is application, otherwise abstraction would become as pointless as applying the concept of apple to chairs. So one must be able to go back to the intensional theory, and ideally to the same situation one started from. That is, one should be able to restore the information which had been forgotten. Sometimes I call this the *forget-restore principle*.

Let us see what this means for logic. At the extensional level we wish to have the standard extensional notion of subset, relation, etc. with which every mathematician is familiar. This means that we need the notion of a proposition A being true, independently of which proof-term makes it true. In fact, this is necessary for defining subsets and their elements (Sambin and Valentini 1998; Maietti and Sambin 2005).

We may add intuitionistic logic to the extensional theory and justify it by the principle of reflection, as I explained above. We can also improve on this and see logic at the extensional level as obtained by abstraction from its formulation at the intensional level, that is with proof-terms. In the opposite direction, a well-known metamathematical result asserts that in every derivation of a judgement A true in intuitionistic logic one can automatically restore proof-terms. One thus obtains a perfect correspondence between logic at the extensional level (with no proof-terms) and its formulation in type theory, with proof-terms. However, to this aim one has to provide type theory with a primitive notion of proposition and with inference rules for all logical constants; then the link with sets is given by the principle that every proposition is a set (propositions-into-sets), but not conversely (Maietti and Sambin 2005). I see no way of obtaining the same result while keeping the full propositions-as-sets interpretation, which is so typical of Martin-Löf type theory.

When applied to all of the extensional theory, restore amounts to implementation. So the extensional theory should be devised in such a way that all pieces of mathematics one develops in it can automatically be implemented in the intensional theory, that is, in a “proof assistant”. Milly Maietti has shown that this is indeed possible (Maietti 2009). One could also start from this very practical requirement and argue backwards: I claim that, assuming one does not ask mathematicians to change their nature, one would reach conclusions very similar to ours.

One should never forget that both directions are necessary. It is common to consider only the direction of abstraction, like when passing to a theory such as ZFC for which application, or restoring information, becomes difficult. But sometimes only the direction of application is considered, when some form of implementation of a piece of mathematics is provided, without knowing how one could go back to mathematics from it. The usefulness of this form of implementation is in my opinion quite debatable.

Besides automatic implementation, there is another good reason for our minimalist foundation. Indeed, its name has been chosen to recall that it is intentionally devised to preserve all possible conceptual distinctions and to respect the nature of all ingredients of mathematics: computational content, geometrical intuition, abstract algebraic manipulations, deductions, ... Technically, all other foundations are obtained by a combination of some basic principles (law of excluded middle, propositions-as-sets, axiom of choice, powerset axiom, extensional equality of functions, etc.) which have to be added either to the extensional or to the intensional theory. In particular, the idea of two levels of abstraction was conceived as a means of satisfying two otherwise apparently incompatible requirements: on the one hand, the possibility of a computational interpretation of mathematics, which means that the axiom of choice `AC` and internal Church thesis `CT` are consistent with the intensional level, and on the other hand its familiar extensional character, which means that two operations with equal values on equal arguments are equal at the extensional level.

One important advantage of our approach is that it allows one to develop mathematics constructively and still retain its ideal aspects along with its real or effective ones. In fact, real and ideal aspects can coexist in the same framework without being confused with each other. To see how this is possible, we need to review some notions of constructive topology as developed in the minimalist foundation.

4.3 Real and Ideal Notions in Constructive Topology

The minimalist foundation has already been tested in the actual development of mathematics. I had something similar in mind since the very beginning of formal topology, even though I was not yet aware of it. Now I am fully aware that the whole Basic Picture relies on it. It is not possible to explain here in detail what the Basic Picture is (the reader is referred to Sambin (2003, 2011) or better to the forthcoming Sambin (2012)). In one sentence, it is the theory resulting from the discovery that the extra information which is necessary to obtain a predicative and implementable version of topology has, contrary to common expectations, a very clear and deep structure, which actually underlies and clarifies traditional topology.

If we keep the information given by a set-indexed family of basic neighbourhoods $\mathsf{ext}(a) \subseteq X (a \in S)$, a topological space on a set of points X is represented by a structure $(X, \Vdash, S)$ where S is a second set and $x \Vdash a \equiv x \,\epsilon\, \mathsf{ext}\, a$ says that x lies in the basic neighbourhood with index a. Then interior and closure of a subset $D \subseteq X$ are defined by

$$x \,\epsilon\, \mathsf{int}\, D \equiv (\exists a \in S)(x \Vdash a \;\&\; \mathsf{ext}\, a \subseteq D)$$

and

$$x \,\epsilon\, \mathsf{cl}\, D \equiv (\forall a \in S)(x \Vdash a \longrightarrow \mathsf{ext}\, a \between D).$$

(I use the sign $\between$ for the relation of overlap between subsets, that is

$$D \between E \equiv (\exists x \in X)(x \,\epsilon\, D \;\&\; x \,\epsilon\, E)$$

for $D, E \subseteq X$). Thus one sees that interior and closure are dual to each other, in a rigorous logical sense. One can further analyse the structure of quantifications and discover that:

$$\begin{array}{lll} \mathsf{int} = \mathsf{ext}\,\Box & \mathcal{J} = \Diamond\,\mathsf{rest} & \exists\forall \\ \mathsf{cl} = \mathsf{rest}\,\Diamond & \mathcal{A} = \Box\,\mathsf{ext} & \forall\exists \end{array}$$

Here ext, $\Diamond$ are the operators on subsets giving the existential image of a subset along the relation $\Vdash$, that is, $x \,\epsilon\, \mathsf{ext}\, U \equiv (\exists a \in S)(x \Vdash a \;\&\; a \,\epsilon\, U)$ for $U \subseteq S$ and $a \,\epsilon\, \Diamond D \equiv (\exists x \in X)(x \Vdash a \;\&\; x \,\epsilon\, D)$ for $D \subseteq X$. Similarly, the operators $\Box$, rest giving universal images are defined by $x \,\epsilon\, \mathsf{rest}\, U \equiv (\forall a \in S)(x \Vdash a \longrightarrow a \,\epsilon\, U)$ and $a \,\epsilon\, \Box D \equiv (\forall x \in X)(x \Vdash a \longrightarrow x \,\epsilon\, D)$.

This analysis leads one to discover that one can define by symmetry two operators $\mathcal{A} \equiv \Box\,\mathsf{ext}$ and $\mathcal{J} \equiv \Diamond\,\mathsf{rest}$ on the set S. The operator $\mathcal{A}$ is well-known, perhaps under its alternative notation $\lhd$ which is defined by $a \lhd U \equiv a \,\epsilon\, \mathcal{A}U$. So $a \lhd U \equiv \mathsf{ext}\, a \subseteq \mathsf{ext}\, U$ says that $\mathsf{ext}\, U$ is a covering of $\mathsf{ext}\, a$. The operator $\mathcal{J}$ is a novelty. The relation $a \ltimes U \equiv a \,\epsilon\, \mathcal{J}U$ says that $\mathsf{ext}\, a \between \mathsf{rest}\, U$, that is, $\mathsf{ext}\, a$ is inhabited by a point all neighbourhoods of which have indices in U.

The discovery of $\ltimes$ and an axiomatic description of the structure induced by $(X, \Vdash, S)$ on S leads to a new definition of pointfree structure $\mathcal{S} = (S, \lhd, \ltimes)$ in which one has both a formal cover $\lhd$ and a positivity relation $\ltimes$, linked by a suitable condition (called compatibility). The positivity relation $\ltimes$ is dual to the cover $\lhd$ in a way which is abstractly similar to that of the closure cl being dual to interior int. Just as $\lhd$ or $\mathcal{A}$ provides a pointfree notion of open subset, namely one for which $U = \mathcal{A}U$, now the addition of $\ltimes$ or $\mathcal{J}$ provides a pointfree notion of closed subset, namely $U = \mathcal{J}U$ or equivalently $a \,\epsilon\, U \leftrightarrow a \ltimes U$ for all a. This is an absolute novelty which considerably enriches the expressive power of constructive pointfree topology. For this reason I recently took the decision to change terminology and now call $\mathcal{S}$ a *positive topology* (and leave formal topology for the notion originally introduced in Sambin (1987)).

Of course, once we add the positivity relation $\ltimes$ to the definition we must take care that all morphisms will preserve it. On the other hand, we can exploit its presence. For instance, following a general principle we define ideal points α of a positive topology $\mathcal{S}$ so that they correspond bijectively to morphisms from $\mathcal{P}1$, the positive topology on a set 1 with only one element, into $\mathcal{S}$. This gives a notion of point which is stronger than the previous notion of formal point, since α is an *ideal point* if it is a formal point according to Sambin (1987) which moreover is formal closed, that is $\alpha = \mathcal{J}\alpha$.

For every positive topology $\mathcal{S} = (S, \lhd, \ltimes)$, $\mathcal{P}t(\mathcal{S})$ is the collection of all ideal points of $\mathcal{S}$. We can then consider the structure $(\mathcal{P}t(\mathcal{S}), \Vdash, S)$, where $x \Vdash \alpha \equiv a \,\epsilon\, \alpha$. This suggests that we can equip $\mathcal{P}t(\mathcal{S})$ with an interior and closure operator. However, since $\mathcal{P}t(\mathcal{S})$ is a collection and not a set, this is possible only through an impredicative definition and thus $\mathcal{P}t(\mathcal{S})$ with the resulting topology is called the *ideal space* associated with $\mathcal{S}$. This is a paradigmatic example of an ideal notion, even when $\mathcal{S}$ is given fully effectively. One could say that the positive topology $\mathcal{S}$ is the real part, and the ideal space $\mathcal{P}t(\mathcal{S})$ associated with it is the ideal part of the same space. Note that, contrary to what one would guess from the classical point of view, it is the pointfree part which is the real one, and that with points is the ideal one.

Following what we did for $(X, \Vdash, S)$, we can define a formal cover $\lhd_{\mathcal{P}t}$ on S by putting

$$a \lhd_{\mathcal{P}t} U \equiv (\forall \alpha : \mathcal{P}t(\mathcal{S}))(\alpha \Vdash a \longrightarrow \alpha \between U).$$

Since it uses a quantification over a proper collection, this too is not a proper constructive definition. We call $\lhd_{\mathcal{P}t}$ the *ideal cover* induced by $\mathcal{P}t(\mathcal{S})$ on S. One can look at it as an ideal aim to be reached.

When $\lhd_{\mathcal{P}t} = \lhd$, that is when the ideal cover coincides with the one given effectively, we say that $\mathcal{S}$ is *spatial*. This notion has been reached here only by a structural, or theoretical argument. Then it is natural to ask ourselves about its meaning in specific cases.

Let us consider the positive topology $\mathcal{T}_{\mathbb{N}} = (\mathbb{N}^*, \lhd, \ltimes)$, where $\mathbb{N}^*$ is the set of lists over natural numbers, or equivalently it is the tree in which at every node one can proceed by choosing any element of $\mathbb{N}$. The cover $\lhd$ and positivity relation $\ltimes$ are generated by the axioms saying that every node is covered by the subset of all its immediate successors or by any of its initial segments. Thus $\mathcal{T}_{\mathbb{N}}$ is a pointfree version of Baire space. One can easily see that ideal points of $\mathcal{T}_{\mathbb{N}}$ can be identified with infinitely proceeding sequences of natural numbers.

In order to maintain a distinction between operations, that is dependent families of elements, and functions, that is total and singlevalued relations, the so-called axiom of unique choice AC! is not assumed to be valid in the minimalist foundation.[2] This move is crucial to obtain as a consequence that one can identify ideal points of $\mathcal{T}_{\mathbb{N}}$ with choice sequences as conceived by Brouwer, rather than with lawlike sequences (see Sambin 2008).

[2]This option is in common with some foundational theories by Solomon Feferman (1979).

Since the implication $k \lhd U \longrightarrow k \lhd_{\mathcal{P}t} U$ holds for all $k \in \mathbb{N}^*, U \subseteq \mathbb{N}^*$ by the definition of ideal point, spatiality of $\mathcal{T}_{\mathbb{N}}$ becomes equivalent to

$$(\forall \alpha : \mathcal{P}t(\mathcal{T}_{\mathbb{N}}))(\alpha \Vdash k \longrightarrow \alpha \between U) \longrightarrow k \lhd U.$$

By the identification of ideal points with choice sequences, this turns out to be exactly the same as Brouwer's principle of bar induction BI. Thus we rediscover by theoretical reasons an important principle which was there before for other more specific reasons. It seems proper to call this an application of our new foundational attitude.

Some more recent applications arise from the presence of the positivity relation $\ltimes$. It is usually said that a subset $H \subseteq \mathbb{N}^*$ is a *spread* if

H is inhabited: $H \between \mathbb{N}^*$
H leaks upwards: $k \in H \longrightarrow \exists a(k * a \in H)$
H is downward closed: $k \in H,\ l \sqsubseteq k \longrightarrow l \in H$

(where $l \sqsubseteq k$ means that l is an initial segment of k). The second condition is a positive way of expressing that H is not a bar (the tree is supposed to have branches going upwards). One should also recall that often in the definition of spread one requires in addition that H is a decidable subset.

It is an immediate consequence of the definitions that spreads are *exactly* inhabited formal closed subsets. Then Brouwer's idea of a choice sequence satisfying the restriction of remaining inside a spread is now expressed by an ideal point α living inside a formal closed subset H, which formally is nothing but $\alpha \subseteq H$.

For every formal closed subset H in an arbitrary positive topology $\mathcal{S} = (S, \lhd, \ltimes)$, the presence of $\ltimes$ and the fact that it is not uniquely determined by $\lhd$ allow us to define a new positivity relation $\ltimes_H$ by putting:

$$a \ltimes_H U \equiv a \ltimes H \cap U.$$

One can prove that $\ltimes_H$ is still compatible with $\lhd$ and hence that $\mathcal{S}_H = (S, \lhd, \ltimes_H)$ is a positive topology. Moreover one can prove that, whenever $\mathcal{S}$ is generated (that is, $\lhd$ is generated by induction and $\ltimes$ by coinduction (Martin-Löf and Sambin 2012)), then also $\mathcal{S}_H$ is generated. Of course, while the axioms for $\lhd$ remain the same, those of $\ltimes_H$ will be different from those of the original $\ltimes$.

Ideal points of $\mathcal{S}_H$ are exactly those points of the original $\mathcal{S}$ which moreover are contained in H:

$$\alpha : \mathcal{P}t(\mathcal{S}_H) \quad \text{iff} \quad \alpha : \mathcal{P}t(\mathcal{S}) \quad \text{and} \quad \alpha \subseteq H.$$

In other words, one has $\mathcal{P}t(\mathcal{S}_H) = \mathcal{P}t(\mathcal{S}) \cap \mathsf{Rest}\, H$, where $\alpha : \mathsf{Rest}\, H \equiv \alpha \subseteq H$. In the specific case of the Baire topology $\mathcal{T}_{\mathbb{N}}$, a choice sequence α lives in a spread H iff α remains an ideal point when passing from $\ltimes$ to $\ltimes_H$, that is if α is an ideal point of $(\mathcal{T}_{\mathbb{N}})_H$.

For every positive topology $\mathcal{S}$, besides the ideal cover $\lhd_{\mathcal{P}t}$ we can define also an ideal positivity relation $\ltimes_{\mathcal{P}t}$, now by means of an impredicative existential quantifier:

$$a \ltimes_{\mathcal{P}t} U \equiv (\exists \alpha : \mathcal{P}t(\mathcal{S}))(\alpha \Vdash a \;\&\; \alpha \subseteq U).$$

Note that in this case it is the opposite direction (with respect to the case of covers) $a \ltimes_{\mathcal{P}t} U \longrightarrow a \ltimes U$ the one which holds in every topology. So the ideal notion implies the pointfree, effective one.

If we require also the other direction $a \ltimes U \longrightarrow a \ltimes_{\mathcal{P}t} U$ to hold, we obtain the notion of a *reducible* positive topology. In a reducible formal topology $\mathcal{S}$ the real positivity relation $\ltimes$, the one which is usually given by coinduction, coincides with the ideal one $\ltimes_{\mathcal{P}t}$, which is defined only improperly.

In the case of the Baire topology $\mathcal{T}_{\mathbb{N}}$, being reducible is exactly the principle that every spread is inhabited by a choice sequence. So again we find something important which had been considered before.

For every commutative ring A, we can define the Zariski positive topology $\mathcal{Z}_A$, whose formal open subsets are (radical) ideals and whose formal closed subsets are coideals (Rinaldi et al. 2012). The associated ideal space $\mathcal{P}t(\mathcal{Z}_A)$ then corresponds to a well known notion, namely the Zariski spectrum of a ring $Spec(A)$. And the statement that $\mathcal{Z}_A$ is reducible is classically equivalent to a very important principle in the classical approach to algebra, namely the prime filter theorem.

These examples remind us that the method of adding ideal notions and of assuming some principles about them is fecund and common in mathematics. Indeed, in the whole history of mathematics ideal notions have proved to be a very convenient artifice to organize our understanding of a certain topic or field in a simple way. Typical examples are complex numbers and points at infinity (with the result, respectively, that every equation has a solution and every two lines have a common point).

Why on earth should we deprive ourselves of such a powerful method? I believe that constructivism should aim at a foundation of *all* of mathematics, with no prejudice, and hence that it should not leave out a treatment of its ideal aspects (and such treatment – it should be understood – should be constructively acceptable).

However, we know that ideal notions are rejected in the computational interpretation of constructivism, by Bishop (1967) and Martin-Löf (1970), because a piece of mathematics is there accepted as meaningful only if it has a direct computational content. I believe that this irreconcilability is only apparent and that it is due mostly to contingent historical misunderstandings.

We know that Hilbert's aim was to justify ideal notions, which in his case means primarily the notion of actual infinite as in "Cantor's paradise", by a proof of formal consistency of an axiomatic theory about them. Leaving aside the metamathematical difficulties which emerged with Gödel's second incompleteness theorem, Hilbert's program has little or no value constructively, as Brouwer and Bishop have illustrated in a forceful and vivid way (an outstanding example is the first dozen pages of Bishop (1967)). In my view, one can locate Hilbert's mistake precisely in the fact that no attention is given to the direction back, from abstractions to reality. In fact, consistency alone is not sufficient to guarantee a clear and reliable contact between the formal system for ideal notions and the real notions from which it started.

Let us consider the two examples above. Nobody would expect the addition of complex numbers (or imaginary numbers, as they were called first) and of points at infinity to modify real numbers and the Euclidean plane, respectively. Indeed, this remark is usually assumed to be so trivial that it is simply taken for granted. It seems to me that the same applies to all other examples of ideal notions introduced in the history of mathematics before Cantor's actual infinite. The use of ideal elements does not modify our knowledge of a previous more concrete notion while embedding that notion in a framework which behaves better. All the results about that concrete notion which can be proved using ideal elements should in principle be provable also without them. A procedure transforming a proof of a real statement which uses ideal elements into one without would be a tangible expression of the direction from ideal back to real.

Using formal systems and their terminology, we should require some form of *conservativity* of the theory T' for ideal elements over the theory T for real elements. Assuming T to be consistent, pure consistency of T' is a consequence. Perhaps one should not expect full conservativity, again by Gödel's theorem: one can imagine that T' could prove a sentence in the language of T which expresses consistency of T and which is therefore unprovable in T. However, the idea of conservativity seems sufficiently clear and I expect it is possible to express it also formally, case by case, in a suitable way which is not jeopardized by Gödel's theorems.

In my opinion, conservativity of T' over T makes the use of ideal elements described in T' perfectly acceptable constructively. In fact, no engagement on the existence of such ideal elements is necessary and their use in the end is only a way of speaking which makes some arguments easier to follow. In particular, there is no claim that one can create facts of T by manipulating ideas of T', which is what caused the reaction by Brouwer and Bishop.

Present day classical mathematics is a "mixture of the real and the ideal, sometimes one, sometime the other, often so presented that it is hard to tell which is which" (Bishop 1967). On the other hand, a necessary condition even to formulate conservativity of the ideal over the real is that both be present and distinct in the same foundation. This is an important peculiarity of the minimalist foundation, which has both a real part (sets, operations, positive topologies) and an ideal part (collections, functions, ideal spaces), and still the nature of each part is preserved (in particular, they are kept well distinct from each other).

The real part admits a computational interpretation; this is a technical way of expressing that it comes through abstraction from some kind of reality. The ideal part is a useful device which is added to organize our knowledge of the real part in a simpler and sometimes more intuitive way. All computational interpretations of mathematics which were possible before should remain possible using the minimalist foundation. This means that the addition of ideal aspects should not change our knowledge of real ones.

I find it very interesting to show, case by case, that one can use ideal notions without loosing effectivity. Indeed, the presence of the real and the ideal in the same framework allows us to formulate a constructive and local version of Hilbert's program, requiring that one should show that each ideal aspect is conservative over the real ones. I here discuss only the case of choice sequences and Bar Induction BI : $(\forall \alpha : \mathcal{P}t(\mathcal{T}_{\mathbb{N}}))(\alpha \Vdash k \longrightarrow \alpha \between U) \longrightarrow k \lhd U$. I expect similar considerations can be carried on for the prime filter theorem and other ideal principles.

Brouwer himself introduced the ideal notion of choice sequence to characterize the continuum. Using BI, or its consequence the Fan Theorem, one obtains a convenient development of intuitionistic analysis. However, Brouwer's explanation of choice sequence rests on the assumption of a creating subject and is still controversial. While, as we have seen, the concept of choice sequence finds a simple and rigorous characterization as ideal points of Baire topology; in particular, the absence of AC! allows us to distinguish between choice sequences, which are functions, from lawlike sequences, which are operations. Informally, a choice sequence is the execution of a program of which we do not know the instructions.

The intuitive argument in favour of BI given by Brouwer can be put in our words as follows. Precisely because the notion of choice sequence/ideal point is so weak and one knows nothing about the αs except that they proceed indefinitely, we can know the antecedent to be true only in case we actually know the consequent $k \lhd U$ to be true.

Conversely, BI ceases to be plausible as soon as one makes some further assumption on the αs. One extreme is to assume α to be given by a recursive operation, in which case BI is contradictory by Kleene's counterexample. But BI is no longer plausible also when α is just an operation, which follows if AC! is assumed to be valid. So plausibility of BI is another way to explain the absolute freedom of the notion of choice sequence.

When I say that BI is plausible I mean more precisely that it is plausible to assume it. In fact, just after considering its logical form, one should give up any hope of proving BI in the minimalist foundation. However, I expect the addition of choice sequences and of Bar Induction BI to be conservative over statements not involving choice sequences, and perhaps only those of a specific form.

A partial result in this direction is that BI is consistent with Church Thesis (Maietti 2012), while they become inconsistent, by Kleene's counterexample, if also AC! is (silently) assumed. The elimination of choice sequences in Kreisel and Troelstra (1970), Troelstra and van Dalen (1988) follows a similar spirit, namely showing that choice sequences are a mere "figure of speech". However, since AC! is assumed, the distinction between lawlike and choice sequences is there due to some ad hoc axioms, and is not reduced to the very general distinction between operation and function, as done here.

A proof of the conjecture of conservativity of BI would show that we can use our spatial intuition and still be sure that computational content is not lost. One can say that spatial intuition is built-in in our physiology of human beings: our vision is "continuous" even if the retina has only a finite number (two to three millions) of receptors. This means that the brain automatically produces a sort of "completion" and it makes us "see" the ideal space associated with the positive topology of visual inputs. If evolution has found this method convenient, why should we abstain from it?

The use of spatial intuition makes many arguments shorter and easier to follow. The purpose of ideal points, and of choice sequences with BI in particular, is precisely that of providing it with a mathematical base. Conservativity would mean that we can freely use (ideal) points in our proofs of pointfree statements and still know that a purely pointfree proof of the same statement can be found automatically.

On the other hand, it should be clear from my general remarks above that the criterion of conservativity of ideal notions over real ones has a priority, so that one should give up assuming BI if it turns out that it is not conservative.

References

Bishop, E. 1967. *Foundations of constructive analysis*. New York: McGraw-Hill.

Feferman, S. 1979. Constructive theories of functions and classes. In *Logic Colloquium '78*, ed. M. Boffa, et al., 159–224. Amsterdam: North-Holland.

Kreisel, G., and A.S. Troelstra. 1970. *Formal systems for some branches of intuitionistic analysis*. Annals of mathematical logic, vol. 1, 229–387.

Maietti, M.E. 2009. *A minimalist two-level foundation for constructive mathematics*. Annals of pure and applied logic, vol. 160, 319–354.

Maietti, M.E. 2012. *Consistency of the minimalist foundation with Church Thesis and Bar Induction*. to appear.

Maietti, M.E., and G. Sambin. 2005. Toward a minimalist foundation for constructive mathematics. In *From sets and types to topology and analysis. Towards practicable foundations for constructive mathematics*. Oxford logic guides, vol. 48, 91–114, ed. L. Crosilla, and P. Schuster. Oxford: Clarendon.

Martin-Löf, P. 1970. *Notes on constructive mathematics*. Stockholm: Almqvist & Wiksell.

Martin-Löf, P. 1984. *Intuitionistic type theory. Notes by G. Sambin of a series of lectures given in Padua, June 1980*. Naples: Bibliopolis.

Martin-Löf, P. and G. Sambin. 2012. *Generating positivity by coinduction*. To appear, privately circulated since 2003.

Nordström, B., Petersson, K., and J. M. Smith. 1990. *Programming in Martin-Löf's Type Theory, an introduction*. Oxford: Oxford University Press.

Rinaldi, D., Sambin, G., and P. Schuster. 2012. *The Zariski basic topology*. Submitted.

Sambin, G. 1987. Intuitionistic formal spaces – a first communication. In *Mathematical logic and its applications*, ed. D. Skordev, 187–204, New York: Plenum.

Sambin, G. 2003. *Some points in formal topology*. Theoretical computer science, vol. 305, 347–408.

Sambin, G. 2008. Two applications of dynamic constructivism: Brouwer's continuity principle and choice sequences in formal topology. In *One hundred years of intuitionism (1907–2007). The cerisy conference*, ed. M. van Atten, P. Boldini, M. Bourdeau, and G. Heinzmann, 301–315. Basel/Boston: Birkhäuser

Sambin, G. 2011. A minimalist foundation at work. In *Logic, Mathematics, Philosophy, Vintage Enthusiasms. Essays in Honour of John L. Bell*. Volume 75 of The Western Ontario series in philosophy of science, ed. D. DeVidi, M. Hallett, and P. Clark, 69–96. Berlin: Springer.

Sambin, G. 2011. Reale e ideale in matematica. In *La ricerca logica in Italia. Studi in onore di Corrado Mangione*. volume 124 of Quaderni di Acme, ed. E. Ballo and C. Cellucci, 425–446. Milano: Cisalpino.

Sambin, G. 2012. *The basic picture and positive topology. New structures for constructive mathematics*, Oxford: Oxford University Press. To appear.
Sambin, G., Battilotti, G., and C. Faggian. 2000. Basic logic: reflection, symmetry, visibility. *Journal of Symbolic Logic* 65: 979–1013.
Sambin, G., and S. Valentini. 1998. Building up a toolbox for Martin-Löf's type theory: subset theory. In *Twenty-five years of constructive type theory, Proceedings of a Congress held in Venice, October 1995*, ed. G. Sambin and J. Smith, 221–244. Oxford: Oxford University Press.
Troelstra, A.S., and D. van Dalen, 1988. *Constructivism in mathematics, an introduction*. Studies in logic and the foundations of mathematics. Amsterdam: North-Holland.

Chapter 5
In the Shadow of Incompleteness: Hilbert and Gentzen*

Wilfried Sieg

5.1 A Puzzle

During the last quarter century or so, the early history of *modern mathematical logic* has been explored in detail and we have gained a much richer perspective of its evolution from nineteenth century roots. Some purely historical insights have been astounding: The beginning of the subject, associated for many with Hilbert and Ackermann's book from 1928, was pushed back by a whole decade to lectures Hilbert gave in the winter term of 1917/18. These lectures also reveal the impact of Whitehead and Russell's *Principia Mathematica* on the logical framework for the investigations in the Hilbert School. Other issues of central importance are coming into sharper historical and systematic focus; among them is a deepened understanding of how (versions of) Gödel's theorems affected proof theory between 1930 and 1934.

1934 witnessed, of course, the publication of the first volume of Hilbert and Bernays's *Grundlagen der Mathematik*. Less publicly, Gerhard Gentzen finished his first consistency proof for full first-order arithmetic. In late 1931, he had already set himself the goal of proving this result as the central theorem of his dissertation. On the long path to his proof, Gentzen made two remarkable discoveries before the end of 1932: (i) he established, independently of Gödel, the consistency of classical arithmetic relative to its intuitionist version, and (ii) he proved the normalization theorem for (a fragment of) *intuitionist* first-order logic and recognized the subformula property of normal derivations. The latter property was extended in his (Gentzen 1934/35) to *classical* logic formulated in

*Dedicated to Per Martin-Löf.

W. Sieg (✉)
Department of Philosophy, Carnegie Mellon University, Pittsburgh, PA 15213, USA
e-mail: sieg@cmu.edu

P. Dybjer et al. (eds.), *Epistemology versus Ontology*, Logic, Epistemology, and the Unity of Science 27, DOI 10.1007/978-94-007-4435-6_5,

the sequent calculus. That paper was literally based on his *actual* dissertation in which he established the consistency of arithmetic with quantifier-free induction. Two questions naturally arise: why did he temporarily give up on his goal, and how had he intended to reach it?

In a draft of his *Urdissertation* from October 1932, still pursuing the goal of a consistency proof for full arithmetic, Gentzen listed the results he had obtained as items (I) through (IV) and formulated the crucial remaining task as item (V):

> The consistency of arithmetic will be proved; in the process, the concept of an infinite sequence of natural numbers will be used, furthermore in one place the principle of the excluded middle. The proof is thus not intuitionist. Perhaps the *tertium non datur* can be eliminated.

How is this to be understood? A student in the Hilbert School plans to use the concept of an infinite sequence of natural numbers in a consistency proof, when only finite mathematical objects are to be appealed to in such a proof? Even more surprising, he thinks of using the principle of tertium non datur *in* a consistency proof, when that principle is viewed as distinctive for classical mathematical practice and has to be secured *through* a consistency proof? – This startling puzzle will be resolved in the end, but not without starting at the beginning.**

5.2 Results, Methods, and Problems

Finitist proof theory originated in lectures Hilbert and Bernays gave in February of 1922. Its gradual emergence can be documented with reference to notes for lectures on the principles of mathematics Hilbert presented between 1917–18 and 1923–24.[1] In the spring of 1922, a finitist consistency proof was obtained for a quantifier-free fragment of arithmetic and a year later for primitive recursive arithmetic. The

---

** In Fall 2009, I wrote an *Introduction* to late Hilbert papers for (Hilbert 2012) and presented, on 6 November 2009, a version in a talk (with the same title as this essay) to the *Workshop on Logical Methods in the Humanities* at Stanford University. The present essay expands that paper.

However, it could not have been written without Menzler-Trott's historical work, von Plato's discovery of Gentzen's *Urdissertation*, and Thiel's efforts to establish a Gentzen Nachlass and to transcribe early manuscripts. I am grateful to all three, but in particular to Thiel for sending me, in the middle of December 2009, the important manuscript INH. Von Plato was deeply involved in its transcription and pointed out to me, why he considers it as very significant. INH and other manuscripts will be made available, I hope very soon, in a full edition of Gentzen's Nachlass.

Many thanks are also due to Sabine Friedrich, Ulrich Majer, Wilfried Nippel, Winfried Schultze, and Marion Sommer for exploring archival questions in Berlin, Göttingen, and Hamburg. For pertinent remarks, suggestions and additional information I thank Wilfried Buchholz, Martin Davis, John Dawson, Heinz-Dieter Ebbinghaus, Eckart Menzler-Trott, Grigori Mints, William Tait, and Christian Thiel.

[1]See my papers (1999) and (2009). All the Hilbert lectures I am referring to are contained in (Hilbert 2012).

proofs used quite novel means and involved, in particular, transformations of formal derivations.[2] The ultimate goal was to turn proofs of *numeric*[3] statements into proofs containing only numeric statements that can then be evaluated as correct. This approach was programmatically extended to theories involving quantifiers via Hilbert's 1923-*Ansatz*, the ε-substitution method. By 1925, Ackermann, expanding Hilbert's method, and von Neumann had obtained stronger results, the extent of which was not perfectly clear. Indeed, in his address *Über das Unendliche* presented in Münster on 4 June 1925, Hilbert was rather vague about the theories that had been proved to be consistent.[4]

No official lecture notes from the decade's second half illuminate this situation.[5] There are, however, four papers of Hilbert's that give insight into the developments in Göttingen. They fall into two distinct groups: his talks in Hamburg (July 1927) and Bologna (September 1928) constitute the first group, whereas the papers he presented in Hamburg (December 1930) and Göttingen (July 1931) belong to the second group. The reason for this grouping will become apparent very soon.

In his Hamburg talk of July 1927, Hilbert describes the status of proof theoretic work as he had done in *Über das Unendliche* (p. 179), but also discusses the "considerable progress in the proof of consistency" made by Ackermann. That remark does *not* refer to (Ackermann 1924), but rather to work Ackermann had done in early 1925 and had communicated to Bernays in a letter of 25 June 1925. Almost a year later, Ackermann tells Bernays in a letter of 31 March 1926 that he has turned his attention to the "ε_f-proof", i.e., the consistency proof for analysis, and that he is trying to finish it with all his might. Hilbert begins his progress report by recalling the idea of his 1923-Ansatz.

> In proving consistency for the ε-function the point is to show that from a given proof of $0 \neq 0$ the ε-function can be eliminated, in the sense that the arrays formed by it can be replaced by numerals in such a way that the formulae resulting from the logical axiom of choice by substitution, the "critical formulae", go over into "true" formulae by virtue of those replacements.[6]

---

[2]The first step in this and later consistency proofs is the transformation of linear derivations into tree-like structures. That is achieved through the "Auflösung in Beweisfäden". The structure of the argument is discussed not only in these lecture notes, but also in (Hilbert 1923, p. 1142) and (Ackermann 1924, section II); it is beautifully presented in (Hilbert and Bernays 1934, pp. 221–228).

[3]A formula is called *numeric* if it contains neither bound nor free variables.

[4]Ackermann's paper was submitted for publication on 30 March 1924 and von Neumann's on 29 July 1925.

[5]Hilbert gave lecture courses on *Grundlagen der Mathematik* in the winter term 1927/28 and on *Mengenlehre* in the summer term of 1929. There are no notes for the 1927/28 lectures, but for the set theory course Menzler-Trott describes in his (2007, Note 8, p. 22) detailed notes that were taken by Lothar Collatz; see (Hilbert *1929). Apart from their mathematical value, these notes are of special interest in the context of this paper, as Gentzen attended these lectures with his friend Collatz; cf. Sect. 5.5.

[6](Hilbert 1927, p. 477). Hilbert continues: Diese Ersetzungen werden nach erfolgter Elimination der freien Variablen durch schrittweises Probieren gefunden, und es muß gezeigt werden, daß

The central idea is to transform, as in the earlier proofs, linear derivations into tree-like ones consisting only of numeric formulae, now also involving ε-terms, all of which can be recognized to be true. Bernays gives details in his *Zusatz zu Hilberts Vortrag*. Both Hilbert and Bernays assert unambiguously that Ackermann's considerations establish the consistency of elementary arithmetic.[7] Hilbert, optimistic as ever, believed that the methods of Ackermann could be extended further:

> For the foundations of ordinary analysis his [Ackermann's, WS] approach has been developed so far that only the task of carrying out a purely mathematical proof of finiteness remains. (Hilbert 1927, p. 479)

The logical calculus underlying the proof theoretic investigations is described in great detail in this paper, but actually goes back to 1922, is used in Ackermann's (1924), is indicated in *Über das Unendliche*, and is investigated most carefully in (Hilbert and Bernays 1934, p. 66 ff). The "axioms for implication" allow the introduction or omission of an assumption, the interchange of assumptions, and the outright elimination of a proposition. The axioms for conjunction and disjunction receive a very special formulation:

$$(A \,\&\, B) \rightarrow A \quad \text{and} \quad (A \,\&\, B) \rightarrow B$$
$$A \rightarrow (B \rightarrow (A \,\&\, B))$$

and

$$A \rightarrow (A \vee B) \quad \text{and} \quad B \rightarrow (A \vee B)$$
$$((A \rightarrow C) \,\&\, (B \rightarrow C)) \rightarrow ((A \vee B) \rightarrow C).$$

Two axioms for negation are formulated in a third group:

$$((A \rightarrow B \,\&\, \neg B) \rightarrow \neg A) \quad \text{and} \quad (\neg\neg A \rightarrow A).$$

The first axiom for negation is called the *principle of contradiction* and the second the *principle of double negation*. To round out the description of the calculus, the *transfinite ε-axiom* is stated as $A(a) \rightarrow A(\varepsilon(A))$. The ε-axiom allows the definition of universal and existential quantifiers as well as the proof of the appropriate principles for them. This formulation goes back to 1923 and is the basis for Hilbert's

dieser Prozeß jedenfalls zu einem Abschluß führt. – *This* is the fact that has to be established by "a purely mathematical finiteness proof". – A careful discussion is given in (Avigad and Zach 2007), and a beautiful contemporary presentation of the method is found in (Tait 2010).

[7] Bernays reemphasizes this point in the later discussion surrounding the second incompleteness theorem, when writing on 20 April 1931 to Gödel: "That proof [of Ackermann] – to which Hilbert referred in his lecture on *The Foundations of Mathematics* [i.e., in (Hilbert 1927)] with the addendum appended by me – I have repeatedly considered and viewed as correct." – Ackermann never published his second proof; it was only presented in the second volume of *Grundlagen der Mathematik* in section 2, see (Hilbert and Bernays 1939, pp. 121–130, and note 1 on p. 121).

ε-substitution method, as indicated above. However, we are already in 1927, when that method had been extended by Ackermann and had been used, presumably, to prove the consistency of elementary arithmetic.

More than a year later, Hilbert gave a talk at the International Congress of Mathematicians in Bologna. His report on the status of proof theoretic research is essentially unchanged: Ackermann and, Hilbert adds on this occasion, von Neumann have secured the consistency of elementary number theory. He insists again that Ackermann has carried out the consistency proof for analysis with just one remaining task, namely, that of proving "a purely arithmetic elementary finiteness theorem".[8] Hilbert gives a wonderfully clear presentation of broad methodological issues and important metamathematical problems. I will focus on the problem of syntactic completeness for elementary number theory and analysis, as that will play a central role in the further developments.

The issue is formulated as Problem IV and, in the different form of Post-completeness, as Problem V. (The republications of the Bologna address list these as problems III and IV, respectively.) Hilbert asserts that completeness of the theories for arithmetic and analysis is generally claimed and thinks, I assume, that such claims are founded on their categoricity. The argument would proceed as follows: as all their models are isomorphic, a statement S is either true in all models or false in all of them; thus, S or ¬S is a logical consequence of the axioms. If logical consequence were captured by derivability in this second-order framework, then the claim would follow immediately.[9]

Hilbert continues, "The usual idea for showing that any two interpretations of number theory, respectively of analysis, must be isomorphic does not meet the demands of finitist rigor." He suggests, as a next step, transforming the standard categoricity proof for number theory into a finitist argument that would establish the following assertion:

> If for some statement S the consistency with the axioms of number theory can be established, then it is impossible to also prove for ¬S ... the consistency with those axioms, and most directly connected with this: If a statement is consistent, then it is also provable.[10]

[8](Bernays 1930, p. 58) gives the same description of the status of proof theory: Durch die von Ackermann und v. Neumann geführten Beweise ist die Widerspruchsfreiheit für das erste Postulat der Arithmetik, d.h. die Anwendbarkeit des existentialen Schließens auf die ganzen Zahlen sichergestellt. Für das weitere Problem der Widerspruchsfreiheit des Allgemeinbegriffs der Zahlenmenge (bzw. der Zahlenfunktion) einschließlich des zugehörigen Auswahlprinzips liegt ein weitgeführter Ansatz von Ackermann vor.

[9]This idea for a syntactic completeness "argument" is made explicit at the end of Gödel's Königsberg talk (1930c, pp. 26–29). It is connected with the unprovability result he had "recently" proved (to be discussed below in Sect. 5.3). Gödel obtained as a consequence of categoricity and syntactic incompleteness of PM the (semantic) incompleteness of calculi for higher-order logics. – Connections to similar considerations in the Introduction to Gödel's thesis (Gödel 1929, pp. 60–64) are detailed in (Kennedy 2010).

[10](Hilbert 1928, p. 6). I indicate negations by prefixing with ¬ instead of by "overlining" as Hilbert does.

This problem is taken up as the central issue in Hilbert's third Hamburg talk that was presented to the local Philosophical Society in December 1930. In this talk, Hilbert describes first the philosophical and mathematical background for proof theory and again formulates the central goal of his foundational work:

> Indeed, I would like to eliminate once and for all the questions concerning the foundations of mathematics as such – by turning every mathematical statement into a formula that can be concretely exhibited and strictly derived, thus recasting mathematical concept formations and inferences in such a way that they are irrefutable and nevertheless provide an adequate image of the whole science. (Hilbert 1931a, p. 489)

Hilbert sketches the formal system for elementary number theory and emphatically reasserts that Ackermann and von Neumann have proved its consistency. Thus, they have validated as admissible all transfinite inferences, in particular, the principle of tertium non datur. Referring back to his Bologna talk, he then formulates as "our most important further task" to find the proof of two theorems:

> 1. If a statement can be shown to be consistent, then it is also provable; and furthermore,
> 2. If for some statement S the consistency with the axioms of number theory can be established, then it is impossible to also prove for ¬S the consistency with those axioms. (l.c., p. 491)

Hilbert asserts that he has succeeded in proving these theorems for "certain simple cases".

This success has been made possible by extending elementary arithmetic with, what Hilbert calls, a *new inference rule*. Hilbert's Rule (*HR*) is viewed as finitist and allows the introduction of universally quantified formulae (x) A(x) as initial ones, just in case the numeric instances A(z) have been established finitistically as correct for arbitrary numerals z. (*HR*) is not a standard inference rule that facilitates a step from one or more premises to a conclusion within a formal theory. It rather introduces *universal claims as axioms* when an appropriate finitist justification has been given for all their instances. For the theory thus extended, Hilbert proves claims 1 and 2, but only when the statements involved are purely universal. For purely existential formulae he proves claim 2 and warns that claim 1 is not a consequence for them. Thus, he has shown that the extended theory is indeed complete for "certain simple cases". His proof is presented as a direct extension of Ackermann's inductive argument treating only the additional case of (*HR*).

As it happened, Gödel reviewed Hilbert's paper for the *Zentralblatt* and wrote the brief, careful, and matter-of-fact report (Gödel 1931b). The talk provides, according to Gödel, "a substantial supplement to the formal steps taken thus far toward laying a foundation for number theory". This "substantial supplement" is obtained by extending the formal system through the "following rule of inference, which, structurally, is of an entirely new kind". Gödel describes (*HR*) as I did in the previous paragraph and formulates the consistency as well as the partial completeness results without further comment. In a letter to Heyting of 15 November 1932 he points out that Hilbert resolved the completeness problem (in spite of the new axioms) for only "a small sub-question" and that the extended system still has undecidable statements.[11] It should be noted that Gödel did not have any qualms about (*HR*): following Herbrand's (1931), he used it in the formulation of elementary arithmetic, when proving the consistency of its classical version relative to its intuitionist one in (Gödel 1933).

In his letter to Gödel of 18 January 1931, Bernays remarks that the formalism Hilbert had used in his Hamburg talk of the preceding December introduces the principle of complete induction in two different ways, namely, (i) through (*HR*) for the quantifier-free statements of finitist arithmetic, and (ii) through the induction axioms of the theory of elementary arithmetic. He proposes a unified formulation through an "infinitary" rule that is not tied to finitist argumentation as (*HR*) is: "If $\mathsf{A}(x_1, \ldots, x_n)$ is a (not necessarily recursive) formula in which only $x_1, \ldots, x_n$ occur as free variables and which is transformed, through the substitution of any numerical values whatsoever in place of $x_1, \ldots, x_n$, into a formula that is derivable by the logical rules from the formal axioms and the formulae already derived, then the formula $(x_1)\ldots(x_n)\mathsf{A}(x_1, \ldots, x_n)$ may be adjoined to the domain of derived formulae." (See Note 11 also for the German text.) In his last paper Hilbert will use a similarly expanded rule, but in a fully constructive context and tied to finitist argumentation; that significantly different move of Hilbert's is discussed in Sect. 5.5.

The central question has been why Hilbert took on the issue of syntactic completeness with such prominence. – One answer may be that he was simply motivated to tackle a significant open problem. After all, in his Bologna talk he considered the issue as important and as difficult, and so did Bernays in his (1930). However, there is one fact that speaks against this understanding: both Hilbert and Bernays conjectured elementary number theory to be complete.[12] One is obviously tempted to ask, what was the reason for Hilbert not only to suspect, but to take for granted *now*, in late 1930, that there are elementary, formally undecidable sentences

[11]See Gödel's *Collected Works V*, p. 60. The detailed argument for the latter claim can be found in Gödel's letter of 2 April 1931; it was given in response to Bernays's letter of 18 January 1931 (ibid., p. 86ff). Bernays discussed (*HR*) and introduced an extended version, which is formulated in the next paragraph. Gödel's analysis applies also to that extension. – A clarifying discussion of (*HR*) and the ω-rule is found in (Feferman 1986) and (Feferman 2003), but also in (Tait 2002, pp. 417–418).

The translation is slightly modified from that in the *Collected Works*; here is the German formulation of the infinitary Bernays Rule: Ist $\mathsf{A}(x_1, \ldots, x_n)$ eine (nicht notwendig rekursive) Formel, in welcher als freie Individuenvariablen nur $x_1, \ldots, x_n$ auftreten und welche bei der Einsetzung von irgendwelchen Zahlwerten anstelle von $x_1, \ldots, x_n$ in eine solche Formel übergeht, die aus den formalen Axiomen und den bereits abgeleiteten Formeln durch die logischen Regeln ableitbar ist, so darf die Formel $(x_1)\ldots(x_n)\mathsf{A}(x_1, \ldots, x_n)$ zum Bereich der abgeleiteten Formeln hinzugenommen werden.

[12]Bernays remarks in his (1930, p. 59): Von der Zahlentheorie, wie sie durch die Peanoschen Axiome, mit Hinzunahme der rekursiven Definition, abgegrenzt wird, glauben wir, daß sie in diesem Sinne deduktiv abgeschlossen ist [i.e., is syntactically complete]; die Aufgabe eines wirklichen Nachweises hierfür ist aber noch völlig ungelöst. Noch schwieriger wird die Frage, wenn wir, über den Bereich der Zahlentheorie hinaus, zu der Analysis und den weiteren mengentheoretischen Begriffsbildungen aufsteigen.

and, in addition, to expand elementary number theory in order to overcome that incompleteness at least partially – by an inference rule that is in Gödel's words "of an entirely new kind"?[13]

A key to answering the question may be the second assertion in the list of objections to proof theory Hilbert discusses in the last third of his talk. Hilbert views the objections naturally as unjustified. Here is his formulation of the second objection:

> It has been said, in criticism of my theory, that the statements are indeed consistent, but that they are not thereby proved. But certainly they are provable, as I have shown here in simple cases. (Hilbert 1931a, p. 492)

Is the first sentence not an allusion to the first incompleteness theorem? But how could Hilbert have known about it? Did he know about the second theorem and von Neumann's related conjecture that the consistency of classical mathematics is unprovable? If he did, how could he take for granted without any hesitation that the consistency of elementary number theory had been established?

5.3 Unprovability in General, First

In order to make explicit the assumptions in these historical questions and to approach answers to them, let me examine, to begin with, what Gödel announced at the roundtable discussion in Königsberg on 7 September 1930. (This discussion was part of the Conference on Epistemology of the Exact Sciences. Hans Hahn was its moderator, and Carnap, Gödel, Heyting and von Neumann were among the participants. For details, see (Dawson 1997, pp. 68–71).) The stenographic transcript of Gödel's remarks was published in *Erkenntnis* together with a later *Postscript* in which he summarized, at the request of the journal's editors, the results of his classical paper (1931). In the transcript, the following theorem is stated:

> (Assuming the consistency of classical mathematics) one can give examples of statements (and in fact statements of the type of Goldbach's or Fermat's) that are in fact contentually true, but are unprovable in the formal system of classical mathematics. Therefore, if one adjoins the negation of such a proposition to the axioms of classical mathematics, one obtains a consistent system in which a contentually false proposition is provable.[14]

---

[13]Feferman makes exactly this point and conjectures in his Introductory Note to Gödel's correspondence with Bernays: "Since Hilbert had previously conjectured the completeness of Z, he would have had to have a reason to propose such an extension, and the only obvious one is the incompleteness of Z." (Feferman 2003, Note 1 on p. 44) – Z is of course elementary number theory.

[14](Gödel 1931a, p. 203 in *Collected Works I*). I modified the translation; the German text is: Man kann (unter Voraussetzung der Widerspruchsfreiheit der klassischen Mathematik) sogar Beispiele für Sätze (und zwar solche von der Art des Goldbachschen oder Fermatschen) angeben, die zwar inhaltlich richtig, aber im formalen System der klassischen Mathematik unbeweisbar sind. Fügt man daher die Negation eines solchen Satzes zu den Axiomen der klassischen Mathematik hinzu, so erhält man ein widerspruchsfreies System, in dem ein inhaltlich falscher

Thus, these transcribed Königsberg remarks do not yet formulate the syntactic incompleteness of elementary number theory and neither do those at the end of the presentation of his thesis work on the completeness of first-order logic, which Gödel had delivered a day earlier, on 6 September; cf. Note 14.

Von Neumann sat at the Königsberg roundtable and talked with Gödel immediately after the session. Gödel's recollection of this conversation and his perspective on subsequent developments are reported in (Wang 1981):

> Von Neumann was very enthusiastic about the result and had a private discussion with Gödel. In this discussion, von Neumann asked whether number-theoretical undecidable propositions could also be constructed in view of the fact that the combinatorial objects can be mapped onto the integers and expressed the belief that it could be done.

Von Neumann's question and conjecture point to an intriguing fact that is clearly formulated in Wang's paper: at the time of the Königsberg meeting, syntax had not been arithmetized. Rather, symbols were directly represented in the formal theory by numerals, sentences by sequences of numerals, and proofs by sequences of sequences of numerals. The crucial syntactic notions and the substitution function are expressible in subsystems of type or set theory and, consequently, so is the undecidable statement. Gödel responded to von Neumann's query by saying: "Of course undecidable propositions about integers could be so constructed, but they would contain concepts quite different from those occurring in number theory like addition and multiplication." Remarks about the subsequent developments follow:

> Shortly afterward Gödel, to his own astonishment, succeeded in turning the undecidable proposition into a polynomial form preceded by quantifiers (over natural numbers). At the same time, but independently of this result, Gödel also discovered his second theorem to the effect that no consistency proof of a reasonably rich system can be formalized in the system itself.[15]

Satz beweisbar ist. – The background is described in Dawson's *Introductory Note*, ibid., pp. 196–199, and in (Dawson 1997, pp. 68–79). – Gödel's *Collected Works III* contains the report on his thesis work he gave to the Königsberg Congress (cf. note 9). He remarks after discussing the connection of categoricity and *Entscheidungsdefinitheit* that his result can be stated as follows: Das Peanosche Axiomensystem mit der Logik der *Principia Mathematica* als Überbau ist nicht entscheidungsdefinit.

[15](Wang 1981, pp. 654–655). Parsons's *Introductory Note* to the correspondence with Wang in Gödel's *Collected Works* describes in Section 3.2 the interaction between Gödel and Wang on which Wang's paper was based. – There are some seeming oddities with this and related other accounts: (i) Gödel parenthetically seems to claim in the published transcript (1931a) that the undecidable statement is of a restricted *number theoretic* form; that would be in striking conflict with his own account in (Wang 1981). However, it is only claimed that the statement is universal with a finitist matrix. Indeed, in the *Nachtrag* Gödel specifies, fully in accord with the report in (Wang 1981), the purely arithmetic character of the undecidable sentence. (ii) Carnap reports from an August 1930 meeting with Gödel that the latter had pointed out undecidability, but also "difficulty with consistency". That has been taken as an indication that Gödel had a version of his second theorem already then. Such an assumption is in direct conflict with the Gödel-Wang account described above; it receives a convincing explanation through Gödel's description in (l.c., p. 654), how he found the result, namely, when running into difficulties in his attempt of proving the consistency of analysis.

So it is clear that Gödel announced only *one* result at the Königsberg meeting; it was, as quoted above from (Gödel 1931a), a quite restricted unprovability result for "classical mathematics" as formulated in type or set theory. He had not yet obtained his second incompleteness theorem.

Now I come back to the question, what Hilbert may have known about Gödel's result(s) before giving his talk in Hamburg or before submitting his paper to *Mathematische Annalen* on 21 December 1930. As von Neumann is often mentioned as a possible conduit to Hilbert, let me recall two facts from von Neumann's correspondence with Gödel: first, von Neumann learned about the formulation of both incompleteness theorems around 25 November 1930 and, second, he got to know the details of Gödel's arguments only at the very beginning of 1931. It is equally crucial to realize that Hilbert delivered his lecture *Naturerkennen und Logik* in Königsberg on 8 September 1930, the very day after the roundtable discussion. His lecture was an invited address at the meeting of the Society of German Scientists and Physicians that took place from 7 to 11 September. We also know that Hilbert's and von Neumann's stays in Königsberg overlapped.[16] Is it then implausible to think that von Neumann (or other colleagues who attended the roundtable discussion like Emmy Noether) talked with Hilbert about Gödel's result as formulated above? – I think not. In any event, it seems implausible that Hilbert was *not* informed then or later about a result that elicited not only von Neumann's deep and immediate response, but also a longer preoccupation (see Note 22). If one assumes that Hilbert knew just the result Gödel had announced in Königsberg, then the metamathematical considerations in his (1931a) take on the completeness issue for statements of the form of Gödel's unprovable sentence.[17]

What can we consider as the immediate effect of Gödel's Königsberg result on Hilbert's finitist consistency program? As that result does not concern consistency, there is no real direct effect – unless syntactic completeness is taken as a crucial

---

[16]Oystein Ore attended the meeting of the Society of German Scientists and Physicians and reports in (Reid 1970, p. 195), "I remember that there was a feeling of excitement and interest both in Hilbert's lecture and in the lecture of von Neumann on the foundations of set theory – a feeling that one now finally was coming to grips with both the axiomatic foundation of mathematics and with the reasons for the applications of mathematics in the natural sciences." Dawson reports that Gödel left Königsberg only on 9 September, and he speculates that "it is very likely that he [Gödel] was in the audience" when Hilbert presented his lecture (l.c., p. 71).

[17]That seems to be in conflict with (Bernays 1935, p. 215): "Noch ehe dieses Gödelsche Resultat bekannt war, hatte Hilbert die ursprüngliche Form seines Vollständigkeitsproblems bereits aufgegeben. In seinem Vortrag *Die Grundlegung der elementaren Zahlenlehre* [i.e., (Hilbert 1931a)] behandelte er dieses Problem für den Spezialfall von Formeln der Gestalt (x)A(x), welche außer x keine gebundene Variable enthalten." The question is of course not, why did Hilbert give up on the *form of that problem* (which he did not), but rather, why did he give up on *the completeness conjecture*? – In Bernays's correspondence with Gödel (*Gödel's Collected Works IV*, p. 84) there is a peculiar and uncharacteristic lack of familiarity with Hilbert's Hamburg talk. Bernays attributes the insight – "Die Widerspruchsfreiheit der neuen Regel folgt aus der Methode des Ackermannschen (oder auch des v. Neumannschen) Nachweises für die Widerspruchsfreiheit von Z." – to an observation of A. Schmidt, when Hilbert's consistency proof for the new rule is explicitly an extension of Ackermann's proof (as discussed above in Sect. 5.2)!

ingredient of the program. Hilbert had formulated, as described above, the syntactic completeness question for arithmetic as well as analysis in his Bologna lecture and conjectured a positive answer. However, there was also a speculation at the Bologna Congress that some formal systems might be *incomplete*. To support this assertion I point to two sources.

1. In the first republication of the Bologna lecture, submitted to *Mathematische Annalen* on 25 March 1929, a remark was added on page 6:

> In higher domains we might conceivably have a situation in which both S and ¬ S are consistent: then the adoption as an axiom of one of the two statements S, ¬ S is to be justified by systematic advantages (principle of the permanence of laws, the possibility of further development, etc.).

It is not clear which "higher domains" are envisioned or why this remark was added. It would be of interest to have a sense, whether Skolem's results were taken into account. Skolem suggested in his (1922, p. 299, note 9) the continuum problem to be undecided by the first-order formulation of Zermelo's axioms and gave as a reason the non-categoricity of those axioms.

2. In a letter of 9 April 1947 to Heinrich Scholz, Bernays wrote:

> The possibility of the underivability of the components of a derivable disjunction of the form
>
> $$\textbf{(A)} \qquad (X)(\phi(X) \rightarrow \gamma(X)) \vee (X)(\phi(X) \rightarrow \neg\gamma(X))$$
>
> was moreover already contemplated a considerable time before the appearance of the Gödel theorem. I discussed such matters at the time with Tarski at the Congress in Bologna.[18]

The proof of **(A)** is presumably based on the categoricity considerations motivating the "usual claim" of syntactic completeness for arithmetic and analysis. That claim was discussed by Hilbert in Bologna and was described above in the middle of Sect. 5.2 as well as in note 9.

Finally, let me add a third perspective. Hilbert frequently emphasized a concept of "quasi-empirical" completeness: an axiomatic theory was to make possible, in an intelligibly structured way, proofs of all the elements of a given collection of mathematical facts. That is articulated in many places, but most directly in the set theory lectures from 1917. Hilbert suggests there that a collection of more or less secure facts should be shaped into a system following this general approach:

> If we have certain statements in front of us and we are unable to assert anything certain about their correctness, we select some that seem to play a distinguished role as a preliminary axiom system – be it that we choose the simplest among them, be it that we prefer those that seem to have the most secure foundation or to be the most intuitive.[19]

---

[18] *Bernays Nachlass* at the ETH Zürich, Hs 975: 4123. See also (Mancosu 1999, p. 33). Like Mancosu I thank Bernd Buldt for pointing me to this letter.

[19] (Hilbert *1917, p. 40). The German text is: Haben wir gewisse Sätze vor uns, über deren Richtigkeit wir nichts Sicheres aussagen können, so greifen wir einige, die uns eine ausgezeichnete

On the next page of these notes Hilbert raises the completeness question in the form, "Is it really the case that all the facts of the collection are logical consequences of the selected particular statements, which as axioms are the basis for the system?" – When answering the question (on p. 48) for the axioms of complete ordered fields, Hilbert tentatively "shows" their syntactic completeness by an argument that reflects the one I sketched in Sect. 5.2 and that exploits the categoricity of the axiom system.

As a consequence of these considerations, I look at matters in the following way. However surprising Gödel's Königsberg announcement was, it was not a complete shock to everyone, as syntactic completeness was not viewed as a *sine qua non* for proof theory or the axiomatic presentation of parts of mathematics. For von Neumann the unprovability result must have been nevertheless striking, and the reason seems to be straightforward. He had formulated the central tasks of proof theory in his talk on Hilbert's Program given just a few days earlier; that talk was published as his (1931). Von Neumann emphasized the quasi-empirical completeness requirement with an interestingly stronger condition, articulated as follows:

> A construction procedure has to be given that allows producing successively all formulae, which correspond to the "provable" claims of classical mathematics. This procedure may thus be called "proving".[20]

This requirement for the proof theoretic program, von Neumann claims, has been secured by the work of Russell and his school and guarantees in particular that every correct finitist statement can be obtained by means of this construction procedure. It does not amount to a decision procedure for particular statements, as von Neumann points out (1931, p. 120), and it is not to be equated with syntactic completeness, as I want to emphasize.[21] Von Neumann's fundamental conviction that finitist mathematics can be formally captured was not undermined by Gödel's unprovability result; rather, that result pointed to limitations of finitist mathematics

---

Rolle zu spielen scheinen, als ein vorläufiges Axiomsystem heraus – sei es dass wir die Einfachsten unter ihnen wählen, sei es dass wir diejenigen bevorzugen, die uns am sichersten fundiert oder auch am anschaulichsten erscheinen.

The German text of the following sentence is: Ist wirklich das gesamte vorliegende Tatsachenmaterial die logische Folgerung aus den herausgegriffenen besonderen Sätzen, die wir als Axiome dem System zu Grunde gelegt haben?

[20](von Neumann 1931, p. 118). The German text is: Es ist ein Konstruktionsverfahren anzugeben, das sukzessiv alle Formeln herzustellen gestattet, welche den "beweisbaren" Behauptungen der klassischen Mathematik entsprechen. Dieses Verfahren heiße darum "Beweisen".

[21]The insight that syntactic completeness and decidability amounted to the same thing had to wait for the mathematical characterization of "formal" theories and decidability. Bernays discusses in his (1930) syntactic completeness and adds on p. 59 in note 19 the remark: Man beachte, daß die Forderung der deduktiven Abgeschlossenheit [i.e., syntactic completeness, WS] noch nicht so weit geht wie die Forderung der *Entscheidbarkeit* einer jeden Frage der Theorie, welche besagt, daß es ein Verfahren geben soll, um von jedem beliebig vorgelegten Paar zweier der Theorie angehöriger, einander kontradiktorisch entgegengesetzter Behauptungen zu entscheiden, welche von beiden beweisbar ("richtig") ist.

and led von Neumann quickly to a dramatic conclusion that is discussed in the next section. There I will also explore what we know about (i) when members of the wider Hilbert circle learned about Gödel's second incompleteness theorem, and (ii) when this theorem's full impact was realized. I will discuss von Neumann, Bernays, and Herbrand.

5.4 Unprovability of Consistency, Second

Von Neumann was captivated at once by Gödel's result and, a couple of weeks later, made the remarkable discovery that, in his own words, "the consistency of mathematics is unprovable". I.e., he had arrived independently at a proof of Gödel's second incompleteness theorem. Taking for granted the co-extensionality of finitist and intuitionist mathematics, he argued for his discovery in two steps. Here is the first step, where W stands for the formula expressing the consistency of the formal system under consideration: "If the consistency [of the system] is established intuitionistically, then it is possible, through a 'translation' of the contentual intuitionistic considerations into the formal [system], to prove W also [in that system]." The possibility of doubting the translatability of finitist arguments because of Gödel's result is considered and rejected; von Neumann believes that "in the present case it [the translatability] must obtain". In the second step he argues, "W is always unprovable in consistent systems, i.e., a putative effective proof of W could certainly be transformed into a [proof of a] contradiction." Thus, the consistency of mathematics is unprovable by intuitionist means. Von Neumann conveyed these considerations to Gödel in a letter of 20 November 1930 and closed by calling Gödel's unprovability result as formulated in Königsberg "the greatest logical discovery in a long time". Gödel responded almost immediately and informed von Neumann of his *new* results and most likely sent him a copy of the abstract (Gödel 1930b), which Hahn had presented to the Vienna Academy of Sciences on 23 October 1930. This abstract contains the classical formulation of both incompleteness theorems. The full text of Gödel's 1931-paper was submitted to *Monatshefte* on 17 November 1930, before von Neumann had formulated his letter to Gödel.

In his next letter of 29 November, von Neumann assured Gödel that he would not publish on the subject "as you have established the theorem on the unprovability of consistency as a natural continuation and deepening of your earlier results".[22]

[22]Gödel's first two letters to von Neumann have not been preserved, it seems. The core content of the letters can be inferred from von Neumann's responses and from Gödel's discussion of von Neumann's perspective at the meeting of the Vienna Circle of 15 January 1931; see also (Gödel 1932). In (Dawson 1997, p. 70) one finds Hempel's report on the course in proof theory he took with von Neumann in Berlin during the winter term 1930–1931: "... in the middle of the course von Neumann came in one day and announced that he had just received a paper from a young mathematician in Vienna ... who showed that the objectives which Hilbert had in mind ... could

However, a disagreement emerged between him and Gödel on how this theorem affects Hilbert's finitist program. Its roots go back, on the one hand, to von Neumann's conviction of the "translatability" of finitist proofs and, on the other hand, to Gödel's view that there might be finitist proofs that are not obtainable in a particular formal theory. The dispute started after von Neumann had received the proof sheets of (Gödel 1931), at the end of which Gödel had expressed that view. After having thanked Gödel for the galleys, von Neumann writes on 12 January 1931, "I absolutely disagree with your view on the formalizability of intuitionism." The fact that for each formal system there is a proof theoretically stronger one, as Gödel's first result shows, does not "touch intuitionism" in von Neumann's view. He defends that belief again, as he had done in his first letter to Gödel, by claiming intuitionist or finitist arguments can be "translated" into formal ones in number theory and, if not there, certainly in analysis or set theory. In this letter, he also indicates a more "abstract" proof of the second incompleteness theorem using conditions on the provability predicate in order to show the equivalence of the unprovable Gödel sentence with the consistency statement W.

Bernays wrote to Gödel on 24 December 1930 from Berlin, where he was spending the Christmas break with his family.[23] He had already earlier received a reprint of (Gödel 1930a) and begins his letter with some comments on Gödel's completeness proof for first-order logic. Then Bernays asks for the galleys of Gödel's new investigations. As the reason for this request he notes that he has learned from Professors Courant and Schur,[24] "that you have recently succeeded in obtaining significant and surprising results in the area of foundational problems, and that you intend to publish them shortly". Let me remark, parenthetically, that I don't see any conflict between this request and the assumption that Bernays already

---

not be achieved at all." In a letter of 3 December 1930 to his friend Chevalley, Herbrand writes about the same period: "... I have been here for two weeks, and every time I have seen von Neumann we talk about the work of a certain Gödel, who has produced very curious functions; and all of this has destroyed some quite solidly anchored ideas." (Details concerning this letter are found in the Appendix of (Sieg 1994).)

[23]The letters I refer to in the following are all contained in volume IV of *Gödel's Collected Works*. The longer excerpts in this paragraph are found on p. 80, respectively on p. 82.

[24]Here and below are some important factual questions. When and where did Bernays learn this from Courant and Schur? How had *they* been informed? What did Bernays do in the winter term 1930–1931? (It seems that he attended neither the Königsberg meeting nor the lecture Hilbert gave in December 1930 in Hamburg.) We know that Gentzen spent this very term in Berlin; did he attend von Neumann's lectures? Did he have contact there with Bernays and Herbrand? (Cf. Note 22.) – As to the latter question, Menzler-Trott is "convinced" that Gentzen talked with all three (Bernays, Herbrand, and von Neumann) while in Berlin. In the same note of 7 December 2010 in which he expressed this conviction about Gentzen, Menzler-Trott reports that Bernays participated in 1930/1931 in the meetings of the *Berliner Gesellschaft für empirische Philosophie* with, among others, Dubislav, Grelling, Hempel, and Reichenbach. It may very well be that Bernays learned here about Gödel's Königsberg result, as Dubislav participated in that meeting and Hempel heard about it in von Neumann's lecture. (Cf. Note 22.) – The name of the Berliner Gesellschaft was changed in 1931, as suggested by Hilbert, to *Berliner Gesellschaft für wissenschaftliche Philosophie.*

knew about Gödel's restricted Königsberg result. In any event, Gödel answered on 31 December 1930, sent Bernays a reprint, presumably the abstract (1930b), and promised to send the galleys of his (1931). Bernays received the galleys on 14 January 1931, studied them and wrote a long, detailed letter to Gödel only four days later. He remarks, "For me that was very interesting and very instructive reading. What you have done is really an important step forward in the investigation of the foundational problems."

It took some time before Bernays saw clearly that Ackermann's and von Neumann's proofs established the consistency of arithmetic only when the induction principle is restricted to quantifier-free formulae. On 3 May 1931, Bernays writes again to Gödel and views Ackermann's considerations as proving the consistency of full arithmetic. He tries to find the reason why that proof cannot be formalized in arithmetic as required by the second incompleteness theorem. Incorrectly, he sees "the explanation of the matter" in the fact that nested recursions cannot be formalized in elementary arithmetic.[25] It was only in the brief note (Bernays 1933) for the International Congress of Mathematicians in Zürich that the limited nature of these consistency proofs was explicitly stated and publicly acknowledged.[26] Here is Bernays's formulation concerning the "method of valuation" (Bewertungsmethode):

> This [method, WS] obtained its essential development by Hilbert's procedure of trial valuation. Using this procedure Ackermann and von Neumann demonstrated the consistency of number theory – admittedly, under the restrictive condition that the application of the inference from n to $(n + 1)$ is only allowed for formulae with just free variables. (Bernays 1933, p. 201)

Bernays remarks, at the end of the note, that the limitation of the proof theoretic results is apparently a fundamental one "because of Gödel's new theorem – and a related conjecture of von Neumann's – on the limits of decidability in formal systems". It seems noteworthy that Hilbert presided over the opening of this Congress; as in Bologna 4 years earlier, he received a standing ovation. That is reported in (Richardson 1932).

Finally, let me make a remark about Herbrand who was also working intensively at that time on the proof theory of elementary arithmetic. He was well aware of the incompleteness results through the contacts he had with both von Neumann and Bernays during his long stay in Berlin, from late November 1930 to the middle of May 1931. The central consistency result in Herbrand's (1931) is formulated for arithmetic with only quantifier-free induction, but including Hilbert's rule (*HR*) and an open characterization of finitist functions. This characterization also covers the

---

[25](*Gödel's Collected Works IV*, p. 104). It is mentioned in (Hilbert and Bernays 1934, p. 422), that nested recursions are formalizable in arithmetic and that Gödel and von Neumann discovered this fact.

[26]In his letter of 15 November 1932 to Heyting, Gödel remarks somewhat indignantly: Bernays hat in seinem Züricher Vortrag (soviel mir bekannt) auch zugegeben, daß man die Widerspruchsfreiheit d[er] Zahlentheorie bisher nur mit einer von Herbrand gegebenen Einschränkung für die vollständige Induktion beweisen kann. (*Gödel's Collected Works V*, p. 60).

functions that are definable by nested recursion, like the Ackermann function. In the last part of his paper, Herbrand discusses Gödel's results. He articulates in almost identical words what he had already claimed on 7 April 1931 in a letter to Gödel, namely, that – contrary to Gödel's opinion, but consonant with von Neumann's view – all finitist (or intuitionist) arguments can be formalized in analysis and possibly in elementary number theory. "If this were so," he concludes, "the consistency of ordinary arithmetic would already be unprovable."

In a letter to Hans Reichenbach that is undated, but was definitely written in the summer of 1931, von Neumann makes a remark on the publication of his contribution to the Königsberg Congress. In a way, he summarizes the broader state of affairs:

> Incidentally, I have decided not to mention Gödel, since the opinion that there still exists a certain hope for proof theory has found champions - *among others*, Bernays and Gödel himself. To be sure, in my opinion this view is erroneous; but a discussion of this question would stray outside the existing boundaries, and so I would rather treat it at another occasion.[27]

So it is perhaps not too surprising that Hilbert himself did not give up on the proof theoretic program, but rather shifted its direction. As a matter of fact, he returned to a perspective that underlies the first part of his (Hilbert 1922), when formulating the base theory in a "quasi-constructive" way with a restricted treatment of negation; see (Sieg 2009, p. 376, note 42) and (Sieg and Tapp 2012). But let me take one step at a time.

5.5 Hilbert's Response

Even in his last paper, *Beweis des Tertium non datur*, Hilbert does not mention Gödel by name, but the proof theoretic considerations react clearly to the second incompleteness theorem and – as Bernays put it in his 1977 interview – try to deal positively with Gödel's results. The approach in this paper is dramatically different from that in the third Hamburg talk. Then, in December 1930, Hilbert concentrated on the syntactic completeness of elementary arithmetic, assuming that its consistency had been secured by Ackermann's and von Neumann's work. Now, in July 1931, there is an almost exclusive focus on consistency and the principle of

[27](Mancosu 1999, p. 49 and note 14). – The German text of von Neumann's note is: Übrigens habe ich mich entschlossen, Gödel nicht zu erwähnen, da die Ansicht, dass noch eine gewisse Hoffnung für die Beweistheorie existiert, Vertreter gefunden hat: u[nter] a[nderen] Bernays und Gödel selbst. Zwar ist m[eines] E[rachtens] diese Ansicht irrig, aber eine Diskussion dieser Frage würde aus dem vorliegenden Ra[h]men hinausführen, ich möchte daher bei einer anderen Gelegenheit darüber sprechen.

tertium non datur, for short: *tnd*.[28] The latter principle postulates, for all statements A(x), that A(x) either holds for all natural numbers or has a counterexample; it is formally expressed by $(x)A(x) \vee (Ex)\neg A(x)$.

The renewed focus on consistency is accompanied, however, by a radical strategic shift. Instead of aiming to prove finitistically the consistency of classical arithmetic by the ε-substitution method, Hilbert formulates a constructive (he thinks, finitist) theory of arithmetic and argues simultaneously for the correctness of its seemingly transfinite inference principles. Let me describe matters in greater detail. The principles for sentential logic – with conditionals, conjunctions and disjunctions as the basic forms – are those of the logical calculus described in Sect. 5.2 above. However, only equations between numerals are negated, and the sole principle involving negations of this restricted sort is the *Axiom des Widerspuchs*, i.e., a contradiction implies any formula whatsoever.[29] For these basic literals Hilbert introduces the concepts "correct" (richtig) and "false" (falsch). After the formulation of the transfinite inference principles these concepts are extended to literals containing recursively defined function symbols and the special symbols connected with the elimination or, as Hilbert says, "application" of the existential quantifier. The concept of "correctness" is extended straightforwardly to conjunctions and disjunctions. Much more problematically, a conditional is viewed as correct "if its antecedent is correct, so is its consequent".

The combination of syntactic and semantic considerations is particularly striking in the case of the "transfinite axioms and inference schemata" for quantifiers. Viewed from a syntactic perspective, existential quantifiers are analyzed essentially by the natural deduction rules for *introducing* and *eliminating* them. Universal quantifiers are *introduced* by the rule (HR^*), the extension of (HR) to formulae of arbitrary complexity all of whose instances are finitistically correct; they are *eliminated* via the axiom $(x)A(x) \rightarrow A(a)$. However, the official partially semantic formulation of these principles is given as follows, extending the concept "correct" to quantified formulae:

> If the statement A(z) is correct as soon as z is a numeral, then the statement (x)A(x) holds; in this case (x)A(x) is called correct. The converse is provided by the axiom $(x)A(x) \rightarrow A(a)$. These stipulations concern the introduction and application of the concept "all". For the introduction and application of the concept "there is" I may apply the following two schemata:
>
> $$\frac{A(a)}{(Ex)A(x)}$$

[28]This focus is fully in accord with the earlier considerations, e.g., in Hilbert's second Hamburg talk of 1927; see p. 471 and p. 479 for consistency, p. 470 and p. 476 for tertium non datur.

[29]The negation for general statements is later defined inductively by using classical equivalences; for example, the negation of the conditional $A \rightarrow B$ is given by $A\&\neg B$, that of $(x)A(x)$ by $(\exists x)\neg A(x)$. – In the earlier Hamburg lecture the *Axiom des Widerspruchs* is the principle $(A \rightarrow (B\&\neg B)) \rightarrow \neg A$. – Note that the law of excluded middle implies *tnd* via the definition of the negation of a universally quantified statement.

> where a is a numeral; in this case, (Ex)A(x) is called correct. Conversely, an expression (Ex)A(x) may be replaced by A(η), where η is a letter that has not been used yet. The following contentual understanding corresponds to this rule: to the formula (Ex)A(x) is associated the definition of a numeral η, according to which [i.e., according to the definition of η, WS] A(η) is correct, whenever (Ex)A(x) is. (Hilbert 1931b, p. 121)

The ultimate goal is to show that this system can be expanded by instances of *tnd* without leading to contradictions.[30]

Before following Hilbert's path of trying to achieve that goal, I want to make a brief remark about the proof theoretic strength of the system I just described. Gentzen's way of interpreting universal quantifiers applied to arbitrary finitistically meaningful statements, given in his (1936) on pp. 526 and 528, can be adapted to argue for the correctness of the standard introduction rule for universal quantifiers, but also for that of the rule of complete induction. Thus, the standard formalization of intuitionist arithmetic, Heyting arithmetic, is contained in Hilbert's constructive system.

To pursue the "proof" of *tnd*, some preparatory work is required. First, the discussion of the transfinite principles is exploited for arguing that "the whole system is consistent", where consistency requires, as usual, that no proof has $1 \neq 1$ as its endformula. Hilbert's reasons for asserting the consistency of the system are compressed into a single sentence: "All transfinite rules and inference schemata are consistent; for they amount to definitions." Second, Hilbert tries to show that consistency and correctness are "identical". Third, defining a statement as "false" in case it leads to a contradiction, he claims that *any statement is either false or correct*. Given the definition of "false" and the identity of consistency and

---

[30]The arguments leading to this goal are barely sketched, and some are deeply problematic. Gödel made a critical remark about Hilbert's paper in a letter to Heyting of 16 May 1933: Ich glaube überhaupt, Sie beurteilen Hilberts letzte Arbeiten etwas zu günstig. Z.B. ist doch in Göttinger Nachr. 1931 [(Hilbert 1931b), WS] kaum irgend etwas bewiesen. – In conversation with Olga Taussky-Todd, so it is reported in (Taussky-Todd 1987, p. 40), Gödel "lashed out against Hilbert's paper [(1931b), WS], saying something like 'how can he write such a paper after what I have done?' Hilbert in fact did not only write this paper in a style irritating Gödel, he gave lectures about it in Göttingen in 1932 and other places."

It is worth noting that (Hilbert 1931b) was neither reprinted in Hilbert's *Gesammelte Abhandlungen* nor mentioned in (Bernays 1967). In contrast, Bernays's paper discusses (Hilbert 1931a) on pp. 215–216: Das Verfahren, durch welches hier Hilbert die positive Lösung des Vollständigkeits problems (für den von ihm betrachteten Spezialfall) sozusagen erzwingt, bedeutet ein Abgehen von dem vorherigen Programm der Beweistheorie. In der Tat wird ja durch die Einführung der zusätzlichen Schlußregel die Forderung einer restlosen Formalisierung der Schlüsse fallen gelassen.

In his contribution to the *Encyclopedia of Philosophy*, Bernays mentions both papers and remarks summarily after a brief discussion of Gentzen's consistency proof: "The broadened methods [of Gentzen, WS] also permitted a loosening of the requirements of formalizing. One step in this direction, made by Hilbert himself, was to replace the schema of complete induction by the stronger rule later called infinite induction ..." (p. 502) – On my reading of (Hilbert 1931b) there is no infinite rule *in* the system, but rather a finitistically justified introduction of universally quantified statements using (HR^*). (Indeed, Hilbert's proof that the system expanded by *tnd* is consistent exploits the finiteness of proof figures; see the discussion below.)

correctness, Hilbert has to prove that any statement either does or does not lead to a contradiction. This metamathematical statement is an instance of *tnd*, and Hilbert views it as "necessary" for the founding of mathematics. He then uses *tnd* to show that correctness, falsity, and the generalized negation of statements (see Note 29) harmonize in the appropriate way. Having taken these preparatory steps, Hilbert proceeds (on p. 124) to argue that adding instances of *tnd* as axioms to the base system does not lead to contradictions. The indirect metamathematical argument for this claim is sketched in the next paragraph.

Consider a proof in the constructive system that uses, for simplicity's sake, a single instance of *tnd*

(1) $$(x)A(x) \vee (Ex)\neg A(x)$$

as an initial formula. Hilbert lists the instances of the axiom schema (expressing for-all elimination) that have been used in the proof, say,

(2) $$(x)A(x) \rightarrow A(a_1), \ldots, (x)A(x) \rightarrow A(a_n).$$

Now he forms the conjunction $A(a_1)$ &...&$A(a_n)$ and, using it, replaces all occurrences of the statement $(x)A(x)$ in the proof. Hilbert claims that all the transformed initial formulae are correct and that the syntactic configuration (resulting from this replacement) is a derivation from the initial formulae. The formulae resulting from those in (2), $A(a_1)\&\ldots\&A(a_n) \rightarrow A(a_i)$, are trivially provable, whereas the formula resulting from (1), namely,

(3) $$A(a_1)\&\ldots\&A(a_n) \vee (Ex)\neg A(x),$$

is provable from correct instances of the law of excluded middle[31]

(4) $$A(a_i) \vee \neg A(a_i).$$

Thus, one has $A(a_i) \vee (Ex)\neg A(x)$ for all i between 1 and n, and distributivity establishes (3). So we have obtained, Hilbert argues, a proof from correct initial formulae; as correctness is preserved by inferences, all of the formulae in the proof are correct and its endformula can't possibly be $1 \neq 1$.

---

[31]Hilbert argues simultaneously for the correctness of these instances of the law of excluded middle by an inductive argument on the complexity of the formulae in the instances of *tnd* used in the proof. It is difficult to see, why the matrix $A(a_i)$ should be of the appropriate form $(y)B(y, a_i)$ to allow the formulation of *tnd* and its use in the induction hypothesis. – Note that Hilbert does not mention a necessary modification in the upper part of the derivation in case axioms of the form (2) are actually used to infer $A(a_i)$ via modus ponens with $(x)A(x)$ as the minor premise. All the instances $A(a_i)$ must be inferred from $(x)A(x)$ and formed into a conjunction – *before* carrying out modus ponens ... to infer the $A(a_i)$. Gentzen will avoid for intuitionist logic such odd detours through the natural deduction formulation of the logical principles and his normalization proof!

Whatever problematic features there are in the overall considerations (and there are many), a central one is Hilbert's use of *tnd* when arguing that any statement is either correct or false. This appeal is openly acknowledged and is, of course, in conflict with the finitist position. In a letter sent from Berlin on 11 October 1931, Bernays reports to Hilbert on progress with the *Grundlagenbuch*: he just finished the presentation of Behmann's decision procedure for monadic predicate logic with identity and is about to begin replacing the consistency proof "for universal and existential quantifiers that is not correct" by another proof. Then he continues:

> As for the relationship to your last article [i.e., (Hilbert 1931b), WS], it can be said in the Preface that the arguments in the book are carried out entirely within the framework of the finitist standpoint (i.e., other considerations are used at most in a heuristic sense), so that your last article, which is based on a different methodological standpoint, does not come within the scope of these considerations.[32]

The phrasing of this remark suggests that Hilbert and Bernays had discussed the issue and clearly agreed that the methods of *Beweis des Tertium non datur* go beyond the finitist standpoint. But why should they be excluded from further exploration? After all, Hilbert's self-conscious use of an instance of *tnd* in the metamathematical argument may be viewed as parallel to the use (and later removal) of that principle in proofs of his early mathematical career, for example, when solving Gordan's problem in invariant theory.[33]

Indeed, it seems that the methods were explored in critical detail only a few months later by a young student, who had been engaged with Bernays on another project. With the explicit goal of proving the consistency of arithmetic, he began working on his thesis in late 1931 and quickly obtained significant results. An outline of his *Urdissertation* from early October 1932 summarizes these results and formulates the remaining central task. The young student is Gerhard Gentzen. The outline of the thesis is organized in five parts.[34] Part I is worked out on

---

[32](Cod. Ms. D. Hilbert 21, 5). The German text is: Was die Beziehung zu Ihrer letzten Note betrifft, so kann ja im Vorwort gesagt werden, dass die Ausführungen des Buches sich ausschliesslich im Rahmen des finiten Standpunktes bewegen (d.h. anderweitige Betrachtungen werden höchstens im heuristischen Sinne angestellt), dass daher Ihre letzte Note, die einen anderen methodischen Standpunkt zugrundelege, nicht in den Bereich dieser Betrachtungen falle. Bernays continues: Auch im Rahmen des finiten Standpunktes wird ja einiges noch ausserhalb bleiben, dessen Behandlung in einem "2. Teil" ja am Schluss des Textes angekündigt, bezw. in Aussicht genommen werden kann; nämlich, 1. die Formalisierung der "zweiten Stufe" (ε_f), 2. die Formalisierung der Metamathematik (Resultate von Gödel).

[33]This connection is never far from Hilbert's considerations as is obvious from his publications. In the manuscript SUB 603 one finds this remark made in a somewhat obscure context in which tertium non datur is discussed: Hiermit ist gezeigt, dass man so schliessen darf. Es ist in der Tat ein Unterschied, ob man hier Schluss mit tertium non datur anwendet oder nicht; v[er]gl[eiche] meine Beweise der Endlichkeit i[n] d[er] Invariantentheorie.

[34]For more details concerning Gentzen's manuscript, see (von Plato 2009, section 5); that section is entitled *A newly discovered proof of normalization by Gentzen*. – The dating of this summary is a *conjecture* of mine that is supported by three facts: (i) all the results described in Parts I through III have been obtained already (and Part IV for a significant sub-calculus), (ii) the detailed

11 handwritten pages and presents the natural deduction calculi for intuitionist and classical first-order logic under the heading "Der Schlussweisenkalkül N1J". Gentzen indicates also the reductions that are the basic steps for normalizing arbitrary intuitionist derivations.

A one-page "Overview of my further results" follows this part, which ended with the remark that a "positive characterization of intuitionist inferences" has been given. In Part II the calculus N1J is shown to be equivalent to the "logistic calculi of intuitionist reasoning" given by Hilbert, Heyting, and Glivenko.[35] In Part III the consistency of classical arithmetic is shown – relative to its intuitionist version. Articulating the gist of the result, Gentzen writes: "It is thus possible to give so-to-speak an 'intuitionist interpretation' to the arithmetic statements." In Part IV Gentzen conjectures the subformula property for normal N1J-derivations of *logical statements*, i.e., statements that do not depend on open assumptions. At the original writing of the summary, the conjecture was established only for the calculus N2J with the Introduction and Elimination rules for conjunction and universal quantification, as well as the rule for negation introduction.

The crucial next task is formulated as Part V. I quoted it in Sect. 5.1 and hope that it is less puzzling now:

> The consistency of arithmetic will be proved; in the process, the concept of an infinite sequence of natural numbers will be used, furthermore in one place the principle of the excluded middle. The proof is thus not intuitionist. Perhaps the *tertium non datur* can be eliminated.[36]

The connection to Hilbert's considerations in *Beweis des tertium non datur* seems unmistakable, as these remarks point exactly to the central features of Hilbert's argument, i.e., the metamathematical use of the rule (HR^*) and *tnd*. What did Gentzen intend to do in order to establish the consistency of arithmetic, i.e., of the part of intuitionist arithmetic that is needed for the interpretation of its classical version? In the next section, let me indicate first some of the historical circumstances and then an essential part of the systematic logical context.

---

normalization of intuitionist derivations (as indicated most clearly in the longer quotation – at the very end of Sect. 5.6 below – that begins with "Some thought") is not being pursued, and (iii) Part V articulates the consistency problem still in the manner of (Hilbert 1931b). The consistency problem as formulated here would naturally be dealt with by detailed semantic considerations – and Gentzen began in October 1932 to write detailed, reflective notes on his approach to the problems he was facing in the manuscript INH discussed below.

[35] The reference is, I assume, to (Heyting 1930) and to (Glivenko 1929); in the paper (Gentzen 1933) that derives from Part III of the *Urdissertation*, Heyting is mentioned, but oddly enough Glivenko is not.

[36] The German is: Die Widerspruchsfreiheit der Arithmetik wird bewiesen; dabei wird der Begriff der unendlichen Folge von natürlichen Zahlen benutzt, ferner an einer Stelle der Satz vom ausgeschlossenen Dritten. Der Beweis ist also nicht intuitionistisch. Vielleicht lässt sich das *tertium non datur* wegschaffen.

5.6 The New Student

In a certain sense Gentzen was not at all a *new* student in Göttingen. After a year of studying mathematics in Greifswald under the tutelage of Hilbert's student Hellmuth Kneser, he went to Göttingen and spent the academic year 1929–30 there, i.e., from late April 1929 to early March 1930. Gentzen and his friend from Greifswald, Lothar Collatz, attended Hilbert's lectures on set theory in the summer term of 1929. From Collatz's careful notes we know that these lectures were divided into three parts. For the purposes here only the third part is of interest where Hilbert discusses, in an elementary way, mathematical logic and his proof theoretic program.[37] He presents the elements of the consistency proof for primitive recursive arithmetic, in particular and with many examples, the "Auflösung in Beweisfäden", i.e., the transformation of linear derivations into tree structures. That is the central step in preparing derivations for further proof theoretic analysis; see Note 2. It was taken in *every* consistency proof given in Göttingen starting in 1922, when Hilbert and Bernays first proved their result, and includes even Gentzen's proof in his (1936).[38]

After spending the summer term of 1930 in Munich (where he read on his own Hilbert and Ackermann's book *Grundzüge der theoretischen Logik*), he went to Berlin and studied there during the winter term 1930–31. Unfortunately, we do not know what courses he took. This remains an intriguing question: von Neumann lectured on proof theory and discussed, as I pointed out in Sects. 5.3 and 5.4, what he had learned about Gödel's theorems. As to Gentzen, Winfried Schultze (Director of the University Archive of the Humboldt-University in Berlin) told me in a letter of 23 February 2010:

---

[37] The first two parts are adapted from the lectures on set theory Hilbert gave in the summer term of 1917, (Hilbert *1917). It would be of interest to examine Hilbert's possibly modified perspective on set theory. What is also of interest is Hilbert's discussion of ordinals, in particular ordinals less than ε_0.

[38] In (Gentzen 1936, p. 513) derivations are defined as sequences of (one-sided) sequents; the transformation into, essentially, tree form is made on p. 542 for the very same reason Hilbert and Bernays made it in 1922, namely, to insure that every sequent, except for the endsequent of course, is used at most once as a premise. Gentzen employed this technique also in the very first step of his relative consistency proof for classical arithmetic when transforming classical into intuitionist proofs, cf. (Gentzen 1933, pp. 126–127, section 4.21). – I mention these matters here, as they seem to answer convincingly, how Gentzen learned about the tree representation of proofs. Von Plato views the issue in a different way. In his (2010) he claims, already in the abstract, "the central component in Gentzen's work on logical calculi was the use of a tree form for derivations." Later, when comparing Gentzen's work with Einstein's in the latter's *annus mirabilis* 1905, he writes: "His [Gentzen's] amazing discovery of natural deduction and sequent calculus in 1932–1933, with its full control over the structure of derivations, followed from the use of a **tree form** [von Plato's emphasis; WS] for derivations. That was the new, simple, and right idea." That idea was prefigured in earlier proof theoretic work; but that fact by no means distracts from the "amazing" character of the insight into the structure of normal proofs, in particular, their subformula property in first-order logic.

> Gerhard Gentzen, coming from Munich, registered on 29 October 1930 with [student] number 1335 of the 121st academic year at the Friedrich-Wilhelms-University in Berlin. He studied here mathematics until 11 March 1931. A final report, unfortunately, has not been preserved; thus, we cannot make any assertions concerning the question, which lectures given by which faculty member he actually attended during this semester.[39]

Gentzen returned to Göttingen for the summer term 1931. During that term Hilbert conducted a seminar on *Grundlagen der Mathematik* and submitted, on 17 July 1931, his (1931b) for publication. It would be of special interest to know some facts about this seminar, from topics treated to who actually attended. Given Hilbert's habit of discussing topics of papers first in lectures or seminars, it is most plausible that he presented aspects of his (1931b) during this term; given the customs of German university institutes and doctoral education, it is almost inconceivable that Gentzen did not attend the seminar. One might also conjecture (at the moment without any concrete archival support) that the letter Richard Courant wrote to his colleague Hermann Nohl in support of Gentzen's application to the *Studienstiftung* refers to that seminar. The letter was written on 31 July 1931:

> As we agreed, I am reporting to you today about Mr. Gentzen on the basis of his seminar talk and a personal consultation. Mr. Gentzen discussed a particularly difficult topic in his seminar talk; he showed through the external as well as the intellectual grasp of the material a superior independence that marks him as a scientifically oriented human being. On account of his talk and after an oral interview I am confident that Mr. Gentzen can complete his doctorate relatively easily and that he can continue afterwards with scientific work. As his inclinations drive him obviously and very strongly in this direction, I can advise the *Studienstiftung* with full responsibility to grant him the doctoral scholarship.[40]

This is a relatively free translation of the full letter. Unfortunately, we don't know anything about the content of Gentzen's application for this scholarship.

Bernays recommended that Gentzen work, during the summer break of 1931, on the "sequent systems" that had been introduced by Paul Hertz; as to the latter's work see (Schröder-Heister 2002), but also (Bernays 1965). The paper

---

[39]Here is the German text: Gerhard Gentzen hat sich – von München kommend – am 29. Oktober 1930 unter der Nummer 1335 des 121. Rektorats in die Matrikel der Friedrich-Wilhelms-Universität zu Berlin eingetragen. Er studierte hier bis zum 11. März 1931 Mathematik. Ein Abgangszeugnis ist hier leider nicht überliefert, so dass wir keine Aussage darüber treffen können, welche Vorlesungen er bei wem in diesem Semester belegt hat.

[40]Majer has explored all the obvious archival issues implicit in this letter and my account above, but without any positive findings. - Here is the German text of Courant's letter: Verabredungsgemäss berichte ich Ihnen heute über Herrn Gentzen auf Grund seines Seminarvortrages und einer persönlichen Rücksprache mit ihm. Herr Gentzen behandelte in seinem Seminarvortrag ein besonders schwieriges Thema und bewies dabei in der äusseren und in der geistigen Durchdringung des Stoffes eine überlegene Selbständigkeit, die ihn durchaus als den Typus eines wissenschaftlichlich gerichteten Menschen kennzeichnet. Ich habe danach wie nach einer mündlichen Besprechung das Zutrauen, dass Herr Gentzen verhältnismässig leicht promovieren und auch dann weiter wissenschaftlich arbeiten kann. Da offenbar seine inneren Neigungen ihn sehr stark auf diese Bahn drängen, so kann ich mit voller Verantwortung der Studienstiftung den Rat geben, ihm die Promotion zu bewilligen.

(Gentzen 1932) would result from this work, and on 6 February 1932 it was submitted to *Mathematische Annalen*. In a letter of 13 December 1932 to Kneser, Gentzen mentions that he had turned, at the beginning of the winter term 1931/32, to more general problems of proof theory and that he had set himself almost a year ago, "the task of finding a proof of the consistency of logical deduction in arithmetic". When writing this letter he was hopeful to "finish soon".[41] By then, indeed somewhat earlier as argued in Note 34, Gentzen had discovered the consistency proof for classical arithmetic relative to intuitionist arithmetic.[42] He sent the resulting paper to Heyting in January or early February 1933 and submitted it to *Mathematische Annalen* on 15 March 1933, but withdrew it when he learned of Gödel's publication (1933). Through the argument in this paper, the principle of *tertium non datur* could be added consistently to intuitionist arithmetic; in fact it had been proved in its interpreted form within the intuitionist theory. In a letter to Heyting written on 25 February 1933, Gentzen suggested investigating the consistency of intuitionist arithmetic, since a consistency proof for classical arithmetic had not been given so far by finitist means, "so that this original aim of Hilbert has not been achieved". He then continued:

> If, on the other hand, one admits the intuitionistic position as a secure basis in itself, i.e., as a consistent one, the consistency of classical arithmetic is secured by my result. If one wished to satisfy Hilbert's requirements, the task would still remain of showing intuitionistic arithmetic consistent. This, however, is not possible by even the formal apparatus of classical arithmetic, on the basis of Gödel's result in combination with my proof. Even so, I am inclined to believe that a consistency proof for intuitionistic arithmetic, from an even more evident position, is possible and desirable. (Quoted in (von Plato 2009a, p. 672).)

Gentzen expressed the hope that he would investigate the consistency of intuitionist arithmetic "next year". "Thus", von Plato concludes, "the hopes of the previous December of 'finishing soon' a consistency proof, as in the letter to Kneser, had faded in a little over a month." Instead of pursuing the consistency proof, Gentzen turned his attention to writing *the* thesis, which was published as *Untersuchungen über das logische Schließen*. Gentzen defended it on 12 July 1933 and submitted it 9 days later for publication in *Mathematische Zeitschrift*.

[41] A longer excerpt from this letter to Kneser is quoted in (von Plato 2009a, p. 670); a translation of the full letter is in (Menzler-Trott, pp. 30–31).

[42] In (Gentzen 1936, note 17, p. 532), this result is attributed to both Bernays and Gentzen himself: Das im Text genannte Ergebnis wurde etwas später, unabhängig von Gödel, auch von P. Bernays und mir gefunden. (In Note 2 to the publication of the German original, i.e., on p. 119, Gentzen describes very concisely in what way Bernays contributed.) It is not clear with respect to which event "etwas später" is to be understood. Gödel presented his result to Menger's Colloquium on 28 June 1932. In his letter to Heyting of 16 May 1933, Gödel explicitly claims that his result should have become known in Göttingen shortly after his presentation. – The proof of this result is given and refined in (Hilbert and Bernays 1939); it is presented there tellingly under the heading *Eliminierbarkeit des 'tertium non datur' für die Untersuchung der Widerspruchsfreiheit des Systems (Z).*

Why did the hopes fade for finishing a consistency proof? Can we get a sense of the difficulties Gentzen ran into? – Based on Gentzen's manuscript INH to be discussed in more detail below, von Plato suggests in his (2009a, p. 677) that Gentzen recognized at the beginning of 1933 "that the planned approach [to the consistency proof for arithmetic in the *Urdissertation*, WS] through an extension of the method of section IV, i.e., normalization and the subformula property, from pure logic to derivations in arithmetic, did not work." In (von Plato 2010, p. 20) the task Gentzen had set himself in Part V is reformulated simply as, "To extend normalization and the subformula property to arithmetic." In contrast to von Plato, I view the *full* statement (V) as a crucial programmatic passage: it reveals, on the one hand, what Gentzen was trying to accomplish, and it indicates, on the other hand, how deeply his investigations were rooted in Hilbert's (1931b). INH and the discussion below will show the significant conceptual difficulties Gentzen had to overcome.

With this perspective we can ask, what had to be done in order to sharpen Hilbert's argument for the correctness of the basic constructive theory and then to show that the addition of *tnd* does not lead to a contradiction. The first step would extend Hilbert's considerations for quantifiers to the sentential logical connectives. That involves the formulation of introduction and elimination rules for those connectives, so that the correctness of logical steps can be viewed as "definitional" in the way Hilbert had done for quantificational inferences.[43] Given Hilbert and Bernays's formulation of the sentential logical axioms and their technique of "Auflösung in Beweisfäden", this first step is not difficult. In the tree representation of proofs the axioms for the connectives are used only at the very top, e.g., in the case of conjunction as follows:

$$\dfrac{\begin{matrix}|\\|\\B\end{matrix}\qquad \dfrac{\begin{matrix}|\\A\end{matrix}\qquad A \rightarrow (B \rightarrow A\&B)}{B \rightarrow A\&B}}{A\&B} \qquad\qquad \dfrac{\begin{matrix}|\\|\\A\&B\end{matrix}\qquad A\&B \rightarrow A\ [B]}{A\ [B]}$$

Why make the absolutely bureaucratic detour through the axioms? Is there any reason not to infer A&B from proofs of A and of B, or not to conclude A [B] from a proof of A&B? – One is led quite directly to Gentzen's calculus N1J as formulated in Part I of the *Urdissertation*. Then it is important to show that the formalization of

---

[43] And as Gentzen explicitly did in his dissertation, (Gentzen 1934/35, p. 189): Die Einführungen stellen sozusagen die "Definitionen" der betreffenden Zeichen dar, und die Beseitigungen sind letzten Endes nur Konsequenzen hiervon, was sich etwa so ausdrücken läßt: Bei der Beseitigung eines Zeichens darf die betreffende Formel, um deren äußerstes Zeichen es sich handelt, nur "als das benutzt werden, was sie auf Grund der Einführung dieses Zeichens bedeutet". A few lines below Gentzen continues: Durch Präzisierung dieser Gedanken dürfte es möglich sein, die *B* (eseitigungs)-Schlüsse auf Grund gewisser Anforderungen als eindeutige Funktionen der zugehörigen *E* (inführungs)-Schlüsse nachzuweisen.

logical principles via this calculus is equivalent to those given by Hilbert, Heyting and Glivenko. That is done with explicit reference to these three authors in Part II of the *Urdissertation.*

Now, the second question can be addressed, namely, whether the addition of the law of excluded middle leads to contradictions. Gentzen (on page 10 of the *Urdissertation*) defines ¬A as A→⊥. Given this definition, one sees readily that an instance of that law in the form A∨(A→⊥) cannot lead to a contradiction, unless intuitionist logic is inconsistent. After all, its double negation is provable intuitionistically and very easily so in Gentzen's N1J.[44] The step from this straightforward argument to the full interpretation of classical into intuitionist arithmetic is clearly a significant one, but it is technically not difficult; in particular not, as Gentzen was presumably familiar with Glivenko's paper that is explicitly mentioned in the *Urdissertation.* The result is nevertheless striking and is contained in Part III.[45]

We come to Part IV and its main results: normalizability and the subformula property of normal derivations for the calculus N2J, a fragment of the calculus N1J for full intuitionist logic. Von Plato conjectures that the results for N1J were proved and added to the manuscript only in March 1933. (An English translation of the proof is found in (von Plato 2008).) At the time of writing Part IV, Gentzen did not view them as crucial for achieving the main goal of his investigation, a consistency proof for intuitionist arithmetic. He noted in the summary after having indicated the reduction steps for the normalization procedure:

> Some thought is needed to recognize that in fact a correct proof is obtained in each case. I am not going to pursue this in detail, as I am not going to use those facts – I just present them for the purpose of illustration. [46]

The results are nevertheless the most surprising logical discovery of Gentzen's, and he must have seen them in a similar light. After all, when facing the impasse in establishing the consistency of arithmetic, he chose to establish his *Hauptsatz* (or cut-elimination theorem) for classical as well as intuitionist sequent calculi. How surprising the result was may also be seen indirectly from observations von Neumann made. Both in his (1927, pp. 11–12) and (1931, p. 120), he argued against

---

[44]Heyting proved this theorem as Theorem 4.8 in his (1930, p. 52) and remarks in a note that the proof was given in (Glivenko 1929); he emphasizes that it is connected to Brouwer's "Satz von der Absurdität der Absurdität des Satzes vom ausgeschlossenen Dritten".

[45]Documents in the Hilbert Nachlass contained in the folder SUB 603 indicate that Hilbert must have been informed about these matters by the fall of 1932; pages 44 through 47 on which pertinent remarks are made stem from *Druckfahnen* (page proofs) dated 15 July 1932. On p. 44, in particular, one finds this remark: Es sei eine Formel vorgelegt. Dann ersetze man A ∨ B durch ¬(¬A&¬B), ferner (Ex)A(x) durch ¬(x)¬A(x), ferner ¬A durch A→1 ≠ 1. ... Wenn dann jene Formel bewiesen ist (Tertium wird zugelassen) dann ist die entstandene Formel inhaltlich richtig. Z.B. (x)¬A(x) ∨ (Ex)A(x), (x)A(x) ∨ ((x)A(x) → w). [w stands for a contradictory formula; WS.]

[46](Gentzen 1932/1933, p. 9). The German text is: Es bedarf einiger Überlegungen, um einzusehen, dass in der Tat jeweils wieder ein richtiger Beweis entsteht. Ich verzichte auf diese genaue Durchführung, da ich von diesen Tatsachen keinen Gebrauch machen werde, sie vielmehr nur zur Veranschaulichung vorführe.

the plausibility of a positive solution of the *Entscheidungsproblem* for provability in first-order logic by pointing out the following fact: the minor premise used in modus ponens and, thus, the antecedent of the major premise are completely unconstrained when trying to determine whether a particular statement can be inferred by this inference. Even in the 1990s, eminent logicians saw that as a decisive obstacle to using natural deduction calculi for automated proof search. For normal proofs the "fact" is, of course, false.

The above reconstruction of *one* path that may have led Gentzen up to Part (V) is most plausible if one focuses on the problem he was addressing and the steps he, in fact, took to solve it, but also those he consciously did not take (most importantly, the normalizability of N1J proofs). At this point, Gentzen had shown that classical arithmetic is consistent relative to intuitionist arithmetic and, in particular, that the addition of *tnd* does not lead to contradictions. The next issue was to establish the consistency of intuitionist arithmetic – following Hilbert's semi-semantic pattern. It is this intention that is signaled through the programmatic statement (V).

5.7 An Impasse

When Gentzen finally confronted the consistency issue head on, he must have recognized that Hilbert's considerations were methodologically unsatisfactory: he had difficulties, broadly, to reconcile them with Gödel's result and, more narrowly, to give a finitist interpretation of the conditional and hence of negation. These concerns shape the exposition in (Gentzen 1936) and actually go back to this point in time, mid-October 1932. Thanks to Gentzen's manuscript INH we don't have to resort to conjectures, as INH directly mirrors his concerns with its detailed reflective observations.[47] Two thirds of this document was written between mid-October and mid-November 1932, a few pages in early 1933, and the rest in October 1934. The manuscript deals with the concept of *contentual correctness in pure number theory* and its relation to consistency proofs. The discussion is intricate and deserves a fuller treatment than I can provide here, as I want to focus on those issues that are directly connected to (Hilbert 1931b) and are explicitly taken up in (Gentzen 1936).

As we saw in Sect. 5.5, the notion of *correctness* for the basic constructive theory was central in Hilbert's considerations. A more restricted semantic component played a role in Hilbert and Bernays's consistency proofs from the very beginning. Recall that in (Hilbert *1921/22) linear derivations were transformed into tree-like ones with Boolean combinations of numeric formulae at their nodes. As

[47] The full title of INH is *Die formale Erfassung des Begriffs der inhaltlichen Richtigkeit in der reinen Zahlentheorie, Beziehungen zum Widerspruchsfreiheitsbeweis* (The formal characterization of the concept of contentual correctness in pure number theory, relations to the consistency proof). – The manuscript consists of 36 pages in shorthand; Thiel's typewritten transcription is 69 pages long.

the (instantiated) axioms are correct and the inferences preserve correctness, all combinations occurring in such a derivation must be correct. Consequently, there cannot be a derivation with an incorrect endformula. In order to extend this *fundamental proof theoretic idea* to derivations involving quantificational logic, Hilbert introduced the ε-substitution method, removing quantifiers from derivations in favor of epsilon terms and requiring also their evaluation to determine correctness (of quantifier-free formulae). Now, in 1931, correctness is defined directly for quantified statements exploiting the understanding of quantifiers (with variables ranging over natural numbers) and "guaranteeing" consistency immediately. As described in Sect. 5.5, Hilbert reasons in (1931b) for the consistency of his constructive theory by asserting, "All transfinite rules and inference schemata are consistent; for they amount to definitions."

A critical reader of Hilbert's paper could ask, how is the notion of correctness defined, and what principles are used to prove the correctness of all theorems? In fact, that is the opening of INH, written on 14 and 16 October 1932. Having defined contentual correctness for an intuitionist theory[48] and the concept "B is an intuitionist consequence of (assumptions) $V_1, \ldots, V_\nu$", Gentzen raises the questions, how far these notions are formal, and why the correctness proof by induction on the length of derivations is, according to Gödel, definitely not formal. In both questions Gentzen takes formal to mean expressible in the formal theory, i.e., he takes the correctness notion to be definable and the correctness proof to be formalizable in the theory. A few days later, on 19 October, he gives a more general articulation of the issues and understands formal in a different way, contrasting *proof theoretic* with *semantic* consistency proofs:

> I seek to clarify the questions: what distinguishes a formal correctness or consistency proof from a contentual one, why is the former for certain inferences not even possible by these same inferences (according to Gödel), is a bridge inference involved then, how secure is that [bridge inference, WS], what are the connections with Gödel's proof, what role do the mathematical axioms play?[49]

He formulates a plan for investigating these questions in clear stages. He intends to treat first a theory consisting only of mathematical axioms (formulated as

---

[48]INH, p. 2: Man definiert inhaltliche Richtigkeit so: die mathematischen Axiome sind richtig. A&B ist richtig, wenn A richtig ist und B richtig ist; AvB ist richtig, wenn mindestens eines richtig; Ax, wenn bei jeder Zahleinsetzung für x dies richtig, ebenso (x)Ax; Aa, wenn eine Zahl angegeben werden kann, so daß Aa gilt, ebenso (Ex)Ax; A $\rightarrow$ B, wenn aus der Richtigkeit von A die von B geschlossen werden kann; $\neg$A, wenn aus A der Widerspruch geschlossen werden kann. - This is in accord with Hilbert's intended definition of correctness. I have made some trivial notational changes from Thiel's transcription for easier comparability, and I separated the clauses for the different connectives by semicolons instead of commas.

[49]INH, p. 3: Ich suche Aufklärung über die Fragen: wie unterscheidet sich ein formaler Richtigkeits- bzw. Widerspruchsfreiheitsbeweis von einem inhaltlichen, wieso ist ersterer bei gewissen Schlußweisen nicht einmal mit Hilfe dieser selbst (nach Gödel) möglich, liegt dann ein Brückenschluß vor, wie groß ist dessen Sicherheit, wie sind die Zusammenhänge mit dem Gödelschen Beweis, welche Rolle spielen die mat[hematischen] Axiome?

inferences), then to add induction, after that to consider the sentential logical connectives &, v as well as the quantifiers ∀, E, only later the conditional and negation, and finally to examine *tnd.*

After some experimentation with contentual correctness proofs for such restricted theories, Gentzen asks whether they involve inferences that are avoided by "purely formal proofs", seeking to explain the seemingly essential difference between the two kinds of correctness proofs.[50] That leads him to the question, "Whether Gödel's result is essentially based on the fact that consistency instead of correctness is being considered." On the next day, Gentzen returns to this question with a definite answer, "consistency is a much more formal property than *correctness*". He also formulates an insight that shapes his subsequent considerations, namely, that consistency is equivalent (under minimal conditions on the formal theory) to what, in contemporary proof theory, is called the *reflection principle* for numeric statements: if such statements are provable, then they hold. On the one hand, this insight captures and reiterates the core idea of the proof theoretic work of Hilbert, Bernays and Ackermann, and, on the other hand, it frames Gentzen's further discussion that addresses Hilbert's issues in an independent and exploratory way.

Calling a proof of a numeric statement a *Normalbeweis* if it contains only numeric statements, Gentzen can now express the difference between (purely) formal and (semi-) contentual correctness proofs by formulating carefully the claim each is to establish. The claim for a *purely formal correctness proof* is, "for every proof of a numeric statement there is a *Normalbeweis* of that statement", and the corresponding claim for the (semi-) contentual correctness proof is, "every proof has a correct result" [where result means endformula, WS]. Calling a formula with free and just universally quantified variables *correct* if every substitution instance has a *Normalbeweis,* Gentzen sketches then the proofs for a simple theory with quantifier-free axioms and rules for universal quantifiers. Comparing the proofs, Gentzen sees the essential difference as follows:

> The semi-contentual proof uses complete induction for a rather complicated statement. This contains Ri erg x [the result of proof x is correct, WS], and this predicate becomes ever more complicated in complicated cases. The formal proof uses complete induction for [the statement] Ey. No y & erg x = *erg* y [there is a *Normalbeweis*y having the same result as the given proof x, WS]; this is also a statement containing logical signs; it is however of a simpler nature, also in more complicated cases.[51]

---

[50] "Purely formal correctness proofs" are understood now in a way that will be sharpened in the next paragraph. – It seems that the difference formulated there between *semi-contentual and contentual*, respectively, *purely formal and formal* can be accounted for *mostly* by the formulation of correctness for numeric statements: if that is done syntactically as below, then the proofs would be semi-contentual and purely formal; if that is done by a semantic evaluation of terms, then they would be contentual and formal. On p. 5 of INH Gentzen calls his definition of "correct" for numeric statements *purely formal*; the definition specifies that the statement can be obtained syntactically by an immediate derivation from the axioms, what is called *Normalbeweis* below.

[51] INH, p. 8: Der halbinhaltliche Beweis macht eine VJ über eine ziemlich komplizierte Aussage. Diese enthält Ri erg x, und dieses Prädikat wird in komplizierten Fällen immer komplizierter.

And yet, Gentzen is still not completely sure where the *Gödel-Punkt* lies. On the next page of INH he writes: "I think I see now clearly, why a consistency proof by a crude contentual interpretation is not formalizable. [It is not formalizable in the usual formalisms, WS] for the very reason that the interpretation itself is not formalizable." He emphasizes that it is necessary to see more clearly that *such* a consistency proof is not formalizable, but also that no consistency proof whatsoever can be carried out in the usual formalisms. In order to gain a clearer view, he extends the earlier comparative methodological experiments to a theory with quantifier-free axioms and complete induction. In this case the formal correctness proof involves a *transfinite* valuation of derivations of the sort used already in Ackermann's proof from 1924; see (Zach 2003). The experimenter Gentzen is prompted to make the following observation:

> In any event, I think there must be some connection between the non-formal element of the "correctness" definition (that is not formalizable) and that of the transfinite induction (that is not formalizable). After all, each of them apparently makes possible a consistency proof that cannot be formalized.[52]

Having gained clarity, at least in a broad sense, about the *Gödel-Punkt* and its location relative to proof theoretic and semantic considerations, he turns on 31 October to the last point of his plan from 19 October, i.e., the investigation of systems expanded by *tnd*. Central is the characterization of "richtig" and "falsch"; thus, Gentzen addressed the very same issue as Hilbert and recognized as well, it seems, that his own attempt of doing so needed *tnd* in the meta-theory.

He decides on 2 November, after quite a bit of reflection concerning a direct treatment of *tnd*, to return to the investigation of the contentual correctness of intuitionist mathematics. After all, relative to intuitionist arithmetic he had already shown that the addition of *tnd* does not lead to a contradiction. He begins by considering the essence of a purely formal consistency proof: "The main thing for this [kind of proof, WS] is to associate with a derivation of a numeric result a *Normalbeweis* of the very same result." (INH, p. 17) This association must be intuitionistically unobjectionable and should be achieved by a finite number of reduction steps similar to those used in the normalization of derivations in intuitionist first-order logic. The stepwise procedure should *simplify* the given derivation until it is a *Normalbeweis*; the initial derivation is then called *reducible*. This idea is reemphasized a little later:

Formale[r] Beweis macht VJ über Ey. No y & erg x = erg y, eine Aussage zwar auch mit logischen Zeichen, doch von einfacherer Natur, auch bei komplizierteren Fällen. – *VJ* stands, of course, for *Vollständige Induktion*.

[52]INH, p. 10: Jedenfalls muß doch meines Erachtens irgendein Zusammenhang zwischen dem unformalen Element bei der nicht formalisierbaren Definition der "Richtigkeit" und bei der nicht formalisierbaren transfiniten Induktion bestehen. Da ja anscheinend jede von beiden einen nicht formalisierbaren Widerspruchsfreiheitsbeweis ermöglicht.

> The intent is: to simplify a given proof with a numeric result in steps until there is a numeric proof for the same result ... I want to get by with a valuation that is as simple as possible.[53]

The discussion in the first part of INH ends abruptly on 8 November, when Gentzen compares a contentual proof with a proof using his reducibility concept. He discovers a difficulty for the latter proof, but emphasizes, "... the difficulty is seemingly not connected with complete induction, but rather with the nature of consequence, thus to the $\rightarrow$ sign." (INH, p. 20,1) The difficulties surrounding the interpretation of the intuitionist conditional, thus of negation, are also emphasized in the writings of Bernays, for example in his (1934, pp. 71–72) and later in *Grundlagen der Mathematik II*, pp. 358–360.

Gentzen takes up the thread, most likely, in February of 1933 and then again in June of that year. Important developments are indicated, and Gentzen ends with general thoughts on consistency proofs and transfinite values of derivations:

> Every proof has a (transfinite) value. Consistency of a system of proofs can only be shown by a proof that has a higher value than all of these. Thus Gödel's theorem. This one has to try to prove. When doing it the values have to be determined. Furthermore, [investigate] whether there are proofs with higher values that have nevertheless greater certainty.[54]

This remark connects and fully reconciles his proof theoretic considerations with Gödel's second theorem. The impasse for a consistency proof he had encountered in November 1932 remains, however, an impasse in June 1933.

5.8 Toward a Solution

Gentzen completed a consistency proof based on his reducibility concept in late 1934, but definitely not later than sometime in October of that year. The reason for this upper bound is simple. On p. 25 of INH, dated "X. 34" (October 1934), Gentzen expresses "Nagging doubts. Concerning the value of the consistency proof". He goes back to the discussion contrasting formal and semi-contentual correctness proofs and locates his proof with the reducibility notion "as a kind of intermediate thing between the other two":

> For & $\forall$ $\vee$ $\exists$ it [the proof via reducibility, WS] runs parallel to the semi-contentual proof recognizing here the contentual [understanding, WS]. For $\supset$ it [the proof, WS] jumps off and moves over to the formal proof, intent on avoiding the contentual $\supset$.[55]

---

[53]INH, p. 19: Die Absicht ist: einen vorliegenden Beweis mit numerischem Ergebnis schrittweise zu vereinfachen, bis ein numerischer Beweis für dasselbe Ergebnis dasteht. ... Ich möchte mit einer möglichst einfachen Wertung auskommen.

[54]INH, p. 23: Jeder Beweis hat einen (transfiniten) Wert. Widerspruchsfreiheit für ein System von Beweisen läßt sich nur zeigen durch einen Beweis von höherem Wert als alle diese. Daher der Satz von Gödel. Dies ist zu beweisen zu versuchen. Dabei die Werte feststellen. Ferner dann, ob es Beweise höheren Wertes und trotzdem größerer Sicherheit gibt.

[55]Note that Gentzen uses here for the first time in INH the horseshoe as the symbol for the conditional.

> That creates the impression: one begins with the semi-contentual. This becomes increasingly ominous, as one obviously assumes exactly what one wants to prove. Finally, for ⊃ one decides to jump off and to take refuge in the formal [considerations, WS]; in that way the proof becomes more complicated and then requires, to be carried through, auxiliary tools of a special sort, for example, the super-ordering.[56]

Intent on putting the nagging doubts aside and "regaining firm ground", Gentzen interprets, (INH, p. 29), universal and existential statements, in a first step, with just finitist matrices. Before expanding the interpretation, he formulates a plan for Part III of what will be his (1936); that part is to discuss the methodological issues he has been grappling with:

> Plan for Part III: first, after the finite, the An-sich-view for the infinite, which fully corresponds to that for the finite. This we reject. Then the development of the intuitionist standpoint in words like those on pp. 21–22 of INH. (Perhaps Gödel's transfer theorem should be formulated right away.) Then probing critical analysis, as still to be developed.[57]

This will remain the blueprint for Part III. Indeed, the structure of the paper's Part III is fully in accord with this plan.[58]

When coming back to the interpretation of universal quantifiers, Gentzen expands on the marginal note from p. 25 of INH (see fn. 55) and claims that for the theory without the conditional, no reduction is achieved, just a confirmation: "No reduction to something more simple, but rather a confirmation of the constructivity [of the

On the margin, he wrote: Richtiger gesagt: bei & ∀ ∨ ∃ sind die finiten Deutungen eben noch fast gleichwertig dem An-sich-sein, bei ⊃ geht das … weniger! – Here is the English translation: More correctly one would say: for & ∀ ∨ ∃ the finitist interpretations are still almost the same as those given by the *An-sich-Auffassung*; for ⊃ that works less.

It should be emphasized that Gentzen, in this late part of INH from October 1934, repeatedly refers back to the INH notes from 1932; for example, on pp. 25, 29, 30 he refers back to pp. 17–18 and 19–20, 21–22, 16 and 21, respectively.

[56]INH, p. 25: Bei & ∀ ∨ ∃ geht er gleichlaufend mit dem halbinhaltlichen Beweis, indem er hier das inhaltliche anerkennt. Bei ⊃ springt er ab und geht über zum formalen Beweis, indem er das inhaltliche ⊃ vermeiden will. Das erweckt den Eindruck: man beginnt mit Halbinhaltlich[em]. Dies wird einem fortschreitend immer unheimlicher, da man offenbar fast genau das schon voraussetzt, was man beweisen will. Bei ⊃ endlich entschließt man sich, abzuspringen und rettet sich ins Formale, wodurch der Beweis kompliziert wird und nachher zu seiner Rettung wieder Hilfsmittel besonderer Art, sagen wir: die Superordnung, benötigt.

[57]INH, p. 29: Plan für den III. Abschnitt: erst, nach dem Endlichen, die An-sich-Auffassung im Unendlichen, welche der im Endlichen ganz entspricht. Diese lehnen wir ab. Dann die Entwicklung des intu[itionistischen] Standpunktes mit Worten wie auf INH 21–22. (Evtl. gleich Angabe des Gödelschen Übertragungssatzes.) Dann tiefergehende Kritik, wie noch zu entwickeln.

[58]Part III is entitled "Bedenkliche und unbedenkliche Schlußweisen in der reinen Zahlentheorie" and the sections have the headings, "Die Mathematik endlicher Gegenstandsbereiche" (section 7), "Entscheidbare Begriffsbildungen und Aussagen im unendlichen Gegenstandsbereich" (section 8), "Die "an-sich"-Auffassung der transfiniten Aussagen" (section 9), "Finite Deutung der Verknüpfungszeichen ∀, &, ∃ und ∨ in transfiniten Aussagen" (section 10), and finally, "Die Verknüpfungszeichen ⊃ and ¬ in transfiniten Aussagen; die intuitionistische Grenzziehung" (section 11).

interpretation, WS].” That constructivity is not confirmed by proof, but is being grasped with words. Here is the reason:

> Because the difference between An-sich and constructive is not formally captured. It is only known by its meaning. And that has to be that way, as it is indeed the last, extra-mathematical foundation.[59]

The interpretation is finally given in greater detail (on p. 33 of INH) under the heading “The correctness proof by finitist interpretation for the theory without $\supset$ and $\neg$”. Here is, paradigmatically and exhibiting the deep parallelism with Hilbert's rule (*HR**), the interpretation of $\forall$x Fx: this statement holds (and has consequently meaning) if for every numeral ν there is a constructible derivation for Fν that has been recognized already as correct. The interconnectedness of the contentual considerations for statements and for derivations is taken on the next page, which is indeed also the last of INH, as a reason for emphasizing the significance of the reducibility notion. As to the circularity for $\supset$ Gentzen writes:

> Perhaps $\supset$ does not involve a circle, or more correctly, the situation is as follows: Also for the previous [logical, WS] signs a kind of “An-sich”-meaning is assumed. But surely this is somehow of a constructive kind ... also e.g. $\forall$ involves contentually an “An-sich”-for-all. The foundation on the “existence of a [correct, WS] derivation” does not seem to be a real foundation. Indeed, it cannot be. After all, the “correctness” presupposes again contentual knowledge of the meaning for the $\forall$, and this is as well an “An-sich”-meaning, if one wishes.[60]

Gentzen re-asserts then that $\forall$ and $\exists$ are explicated by a contentual for-all and there-is. But this contentual understanding is not that of the “An-sich”-for-all or the “An-sich”-there-is. So for him the real question is:

> What differentiates the “An-sich”-$\forall$ from the contructive $\forall$, though the same inferences hold for both? And what about $\exists$? The constructive interpretation reduces to something “conceptually simple”. For $\supset$ one can argue, whether there is something simpler or not. Also, the “idea of reducibility” is important.[61]

[59]INH, p. 29: Weil eben der Unterschied von An-sich und konstruktiv nicht formal erfaßt ist. Eben nur dem Sinn nach bekannt. Und das muß auch so sein, dies ist eben das letzte, außermathematische Fundament.

[60]INH, p. 33: Vielleicht liegt doch kein Zirkel vor beim $\supset$, oder richtiger, liegt die Sache so: Auch bei den vorigen Zeichen setzt man eine Art “An-sich”-Sinn schon voraus. Freilich ist dieser irgendwie konstruktiver Art. ... auch bei $\forall$ z.B. liegt doch inhaltlich ein “An-sich”-Alle vor. Die Begründung auf das “Bestehen einer [korrekten, WS] Herleitung” scheint doch auch keine wirkliche Begründung zu sein. Kann es ja gar nicht. Die “Korrektheit” setzt eben wieder inhaltliches Wissen über den Sinn des $\forall$ voraus, und dies ist ebenso gut ein “An-sich”-Sinn, wenn man will.

[61]INH, p. 33: Wodurch unterscheidet sich das”An-sich”-$\forall$ von dem konstruktiven $\forall$, obwohl für beide die gleichen Schlußweisen gelten? Und bei $\exists$? Die konstruktive Deutung führt zurück auf etwas “begrifflich Einfaches”. Bei $\supset$ kann man darüber streiten, ob etwas Einfacheres vorliegt oder nicht. Auch die “Idee der Reduzierbarkeit” ist wichtig.

These considerations are systematically taken up in the 1936-paper. In particular, the interpretation of the conditional remains (on p. 530) a main task for the consistency proof. When discussing the interpretation of statements in the language of first-order arithmetic, Gentzen formulates at the end of section 9 two explicit goals, namely, (i) to provide a finitist meaning for these statements, i.e., "to *interpret* each such statement as expressing a determinate, finitely representable fact", and (ii) to ensure that the logical inferences are in harmony with the finitist meaning of the statements involved. Here is the last paragraph of section 9 outlining the work that is related to (i) and (ii) and that is to be accomplished in the next two sections:

> In the following section 10 that [program indicated by (i) and (ii), WS] is to be carried out for a considerable class of transfinite statements and the associated inferences. In section 11, I will treat the remaining statement forms and inferences; there the method will encounter difficulties, and the significance of the *intuitionist* (1.8) *boundary* between permissible and impermissible inferences will emerge; furthermore, another even more restrictive boundary can be seen to be defensible.[62]

Indeed, Gentzen discusses in section 10 the finitist interpretation for some of the connectives in transfinite statements, namely, for universal and existential quantifiers, as well as for conjunctions and disjunctions. Before turning in section 11 to the problematic connectives (conditional and negation) and to the discussion of the intuitionist boundary, he concludes section 10 by asserting what he had indicated briefly in INH:

> One could then, proceeding from these considerations, develop a purely formal consistency proof for this part of number theory.[63] But such a proof would have little value, for in the proof itself one would have to *use* transfinite statements and the accompanying modes of inference that one wants to "*ground*" by the proof. So the proof would not be a proper *reduction*, but rather a *confirmation* of the *finitist* character of the formalized rules of inference. But one must already be clear in advance *what* counts as *finitist* (in order then to be able to carry out the consistency proof itself with finitist means of proof).[64]

---

[62](Gentzen 1936, p. 525). The German text is: Das soll im folgenden § 10 für einen beträchtlichen Teil der transfiniten Aussagen und der zugehörigen Schlußweisen durchgeführt werden. In § 11 behandle ich dann die restlichen Aussageformen und Schlußweisen; dabei stößt die Methode auf Schwierigkeiten, und es zeigt sich die Bedeutung der *intuitionistischen* (1.8) *Grenzziehung* zwischen erlaubten und unerlaubten Schlußweisen innerhalb der Zahlentheorie; ferner ergibt sich eine andere, noch engere Grenzziehung als ebenfalls verfechtbar. – On p. 532 Gentzen makes this narrower boundary explicit, namely, no general use of the conditional.

[63]"Purely formal" has to be understood here differently from INH, namely, as emphasizing, "informally" so-to-speak, that the proof does not have any real content and does not provide a meaningful reduction.

[64](Gentzen 1936, p. 529). The German text is: Man könnte dann, von diesen Überlegungen ausgehend, einen rein formalen Widerspruchsfreiheitsbeweis für diesen Teil der Zahlentheorie entwickeln. Ein solcher hätte aber nicht viel Wert, denn man müßte in dem Beweis selbst transfinite Aussagen und dieselben zugehörigen Schlußweisen *benutzen*, die man durch ihn '*begründen*' will. Der Beweis würde also keine eigentliche *Zurückführung* bedeuten, wohl aber eine *Bestätigung* des *finiten* Charakters der formalisierten Schlußregeln. *Was* aber *finit* ist, darüber müßte man sich zuvor im klaren sein (um dann den Widerspruchsfreiheitsbeweis selbst mit finiten Beweismitteln führen zu können).

That outlines a consistency proof using an appropriate truth definition or the contentual concept of *Richtigkeit*. Hilbert's way of proceeding, not clearly respecting the line between syntactic and semantic considerations, has been made blindingly clear by Gentzen.

Hilbert and Bernays, in their *Grundlagen der Mathematik* (1939, p. 390), consider a consistency proof that rests ultimately or mainly on semantic considerations as unsatisfactory for proof theory. Indeed, they recall the "formalism of recursive number theory" and their proof theoretic treatment establishing its consistency: the latter was central despite the fact that, in principle, one could simply have pointed to its finitist interpretation. (See Sect. 5.2 above.) In their judgment, Gentzen's consistency proof for elementary number theory does justice to this concern. They consider as a very serious possibility that the fundamental idea of Gentzen's consistency proof (using transfinite induction up to ε_0) can be extended beyond elementary number theory to more comprehensive formalisms, involving then of course larger ordinals. Their book ends with the remark:

> If this perspective should prove accurate, Gentzen's consistency proof would open a new era for proof theory.[65]

5.9 New Perspectives

Gentzen gives, in his 1936-paper, not only a consistency proof, but provides a formal substitute for the contentual correctness notion, namely that of "statability of a *reduction procedure*" (1936, p. 536). This concept, together with the translation from classical into intuitionist arithmetic, Gentzen views as giving "a particular finitist interpretation[66] of the statements [of classical arithmetic, WS], which

[65](Hilbert and Bernays 1939), p. 374. The German text is: Falls diese Perspektive sich bewähren sollte, so würde mit dem Gentzenschen Widerspruchsfreiheitsbeweis ein neuer Abschnitt der Beweistheorie eröffnet.

[66]In his (1941), Bernays considers the principle of transfinite induction up to ε_0 as non-finitist, but remarks that his position should not be considered "as the standpoint of the Hilbert School". In the quotation here, but also in other places of his paper, for example on p. 564, Gentzen speaks quite clearly as providing a *finitist* interpretation. May that also be in the background for Hilbert's famous remark in the Preface to the first volume of *Grundlagen der Mathematik*, that Gödel's results do not show that proof theory cannot be carried through, but rather that the finitist standpoint has to be exploited in a sharper way in order to obtain consistency proofs for more complex formalisms? – It would be of great biographical interest, but also of significance for our understanding of the systematic proof theoretic developments, if we would know more about the relation between Gentzen and Hilbert, most importantly, during the time from 1931 to 1934, but also after November 1935, when Gentzen had been appointed as Hilbert's "Special Assistant". Menzler-Trott conjectures that, during the later period, they only talked about "newspapers, poems and popular science".

replaces their interpretation via the *an-sich-Auffassung*".[67] In a letter to Kneser written on 5 December 1934, Gentzen reports that he is in the process of polishing his paper for publication in *Mathematische Annalen*.[68] The paper was submitted on 11 August 1935. After correspondence with Bernays and, indirectly, Gödel (with whom Bernays had discussed Gentzen's proof), Gentzen inserted sections 14.1 to 16.11 that replaced the earlier treatment by one involving transfinite induction up to ε_0. (No changes were made in the parts of the paper to which I have been referring.) That replacement was made in February 1936. The original argument, together with an introduction by Bernays, was published in 1974. Here remains the task of joining Gentzen's methodological goals with the details of his original argument, with the mathematical analyses of Coquand and Tait 1995, as well as with contemporary directions of proof theory.

Ironically, Gödel was deeply influenced by Gentzen's and Hilbert's considerations. That is clear from his Lecture at Zilsel's, (Gödel 1938), in which he explored various extensions of finitist mathematics and analyzed, in particular, Gentzen's first consistency proof. One of these extensions, via finite type theories for computable functionals, was pursued in lectures Gödel gave in 1941 at Princeton and at Yale. That work led ultimately to the Dialectica interpretation and was published in (Gödel 1958); its mathematical and foundational aspects are discussed in Troelstra's most informative *Introductory Notes*, (Troelstra 1990, 1995). In the Yale lecture, (Gödel 1941), Gödel viewed his project in the same general way as Gentzen saw his, namely, as giving an interpretation of intuitionist arithmetic from a more strictly constructive standpoint. He articulates three requirements such a position should satisfy and writes then:

> Let me call a system strictly constructive or finitistic if it satisfies these three requirements (relations and functions decidable, respectively, calculable, no existential quantifiers at all, and no propositional operations applied to universal propositions). I don't know if the name "finitistic" is very well chosen, but there is certainly a close relationship between these systems and what Hilbert called the "finite Einstellung". (Gödel 1941, p. 191)

Coming back to Hilbert's last paper, I quote an important passage that occurs right after the discussion of the concept of *Richtigkeit*, which – as we saw – is also central for Gentzen:

> Finally the important and, for our investigation, decisive fact has to be emphasized, namely, that all the axioms and inference schemata VI, which I have called "transfinite", have a strictly finitist character: the prescriptions they contain can be executed in the finite.[69]

---

[67]This connection remains to be explored; here I point to the summary in Gentzen's paper, pp. 564–565, and the related discussion in (Sieg & Parsons, pp. 83–85). The most significant connection of Gödel's to Gentzen's first consistency proof, revealed in his Zilsel Lecture, has been detailed in (Tait 2005).

[68](Menzler-Trott, p. 55). The more extended text is: "At the moment I am preparing a consistency proof for pure (i.e. no analytic means employed) number theory, which I have finished, for publication in *Mathematische Annalen*."

[69](Hilbert 1931b, p. 121). The German text is: Endlich werde noch die wichtige und für unsere Untersuchungen entscheidende Tatsache hervorgehoben, die darin besteht, daß die sämtlichen

Hilbert's considerations in (1931b) were a crucial germ for Gentzen's work on consistency, presenting – as they did – a new perspective and pointing in novel directions. The concrete problems arising in part from difficulties that had been pushed aside, Gentzen resolved in surprising and ingenious ways, but fully in the spirit of Hilbert's view that true contentual thinking consists in operations on proofs.

There are additional Hilbert manuscripts from around this time; for example, folder SUB 603 contains a copy of Hilbert's last paper with notes in his own hand (not helpful, it appears, for understanding the paper) and many seemingly unconnected pages. On one of these loose pages one finds a remarkable general statement that conveys the impact of his reflections on Gödel's second theorem: "Consistency is naturally a relative notion; that is not an objection to my theory, but rather a necessity." Hilbert drew an arrow to "relative" and wrote at the end of the arrow "New!"[70]

Appendix

Wilfried Buchholz pointed to a paper of Kurt Schütte that appeared in an obscure journal and was based on a talk Schütte had given on 13 May 1993 to a meeting in Munich to memorialize Hilbert who had died 50 years ago. The paper's title is *Bemerkungen zur Hilbertschen Beweistheorie*. Here are two quotations that are of logical and deep human interest:

> Hilbert hat selbst keine Widerspruchsfreiheitsbeweise durchgeführt, sondern nur die Anregung dazu gegeben. Er hat bald nach Erscheinen der Gödelschen Unvollständigkeitssätze und noch vor dem Gentzenschen Widerspruchsfreiheitsbeweis in einem Göttinger Kolloquiumsvortrag vorgeschlagen, das Axiomenschema der vollständigen Induktion in Erweiterung des finiten Standpunktes durch die konstruktive Verwendung einer Schlussregel mit unendlich vielen Prämissen zu ersetzen. Diese von Hilbert vorgeschlagene Methode wurde zunächst überhaupt nicht beachtet, sondern erst fast 20 Jahre später aufgegriffen und ist heute für beweistheoretische Untersuchungen von starken Teilsystemen der Analysis unentbehrlich geworden. ... Ich selbst hatte nur mit Hilberts Mitarbeiter Paul Bernays, der bis 1933 ausserordentlicher Professor in Göttingen war, einen engen wissenschaftlichen Kontakt gewonnen. Mit Hilbert bin ich persönlich nur zweimal zusammengekommen, nämlich erstens bei meinem Besuch in seiner Wohnung vor meiner Doktorprüfung, und zweitens in der mündlichen Prüfung vor genau 60 Jahren als letzter Doktorand von Hilbert, wobei ich beide Male seine menschliche Güte erfahren habe.

Axiome und Schlußschemata VI, die ich transfinit genannt habe, doch ihrerseits streng finiten Charakter haben: die in ihnen enthaltenen Vorschriften sind im Endlichen ausführbar.

[70]That is in conflict with the earlier perspective on the finitist standpoint. – Even in *Über die Grundlagen des Denkens*, SUB 604, Hilbert explicitly asserts that the concept "widerspruchsfrei" is "absolut" (on p. 6 of the original manuscript); but that particular use may not be in conflict with the new perspective, as absolute is used here in a different sense: 'Widerspruchsfrei' – ebenso wie 'Richtig' – ist ein absoluter Begriff; denn wir setzen ausdrücklich fest, dass beim Beweisen als zulässige Begriffs- und Schlussmethoden jedesmal nur diejenigen anzusehen sind, die das Verständnis der Aussage A nötig macht, so dass hinsichtlich der Bedeutung unserer Definitionen von 'widerspruchsfrei' und 'richtig' keine Unbestimmtheit eintritt.

References

The translations in the text are mine, except when quoted explicitly from particular publications.

Ackermann, W. 1924. Begründung des "tertium non datur" mittels der Hilbertschen Theorie der Widerspruchsfreiheit. Mathematische Annalen 93: 1–36.

Ackermann, W. 1925–1926. Letters to Bernays of 25 June 1925 and 31 March 1926: Wissenschafts historische Sammlung, ETH Zürich, Bernays Nachlass, HS 975: 96 and 97.

Avigad, J., and R. Zach. 2007. The epsilon calculus, Stanford encyclopedia of philosophy (version of July 2007). http://plato.stanford.edu/entries/epsilon-calculus/#6.

Benacerraf, P., and H. Putnam (eds.). 1983. Philosophy of mathematics, Cambridge: Cambridge University Press.

Bernays, P. 1930. Die Philosophie der Mathematik und die Hilbertsche Beweistheorie. Blätter für Deutsche Philosophie 4: 326–367. Reprinted in (Bernays 1976, 17–61).

Bernays, P. 1933. Methoden des Nachweises von Widerspruchsfreiheit und ihre Grenzen, in (Saxer 1933, 342–343).

Bernays, P. 1934. Über den Platonismus in der Mathematik. Reprinted in (Bernays 1976, 62–78). Trans. in (Benacerraf and Putnam 1983, 258–271).

Bernays, P. 1935. Hilberts Untersuchungen über die Grundlagen der Arithmetik, in (Hilbert 1935, third volume, 196–216).

Bernays, P. 1941. Sur les questions méthodologiques actuelles de la théorie Hilbertienne de la démonstration. In Les entretiens de Zürich sur les fondements et la méthode des sciences mathématiques, ed. F. Gonseth, 144–152. Zürich: Leemann & Co., Discussion, 153–161.

Bernays, P. 1965. Betrachtungen zum Sequenzenkalkül. In Contributions to logic and methodology in honor of J.M. Bochenski, ed. A-T. Tymieniecka and C. Parsons, 1–44. Amsterdam: North-Holland Publishing Company.

Bernays, P. 1967. Hilbert, David. In Encyclopedia of philosophy, vol. 3, ed. P. Edwards, 496–504. New York: Macmillian and Co.

Bernays, P. 1976. Abhandlungen zur Philosophie der Mathematik. Darmstadt: Wissenschaftliche Buchgesellschaft.

Bernays, P. 1977. Three recorded interviews, given on 25.7.1977, 13.8.1977, 27.8.1977. Wissenschaftshistorische Sammlung, ETH Zürich, Bernays Nachlass, Cod. Ms. P. Bernays T 1285.

Coquand, T. 1995. A semantics of evidence for classical arithmetic. Journal of Symbolic Logic 60: 325–337.

Dawson, J. 1986. Introductory note to (Gödel 1931a), in (Gödel 1986, 196–199).

Dawson, J. 1997. Logical dilemmas: The life and work of Kurt Gödel. Wellesley: A. K. Peters.

Ewald, W.B. 1996. From Kant to Hilbert: A source book in the foundations of mathematics, 2 vols. Oxford: Oxford University Press.

Feferman, S. 1986. Introduction to (Gödel 1931b), in (Gödel 1986), 208–213.

Feferman, S. 2003. Introductory note to Gödel's correspondence with Bernays, in (Gödel 2003a, 41–79).

Feferman, S., and W. Sieg (eds.). 2010. Proofs, categories and computations – Essays in honor of Grigori Mints. London: College Publications.

Gentzen, G. 1932. Über die Existenz unabhängiger Axiomensysteme zu unendlichen Satzsystemen. Mathematische Annalen 107: 329–350.

Gentzen, G. 1932/1933. "Urdissertation", Wissenschaftshistorische Sammlung, Eidgenössische Technische Hochschule, Zürich, Bernays Nachlass, Ms. ULS. (A detailed description of the manuscript is found in (von Plato 2009a, 675–680)).

Gentzen, G. 1933. Über das Verhältnis zwischen intuitionistischer und klassischer Arithmetik. Archiv für mathematische Logik und Grundlagenforschung 16 (1974): 119–132.

Gentzen, G. 1934/35. Untersuchungen über das logische Schließen I, II. Mathematische Zeitschrift 39: 176–210, 405–431.

Gentzen, G. 1935. Der erste Widerspruchsfreiheitsbeweis für die klassische Zahlentheorie. Archiv für mathematische Logik und Grundlagenforschung 16 (1974): 97–118.
Gentzen, G. 1936. Die Widerspruchsfreiheit der reinen Zahlentheorie, Mathematische Annalen 112: 493–565. Trans. in (Gentzen 1969).
Gentzen, G. 1969. The collected papers of Gerhard Gentzen. Amsterdam: North-Holland Publishing Company. Edited and translated by M. E. Szabo.
Glivenko, V. 1929. Sur quelques points de la logique de M. Brouwer, Académie Royale de Belgique. Bulletins de la classe des sciences, ser. 5, 15: 183–188.
Gödel, K. 1929. Über die Vollständigkeit des Logikkalküls, Dissertation, Vienna, in (Gödel 1986, 60–101).
Gödel, K. 1930a. Die Vollständigkeit der Axiome des logischen Funktionenkalküls, in (Gödel 1986), 102–123.
Gödel, K. 1930b. Einige metamathematische Resultate über Entscheidungsdefinitheit und Widerspruchsfreiheit, in (Gödel 1986, 140–143).
Gödel, K. 1930c. Vortrag über Vollständigkeit des Funktionenkalküls, in (Gödel 1995, 16–29).
Gödel, K. 1931. Über formal unentscheidbare Sätze der Principia Mathematica und verwandter Systeme I, in (Gödel 1986, 126–195).
Gödel, K. 1931a. Diskussion zur Grundlegung der Mathematik, in (Gödel 1986, 200–205).
Gödel, K. 1931b. Besprechung von Hilberts Die Grundlegung der elementaren Zahlentheorie. Zentralblatt für Mathematik und ihre Grenzgebiete 1: 260. Reprinted and translated in (Gödel 1986, 212–214).
Gödel, K. 1932. Über Vollständigkeit und Widerspruchsfreiheit, Mathematisches Colloquium, dated 22 January 1931. Ergebnisse eines mathematischen Kolloquiums 3: 12–13. Reprinted and translated in (Gödel 1986, 234–237).
Gödel, K. 1933. Zur intuitionistischen Arithmetik und Zahlentheorie, in (Gödel 1986, 286–295).
Gödel, K. 1938. Vortrag bei Zilsel, in (Gödel 1995, 85–113).
Gödel, K. 1941. In what sense is intuitionistic logic constructive?, in (Gödel 1995, 189–200).
Gödel, K. 1958/1972. Über eine bisher noch nicht benützte Erweiterung des finiten Standpunktes. Dialectica 12: 280–287. Revised and expanded as (Gödel 1972). The German original was reprinted and translated in (Gödel 1990, 240–251). The expanded English version is in (ibid., 271–280).
Gödel, K. 1986. Collected works, vol. I. New York: Oxford University Press.
Gödel, K. 1990. Collected works, vol. II. New York: Oxford University Press.
Gödel, K. 1995. Collected works, vol. III. New York: Oxford University Press.
Gödel, K. 2003a. Collected works, vol. IV. New York: Oxford University Press.
Gödel, K. 2003b. Collected works, vol. V. New York: Oxford University Press.
Herbrand, J. 1931. Sur la non-contradiction de l'arithmétique. Crelles Journal für die reine und angewandte Mathematik 166: 1–8. Trans. in (Herbrand 1971, 282–298).
Herbrand, J. 1971. Logical writings, ed. Warren Goldfarb. Cambridge: Harvard University Press.
Heyting, A. 1930. Die formalen Regeln der intuitionistischen Logik I, Sitzungsberichte der Preussischen Akademie der Wissenschaften, 42–56.
Hilbert, D. *1917. Mengenlehre. Lecture notes by M. Goeb, MI.
Hilbert, D. *1921/22. Grundlagen der Mathematik. Lecture notes by P. Bernays, MI.
Hilbert, D. 1922. Neubegründung der Mathematik. Abhandlungen aus dem mathematischen Seminar der Hamburgischen Universität 1, 157–177. Trans. in (Ewald 1996, 1117–1134).
Hilbert, D. 1923. Die logischen Grundlagen der Mathematik. Mathematische Annalen 88: 151–165. Trans. in (Ewald 1996, 1136–1148).
Hilbert, D. 1927. Die Grundlagen der Mathematik. Abhandlungen aus dem mathematischen Seminar der Hamburgischen Universität 6(1/2), 65–85. Trans. in (van Heijenoort 1967, 464–479).
Hilbert, D. 1928. Probleme der Grundlegung der Mathematik. Mathematische Annalen 102: 1–9. Reprint, with emendations and additions, of paper with the same title, published in Atti del Congresso internazionale dei matematici, Bologna 1928, 135–141.

Hilbert, D. *1929. Mengenlehre, Lecture notes by Lothar Collatz, Staats- und Universitätsbibliothek Hamburg, Signatur: NL Collatz, 82 pages.

Hilbert, D. 1930. Naturerkennen und Logik. Die Naturwissenschaften 18: 959–963. Reprinted in (Hilbert 1935, 378–387). Trans. in (Ewald 1996, 1157–1165).

Hilbert, D. 1931a. Die Grundlegung der elementaren Zahlenlehre. Mathematische Annalen 104: 485–494. Partially reprinted in (Hilbert 1935, 192–195). Trans. in (Ewald 1996, 1148–1157).

Hilbert, D. 1931b. Beweis des tertium non datur, Nachrichten von der Gesellschaft der Wissenschaften zu Göttingen, Mathematisch-physikalische Klasse, 120–125.

Hilbert, D. 1935. Gesammelte Abhandlungen, 3 vols. Berlin: Springer.

Hilbert, D. 2012. David Hilbert's lectures on the foundations of mathematics and physics, 1891–1933, vol. 3, ed. W.B. Ewald and W. Sieg. Berlin: Springer. To appear.

Hilbert, D., and P. Bernays. 1934. Grundlagen der Mathematik, vol. I. Berlin: Springer. Second edition, 1968, with revisions detailed in foreword by Bernays.

Hilbert, D., and P. Bernays. 1939. Grundlagen der Mathematik. vol. II. Berlin: Springer. Second edition, 1970, with revisions detailed in foreword by Bernays.

Kennedy, J. 2010. Gödel and "Formalism freeness": Manuscript, February 25, 2010.

Mancosu, P. 1999. Between Vienna and Berlin: The immediate reception of Gödel's incompleteness theorems. History and Philosophy of Logic 20: 33–45.

Menzler-Trott, E. 2007. Logic's lost genius – The life of Gerhard Gentzen. Providence/London: American Mathematical Society/London Mathematical Society.

Menzler-Trott, E. 2010. Personal communication, December 7, 2010.

Parsons, C.D. 2003. Introductory note to the Wang correspondence, in (Gödel 2003b, 379–397).

Reid, C. 1970. Hilbert. New York: Springer.

Richardson, R.G.D. 1932. Report on the congress. Bulletin of the AMS 38(11): 769–774.

Saxer, W. (ed.). 1933. Verhandlungen des Internationalen Mathematiker-Kongresses Zürich 1932, II. Band: Sektionsvorträge, Orell Füssli Verlag.

Schröder-Heister, P. 2002. Resolution and the origins of structural reasoning: Early proof-theoretic ideas of Hertz and Gentzen. Bulletin of Symbolic Logic 8: 246–265.

Schütte, K. 1993. Bemerkungen zur Hilbertschen Beweistheorie, ACTA BORUSSICA V, Beiträge zur ost- und westdeutschen Landeskunde 1991–1995.

Sieg, W. 1994. Mechanical procedures and mathematical experience. In Mathematics and mind, ed. A. George, 71–117. Oxford: Oxford University Press.

Sieg, W. 1999. Hilbert's programs: 1917–1922. Bulletin of Symbolic Logic 5: 1–44.

Sieg, W. 2009. Hilbert's proof theory. In Handbook of the history of logic, ed. D.M. Gabbay and J. Woods, 321–384. Amsterdam: Elsevier.

Sieg, W., and C.D. Parsons. 1995. Introductory note to (Gödel 1938), in (Gödel 1995), 62–84.

Sieg, W., and C. Tapp. 2012. Introduction to the Undated Draft: The "Undated Draft" is a manuscript most likely written in late 1920/early 1921, in (Hilbert 2012).

Skolem, T. 1922. Einige Bemerkungen zur axiomatischen Begründung der Mengenlehre. Trans. in (van Heijenoort 1967, 290–301).

Tait, W.W. 2002. Remarks on finitism. In Reflections on the foundations of mathematics, Lecture notes in logic, vol. 15, ed. W. Sieg, R. Sommers, and C. Talcott, 410–419. Urbana: Association for Symbolic Logic.

Tait, W.W. 2005. Gödel's reformulation of Gentzen's first consistency proof of arithmetic. The Bulletin of Symbolic Logic 11: 225–238.

Tait, W.W. 2010. The substitution method revisited, in (Feferman and Sieg 2010, 231–241).

Taussky-Todd, O. 1987. Remembrances of Kurt Gödel. In Gödel-Symposium in Salzburg, July 10–12, 1983. 31–41. Naples: Bibliopolis.

Troelstra, A.S. 1990. Introductory note to (Gödel 1958/1972), in (Gödel 1990, 217–240).

Troelstra, A.S. 1995. Introductory note to (Gödel 1941), in (Gödel 1995, 186–189).

Van Heijenoort, J. (ed.). 1967. From Frege to Gödel: A sourcebook of mathematical logic, 1879–1931. Cambridge: Harvard University.

Von Neumann, J. 1927. Zur Hilbertschen Beweistheorie. Reprinted in (von Neumann 1961, 256–300).

Von Neumann, J. 1931. Die formalistische Grundlegung der Mathematik. Erkenntnis 2: 116–121. Trans. in (Benacerraf and Putnam 1983, 61–65).
Von Neumann, J. 1961. Collected works, vol. I, ed. A.H. Taub. New York: Pergamon.
von Plato, J. 2008. Gentzen's proof of normalization for natural deduction. The Bulletin of Symbolic Logic 14: 240–257.
von Plato, J. 2009a. Gentzen's logic. In Handbook of the history of logic, vol. 5, ed. D.M. Gabbay and J. Woods, 667–721. Amsterdam: Elsevier.
von Plato, J. 2009b. Gentzen's INH: A brief summary of its main ideas: Personal communication, November 24, 2009.
von Plato, J. 2010. Gentzen's logical calculi: Aspects of a work of genius: Manuscript.
Wang, H. 1981. Some facts about Kurt Gödel. Journal of Symbolic Logic 46: 653–659.
Zach, R. 2003. The practice of finitism: Epsilon calculus and consistency proofs in Hilbert's program. Synthese 137: 211–259.

Chapter 6
Evolution and Logic

Jan M. Smith

There is more wisdom in your body than in your deepest philosophy. Nietzsche

Biological evolution is perhaps the most revolutionary discovery ever made. For many of us it eliminates questions about our origin which otherwise would have been troubling. During the last decades a deeper understanding of the mechanisms of evolution, like the gene-perspective, has broadened the explanations to include, for instance, altruism and social structures. Edward O. Wilson, one of the leading evolutionary biologists, has even proposed a unification of all knowledge where the social sciences and humanities are integrated with the natural sciences by the theory of biological evolution (Wilson 1998).

But still, does evolution help us in answering the fundamental questions of philosophy like "What can we know?" and "What exists?"; more specifically, can evolution say something about the foundations of mathematics and logic? I will here give some personal thoughts on the subject and for me the starting point is David Hume's sceptical empiricism and in particular his view on causality. Hume's analysis, rather straightforwardly, opens up for an evolutionary understanding of our ability to form causal relations; my view is that very basic logic in a similar way can be given an evolutionary explanation.

Hume gives a restrictive limit for the knowledge we can obtain through direct empirical evidence and hence also a border which will involve metaphysical speculation to pass; and it is mainly on this other side of the border that the foundational questions of mathematics reside. Hume himself did not speculate beyond direct empirical evidence, but Kant's *Copernican revolution* can be seen

J.M. Smith (✉)
Department of Computer Science and Engineering, Chalmers University of Technology and University of Gothenburg, S-412 96 Gothenburg, Sweden
e-mail: smith@chalmers.se

P. Dybjer et al. (eds.), *Epistemology versus Ontology*, Logic, Epistemology, and the Unity of Science 27, DOI 10.1007/978-94-007-4435-6_6,

as an answer to the questions asked by Hume's analysis. As has been observed by many people, some already in the nineteenth century, Kant's categories and forms of intuition are very much in coherence with biological evolution.

To connect Hume, evolution, and logic might seem a bit far-fetched from a traditional view on the foundations of mathematics and logic. However, this path is in consonance with philosophy developed in the perspective of biological evolution; I especially want to refer to Michael Ruse's book *Taking Darwin seriously* (Ruse 1986).

6.1 Hume's Analysis of Causality

According to Hume's famous analysis, we have no perceptual evidence for causality:

> Motion in one body is regarded upon impulse as the cause of motion in another. When we consider these objects with utmost attention, we find only that the one body approaches the other; and that the motion of it precedes that of the other, but without any, sensible interval. It is in vain to rack ourselves with farther thought and reflection upon this subject. We can go no farther in considering this particular instance.

Causality has played a crucial role in western philosophy since at least Aristotle. By Hume's analysis we must put that central position in doubt:

> It is a general maxim in philosophy, that whatever begins to exist, must have a cause of existence. This is commonly taken for granted in all reasoning, without any proof given or demanded. It is supposed to be founded on intuition, and to be one of those maxims, which though they may be denied with the lips, it is impossible for men in their hearts really to doubt of. But if we examine this maxim by the idea of knowledge above-explained, we shall discover in it no mark of any such intuitive certainty; but on the contrary shall find, that it is of a nature quite foreign to that species of conviction.

What remains is just that

> All our reasoning concerning causes and effects are deriv'd from nothing but custom; and that belief is more properly an act of the sensitive, than of the cogitative part of our natures.

These quotations are from Hume's *A Treatise of Human Nature* (1739–1740), his magnum opus in philosophy.

Hume's conclusions may be compared to the rationalist Descartes who, a 100 years earlier, begins with even less than what the empirist Hume accepts, *Cogito ergo sum*, but from that derives knowledge about the world. How is that possible? The answer is that Descartes had an absolute belief in the existence of God. He does give a proof of the existence of God, but, as all such proofs, it is not particularly convincing: it is clear that he is constructing an argument for something which for him is obviously true. Once the omnipotent and benevolent God is present, Descartes can proceed to deduce reality. The existence of God was not questioned by any of the rationalist of the seventeenth century, Leibniz even has it among the obvious truths on pair with $1 + 1 = 2$.

Hume had a profound impact on Kant. In the *Prolegomena* (1783) to his *Critique of Pure Reasoning* (1781), Kant writes

> It was Hume that first roused me from a dogmatic slumber of many years, and gave quite a new direction to my researches in the field of speculative philosophy.

and without reservation he accepts Hume's analysis of causality. Kant's *Copernican revolution* in the *Critique* puts the basic concepts space and time inside ourselves. They are forms of intuition and we cannot give them any existence independent of us. We cannot know anything about *Das Ding an sich* which does not exist in space and time and does not enter into causal relations. This radical solution puts causality on pair with time and space as a mean for us to organize the outside world. It is only through our innate forms of intuition that we, according to Kant, can have knowledge of the external world.

Hume's empiricism and scepticism were important to British science in the early nineteenth century. According to his unpublished notebooks, Darwin read Hume "during a month to six weeks" in 1838 during the time when his first ideas on evolution were conceived (Huntley 1972). Another striking example of Hume's influence on opening up for new ideas can be found in a letter by Einstein to Schlick (Einstein 1915):

> You have also correctly seen that this trend of thought was of great influence on my efforts, and specifically E. Mach and still much more Hume, whose treatise on understanding I studied with fervor and admiration shortly before the discovery of the theory of relativity. It is very well possible that without these philosophical studies I would not have arrived at the solution.

Behind these influences is surely Hume's general critical attitude of questioning what might seem obvious. But in the case of at least Kant, the influence is definitely deeper, resulting in a fundamentally new perspective on what can be known.

6.2 Evolutionistic Understanding of Causality

It is obvious that structuring the world by causality is extremely important for survival and that there, hence, must have been a very strong evolutionary pressure to develop this ability; as Ruse expresses it in *Taking Darwin Seriously* (Op. cit., p. 174):

> The world works in a regular way. It is in our biological interests to take note of this, and so as an adaptive response we tend to make something of the regularities. But, as philosophers, we should not try to make more of the regularities than what they are. Causes are projected into the world by us ... The human who believes in real connections has the biological edge over the human who only sees contingency.

This is reflecting Hume's analysis in an evolutionary perspective. Of course, Hume, who was born almost 100 years before Darwin, cannot have had ideas in this direction, but still he has formulations which can be put in that context (*Treates*):

> Nature, by an absolute and uncountroulable necessity has determin'd us to judge as well as to breathe and feel.

Kant's *Copernican revolution* offers an explanation of causality which is in accordance with evolution in the sense that the explanation in terms of categories and forms of intuition puts the concept inside ourselves as biological beings. An evolutionary interpretation of Kant's ideas was given by Konrad Lorenz in *Kants Lehre vom apriorischen im Lichte gegenwärtiger Biologie* (Lorenz 1941) and it also appears later in *Behind the Mirror* (Lorenz 1973) where he writes (p. 37):

> ...the categories and modes of perception of man's cognitive apparatus are the natural products of phylogeny and thus adapted to the parameters of external reality in the same way, and for the same reason, as the horse's hooves are adapted to the prairie, or the fish's fins to the water.

Adaption through evolution thus becomes a biological description of the world, in Kant's terminology *Das Ding an sich*. That our cognitive apparatus developed by evolution should give us such a direct, although not complete, image of reality might be looked upon as a rather naïve form of realism.

In Donald T. Campbell's influential paper *Evolutionary epistemology* (Campbell 1974) evolution is understood in a broader sense to not only include biology (p. 413):

> ...evolution – even in its biological aspects – is a knowledge process, and ... the natural-selection paradigm for such knowledge increments can be generalized to other epistemic activities, such as learning, thought, and science.

Campbell's view on Kant's connection to evolution focuses on psychology (Op. cit., p. 441):

> The evolutionary perspective is of course at odds with any view of an *ipso facto* necessarily synthetic a priori. But it provides a perspective under which Kant's categories of thought and intuition can be seen as descriptive contribution to psychological epistemology. Though we reject Kant's claim of a necessary a priori for these categories, we can in evolutionary perspective see the categories as highly edited, much tested presumptions, "validated" only as scientific truth is validated, synthetic a posteriori from the point of view of species-history, synthetic and in several way a priori (but not in terms of necessary validity) from the point of view of an individual organism.

Obviously, the forms of intuition have changed during evolution, they must have been very different in the early animals we descend from; in this sense they are a posteriori and not static. But taking the evolutionary perspective seriously and regarding our biology as a prerequisite and frame for our thoughts, then it is the biology of the human species as it is now that is relevant; hence, in our context, we cannot agree with Campbell's statement that evolution is at odds with any necessarily synthetic a priori.

6.3 Evolutionistic Understanding of Logic

As for causality, it is obvious that the ability to perform simple logical reasoning is important for survival. For instance, by *modus ponens* one can retrieve information from earlier experiences when getting into a similar situation, *or-elimination* is case analysis of importance when figuring what is best to do next, etc. Of course, I do not claim that a particular logical system, like Gentzen's *Natural deduction*, is there by evolution; rather that there is some innate capacity for very basic logic. Already a simple argument consisting of more than one step will involve some cognitive activity, and more complex reasoning will certainly need language. I fully agree with Ruse (Op. cit., p. 169):

> My suspicion is that, far from being useless, logic is so necessary and deeply ingrained in our nature that we cannot imagine ourselves thus structure and inform our experiences.

Hume's analysis of causality makes the explanation of causality by evolution rather straightforward, but in the case of logical reasoning it might be less obvious. I think that a reason is that logical truth, since Frege, is often seen as the prime example of analytic truth and hence not in need of any further understanding. But one must not confuse the validity of logic with that of a derivation using logic; in my opinion, the underlying basic logic is not analytic but it is there because of evolution.

If we go back to the early history of logic, there is a connection between causality and rules of logic. Aristotle writes in *Posterior Analytics*:

> We think we have scientific knowledge when we know the cause.

> By demonstration I mean a syllogism productive of scientific knowledge.

> The premisses must be the causes of the conclusion, better known than it, and prior to it; its causes, since we possess scientific knowledge of a thing only when we know its cause; prior, in order to be causes; antecedently known, this antecedent knowledge being not our mere understanding of the meaning, but knowledge of the fact as well.

I do not want to make too much out of Aristotle's use of the word "cause" in this context, just remark that the cause-effect of the rules of logic might have been more natural for Aristotle with the practical use of logic he had in mind than for us today with our more mathematical and formal perspective on logic.

From the evolutionary point of view, elementary arithmetic is similar to logic; as Ruse puts it (Op. cit., p. 162):

> The proto-human who innately preferred '$2 + 2 = 4$' to '$2 + 2 = 5$' was at a selective advantage over his/her less discriminating cousin.

However, I do not agree entirely with Ruse's formulation here. I believe that very basic properties of arithmetic are in inborn, like the insight that '$1 + 1 = 2$' and not '$= 1$' or '$= 3$', but already to see that '$2 + 2 = 4$' might involve some form reasoning.

Kant considered arithmetical truths to be synthetic a priori which is in accordance with an evolutionary explanation. This is in contrast with his view on logic, which he held to be analytic. So, I deviate here from Kant since the evolutionary explanation of logic entails that logic is synthetic a priori. Basic logic is a fundamental prerequisite for our cognitive abilities and I would claim that basic logic is a form of intuition equally fundamental as time and space.

6.4 Foundations of Mathematics

Given that biological evolution explains why there is mathematics, can it single out any of the different views on the foundations of mathematics as the correct one? I do not think that is the case: although Formalism, Platonism, and Intuitionism have very different explanations of mathematics, it seems to me be possible for a devotee of any of them to argue for an evolutionary origin.

Formalism is rather straightforward: you may accept that mathematics has an evolutionary origin but that the practise of mathematics nevertheless is a formal game. Although I am not an adherent of formalism, my point would here be that the very basic rules of the game, those of logic, are set up by evolution.

Platonism seems to me more difficult to connect to evolution, but even if you have a naïve realistic view that the mathematical objects exist on the same level as material objects and exist independently of us humans, you may perhaps believe in an evolutionary explanation: in the spirit of Lorenz you might argue that our conception of mathematical objects comes from an adaption to the real mathematical world you believe in. I find this hard to imagine, but I already find naïve Platonism in itself a stance difficult to understand.

Intuitionism, however, is more directly coherent with evolution since it puts constructions by us humans in the centre. In Brouwer's writing, there are many references to Kant and time is for Brouwer the basic intuition behind all of mathematics, beginning with the natural and real numbers. But he did not agree with Kant on space because of the discovery of non-Euclidean geometries; in *Intuitionism and Formalism* (Brouwer 1912) he writes (p. 127):

> However weak the position of intuitionism seemed to be after this period of mathematical development; it has recovered by abandoning Kant's apriority of space but adhering more resolutely to the apriority of time.

Space is by Brouwer understood through analytic geometry and, hence, reduced to real numbers and thereby to the intuition of time.

Also Poincaré was inspired by Kant and in *The Value of Science* (Poincaré 1905) he even refers to evolution; selection will favour those that find regularities and have ideals beneficial for their survival (pp. 5, 9):

> [In a world without regularities] there would be no science; perhaps thought and even life would be impossible since evolution could not there develop the preservational instincts.

> ... is there a play of evolution and natural selection? Have the peoples whose ideal most conformed to their highest interest exterminated the other and taken their place? All pursued their ideals without reference to consequences, but while this quest led some to destruction, to others it gave empire. One is tempted to believe it.

According to Poincaré, mathematics is a cognitive result of evolution because of our search for harmony and regularities. Poincaré is aiming at explaining abstract mathematics holistically, while I am arguing that very basic mathematics, and in particular logic, is a direct consequence of evolution. However, neither Brouwer nor Poincaré paid much attention to logic.

6.5 Martin-Löf Type Theory and the Synthetic A Priori

In the paper *Analytic and synthetic judgements in type theory* (Martin-Löf 1994), Martin-Löf has connected his semantics of type theory to Kant. The judgement that a is an object of type A is, according to Martin-Löf, analytic since "the judgement is evident solely by virtue of the meanings of the terms that occur in it". However, the judgement that the proposition A is true, written A *true*, is synthetic a priori since a construction is needed to get an object a of A which makes A *true* evident; for instance, $(A\&B \supset A)$ *true* is synthetic since it comes without a proof (although in this case it is trivial to find one).

Analytical judgements are not the focus of Kant, most of his *Critique* is concerned with the synthetic a priori, and his definition of analytic is confusing. In the *Prolegomena* he says

> Analytical judgments express nothing in the predicate but what has been already actually thought in the concept of the subject, though not so distinctly or with the same (full) consciousness.

but on the same page continues with

> All analytical judgments depend wholly on the law of Contradiction ... the predicate of an affirmative analytical judgment is already contained in the concept of the subject, of which it cannot be denied without contradiction.

The second characterization may lead to a logical understanding of analytic which is different from the first one, a fact which Frege was well aware of. In *The Foundations of Arithmetic* (Frege 1884) he explicates Kant's second definition and defines analytic as logical truth (p. 4):

> If in carrying out this process [of finding a proof], we come only on general logical laws and on definitions, then the truth is an analytic one, ... If, however, it is impossible to give the proof without making use of truths which are not of a general logical nature, but belong to the sphere of some special science, then the proposition is a synthetic one.

Martin-Löf's interpretation of analytic is the first one of containment of the subject in the predicate and it is different from Frege's by which all logical truths are analytic. Martin-Löf's understanding of synthetic a priori is also different from the one I am advocating and which comes from the forms of intuition; as Kant expresses it in the *Prolegomena*:

> ...the intuitions which pure mathematics lays at the foundation of all its cognitions and judgments which appear at once apodictic and necessary are space and time. For mathematics must first have all its concepts in intuition, and pure mathematics in pure intuition, that is, it must construct them.

Hence, according to Kant, mathematics is constructed from the forms of intuition. In the evolutionary interpretation of the forms of intuition, this means that mathematics is there because of evolution. I see the word "intuition" as the important part of Kant's characterisation of mathematics, while Martin-Löf puts emphasis on "construction" which, hence, leads to his view that *A true* is synthetic since a construction is needed to see that the judgement is valid. But I agree with Martin-Löf that formal derivations are analytic, albeit with an underlying evidence which is synthetic a priori.

It is worth mentioning that, in spite of Frege's strong influence on analytic philosophy, Russell in *The Principles of Mathematics* (Russell 1903) considered logic to be synthetic (p. 457):

> Kant never doubted for a moment that the propositions of logic are analytic, whereas he rightly perceived that those of mathematics are synthetic. It has since appeared that logic is just as synthetic as all other kinds of truth; but this is a purely philosophical question, which I shall here pass by.

6.6 Ontology

Accepting Kant's forms of intuition does not mean that you also have to accept his view that *Das ding an sich* exists nor that, if you believe that it exists, we cannot know anything about it.

The idealistic view that *Das ding an sich* does not exist, held by both Fichte and Mach, must be ruled out in an evolutionary perspective since the very idea of biological evolution is adaption to some kind of reality. The opposite view, held by Lorenz, does presuppose a reality which we get to know directly by our adoption to it; ontologically Lorenz agrees with Kant but not epistemically since Kant denies any knowledge of *Das ding an sich*.

Modern physics reveals a reality which in many respects is fundamentally different from our everyday experience of it. The Copenhagen interpretation of quantum mechanics in terms of measurements and probabilities makes it difficult to transfer our macro experiences of how particles behave to elementary particles. Although Hume did not question causality per se but rather the evidence for it, it is clear that in quantum mechanics basic concepts like causality and determinism must be seen in a different perspective than our common sense view of these concepts. Also, the four-dimensional space-time of the theory of relativity is, especially on the cosmic scale, very different from our everyday experience of time and space.

From the evolutionary perspective, it is a striking fact that mathematics is able to describe physics far from the everyday reality we are adopted to. The explanation must be, I guess, that the world functions in a very regular way so that our cognitive abilities, developed for survival, can be used in domains of no relevance to our adaption. However, one should not exclude that our ever more advanced physical experiments and observations could come to a limit where our logic and mathematics do not function anymore and, hence, also a limit beyond which knowledge is not possible.

Hume's sceptical analysis together with modern physics make me reluctant to make any strong ontological commitments. There is an evolutionary adaption but it is doubtful if it is possible for us to have any knowledge about the nature of what we are adopted to; after all, our cognitive apparatus is there for survival and not for metaphysical knowledge (taking an evolutionary perspective also on this question itself). What we can be sure of are the forms of intuition, although probably not precisely in the way Kant envisioned them, and evolution explains why they are there. This is clearly close to Kant's view of *Das ding an sich* but from a very different point of view.

6.7 Concluding Remarks

I believe that very little, if anything, can be said objectively in metaphysics, in Kant's terminology "der spekulativen Philosophie"; rather, metaphysical stances are inevitably dependent on ones general convictions. The frame of mind of the rationalists of the seventeenth century is the scientific revolution of this period together with the belief of the existence of God. Hume is part of the British empiricism which, together with his scepticism, led to his radical analysis of causality as well as to his reluctance to metaphysical speculation.

Hume's analysis and the explanatory power of evolutionary adaption have convinced me that we possess an innate ability for forming causal relations; and that insight leads me to a similar view on logic. I am not claiming precise knowledge of our innate abilities for logic, I cannot tell whether there is a rule structure or if logic is there in some other way; if we had deeper knowledge about human evolution and how the brain functions, we could say more here. But I do not think that the details, important and interesting as they are, matters for general philosophical conclusions coming from the mindset of biological evolution and Humean scepticism; and in particular not for the conclusion that basic logic is synthetic a priori.

References

The texts by Aristotle, Hume, and Kant are available from the internet at The Online Books Page. http://onlinebooks.library.upenn.edu/.

Brouwer, L. 1912. Intuitionism and formalism. In *L.E.J. Brouwer collected works*, ed. A. Heyting, 123–138. Amsterdam and Oxford: North Holland, 1975.

Campbell, D.T. 1974. Evolutionary epistemology. In *The philosophy of Karl Popper*, ed. P. Schlipp, 413–463. Illinois: La Salle.

Einstein, A. 1915. *The collected papers of Albert Einstein, volume 8: the Berlin years: correspondence, 1914–1918*. Princeton: Princeton University Press, 1998.
Frege, G. 1884. *The foundations of arithmetic*. Oxford: Basil Blackwell, 1953. Translated by J.L. Austin.
Huntley, W. B. 1972. David Hume and Charles Darwin. *Journal of the History of Ideas* 33(3): 457–470.
Lorenz, K. 1941. Kants lehre vom apriorischen im lichte gegenwärtiger biologie. *Blätter für Deutsche Philosophie* 15: 94–125.
Lorenz, K. 1973. *Behind the mirror*. Mariner Books: New York and London, 1978.
Martin-Löf, P. 1994. Analytic and synthetic judgements in type theory. In *Kant and contemporary epistemology*, ed. P. Parrini, 87–99. Dordrecht/Boston: Kluwer.
Poincaré, H. 1905. *The value of science*. New York: Dover, 1958. Translated by George Bruce Halsted.
Ruse, M. 1986. *Taking Darwin seriously*, 2nd. ed. Prometheus books: New York, 1998.
Russell, B. 1903. *The principles of mathematics*. Merchant Books: New York, 2008.
Wilson, E.O. 1998. *Consilience: the unity of knowledge*. Random House: New York.

Chapter 7
The "Middle Wittgenstein" and Modern Mathematics

Sören Stenlund

7.1 Introduction

Wittgenstein's work in the philosophy of mathematics has a much more central position in his overall philosophical thinking, and for the continuity of his thought, than what is generally recognized in the so-called "two-Wittgenstein view", where we have the earlier Wittgenstein (the author of the *Tractatus*) and the later Wittgenstein (the author of the *Philosophical Investigations*). In order to see this, a closer look at Wittgenstein's work in the 'Middle Period" (roughly 1929–1936) is needed.[1]

In this essay I will concentrate on the change in Wittgenstein's thinking that takes place mainly in the beginning of the 1930s. By examining certain crucial features in this change I will try to show that he received decisive impulses and ideas from new developments in mathematics and natural science at the time. These ideas affected not only his thinking about mathematics, but also his thinking about language and the nature of philosophy in general.

There is one feature of Wittgenstein's philosophy of mathematics that is present from the beginning to the end, and that is his emphasis of the *difference* between mathematical propositions and propositions of experience. Mathematical propositions, propositions of pure mathematics, do not have a descriptive content according to Wittgenstein; they are not 'about anything', in the sense in which the propositions of physics or biology are so.[2] The difference is not just that physics and biology

[1]The importance of the 'Middle Period' for understanding Wittgenstein's philosophy of mathematics was emphasized by Gerrard (1991).

[2]In PR, p. 186 we find the following sharpened statement of this feature of mathematical symbolism: "Let's remember that in mathematics, the signs themselves *do* mathematics, they

S. Stenlund (✉)
Department of Philosophy, Uppsala University, Uppsala, Sweden
e-mail: Soren.Stenlund@filosofi.uu.se

P. Dybjer et al. (eds.), *Epistemology versus Ontology*, Logic, Epistemology, and the Unity of Science 27, DOI 10.1007/978-94-007-4435-6_7,

are about observable empirical phenomena, which mathematics is not. It is rather that proved mathematical propositions are not propositions, if being about facts of a realm given in advance is taken as an essential feature of propositions. Thus, for instance, the propositions of the *Tractatus* do not include the mathematical propositions.

Now, there is a very common view that there is a realm of entities that mathematical propositions are about and do describe. Wittgenstein expressed the source of what he took to be problematic about this view as follows in a lecture in 1939:

> The difficulty in looking at mathematics as we do is to make one particular section – to cut pure mathematics off from its application. It is particularly difficult to know where to make this cut because certain branches of mathematics have been developed in which the charm consists in the fact that pure mathematics looks as though it were applied mathematics – applied to itself. And so we have the business of a mathematical realm.[3]

As Wittgenstein uses the words 'apply' and 'application' here, an applied mathematical system is always applied to something existing in advance outside the system that is being applied. It is applied to a subject matter that exists independently of the mathematical system, as when we say that the system of pure Euclidean geometry is applied in astronomy for the description of the movements of the planets. The planets and their movements as empirical phenomena are there in advance, independently of geometry and its application. When a mathematical notion is applied *inside* mathematics, and this sense of application is retained (which the 'prose' of mathematics invites us to), there appears to be a mathematical realm existing in advance.[4] For instance, when we talk about the number of roots of an equation, it appears as though this number is there to be discovered in a realm outside the arithmetic-algebraic system in which the notion of the root of an equation is invented and determined.

To cut a system of pure mathematics off from its application, in the sense described in this quotation, is to present the system as an *autonomous* system, a conceptually self-standing system. This autonomy idea and its influence on Wittgenstein's conception of grammar since the beginning of the 1930s, is the main theme of this essay. The idea has affinity with important trends in modern mathematics and natural science around beginning of the twentieth century, such

---

don't describe it. [...] You can't write mathematics [in the sense in which you write history], you can only do it." (References to printed editions of Wittgenstein's works are given using the abbreviations indicated in the bibliography. Such an abbreviation followed by a decimal number refers to the remark or section with that number. An abbreviation followed by a roman numeral indicates a part of the abbreviated work. References by MS and TS number are from the Bergen electronic edition of *Wittgenstein's Nachlass,* Oxford University Press, Oxford, 2000).

[3]LFM, p. 150.

[4]The nature of this realm is the subject matter of the 'ontology of mathematics', in the philosophical discourse.

as for instance Heinrich Hertz' work in mechanics and David Hilbert's work in geometry, and, generally, Hilbert's axiomatic method.[5]

One difficulty with seeing the significance of the autonomy of grammar is the following: if it was inspired by trends in modern mathematics and natural science (which I will say more about in the following sections), why didn't that bring Wittgenstein's thinking closer to scientism or naturalism (as one would perhaps expect), but on the contrary made him more averse to the use of scientific methods in philosophy. For, as already mentioned, the logical positivists were also inspired by the same trends in modern mathematics and natural science.

The inspiration Wittgenstein got from mathematics and science was not something similar to the ideas of 'explication' and 'rational reconstruction', i.e., the use of mathematical tools for the construction of new concepts in science that were designed to replace philosophically problematic notions. Wittgenstein's grammatical method works in the opposite direction, towards the source of problematic notions, and the clarification is for its own sake and not in the service of science. Wittgenstein took particular interest in trends in modern mathematics and physics that *were motivated by a desire to overcome philosophical dogmatism* (without introducing a new dogmatism). He was inspired and stimulated by the efforts to liberate science from dogmatic attitudes and prejudices which had their roots in traditional metaphysical philosophy. We find such efforts, for instance, in the work of Ernst Mach, Ludwig Boltzmann and Heinrich Hertz as well as in the work of the mathematical formalists, such as for instance, Johannes Thomae.[6]

[5]By 'modern mathematics' will be meant, in this paper, the new trends in mathematics that emerge in the late nineteenth and the beginning of the twentieth century when the old conception of mathematics as the science of quantity is experienced as outdistanced and inadequate in regard to the internal change and development of the discipline of mathematics, in particular in regard to the new emphasis of formal rigor, and of pure mathematics as an independent and autonomous discipline. See Mehrtens (1990), Epple (2003), and Gray (2006). Mehrtens focuses in particular on the foundational crisis as a symptom of this change and of the need to revise and redefine the identity of mathematics as a discipline. He presents Hilbert as the main advocate of the modern mathematics, and as the main champion for the autonomy of mathematics and mathematical discourse. In the beginning of the 1930s Wittgenstein (unlike many mathematicians and logicians) was obviously sensitive to this cultural aspect of mathematics and aware of this scientific change as being to some extent part of a broader cultural change.

[6]Thomae writes: "The formal conception of numbers sets itself more modest limitations than does the logical conception. It does not ask, what are and what shall the numbers be, but it asks, what does one need about numbers in arithmetic. For the formal conception, arithmetic is a game with signs which one may call empty; by this one wants to say that (in the game of calculation) they have no other content than that which has been attributed to them concerning their behaviour with respect to certain rules of combination (rules of the game). Similarly, a chess player uses his pieces, he attributes to them certain properties which condition their behaviour in the game, and the pieces themselves are only external signs for this behaviour. To be sure, there is an important difference between the game of chess and arithmetic. The rules of chess are arbitrary, the system of rules for arithmetic is such that by means of simple axioms the numbers can be related to intuitive manifolds, so that they are of essential service in the knowledge of nature. – The formal standpoint relieves us of all metaphysical difficulties, this is the benefit it offers us." Quoted from Epple (2003), p. 301.

The work of Mach, Boltzmann, Thomae and Hertz was done in the service of science, but Wittgenstein saw its significance for language in general: the attitudes and prejudices with roots in metaphysical dogmatism had their counterparts and parallels in our conception of language in general.

In the rest of this essay we will examine some more specific examples of the affinity between Wittgenstein's philosophical method in the beginning of the 1930s and certain trends in modern mathematics and natural science.

7.2 Grammar and Geometry

Already in a notebook in 1913 Wittgenstein makes the following remark about grammar and its significance for philosophy:

> Distrust of grammar is the first requisite for philosophizing.

When he returns to philosophy by the end of the 1920s, questions of the nature of grammar is a recurrent issue in his writings and his work on this issue is motivated by his desire to remove the mistakes he had found in the *Tractatus*, especially the idea that the logical nature of the proposition should be accounted for in one calculus of truth-functions. The idea he is concerned to explore instead is expressed in the *Philosophical Remarks* as follows:

> A proposition is completely logically analysed if its grammar is made completely clear: no matter what idiom it may be written or expressed in.
>
> [...] All that is possible and necessary is to separate what is essential from what is inessential in our language.[7]

He says here 'our language' not only as opposed to the idea of a logically ideal language, but also against the notion of a 'phenomenological language' or a 'primary language', which were ideas he had tried before but rejected.

By this time he is still very much within the framework of the *Tractatus* and concerned with revising it, rather than leaving it behind and starting anew. He is still under the influence of the distinction between language as representation and reality as that which is represented, and that it is of the essence of propositions to represent states of affairs of reality. Language and reality must therefore share a logical form. They must have the same 'multiplicity', (a notion he owes to Hertz). For instance, that a fact is a combination of seven simple objects is shown in the proposition representing this fact in its containing seven distinct simple names. In a lecture in the lent term of 1930 he said:

> The multiplicity of language is given by grammar. A proposition must have the same multiplicity as the fact which it expresses: it must have the same degree of freedom. We must be able to do as much with language as can happen in fact. Grammar lets us do some things with language and not others; it fixes the degree of freedom.

---

[7] PR, p. 51.

> [...] grammar is not entirely a matter of arbitrary choice. It must enable us to express the multiplicity of facts, give us the same degree of freedom as do the facts.[8]

As Wittgenstein still conceived it here, grammar is not autonomous.

Sometime in 1930–1931 Wittgenstein explained some of the first remarks in the *Tractatus* to a student and made the following remark:

> 1.13. "The facts in logical space are the world". Logical space has the same meaning as grammatical space. Geometry is a kind of grammar: there is an analogy between grammar and geometry. Grammatical space includes all possibilities. "Logic treats of every possibility" 2.0121.[9]

The connection between grammar and geometry expressed here is of great importance for how his notion of grammar develops during the following years, and he spends a lot of work exploring this analogy in these years. His thoughts on the nature of grammar influenced his views on the nature of geometry, but the influence and stimulus in the opposite direction is even more important: geometry is often used as a model for grammar. In the *Philosophical Remarks* he says: "Geometry and Grammar always corresponds with one another."[10] The following remark indicates that, in this work, he took notice of and was influenced by the lively discussion about the nature of geometry and its role in physics during the first decades of the twentieth century:

> Euclidean geometry is a part of grammar. It is a convention of expression and so part of grammar. (Minkowski accounts for the result of the Michelson-Morley experiment by a new geometry, Fitzgerald by contraction. These are merely two expressions of the same fact; we can adopt either, unless a decisive experiment is possible between them.)[11]

In the *Philosophical Grammar* we find the following remark on the nature of geometry:

> Geometry isn't the science (natural science) of geometric planes, lines and points, as opposed to some other science of gross physical lines, stripes and surfaces and *their* properties. The relation between geometry and propositions of practical life, about stripes, colour boundaries, edges and corners, etc. isn't that the things geometry speaks of, though *ideal* edges and corners, resemble those spoken of in practical propositions; it is the relation between those propositions and their grammar. Applied geometry is the grammar of statements about spatial objects. The relation between what is called a geometrical line and a boundary between two colours isn't like the relation between something fine and something coarse, but like the relation between possibility and actuality.[12]

What Wittgenstein is questioning here is the common and popular view of geometry (even among prominent mathematicians, such as for instance Felix Klein) according to which the figures and diagrams we write on paper or on the blackboard are only

---

[8] DL, p. 8.

[9] DL, p. 119.

[10] PR, p. 186.

[11] DL, p. 8.

[12] PG, p. 319.

inexact approximations of the ideal or exact geometrical objects that pure geometry is really about, as when it is said "we can never write a perfect circle or find one in physical space, they will always be inexact, while the objects that the theorems of geometry are about are exact". This is to hypostasize possibilities as a kind of objects, which are merely 'ideal'. But, for Wittgenstein, the 'ideal circle' is not literally a circle, it is a rule of grammar, a grammatical picture, that determines what it makes sense and does not make sense to say, within the conceptual framework of Euclidean geometry, about an actual circle on the paper or in physical space.[13]

The circle we draw on paper in the statement and proof of a theorem about circles in pure geometry has the role of a *variable* in the proof. It is, as we say, an 'arbitrary circle'. It is the *use* we make of the written circle as a sign in the proof that makes it an 'arbitrary circle'. It's not that the written circle refers to some invisible 'ideal object', which the theorem is 'about'. In applications of the theorem to actual circles the variable is assigned an actual circle as a value. But the circles we draw in the proofs and constructions in pure geometry are at the same time actual circles, to which already proved theorems about circles apply. This double role of the drawn figures and diagrams of geometry as expressions for a rule and, at the same time, as instances of the rule is one basis for the 'perspicuousness' of geometrical proofs. Wittgenstein's conception of the perspicuousness of grammar was clearly influenced by this perspicuousness of geometrical proofs, for instance in his frequent use by this time of formal grammatical remarks such as 'any rod has a length'. If this is said while you are showing a specific rod, this rod has the double role of an arbitrary rod and a specific instance of a rod.

Suppose that we are doing Euclidian geometry and two distinct points A and B are marked out on the blackboard. Then we may say (correctly) "there is exactly one straight line going between the points A and B". Someone, not very used to the discourse of geometry, might say (also, in a sense, correctly): "I see no straight line going between the points A and B. There isn't one until it has been drawn." The former statement, as a rule of grammar, means for instance that it makes good sense to talk about *the* straight line through the points A and B, but that it does *not* make sense to talk about 'the two distinct straight lines going through the points A and B'.

The counterpart in the grammar of language in general, to the misleading reification of possibilities in geometry (that invites the talk of 'ideal objects'), is the view of grammatical statements as a kind of material statements about states of affairs in some (hidden) conceptual realm, a realm of conceptual states of affairs that would be a ground for, or a reality against which grammar is responsible. But grammar is not responsible to any reality, in that sense.

[13]Wittgenstein knew of course that the picture and the talk of geometry as being about 'ideal objects' is very common in mathematical work, where it may be harmless and even a stimulus for mathematical inventions. But he is concerned with the way in which this pictorial talk tends to deceives us in the philosophy of mathematics. This kind of deception often takes place when philosophers are inclined to talk about "possibilities in principle".

When Wittgenstein said that "applied geometry is the grammar of statements about spatial objects" in the quotation given before, he had surely become aware of the fact that geometry had been axiomatized as an autonomous axiom system, in which the geometrical concepts are detached, or 'cut off', from their applications. This might be why he is saying 'applied geometry' rather than just 'geometry'. Anyway, this result in geometry due to Hilbert was, no doubt, one of the things that helped Wittgenstein to overcome the prejudice that forced him to think as though grammar must somehow be responsible to a reality outside language, rather than being an autonomous calculus.[14] That this was a radical and deep-reaching change in his thinking that affected the heart of the conception of logical analysis he had had before, is clear from the following remarks:

> What is incorrect is the idea that the application of a calculus in the grammar of real language correlates it to a reality or gives it a reality that it did not have before.
>
> [...] So what happens when the 6-place relation is found? Is it like the discovery of a metal that has the desired (and previously described) properties (the right specific weight, strength, etc)?
>
> [...] This is all connected with the false concept of logical analysis that Russell, Ramsey and I used to have, according to which we are writing for an ultimate logical analysis of compounds – an analysis which will enable us really to discover a 7-place relation, like an element that really has the specific weight 7.[15]

Against this "looking in the direction of a misleading analogy" he now declares:

> Grammar is for us a pure calculus (not the application of a calculus to reality).[...] There isn't any question here of a connection with reality which keeps grammar on the rails. The "connection of language with reality", by means of ostensive definitions and the like, doesn't make the grammar inevitable or provide a justification for the grammar. The grammar remains a free-floating calculus which can only be extended and never supported.[16]

Having seen that the idea of a correlation, an agreement between language and reality that fixes the degree of freedom of grammar rests on a false analogy, the task for Wittgenstein is no longer one of correcting or revising the *Tractatus*. Too many notions and ideas in the *Tractatus*, including the very dogmatic spirit in which the book is written, are bound up with the idea of a common form of language and reality. He must have felt that the dogmatic spirit of the *Tractatus* was no longer the spirit in which he could do philosophy. It strikes him now as unfair and dishonest.

---

[14]Hilbert (1902).

[15]PG, pp. 311–312. The question of finding the six- or seven-place relation is connected with the problem stated in TLP 5.554–5.5542.

[16]PG, pp. 312–313.

7.3 Grammar and the Axiomatic Method

Despite Hilbert's and Wittgenstein's different views of how the problems of the foundations of mathematics should be dealt with, there are some interesting similarities between Wittgenstein's grammar or method of grammatical investigation and Hilbert's new axiomatic method that Hilbert first used in his work on the foundations of geometry.[17] Hilbert explicitly stated that the aim of his axiomatic method is that of displaying the logical structure of *existing* mathematical and scientific theories, and he was mainly concerned with existing theories of natural science (among which he counted geometry). The feature of this method that Hilbert called *formal*, and opposed to the *genetic* method, made it possible to conceive of the axioms as formal stipulations that 'implicitly define' the meaning of the words and signs involved. This is not far from what Wittgenstein says about the rules of grammar in the beginning of the 1930s. He says for instance that the law of double negation $\neg\neg p = p$ "does not follow from the meaning of the word 'not' but 'constitutes it''.[18] He elaborates this point again with a comparison with geometry: "Geometry no more speaks about cubes than logic does about negation. Geometry defines the *form* of a cube but does not describe it."[19] To express the logical form of something is not to describe it. The rule of double negation is not an assertion about negation, it is a rule for the replacement of one sign with another, or a rule for how we operate with the sign for negation.

But this also shows a difference between Hilbert's and Wittgenstein's methods. Wittgenstein did not want to approach the axioms of geometry as propositions about a certain domain of (unspecified) things or objects, which was Hilbert's approach. Wittgenstein probably felt that this view is based too much on the problematic analogy between mathematical and empirical propositions, and in geometry it tends to invite the view of geometrical propositions as descriptive statements about 'ideal objects'. Euclidian geometry as an axiomatized system was, for Wittgenstein, rather a pure calculus, a calculus for *operating* with signs and expressions, including figures and diagrams, (not for making assertions *about* figures and diagrams!).

Consider for instance the group of axioms in Hilbert's axiomatization of Euclidean geometry that he calls *Axioms of Order*, which have to do with the Euclidean notion of 'between'.[20] There we find the axiom:

---

[17]See Mülhölzer (2005, 2008), for other interesting similarities (and differences) between Hilbert's and Wittgenstein's thinking about issues in the foundations of mathematics. On the whole one might say that Wittgenstein was much more stimulated by Hilbert's ideas than what has been generally recognized among Wittgenstein scholars. This is not to say that Wittgenstein endorsed Hilbert's foundational program but I think one could say that he found it interesting and useful, not least for the philosophical mistakes it contained.

[18]AL, p. 4.

[19]PG, p. 52.

[20]Hilbert (1902, pp. 5–6).

> If A, B, and C are points of a straight line and B lies between A and C, then B lies also between C and A.

Using Wittgenstein's notion of a 'rule of grammar', it is very natural to call this axiom a rule of grammar for the word 'between' in the language of Euclidean geometry, or a constitutive rule for the Euclidean *concept* of lying between. As a rule of grammar it asserts that there is no significant difference between the expressions 'the point B lies between A and C' and 'the point B lies between C and A', in the system of Euclidean geometry. The expressions can be substituted for one another.

This gives us a simple example of what Wittgenstein meant when he was talking about *breaking the spell of notions we are accustomed to*.[21] This rule for the Euclidean notion of 'between' was not stated as an axiom in the original formulation of Euclidean geometry, but was presumably taken for granted by the 'intuitive meaning' of the word 'between' that one was accustomed to, even in non-mathematical discourse, and that one therefore was inclined to think of as a *unique* notion, that cannot be otherwise. As though the word 'between' not only happen to, but *must* signify a symmetric relation; as though symmetry was implicit in the meaning of the word 'between' as an 'intrinsic property'. So we have a simple example of the kind of dogmatism about what a world *must* mean, and with it the temptation to justify this 'must' by reference to some sort of reality, a substantial and non-temporal realm such as the a priori nature of physical space or the a priori forms of spatial intuition.By making the grammar of the Euclidian notion of 'between' explicit as a rule of grammar of a delimited system, the attitude of the uniqueness and immutability of the notion dissolves, and so does the temptation to give its meaning a metaphysical justification.

Another of Hilbert's examples of an axiom that makes explicit something that is usually taken for granted in scientific practice is the axiom of continuity in physics. He states it as follows: "If for the validity of a proposition of physics we prescribe any degree of accuracy whatsoever, then it is possible to indicate small regions, within which the presuppositions that have been made for the proposition may vary freely, without the deviation of the proposition exceeding the prescribed degree of accuracy"[22].

About this axiom Hilbert says that:

> This axiom basically does nothing more than express something that already lies in the essence of experiment; it is constantly presupposed by the physicists, although it has not previously been formulated. [23]

This axiom also expresses what Wittgenstein would call a rule of grammar for the language of experimental physics. When Hilbert gives it the status of an axiom that expresses the essence of experiment, he indicates that this axiom has a function that

---

[21] In *The Blue Book* Wittgenstein writes: "We shall also try to construct new notations, in order to break the spell of those which we are accustomed to." (BB, p. 23).

[22] Hilbert (1918). The quotation is from Ewald (1996, p. 1110).

[23] Hilbert (1918), in Ewald (1996, p. 1110).

is different from the function of the empirical propositions of physics; its role as a rule of grammar is emphasised.

If we disregard Hilbert's (neo-kantian) philosophical conception of what he was doing in his axiomatics, i.e., his idea of "making explicit a deeper layer of axioms", as he expressed it,[24] and instead pay attention to how he was working, it seems to me that one essential feature of Hilbert's method was that he allowed himself to approach theories with the aim of representing them as *autonomous systems*: a theory has been fully axiomatized only when it has been represented as an autonomous system, as a network of concepts *detached from their applications*. Here we have an example of the 'section' Wittgenstein talks about when he says: "[...] looking at mathematics as we do is to make one particular section – to cut pure mathematics off from its application."[25] This approach can obviously be understood as though the axioms in the fully axiomatized system have the status of arbitrary rules of grammar in Wittgenstein's sense. The axioms determine how we operate with the signs.

In a letter to Frege, Hilbert wrote:

> ... every theory is only a scaffolding or schema of concepts together with their necessary relations to one another, and the basic elements can be thought of in any way one likes.[26]

If the 'basic elements', for instance the words 'point', 'line' and 'plane' in geometry, can be thought of 'in any way one likes', they are treated as mere signs in the axiomatic theory in abstraction from the specific meaning the signs might have in some application of the system. The only 'sense' that the signs have within the axiomatic system, is the formal properties and relations conferred upon them by the axioms, and which determine how we operate with the signs in the system.

But Hilbert is here talking about 'what a theory is' from a logical or formal point of view, as a system of 'pure mathematics', as he also sometimes said. He is not talking about geometry as the scientific discipline it is. He did not mean to deny, that geometry is a science of physical space. Hilbert has been called a formalist in a vulgar sense according to which 'mathematics is nothing but games with meaningless signs'. That Hilbert was erroneously described as a formalist in that sense (e.g., by Brouwer, Bourbaki and others) was presumably the result of a misunderstanding of the nature and aim of his axiomatic method: as a methodological approach for displaying logical structure of *existing* scientific theories, where the signs already have meanings (but ones that are in need of clarification), it was misunderstood as a method or technique for the formulation of new abstract axiomatic systems as an end in itself.[27]

---

[24]Hilbert (1918), in Ewald (1996, p. 1109).

[25]LFM, p. 150.

[26]Frege (1980, p. 40).

[27]About this abstract algebraic trend, inspired by Hilbert's axiomatics, Corry (2000, p. 52), writes: "[..] in the years immediately following the publication of the *Grundlagen,* several mathematicians, especially in the USA, undertook an analysis of the systems of abstract postulates for algebraic concepts such as groups, fields, Boolean algebras, etc., based on the application of techniques and conceptions similar to those developed by Hilbert in his study of the foundations of geometry.

There is an analogous misunderstanding of Wittgenstein'sgrammatical method as a 'linguistic philosophy' or an 'ordinary language philosophy' (in the spirit of the so-called Oxford School), rather than a method for conceptual investigation of concepts *we already have and use*, but which are in need of clarification, and where this need is motivated by a specific conceptual puzzle or confusion.

Hilbert was not a formalist about geometry as a scientific discipline, and neither was Wittgenstein.[28] In a discussion with Waismann, Wittgenstein made the following remark: "Geometry is not something self-sufficient, it is brought to completion by physics. It is part of a hypothesis."[29]

7.4 Grammar and the Theory of Relativity

Even if neither Hilbert nor Wittgenstein were formalists, it seems clear to me that they had both found the formalist perspective of mathematics very useful from a methodological point of view in their work on their respective methods for conceptual clarification. The feature of Hilbert's axiomatic method that is most interesting in a comparison with Wittgenstein's method of grammatical investigation was stated by Einstein as follows:

> The progress achieved by axiomatics consists in its having neatly separated the logical-formal from its objective or intuitive content.[30]

This remark should be compared with Wittgenstein's remark, mentioned before: "[...] looking at mathematics as we do is to make one particular section – to cut pure mathematics off from its application." What Wittgenstein describes with the words 'cut off' is what Einstein describes with the word 'separate', if "the objective content" is thought of as something 'given in advance' to which the logical-formal notions are applied (as suggested in the introduction).

There is no evidence that Hilbert showed any interest in this kind of work, and in fact there are reasons to believe that they implied a direction of research that Hilbert did not contemplate when putting forward his axiomatic program. It seems safe to assert that Hilbert even thought of this direction of research as mathematically ill-conceived." I am inclined to agree with Corry here, and it seems to me that the view of Hilbert as a formalist, has been based very much on the view that this abstract algebraic trend was in harmony with aims of Hilbert's axiomatic program. This trend did not have the epistemological and clarificatory aims of Hilbert's axiomatics. Alfred Tarski's conception of metamathematics as 'methodology of the deductive sciences' is also an example of a work in the spirit of this abstract algebraic trend, but which is incompatible with the epistemological aims of Hilbert's program. Van der Waerden's book *Moderne Algebra* is another work in the abstract algebraic trend. It appears to have been an important source of inspiration for the Bourbaki programme.

[28]If Hilbert was a formalist, I think that the most appropriate sense of the word 'formalism' is the one explained by Detlefsen (1993a, b).

[29]WVC, p. 162.

[30]Einstein (1973, p. 233).

Hilbert's axiomatic method (which was also called 'axiomatics') and the new axiomatization of geometry appears to have had a decisive role for the formulation of the theory of relativity. After having explained what the quoted passage means for geometry, Einstein goes so far as to say: "I attach special importance to the view of geometry which I have just set forth, because without it I should have been unable to formulate the theory of relativity".[31]

It was against this background that Einstein made the famous statement:[32]

> Insofar as the theorems of mathematics are related to reality, they are not certain; and insofar as they are certain, they are not related to reality.

Through the separation (as Einstein calls it), the traditional notion of an axiom of mathematics (as a proposition or judgement expressing a necessary and universal truth about some subject matter) was transformed into one of the rules defining a system, a rule that we follow in that system, a rule having only an internal necessity in the system. There is no sense or meaning that the signs *must* have except within a certain system. We are free to change the sense of signs by changing the axioms, thereby getting a different mathematical system. The non-Euclidian geometries were constructions of such new mathematical systems.

One feature of the axiomatic method that Einstein stressed in particular (and perhaps more than what Hilbert would have wanted) was our freedom to change the meaning of signs by changing the rules for their employment. He thereby avoided the attitude that words such as 'space', 'time', and 'simultaneity' have a traditional, absolute and unique core-meaning that these words *must* have in any use. He even spoke of "the fictitious character of the fundamental principles", which means that the status of 'fundamental principles' was transformed from that of 'eternal laws of nature' to arbitrary rules of grammar (even if Einstein did not express himself in those words), where the essential thing is how we operate with the notions, for instance, how we actually determine the truth of a statement about the simultaneity of two events. This freedom of the invented conceptual systems is presented by Einstein as the decisive difference between the old philosophy of nature and the way of thinking that results in the theory of relativity:

> The natural philosophers of those days were, on the contrary, most of them possessed with the idea that the fundamental concepts and postulates of physics were not in the logical sense free inventions of the human mind but could be deduced from experience by "abstraction" – that is to say by logical means. A clear recognition of the erroneousness of this notion really only came with the theory of relativity.[33]

---

[31]Einstein (1973, p. 235).

[32]In his talk "Geometry and Experience" given at the Berlin Academy of Sciences in 1920. The 'certainty' that Einstein speaks of here is presumably the certainty that comes from the *necessity* of a proved mathematical proposition.

[33]Einstein (1973, p. 273).

Wittgenstein had surely learned about Einstein's work, even if not in some great detail.[34] Einstein's views of geometry were discussed in the meetings of the *Vienna Circle* when Wittgenstein was participating. He was no doubt impressed by certain features of Einstein's work, and in particular the ones I have mentioned. In the *Philosophical Grammar* he writes with approval: "Einstein: how a magnitude is measured is what it is"[35]. And already in 1929–1930, in an investigation of the relationship between expectation and fulfilment, Wittgenstein makes the following explicit reference to relativity theory:

> My whole idea is always that if someone could see the expectation he would necessarily be seeing what was expected. [...]
>
> The anticipation is as it were itself language and cannot go outside itself. (In the 'not being able to go outside of itself' lies the similarity of my views and that of relativity theory.)[36]

The 'not being able to go outside of itself' is a main feature of the autonomy of a calculus, a language game or (what Wittgenstein also calls) a system. Relativity theory is also mentioned in a context in which Wittgenstein discusses Frege's assertion sign. He concludes his argument by saying: "The way of thinking [*Denkbewegung*] that is necessary here is again the way of thinking of relativity theory."[37]. And as late as 1943–1944, we find him saying:

> Following according to the rule is FUNDAMENTAL to our language-game. It characterizes what we call description.
>
> This is the similarity of my treatment with relativity-theory, that it is so to speak a consideration about the clocks with which we compare events.[38]

There is a clear affinity between Wittgenstein's uses of the idea of a system (game, calculus) and Einstein's uses of the idea of a system of reference.

By pointing out these similarities between Hilbert's axiomatic method, Einstein's relativity theory and Wittgenstein's original conception of grammar, I do not want to suggest that these three approaches had the same motivation and purpose. One might say that they all aimed at achieving conceptual clarity, but Hilbert's notion of clarity, in contrast to Wittgenstein's, was strongly conditioned by the endeavour towards the progress of modern science. Clarity for Hilbert, as well as for Einstein, was no doubt a means to that end, while for Wittgenstein clarity was an end in itself. For Hilbert as for Frege and Russell, logic was a scientific discipline, and Hilbert used his axiomatic method not for the purpose of philosophical clarification in Wittgenstein's sense, but rather for contributing to the development of the modern algebraic trend in mathematics in the wake of Dirichlet, Riemann, Dedekind, Weierstrass and Cantor, among others.

---

[34]See Penco (2010).

[35]PG, p. 459.

[36]MS 108, p. 271. Quoted from Hilmy (1987, p. 146). The sentence on relativity theory is repeated in *The Big Typescript* (TS 213, pp.355–356).

[37]MS 109, p. 199.

[38]RFM VI, 28.

If their work does not derive from a common motive or aim, are not the similarities I have suggested quite superficial and far-fetched? – Well, what I want to suggest is that in order to understand, appreciate and approve of this modern orientation in mathematics and physics, Hilbert felt (like Mach, Boltzmann, Hertz and Einstein) that *it was necessary to overcome prevailing dogmatic attitudes* to traditional notions and methods that many mathematicians and scientists were accustomed to. It is in the effort to overcome residues from such traditional attitudes, in a kind of modernist spirit, that we can find an obvious affinity between Hilbert's and Wittgenstein's work. We find this spirit, for instance, in their rejection of the kantian *apriorism*. Wittgenstein asserts already in the *Tractatus* that "Whatever we can describe at all could be other than it is. There is no *apriori* order of things."[39] And about Newton's notion of absolute time, Hilbert asserts in a polemic tone that: "Kant, the critical philosopher, is here not critical at all, for he simply accepted Newton. Only Einstein definitively liberated us from this prejudice [...] the Kantian theory of the *apriori* still contains anthropological dross from which it must be liberated [...]".[40]

7.5 Mental Verbs and the Method of Ideal Elements

Let me take another example that shows, I think, that Wittgenstein found inspiration in modern mathematics in his grammatical investigations of notions outside the field of mathematics. But this does not mean that he was doing or suggesting a certain kind of applied mathematics as in physics or mathematical psychology; no, he is using certain *analogies* in order to "break the spell" of notions we are accustomed to.

There is a passage in the *Blue Book*[41] where Wittgenstein reminds us that a psychological verb such as, for instance, 'being afraid of' are used both in a transitive and an intransitive way. We can be afraid of something or someone, but it also happens that we just have a feeling of being afraid without being able to say what we are afraid of. Wittgenstein suggests, as a kind of thought experiment, that we could introduce a new, more simple and unified, terminology according to which psychological verbs are always used transitively. According to this proposal, when someone just feels afraid, without knowing what she is afraid of, it is always something that she is afraid of even if she can't say what it is. In the same vein he

[39]TLP 5.634.

[40]But Hilbert did not reject Kant's apriorism altogether. I think he was very serious when he continued the same sentence by saying: "afterwards only the *apriori* attitude is left over which also underlies pure mathematical knowledge: essentially that is the finite attitude which I have characterized in several works." (Hilbert (1930), the quotation is from Ewald 1996, p. 1163). Commentators of Hilbert's foundational program tend to ignore this remark and similar ones. I don't think that he insisted on this point so often only in order to "pay a due tribute to the towering figure of his fellow Königberger", Corry (2006a, p. 159).

[41]BB, pp. 57–66. See also PR 58.

suggests that one could imagine situations in which it would be practical to use a terminology in which we could speak of unconscious pain, such as 'unconscious toothache'. He claims that this is the sense in which psychoanalysis talks of unconscious thoughts, unconscious acts of volition, etc. and that there is nothing wrong with such a terminological convention as long as one is capable of going through with it without confusing it with the terminology we are accustomed to. Such confusion would come out in philosophical puzzles such as "Does unconscious pain really exist?", "How is unconscious toothache possible at all?"

There is a striking analogy between Wittgenstein's construction of these new terminologies and the method for construction of new notational systems in the modern mathematics that Hilbert was using and promoting. Hilbert called it *the method of ideal elements* and one of the most well-known examples is the system of complex numbers, in which the complex-imaginary numbers are the ideal elements while the complex-real numbers are the real elements. One point and advantage of this notational system is that it simplifies theorems about the existence and number of roots of an equation. Another example is a notational system of geometry in which infinitely long lines and points at infinity are introduced as ideal elements, and the point of this construction is that it simplifies the algebraic treatment of geometry.[42]

In the similarity that I am suggesting here (between these new mathematical notational systems and Wittgenstein's suggested psychological notational systems), 'unconscious pains', 'unconscious thoughts and feelings', would be the 'ideal elements', and as there may arise philosophical puzzles if one confuses the new and the ordinary psychological terminologies, so there may arise analogous puzzles about the existence and possibility of imaginary numbers and points at infinity in the philosophy of mathematics, as Hilbert was aware of. That's why he called them *ideal* elements, as opposed to real entities. The use of the method of ideal elements was actually a central idea in Hilbert's foundational program, not, however, as a mere analogy, but for defining his conception of metamathematics which is a form of mathematics.

Wittgenstein uses this analogy when he deals with the problem of solipsism in the *Blue Book*. Rather than just dogmatically rejecting the solipsist's statement "only I have real pain" as nonsense or absurd, he proposes that we can make sense of the statement, by taking it as a proposal to *a new grammar*, i.e., as a grammatical rule for a new terminology for the use of mental words, a terminology in which the solipsist is given a central position. The solipsist's confusion appears when he thinks that he can justify his new terminology as being 'more correct' than our ordinary use of mental words – as though the new grammar is more 'true to reality'. But then he is conflating the two terminologies, and takes a grammatical statement to be a kind of material statement.

It is not unlikely that Wittgenstein was inspired by what Hilbert called the method of ideal elements in proposing these psychological notational systems, but of course

[42]Hilbert (1926, p. 195).

this method was not Hilbert's invention. Wittgenstein may have learned it from someone else.[43] My main interest here is, however, the *ideas* in modern mathematics that inspired him.

7.6 Wittgenstein and Hertz

As methods of conceptual clarification, there are important differences between Hilbert's axiomatic method and Wittgenstein's grammar due to the fact that Wittgenstein was neither a mathematician nor a scientist, and as a philosopher he did not share Hilbert's neo-kantian outlook.[44] But there seems to be enough similarities to raise the question: Had Wittgenstein read Hilbert? He seems to have read several of Hilbert's texts on the foundations of mathematics that were used in the discussions within the *Vienna Circle* that Wittgenstein took part in. Hilbert's axiomatic method was fashionable at the time, not least through Einstein's work, so Wittgenstein was certainly also familiar with it from conversations with Waismann, Schlick, Ramsey and others. But I think that the most important explanation of the similarities is that Hilbert and Wittgenstein had common sources of inspiration in their work on conceptual clarification: Ludwig Boltzmann and in particular Heinrich Hertz. Hilbert was influenced, in several respects, by Heinrich Hertz' book "The Principle of Mechanics" in the development and use of his axiomatic method in his work in mathematics and theoretical physics, as has been shown in great detail by Leo Corry.[45] That Wittgenstein was influenced by Hertz is more well-known since the reference to Hertz' book is one of the very few references there are in the *Tractatus*. But the extent of this influence tends to be underestimated by many Wittgenstein scholars. I am inclined to agree with Gordon Baker and others, who have argued that the influence from Hertz is present also in Wittgenstein's later philosophy.[46]

---

[43]Jahnke (1993, p. 279) reports that even the romantic poet Novalis alludes to the method of ideal elements in an attempt to characterise the mathematical genius. Another somewhat surprising allusion to modern mathematics is the following, written by Wittgenstein in his diary in 1947: "Weierstrass introduces a string of new concepts to bring about order in the thinking about the differential calculus. And in that way on the whole, it seems to me, I must bring about order in psychological thinking through *new* concepts. (That it concerns a calculus in the first case, but not in the second, is *not* important.) (MS 135, p. 115–116, my translation).

[44]As the preface to the *Philosophical Remarks* indicates, Wittgenstein did not share the strong faith in modern science that Hilbert often expressed.

[45]Corry (2006b).

[46]Baker (1988). See also Barker (1980), Wilson (1989), Visser (1999) and Kjaergaard (2002).

In November 1946 Wittgenstein spoke to the Cambridge Moral Science Club. His talk had the title *Philosophy*, and was occasioned by a talk three weeks earlier by Karl Popper in which Popper criticized Wittgenstein as the leading figure of the "Cambridge School of linguistic philosophy". In a letter to G.E. Moore Wittgenstein described the content of his talk as follows: "I'm giving a talk, roughly, on what I believe philosophy is, or what the method of philosophy is" (PPO, p. 338).

That Wittgenstein intended to have a quotation from Hertz' book as an epigraph to the first version of the *Philosophical Investigations* lends support to this view. The quotation from Hertz was the following:

> When these painful contradictions are removed, the question as to the nature of force will not have been answered; but our minds, no longer vexed, will cease to ask illegitimate questions.[47]

The expression 'painful contradictions' refers to controversies about the nature of the concept of force in Newtonian mechanics, a basic notion of classical mechanics. As Peter Barker points out, the fundamental mistake of Newton's defenders was according to Hertz to think that every term in a scientific theory corresponds to an entity in the world.[48] Hertz argued that the question of the nature of force is confused since it is based on the attitude that there is a basic content that the concept of force must have and which is justified by physical reality; a content which is still waiting to be found. But Hertz argued that the confusion "must lie in the unessential characteristics which we have ourselves arbitrarily worked into the essential content given by nature." The problems do not relate to the content of the notion, "but only to the form in which the content is represented"[49]. Hertz showed how to reformulate classical mechanics without using force as a basic notion at all. No entity in the world corresponds to the word 'force'. Hertz calls it an 'empty relation' or an 'idle wheel' – a metaphor that Wittgenstein used in his writings even in his later work.

What Wittgenstein learned from Hertz was not only the general idea of resolving philosophical puzzles by dissolving them; he was also influenced by the way in which Hertz accomplished this through reformulation and rearranging notions and expressions of classical mechanics. Hertz expressed this aspect of his work as follows:

> ... the existing defects are only defects in form; [...] all indistinctness and uncertainty can be avoided by suitable arrangement of definitions and notations, and by due care in the mode of expression.[50]

In the *Big Typescript* Wittgenstein writes:

> In my way of doing philosophy, its whole aim is to give an expression such a form that certain disquietudes disappear. (Hertz)[51]

According to the Minutes of the meeting, Wittgenstein was not happy about the description of his philosophy as 'linguistic philosophy'. He quoted Hertz with great approval, and presented Hertz as his main source of inspiration for his view of philosophical problems and his method for dealing with them. He also mentioned Mach. But no other philosopher, not even Frege, appears to have been mentioned in his talk. (PPO, p. 342)

[47] Hertz (1956, pp. 7–8).

[48] Barker (1988, p. 245).

[49] Hertz (1956, p. 8).

[50] Hertz (1956, p. 9).

[51] MS 213, p. 421.

The idea of the autonomy of logical structure makes this way of working possible since it gives us control over our representations and freedom to change and rearrange the conceptual system. It was clearly inspired by Hertz' view of the autonomy of the conceptual system of mechanics, and by Hertz' general view of a scientific theory as an image, a symbolic system, which is an autonomous entity in its formal aspect, independent of the empirical phenomena it is used to explain. By "the principles of mechanics", will be meant, says Hertz, "any selection of such and similar propositions, which satisfies the requirement that the whole of mechanics can be developed from it by purely deductive reasoning *without any further appeal to experience*.[52] This view is obviously echoed in Hilbert's idea of a theory as a network of concepts, connected by purely logical relationships. And the feature of Hilbert's axiomatics that Einstein gives prominence to (i.e., that it has "separated the logical-formal from its objective or intuitive content") is suggested already in Hertz' work.

Through the influence from Hertz, the autonomy-idea is already present in the *Tractatus* notion that the logic of propositions does not represent, that logical syntax must not refer to the meaning (*Bedeutung*) of expressions, and, in general, the idea that 'Logic must take care of itself.'[53]

But there is also one respect in which one might say that Wittgenstein, in his early philosophy, was too much influenced, or even mislead, by Hertz' work. A scientific theory for Hertz is an image, a symbolic system, which is an autonomous entity *only in its formal aspect*. It is independent of the empirical phenomena it is used to explain only in its formal or logical aspect. Classical mechanics is, however, a theory about empirical phenomena that are independently given. It is an arrangement designed to serve a purpose, namely representing and explaining experimental observations, and like cookery (which is also designed to serve a purpose) it is in that respect not autonomous. The requirements of the conceptual system of mechanics that Hertz calls its *correctness* and its *appropriateness* concern ways in which the system serves its purpose. In his early philosophy Wittgenstein seems to have been led to think that there must be something similar to these requirements even for human language in general. In the *Tractatus*, the requirements take the form of the logical structure and the multiplicity that is common to language and reality.[54] As late as in the lent term of 1930 Wittgenstein was still under the influence of Hertz on this point. He tended to see human language as a system of representation and said that "grammar is not entirely a matter of arbitrary choice...". But in less than

---

[52]Hertz (1956, p. 4), my emphasis.

[53]I agree with Marie McGinn when she writes: "One of the central themes of the *Investigations* is to try to show that, insofar as a sentence is used to express a grammatical proposition, it does not represent (cannot be compared with reality for truth or falsity): there is nothing that grounds or justifies grammar. [...] the concern to reveal the autonomy of grammar echoes Wittgenstein's earlier concern to show that logic does not represent – that 'logic takes care of itself' – and it represents a deep continuity of philosophical purpose". McGinn (2006, p. 295).

[54]See Barker (1980), for an elaboration of this similarity between the *Tractatus* and Hertz' *Principes.*

one year, when he had got clearer about the autonomy of the grammar of ordinary language, he rejects this idea as mistaken.

One might think that Hertz' book was not a philosophical work at all but a scientific work, a treatise on mechanics, but it is clear that Hertz himself saw its main purpose as philosophical. In the preface he writes:

> What I hope is new, and to this alone I attach value, is the arrangement and collocation of the whole – the logical or philosophical aspect of the matter.

It was obvious to Wittgenstein that Hertz' idea of how to remove 'painful contradictions' that compel us to raise questions of the form "What is the nature of X?", has a much wider applicability than to problematic notions of mechanics. But not all questions of this form are raised as painful puzzles or mysteries. Hertz asks rhetorically: "Now, why is it that people never in this way ask what is the nature of gold, or what is the nature of velocity"[55].

In traditional philosophy, philosophical mysteries have often been dealt with by persuading oneself to have found *the* final and definitive answer to the question of the nature or essence of X. But many of the answers to such questions that have been proposed in the past can be seen to depend on specific forms of dogmatism, and history shows that these do not tend to be final and definitive. Hertz' idea that the answer we need is *not* really an answer to a question of this form, but an answer that makes the question disappear, made it possible for Wittgenstein to make sense of the idea that the clarity that philosophy is aiming at is *complete* clarity.[56]

An important part of Hertz' account is also his description of how the problematic nature of an alleged fundamental notion of mechanics such as force is experienced: with painfulness, doubt, embarrassment, disquietude. He mentions "the experience that it is exceedingly difficult to expound to thoughtful hearers the very introduction to mechanics without being occasionally embarrassed, without feeling now and then tempted to apologize, without wishing to get as quickly as possible over the rudiments, and on to examples which speak for themselves."[57] This must have made a strong impression on Wittgenstein. He could surely recognize similar experiences in lots of situations as a teacher of philosophy. Hertz' message is of course that we must take our conceptual sensitivity seriously, even when it offends the authority of tradition.

Acknowledgements This article is a revised and abridged version of my original article *Le "Wittgenstein-intermédiaire" et les mathématiques modernes* to be published in the Canadian journal *Philosophiques*. The article appears here by permission of the editors of *Philosophiques*. I am indebted to Kim-Erik Berts, Juliet Floyd and Kim Solin for helpful comments on an earlier version of this article.

[55]Hertz (1956, p. 7).

[56]PI 132.

[57]Hertz (1956, pp. 6–7).

References

Barker, P. 1980. Hertz and Wittgenstein. *Studies in History and Philosophy of Science* 11: 243–56.

Baker, G. 1988. *Wittgenstein, Frege and the Vienna circle*. Oxford: Basil Blackwell.

Benacerraf, P., and Putnam, H. (eds.). 1983. *Philosophy of mathematics, selected readings*, 2nd ed. Cambridge: Cambridge U.P.

Corry, L. 2000. The empiricist roots of Hilbert's axiomatic approach. In *Proof theory: History and philosophical significance*, ed. Vincent F. Henricks et al., 35–54. Dordrecht: Kluwer.

Corry, L. 2006a. Axiomatics, empiricism, and Anschauung in Hilbert's conception of geometry: Between arithmetic and general relativity. In *The architecture of modern mathematics: Essays in history and philosophy*, ed. J.J. Gray and J. Ferreirós, 155–176. Oxford: Oxford University Press.

Corry, L. 2006b. The origin of Hilbert's axiomatic method. In *The genesis of general relativity, vol. 4. Theories of gravitation in the twilight of classical physics: The promise of mathematics and the dream of a unified theory*, ed. Jürgen Renn et al., 139–236. New York: Springer.

Detlefsen, M. 1993a The Kantian character of Hilbert's formalism. In *Proceedings of the 15th International Wittgenstein-Symposium*, ed. J. Czermak, 195–205. Vienna: Verlag Hölder-Pichler-Temsky.

Detlefsen, M. 1993b. Hilbert's formalism. *Revue Internationale de Philosophie* 47: 285–304.

Einstein, A. 1973. *Ideas and opinions*. London: Souvenir Press.

Epple, M. 2003. The end of the science of quantity: Foundations of analysis, 1860–1910. In *A history of analysis*, ed. H.N. Jahnke, 291–323. Providence/London: American Mathematical Society/London Mathematical Society.

Ewald, W. 1996. *From Kant to Hilbert. Readings in the foundations of mathematics*. Oxford: Oxford University Press.

Frege, G. 1980. *Philosophical and mathematical correspondence*, ed. B. McGuinness. Chicago: University of Chicago Press.

Gerrard, S. 1991. Wittgenstein's philosophies of mathematics. *Synthese* 87: 125–142.

Gray, J.J. 2006. Modern mathematics as a cultural phenomenon. In *The architecture of modern mathematics*, ed. J. Ferreirós and J.J. Gray, 371–396. New York: Oxford University Press.

Hertz, H. 1956. *The principles of mechanics presented in a new form*. New York: Dover. English translation of *Die Prinzipen der Mechanik in neuem Zusammanhange dargestellt*, Leipzig, 1894.

Hilbert, D. 1902. *Die Grundlagen der Geometrie*. Trans. as the foundations of geometry, by E.J. Townsend. Chicago: The Open Court Publishing Company.

Hilbert, D. 1918. Axiomatisches Denken. *Mathematische Annalen* 78: 405–415. English translation as "Axiomatic thought" in Ewald (1996).

Hilbert, D. 1926. Über das Unendliche. *Mathematische Annalen* 95: 161–190. English translation as "On the infinite" in Benacerraf, P. and Putnam, H., (1983).

Hilbert, D. 1930. Naturerkennen und Logik In Hilbert: *Gesammelte Adhandlungen*, ed. Dritter Band. New York: Chelsea Publishing Company, 1965. English translation as "Logic and the knowledge of nature" in Ewald (1996, 1157–1165).

Hilmy, S.S. 1987. *The later Wittgenstein, the emergence of a new philosophical method*. Oxford: Basil Blackwell.

Jahnke, H.N. 1993. Algebraic analysis in Germany, 1780–1840: Some mathematical and philosophical issues. *Historia Mathematica* 20: 265–284.

Jahnke, H.N. (ed.). 2003. *A history of analysis*. Providence: American Mathematical Society.

Kjaergaard, P.C. 2002. Hertz' and Wittgenstein's philosophy of science. *Journal for General Philosophy of Science* 33: 121–149.

McGinn, M. 2006. *Elucidating the tractatus, Wittgenstein's early philosophy of logic & language*. Oxford: Oxford University Press.

Mehrtens, H. 1990. *Moderne – Sprache – Mathematik: Eine Geschichte des Streits um die Grundlagen der Disziplin und des Subjekts formaler Systeme*. Frankfurt: Suhrkamp.
Mülhölzer, F. 2005. "A mathematical proof must be surveyable" – What Wittgenstein meant by this and what it implies. *Grazer Philosophische Studien* 71: 57–86
Mülhölzer, F. 2008. Wittgenstein und der Formalismus. In *"Ein Netz von Normen": Wittgenstein und die Mathematik*, ed. M. Kroß, 107–148. Berlin: Parerga Verlag.
Penco, C. 2010. The influence of Einstein on Wittgenstein's philosophy. *Philosophical Investigations* 33: 360–379.
Visser, H. 1999. Boltzmann and Wittgenstein on how pictures became linguistic. *Synthese* 119: 135–156.
Wilson, A.D. 1989. Hertz, Boltzmann and Wittgenstein Reconsidered. *Studies in History and Philosophy of Science* 20: 245–263.
Wittgenstein, L. 1958a. [BB]: *The Blue and Brown books: Preliminary studies for the 'philosophical investigations'*, ed. R. Rhees. Oxford: Blackwell.
Wittgenstein, L. 1958b. [PI]: *Philosophical investigations*, ed.G.E.M. Anscombe and R. Rhees, 2nd ed. Trans. G.E.M. Anscombe. Oxford: Blackwell.
Wittgenstein, L. 1969. [TLP]: *Tractatus Logico-philosophicus*. Trans. D.F. Pears and B.F. McGuinnes. London: Routledge & Kegan Paul.
Wittgenstein, L. 1974. [PG]: *Philosophical grammar*, ed. R. Rhees. Trans. A.J.P. Kenny. Oxford: Blackwell.
Wittgenstein, L. 1975. [PR]: *Philosophical remarks*, ed.R. Rhees, 2nd ed. Trans. R. Hargreaves and R. White. Oxford: Blackwell.
Wittgenstein, L. 1976. [LFM]: *Wittgenstein's lectures on the foundations of mathematics Cambridge, 1939*, ed. Cora Diamond. Hassocks, Sussex: The Harvester Press, Ltd.
Wittgenstein, L. 1978. [RFM]: *Remarks on the foundations of mathematics*, ed. G.H. von Wright, R. Rhees and G.E.M. Anscombe, 3rd ed. Trans. G.E.M. Anscombe. Oxford: Blackwell.
Wittgenstein, L. 1979a. [AL]: *Wittgenstein's lectures, Cambridge, 1932–1935: From the notes of Alice Ambrose and Margaret Macdonald*, ed. Alice Ambrose. Chicago: University of Chicago Press.
Wittgenstein, L. 1979b. [WVC]: *Ludvig Wittgenstein and the Wienna circle: Conversations recorded by Friedrich Waismann*, ed. B.F. McGuiness. Trans. J. Schulte and B. McGuiness. Oxford: Blackwell.
Wittgenstein, L. 1980a. [DL]: *Wittgenstein's lectures, Cambridge 1930–1932, from the notes of J. King and D. Lee*, ed. D. Lee. Oxford: Blackwell.
Wittgenstein, L. 1980b. [RPP I]: *Remarks on the philosophy of psychology*, vol. I, ed. G.E.M. Anscombe and G.H. von Wright. Trans. G.E.M. Anscombe. Oxford: Blackwell.
Wittgenstein, L. 1998. [CV]: *Culture and value*, ed. G.H. von Wright. Trans. Peter Winch. Oxford: Blackwell.
Wittgenstein, L. 2003. [PPO]: *Ludwig Wittgenstein, public and private occasions*, ed. J.C. Klagge and A. Nordmann. Oxford: Rowman & Littlefield Publishers.

Chapter 8
Primitive Recursive Arithmetic and Its Role in the Foundations of Arithmetic: Historical and Philosophical Reflections*

In Honor of Per Martin-Löf on the Occasion of His Retirement

William W. Tait

1. Skolem tells us in the Concluding Remark of his seminal paper on primitive recursive arithmetic (*PRA*), "The foundations of arithmetic established by means of the recursive mode of thought, without use of apparent variables ranging over infinite domains" (1923), that the paper was written in 1919 after he had studied Whitehead and Russell's *Principia Mathematica* and in reaction to that work. His specific complaint about the foundations of arithmetic (i.e. number theory) in that work was, as implied by his title, the essential role in it of logic and in particular quantification over infinite domains, even for the understanding of the most elementary propositions of arithmetic such as polynomial equations; and he set about to eliminate these infinitary quantifications by means of the "recursive mode of thought." On this ground, not only polynomial equations, but all primitive recursive formulas stand on their own feet without logical underpinning.
2. Skolem's 1923 paper did not include a formal system of arithmetic, but as he noted in his 1946 address, "The development of recursive arithmetic" (1947), formalization of the methods used in that paper results in one of the many equivalent systems we refer to as *PRA*. Let me stop here and briefly describe one such system.

---

*This paper is loosely based on the Skolem Lecture that I gave at the University of Oslo in June, 2010. The present paper has profited, both with respect to what it now contains and with respect to what it no longer contains, from the discussion following that lecture.

W.W. Tait (✉)
Professor Emeritus, Department of Philosophy, University of Chicago
e-mail: williamtait@mac.com

P. Dybjer et al. (eds.), *Epistemology versus Ontology*, Logic, Epistemology, and the Unity of Science 27, DOI 10.1007/978-94-007-4435-6_8,

We admit the following *finitist types*[1] of objects:

$$\mathbb{N} \qquad\qquad A \times B$$

when A and B are finitist types. We also admit as *terminal types* the types

$$A \to B$$

of functions from the finitist type A to the finitist type B. The terms of each type are:

Variables $x^A = x$ of each finitist type A

$$x : A.$$

$$0 : \mathbb{N} \qquad\qquad t : \mathbb{N} \Rightarrow t' : \mathbb{N}$$

Primitive recursion

$$s : A \,\&\, t : \mathbb{N} \times A \to A \,\&\, r : \mathbb{N} \Rightarrow PR(s,t,r) : A$$

where A is a finitist type, with the defining axioms

$$PR(s,t,0) = s \qquad\qquad PR(s,t,r') = t(r, PR(s,t,r)).$$

Corresponding to the cartesian product $A \times B$ we have

$$s : A \,\&\, t : B \Rightarrow (s,t) : A \times B$$
$$p : A \times B \Rightarrow pL : A \,\&\, pR : B$$

with the defining axioms

$$(s,t)L = s \qquad\qquad (s,t)R = t$$

and

$$(pL, pR) = p.$$

Finally, for terminal types $A \to B$ we have

$$s : A \,\&\, t : A \to B \Rightarrow ts : B$$

and

$$x : A \,\&\, t : B \Rightarrow \lambda x.t : A \to B$$

[1]The finitist types are called the *finitist types of the first kind* in "Finitism" (Tait 1981). The finitist types of the second kind are the types, corresponding to a constant equation, of the computations proving the equation. We need not discuss these here.

with the defining axiom (lambda-conversion)

$$(\lambda x.t)s = t[s/x]$$

The formulas are built up from equations between terms of the same finitist type by means of implication. The logical axioms and rules of inference are those of identity, implication and mathematical induction

$$\frac{\phi(0) \qquad \phi(x) \rightarrow \phi(x')}{\phi(t)}$$

It is easy to show that

$$0 = 0' \rightarrow \phi$$

for any formula ϕ. So we may abbreviate

$$\neg\phi := \phi \rightarrow 0 = 0'.$$

The axiom of double negation elimination

$$\neg\neg\phi \rightarrow \phi$$

is then also derivable.

The Dedekind axioms

$$\neg 0 = t' \qquad\qquad s' = t' \rightarrow s = t$$

express that the iteration of successor, starting with 0, is 'free'—there are no loops. It is in virtue of that definition by primitive recursion is valid. Indeed, given definition by primitive recursion, we can derive the Dedekind axioms. Define *sgn* and *pred* by

$$sgn\, 0 = 0 \qquad sgn\, t' = 0' \qquad pred\, 0 = 0 \qquad pred\, t' = t.$$

Then

$$0 = t' \rightarrow sgn\, 0 = sgn\, t' \qquad\qquad s' = t' \rightarrow pred\, s' = pred\, t'.$$

Using mathematical induction one easily proves the *uniqueness of definition by primitive recursion*: let $s : A, t : \mathbb{N} \times A \rightarrow A$ and $u : \mathbb{N} \rightarrow A$. Then

$$\frac{u0 = s \qquad un' = t(n, un)}{ur = R(s, t, r)}$$

The converse is also true: mathematical induction can be derived from uniqueness of primitive recursion. (See Goodstein 1945; Skolem 1956; Goodstein 1957.)

3. Concerning his general philosophy of mathematics or at least of arithmetic, Skolem, again in the Concluding Remark of Skolem (1923), expresses his allegiance to Kronecker's finitism. In particular, he refers to *Kronecker's principle* that a mathematical definition (*Bestimmung*), say of a numerical function or property, is genuine if and only if it supplies an algorithm for determining its values. In its consequent rejection of infinite quantifiers, this position also stands with Weyl's later 'intuitionism' (1921) and Hilbert's still later 'methodological' finitism (1922, 1923, 1926).

   It is interesting to note that Hilbert, too, studied the foundations of arithmetic in *Principia Mathematica*[2] and, at roughly the same time as Skolem, rejected it, but for a quite opposite reason, namely because with the axioms of infinity and of reducibility, it could no longer claim to be a *logical* foundation. Indeed, it is just as subject to the demand for a consistency proof as the axiomatic theory of numbers, to which Hilbert then returned. The lingering problem for him was to avoid the circle of a proof of consistency of the axiomatic theory that is itself founded on an axiomatic theory. Hilbert and Bernays felt that they solved that problem around 1922 by restricting the methods of proof theory, in which consistency is to be established, to finitist mathematics.
4. For most of us, constructivists or otherwise, arithmetic does not end with Kronecker's principle or with PRA. As Gödel (1958) has taught us, even without the introduction of logic we can extend PRA non-conservatively but constructively by allowing definition by recursion of higher type objects, such as numerical-valued functions of numerical functions, numerical-valued functions of these, etc.[3] If in the definition of the formal system PRA above we abolish the distinction between finitist and terminal types and admit $A \to B$ as well as $A \times B$ to be a type when A and B are, then the resulting system is Gödel's theory T of primitive recursive functions of finite type, provided we add the axioms (so-called *η-conversion*)

$$\lambda x.tx = t$$

   when $t : A \to B$ and t does not contain x, and the rule of inference

$$\frac{\phi \to s = t}{\phi \to \lambda x.s = \lambda x.t}$$

   when x is not in ϕ.

   Actually, this is not quite Gödel's theory, but it agrees with that theory in its arithmetic consequences (i.e. theorems ϕ containing only equations between terms of type $\mathbb{N}$). Objects of types $A \to B$ are interpreted by Gödel to be

[2]Concerning Hilbert's flirtation with logicism, see Hilbert (1918), his lecture "Prinzipien der Mathematik" in his 1917–1918 lectures on logic Hilbert (2011, Chap. 1) and Mancosu (1999), Sieg (1999a). Concerning his rejection of it, see his lecture "Probleme der Mathematischen Logik" in Hilbert (2011, Chap. 2).

[3]The fact that non-primitive recursive functions such as the Ackermann function, defined by twofold nested recursions, are definable using primitive recursion of higher type was already shown by Hilbert in (1926).

computable (*bereckenbar*) functions from A to B and he interprets equations between objects of this type as expressions of 'intensional' or 'definitional equality'. But, for example, one might prove the equation $sx = tx$, where x is of type $\mathbb{N}$ by induction on x. By the above inference we then have $\lambda xsx = \lambda xtx$ and so, by η-conversion, $s = t$. An application of this yields the higher-type form of uniqueness of primitive recursive definition:

$$\frac{r0 = s \qquad rn' = t(n, rn)}{r = \lambda nPR(s, t, n)}$$

Certainly a notion of equality admitting this inference is not decidable.[4]

But Gödel's conception of T is not entirely satisfactory. Although T is a quantifier-free theory, on his interpretation of it the (domain of objects of) type $A \rightarrow B$ is defined only by means of quantification over infinite domains; namely, a computable function of type $A \rightarrow B$ is one which, applied to any computable object of type A yields a computable object of type B. If one takes the function to be given by a Turing machine, then the statement that it is computable becomes an arithmetic statement whose complexity grows with the type of the function.[5] It seems preferable to accept the notion of function as *sui generis*, to interpret $A \rightarrow B$ simply as the domain of functions from A to B, and to understand equations between objects of such a type to mean equality in the usual sense of extensional equality of functions. What makes T constructive is not that it concerns special domains of 'constructive objects' of higher type, but rather that it treats the higher types constructively.[6] This means that Kronecker's principle will be violated and the law of double negation elimination (i.e. classical propositional logic) will not in general be valid for formulas containing equations between terms of higher type; but Gödel's application of T in the *Dialectica* interpretation does not depend on that anyway.

5. Likewise, the requirements of a constructivism more liberal than Kronecker's can be met even with the (non-conservative) introduction of infinite quantification, as in Heyting arithmetic. In this case again decidability of formulas ($\neg\neg\phi \rightarrow \phi$) cannot in general be proved, as Kronecker demands; but it is not assumed as an axiom either. As is well-known, this constructive extension of Kronecker's finitism is from another point of view—a proof-theorist's point of view—just a generalization of Gödel's. For the basic notion of constructive

[4] The difficulty of dealing with equations of higher type was avoided by Spector in (1962) by restricting equations to those between numerical terms and restricting formulas to equations. Equations $s = t$, standing alone, between terms of type $A \rightarrow B$ are taken to be abbreviations for $sx = tx$, where the variable x of type A does not occur in s or t. $s = t \rightarrow u = v$ is then taken to be an abbreviation for $(1 - sgn|s - t|) + sgn|u - v| = 0$, where $|x - y|$ denotes absolute difference.

[5] Gödel seems to have tried over and over again (see Gödel 1972, footnote h), but I think unsuccessfully, to eliminate this logical complexity.

[6] See Tait (2006) for further discussion.

logic is that of a *proof* of a sentence: a sentence is given by stating what counts as a proof of it. So we may think of the sentence as (or at least having associated with it) the *type* of its proofs and we are led to the Curry-Howard theory of dependent types. And when we consider what is to count as a proof of the implication $A \rightarrow B$ or quantification $\forall x{:}A\phi(x)$, we are led to the types $A \rightarrow B$ and $\Pi_{x:A}\phi(x)$, respectively. And when we analyze mathematical induction

$$\frac{\phi(0) \qquad \forall x[\phi(x) \rightarrow \phi(x')]}{\phi(r)}$$

from this point of view, if s is a proof of the first premise and t is a proof of the second, then the proof p of the conclusion is obtained by primitive recursive definition: $p = PR(s,t,r)$. So, as in the case of Gödel's theory T, we may think of Heyting arithmetic, i.e. the extension of PRA by means of adding infinite quantification, as application of definition and proof by induction to functions and formulas of higher types. This line of thought has of course been developed by Per Martin-Löf, e.g. in 1973 and 1998, into a foundation for constructive mathematics. But that is not the direction that I want to take here; rather I want to reflect on the lowest level of the hierarchy of types, the finitist types.

But, before passing on to that, I want to at least mention a difficulty that arises for the theory of dependent types (as opposed to the types in Gödel's theory T) from the treatment of the concepts of function and higher-order equality that I advocated above in connection with the theory T and say something about its remedy. The relation $s =_{A,B} t$ or for simplicity $s = t$ of extensional equality between objects respectively of type A and type B (unlike that of intensional equality) is definable in the theory itself (see Tait 2005a), but it can happen that the extensionally equal terms s and t are of different (but extensionally equal) types A and B, respectively. For example, it may be that $s = ru$ and $t = rv$, where $r : \forall x : A.\phi(x)$ where x occurs in $\phi(x)$ and u and v are distinct normal terms of type A that are extensionally equal. In that case s and t are of the distinct (but extensionally equal) types $\phi(u)$ and $\phi(v)$, respectively. Because of this, 'substitution of equals for equals' fails for extensionally equality in a strong sense that the substitution may not even be meaningful. For example, if $s : A \rightarrow B$, $t : A$ and $t = r$, where r is not of type A, then $st : B$, but sr is not meaningful. However, we can restrict extensional equality to the relations $=_{A,A}$ between terms of the same type and in that case, substitution of equals for equals holds.

6. Let me turn back to the historical question about the source of the idea of founding arithmetic on the 'recursive mode of thought.' Skolem speaks only of Kronecker as a positive influence, but certainly in the passages in which Kronecker expresses his principle concerning mathematical definitions, namely in footnotes in his papers "*Grundzüge einer arithmetischen Theorie der algebraischen Grössen*" (1881, p. 257) and "*Über einige Anwendungen der Modelsysteme*" (1886, p. 156, footnote*), there is no specific reference to the recursive mode of thought or, in general, to the systematic foundations of

arithmetic. And in his paper "*Über den Zahlbegriff*" (1887) there is no explicit discussion of the principle of proof or definition by mathematical induction at all. In his introductory essay in Skolem's *Selected Works in Logic* (Skolem 1970, pp. 17–52), Wang speculates (p. 48) that Skolem might have been at least unconsciously influenced by Grassmann's 1861 *Lehrbuch der Arithmetik* (1904, p. XXIII); but although Grassmann, in his axiomatic treatment of the theory of the integers, states and uses the principle of *proof* by mathematical induction for propositions about the positive integers and in effect gives the recursive definitions of addition and multiplication of the integers, he does not mention the *general* principle of definition by primitive recursion of arithmetic functions, which is surely the basis of Skolem's paper.[7]

In (1947) Skolem himself states that, as far as he knows, his 1923 paper was the first investigation of "recursive number theory" and he criticizes with some justification Curry's assertion (Curry 1940) that it had its roots in work of Dedekind and Peano. Peano added to Dedekind's concept of a simply infinite system a framework of formal logic, but he did not even state, much less derive, the principle of definition by recursion as did Dedekind. Like Grassmann, he simply introduced as axioms the primitive recursive definitions of the arithmetic operations. As for Dedekind himself, Skolem points out that his foundation for arithmetic had a quite different goal from his own. Indeed, his own motivation, "to avoid the use of quantifiers", was the exact opposite of that of Dedekind who in his monograph *Was sind und was sollen die Zahlen?* (Dedekind 1888) and along with Frege in his *Begriffsschrift* (Frege 1879) aimed at the reduction of the concept of finite iteration to logic, employing what we now recognize as second-order logic. Indeed, Dedekind and Frege, rather than Russell and Whitehead, were the real source of the foundation of arithmetic that Skolem was opposing. And Dedekind's set-theoretic and non-algorithmic approach to mathematics was the explicit target of Kronecker's footnotes.

7. Nevertheless, there is an element of truth in Curry's assessment, for there are two themes in Dedekind's monograph. One is indeed the reduction of definition and proof by induction to logic. That reduction can be criticized not only by extreme constructivists such as Kronecker or more moderate constructivists such as Brouwer, but also by those, like Poincaré and (in his predicativist phase) Weyl, who rejected impredicative definitions. For the core of Dedekind's (and Frege's) reduction is his definition of the least set containing a given object and closed under a given function as the intersection of all such sets; and this is the paradigm of an impredicative definition.[8] But the other theme of Dedekind's

[7] Jens Erik Fenstad suggested in conversation that it was perhaps Skolem's close study of E. Netto's *Lehrbuch der Combinatorik* (1901), a work of which he later collaborated in producing a second edition, that led him to his conception of foundations of arithmetic. Again, it is true that the work abounds in proofs by mathematical induction, but nowhere is a function or concept *defined* by induction.

[8] It is worth noting that in Dedekind's case the impredicativity is limited to (1) constructing from a Dedekind infinite system $\langle D, 0, ' \rangle$ (i.e. where $0 \in D$ and $x \mapsto x'$ is an injective operation on

monograph makes it, I believe, a legitimate ancestor of Skolem's paper. For Dedekind was first to explicitly recognize the central role in the foundations of arithmetic of proof and definition by mathematical induction, the first of which he built into his notion of a simply infinite system and the latter of which (in the form of definition by iteration) he derived from it and then used both to define the arithmetic operations and to prove the categoricity of the axioms for a simply infinite set. Indeed, if one replaces the second-order axiom of mathematical induction in his theory of a simply infinite system by the rule of induction in the form

$$\frac{\phi(0) \qquad \phi(x) \rightarrow \phi(x')}{\phi(t)}$$

for arbitrary numerical predicates $\phi(x)$ of numbers and the principle of definition by iteration

$$I(s,t,0) = s \qquad\qquad I(s,t,r') = t(I(s,t,r))$$

where the operation $I(a, f, n)$ is defined for an arbitrary domain of objects D with $a \in D, f{:}D \rightarrow D$ and $n \in \mathbb{N}$, then we have a logic-free foundation of arithmetic for which Dedekind's proof of categoricity goes through exactly as stated. (We will see that the general principle of definition by primitive recursion is derivable in this framework.) The 'open-endedness' of this system—*arbitrary* numerical predicate $\phi(x)$ and *arbitrary* domain D—may seem to contrast with Dedekind's characterization of the numbers in second-order logic. But, as we know, the range of the second-order quantifier is itself open-ended: by going to ever higher orders, we introduce new sets of numbers into its scope.

An aside: Although Dedekind and Frege shared in the second-order analysis of the notion of a finite iterate of a function (or, in Frege's case, of a binary relation),[9] their conceptions of what would constitute a foundations of arithmetic were profoundly different. In answer to the question "What are the numbers?" Dedekind recognized their *sui generis* character and sought only to uniquely characterize the *system* of numbers, while Frege, focusing on the role

D whose range does not contain 0) the simply infinite subsystem $\langle \mathbb{N}, 0, ' \rangle$, and (2) defining the ordering $<$ of $\mathbb{N}$. According to his well-known letter to Keferstein (1890), (1) was simply part of his proof that the theory of simply infinite systems is consistent. His development of arithmetic from the axioms of a simply infinite system together with the axioms of order ($<$), in particular his familiar bottom-up derivation of the principle of definition by recursion (as opposed to Frege's (1893) top-down derivation), is predicative. (2) is worth noting because $<$ is primitive recursive, but the natural definition of its characteristic function is not by iteration; and the principle of definition by induction that Dedekind proved was restricted to definition by iteration. Dedekind defined the ordering $n < x$ essentially as the least set which contains n' and is closed under $x \mapsto x'$. Dedekind needed the relation $<$ in his derivation of the principle of definition by iteration.

[9]Frege (1879) and Dedekind (1888). It is worth noting that a version of Dedekind's monograph under the same title can be found in his notebooks dating, as he himself indicates in the Preface to the first edition, from 1872–1878. A reference to this manuscript can also be found in a letter to Dedekind from Heinrich Weber, dated 13 November 1878. (See Dugac (1976).)

of the numbers as finite cardinals, sought to *reduce* them to something else, namely to extensions of concepts. One consequence of his approach was that the principle of definition by recursion does not play an essential role in Frege's foundation, since he defined the arithmetic operations on numbers in terms of the corresponding operations on extensions. It was only in Frege (1903) that Frege derived this principle and in effect proved the categoricity of the theory of simply infinite systems.[10]

8. As we just noted, Dedekind proved the principle of definition by iteration, not the general principle of primitive recursive definition. As Peter Aczel pointed out in conversation, the latter principle does seem to have appeared explicitly for the first time in Skolem (1923). It is the special case of iteration $I(s, t, n)$ that is immediately justified on the basis of the notion of the numbers representing the 'free' finite iterations (i.e. without loops): the iteration

$$0, 1 \dots, n$$

of $'$ starting with 0 is imaged by the iteration

$$s, ts, \dots t^n s$$

$= I(s, t, n)$ of $t : D \to D$ starting with $s : D$. In (1981) I argued directly that definition by primitive recursion followed from the notion of number as the form of finite sequences; but behind the argument was the following reduction of primitive recursion to pure iteration.[11] To carry out the reduction, we need to use the uniqueness of the recursion equations for the identity function $\lambda x : \mathbb{N}.x$ on $\mathbb{N}$ in the form.

$$\frac{r0 = 0 \qquad rn' = (rn)'}{rs = s}$$

For the reduction of primitive recursion to iteration the admission of types $A \times B$ may also be essential. As far as I know, this question remains open. Let $s : A$ and $t : \mathbb{N} \times A \to A$. Define $\bar{s} : \mathbb{N} \times A$ and $\bar{t} : \mathbb{N} \times A \to \mathbb{N} \times A$ by

---

[10] It is usual to say that, whereas Frege understood the numbers as cardinals, Dedekind took them to be ordinals. Although there is some justice in this, there is also an objection: when Cantor introduced the concept of ordinal number, it was as the isomorphism types of well-ordered sets, just as he introduced the cardinals as isomorphism types of abstract sets. Just as Cantor did *not* refer to his transfinite numbers in (1883a) as ordinal numbers (see Tait (2000) for further discussion of this), Dedekind did not define the (finite) numbers as order types nor did he refer to them as ordinals. A significant thing about both systems of numbers, Cantor's transfinite numbers and Dedekind's finite numbers, and as opposed to Frege's, is that they are defined intrinsically, without reference to the domain of either sets or well-ordered sets.

[11] Such a reduction is carried out in Robinson (1947) for the case in which D is finitary type, without the introduction of product types $A \times B$. Instead, she introduced the primitive recursive coding of pairs of natural numbers and then showed that all other primitive recursions could be reduced, using this coding, to iteration.

$$\bar{s} = (0, s) \qquad \bar{t}(n, x) = (n', t(n, x)).$$

and write $f = \lambda n I(\bar{s}, \bar{t}, n)$. ($f$ is a term of terminal type containing any variables that might be in s or t.) Thus

$$f(0) = \bar{s} \qquad f(n') = \bar{t}(f(n)).$$

Note that

$$f0L = 0$$

$$fn'L = \bar{t}(fn)L = \bar{t}(fnL, fnR)L = ((fnL)', t(fnL, fnR)L = (fnL)'.$$

So $\lambda n fnL$ satisfies the same iteration equations as the identity function $\lambda n : \mathbb{N}.n$ on $\mathbb{N}$ and so

$$fnL = n$$

and

$$\begin{aligned} fn'R &= \bar{t}(fn)R = \bar{t}(fnL, fnR)R = \bar{t}(n, fnR)R \\ &= (n', t(n, fnR))R = t(n, fnR). \end{aligned}$$

So $PR(s, t, n)$ can be defined to be $gn = fnR = I(\bar{s}, \bar{t}, n)R$.

9. This might be a good place to mention the axiom

$$(pL, pR) = p$$

when $p : A \times B$ or $p : \exists x : A.B(x)$, which we used in the above reduction of primitive recursion to iteration with $p = fn$. Like the corresponding

$$\lambda x f x = f$$

when $f : A \to B$ or $f : \forall x{:}A.B(x)$, it is not usually counted as following from the notion of *definitional equality*. This is certainly right historically, but I am not clear on the principle by which it is excluded. Shouldn't it be part of the characterization of $A \times B$, for example, that every object $p : A \times B$ has the form (a, b) with $a : A$ and $b : B$? If so then the equation $p = (pL, pR)$ seems mandatory. (On the other hand, the analogous argument that every object of type $A \to B$ should be of the form $\lambda x.t$ is perhaps not so compelling.) I would welcome some insight on this. The notion of definitional equality derives from Gödel's *Dialectica* paper (Gödel 1958), where in footnote 7 he states that "identity (*Identität*) between functions is to be understood as intensional or definitional equality." Clearly that statement in itself does not force the narrow meaning of 'definitional equality' that was later established (by me as well as others). If, as Gödel intended, classical propositional logic is to be applied to equations of higher type, this notion of equality or identity needs to be decidable, but that certainly leaves latitude for our equation $(pL, pR) = p$ as well as the corresponding $\lambda x.tx = t$.[12] Is there a natural notion of intensional equality? Whatever answer this question deserves, I don't see any principled grounds for rejecting the equations in question as intensional equations.

However, there is another problematic ingredient in the reduction of primitive recursive definition to iteration, namely the uniqueness of the iteration equations for $\lambda x:\mathbb{N}.x$, and this is a different matter. Equations obtained from uniqueness of iteration are no more intensional than equations obtained by mathematical induction; indeed, as we noted, the two principles are equivalent. So the equation

$$PR(s,t,.n) = I(\bar{s},\bar{t},n)R$$

must be understood as an extensional equality.

Of course, in the case of our present concern, *PRA*, which deals only with numerical terms, the intensional meaning of an equation $s = t$ between closed terms, namely that s and t compute to the same numeral, and the extensional meaning, that they denote the same number, agree.

10. Closest to Skolem's conception of the foundations of arithmetic among his predecessors seems to be that of Poincaré, especially in (1894), although Skolem makes no mention of him in this context and Poincaré stops far short of Skolem's detailed development of the subject.[13] But as Skolem subsequently did, he explicitly founds arithmetic on what he calls 'reasoning by recurrence', i.e. definition and proof by induction. He doesn't state a general principle of definition by recursion and the only definitions by recursion he actually gives are of the operations of addition and multiplication; so it is not clear whether he had in mind definition by iteration or the more general principle of definition by primitive recursion. Indeed, given his attitude towards formality, it seems possible, even likely, that he never explicitly considered the difference.

    Interestingly, both Poincaré and Skolem simply take the initial number, lets say 0,[14] and the successor function $n \mapsto n'$ to be given. "The notions 'natural number' and 'the number $n + 1$ following the number n' (thus the descriptive function $n + 1$) as well as the recursive mode of thought are taken as basic." (Skolem 1923). Poincaré in fact assumes that they are defined and then observes that "these definitions, whatever they may be, do not enter into the course of the reasoning." (Ewald 1996, p. 974). What is striking about this is that both of them were reacting to the logicism of Russell. But for Russell, following Frege, giving a foundation for arithmetic required *defining* the natural numbers, and

---

[12] By the Strong Normalization Theorem, definitional equality means having the same normal form. This theorem is preserved when the conversions of (pL, pR) and λxtx to p and t, respectively, are admitted.

[13] "Here I stop this monotonous series of reasonings." Sect. IV.

[14] Many of the writers that I mention, including Poincaré and Dedekind, in fact take the least natural number to be 1. For the sake of simplicity, since nothing of relevance for us is really at stake in the choice, I will pretend that everyone starts with 0.

the logical complexity and in particular the need for the infinitary quantification that so displeased Skolem and enraged Poincaré resulted precisely from the attempt to define them (namely as the extensions of concepts).

For Poincaré the principle of reasoning by recursion is a synthetic a priori truth. This thesis was explicitly a rejection of both the logicism of Russell (and Couturat) and Hilbert's axiomatic conception of mathematics. As far as I know, he makes no explicit mention in his discussions of philosophy of mathematics of Dedekind's foundations of arithmetic or of Frege's; but in the first decade of the twentieth century (and so prior to the publication of *Principia Mathematica*), he devoted a series of papers (1905, 1906a, 1906b, 1906c, 1909) to an attack on Russell's logical foundations of arithmetic. Unlike the later rejection by Hilbert and Skolem, many of his objections, especially in the earlier papers, were based on a faulty understanding of the new logic which formed the framework for attempts at logical foundations[15] and added nothing useful to the discussion. But that is not so of his critique in general: two of his objections, in conjunction, are quite telling and indeed were taken over some years later by Hilbert. The first is his critique of impredicative reasoning and his recognition that Russell's foundation of arithmetic, once the Axiom of Reducibility is introduced (as it must be), involves impredicative definitions. As we have already noted, Hilbert was later to conclude that this means that the foundation is no longer a *logical* foundation: quantification over sets of numbers cannot be eliminated in the way that Russell claimed in his 'no-class' theory. The second objection now comes into play and it was equally applied by Poincaré as a criticism of Hilbert's initial foray (1905) into proof theory: The axioms of Russell's theory, like Hilbert's, are subject to the demand for a consistency proof. But Poincaré recognized, as Hilbert at that point did not, a circularity problem: for example, a proof of syntactic consistency of a system containing mathematical induction would itself inevitably need to employ mathematical induction. This is the circle that Hilbert only confronted many years later, e.g. in (1922), and, as he at least felt, avoided by restricting the proof theory to finitist methods of reasoning.

11. Poincaré's critique applies also to Dedekind, although as we noted, he makes no explicit reference to him. Dedekind's foundation starts with the assumption that there is a Dedekind infinite system, i.e. a set D with an injective function $f : D \to D$ and an element $e \in D$ that is not in the range of f. His claim to be giving a logical foundation of arithmetic was based upon the fact that, in constructing a simply infinite system from a Dedekind infinite system and developing arithmetic in a simply infinite system, he eliminated the role of inner *Anschauung*. But, as he made explicit in Dedekind (1890), he believed, with Hilbert and Poincaré, that a proof that the concept of a Dedekind infinite system is consistent is required.

[15]For a discussion of this see Goldfarb (1988).

We have with Dedekind and Poincaré an interesting contrast and, perhaps, the polar opposites in foundations of arithmetic. For the latter, the concept of finite iteration is unanalyzable, given to us in intuition. For Dedekind it has a logical analysis and he believes, contrary to Poincaré and Skolem, that this logical analysis is not just a transformation of intuitive truths into grotesque 'logical' constructions, but that, when we reason by recursion, his logical analysis actually plays out in our minds. In the Preface to the first edition of "*Was sind und was sollen die Zahlen?*" he writes

> ...I feel conscious that many a reader will scarcely recognize in the shadowy forms which I bring before him his numbers which all his life long have accompanied him as faithful and familiar friends; he will feel frightened by the long series of simple inferences corresponding to our step-by-step understanding, by the matter-of-fact dissection of the chains of reasoning on which the laws of numbers depend, and will become impatient at being compelled to follow out proofs for truths which to his supposed inner intuition (*Anschaaung*) seem at once evident and certain. On the contrary in just this possibility of reducing such truths to others more simple, no matter how long and apparently artificial the series of inferences, I recognize a convincing proof that their possession or belief in them is never given by inner intuition but is always gained only by more or less complete repetition of the individual inferences. (Dedekind 1963, p. 33)

The first sentence of the preface is

> In science, nothing capable of proof ought to be accepted without proof.

But presumably Poincaré's answer would be that an argument based on impredicative definitions is not a proof.

12. We have mentioned Poincaré's (well-vindicated) belief that arithmetic is founded on a synthetic a priori principle, reasoning by recurrence:

> This rule, inaccessible to analytic demonstration and to experience, is the veritable type of a synthetic a priori judgement. (Poincaré 1894, Sect. VI; Ewald 1996, p. 979)

He goes on to write

> Why then does this judgement force itself upon us with an irresistible evidence? It is because it is only the affirmation of the power of the mind which knows itself capable of conceiving the indefinite repetition of the same act when once this act is possible. The mind has a direct intuition of this power, and experience can only give occasion for using it and thereby becoming conscious of it.

And he continues two paragraphs later:

> Mathematical induction, that is, demonstration by recurrence ... imposes itself necessarily because it is only the affirmation of a property of the mind itself.

Poincaré's use of the term "intuition" is in general quite broad (see Poincaré 1900) and it is not entirely clear that he intends his usage in connection with the principle of reasoning by recurrence to coincide with Kant's. He certainly understands himself to be defending a general Kantian point of view: for example

> This is what M. Couturat has set forth in the work just cited; this he says still more explicitly in his Kant jubilee discourse, so that I heard my neighbor whisper: "I well see this is the centenary of Kant's *death*.
>
> Can we subscribe to this conclusive condemnation? I think not, ... (Poincaré 1905; Ewald 1996, p. 1023)

But it is not clear whether he believed himself to be following Kant in his use of the term "intuition", or even how aware he was of Kant's precise doctrine. There is indeed a difference, but I think one can argue that, at the end of the day and in the case of foundations of arithmetic at least, the difference doesn't matter.

The most important distinction in the usage of "intuition" is between Kant's, according to which intuition is the (non-propositional) *intuition of*, and the meaning according to which it is (propositional) *intuition that*.[16] In the latter sense, intuitive truths historically are those with which reasoning must begin, when premises have been pushed back until no further reduction is possible. It is in this sense, for example, that Leibniz used the term.[17] Poincaré clearly uses the term in the latter, propositional, sense and, on this understanding, given his rejection of impredicative reasoning, it would seem that he is absolutely correct on his own terms in calling the principle of reasoning by recurrence an intuitive truth.

But this is a case of intuition *that*. However much Kant may have on occasion used the term "*inuitus*" or "*Anschauung*" in the propositional sense of *intuition that*, it is a fundament of his philosophy to distinguish sensibility, the faculty of intuition, from understanding, the faculty of concepts, and there is no doubt but that, in this context, intuition is intuition *of*, the unique immediate mode of our acquaintance with objects: All objects are represented in sensible intuition. Abstracted from its empirical content the intuition is just space (pure outer intuition) and time (pure inner intuition). He also speaks of (sensible) intuitions of objects to refer to their representations in intuition. But an intuition by itself is not knowledge: The latter requires recognizing that an object represented in intuition falls under a certain concept or that one concept entails another. A priori knowledge of the latter sort, that all S are P, may be analytic, namely when P is contained in S. But, although the truths of mathematics can be known a priori, they are not in general analytic. When they are not analytic,

---

[16]It has sometimes been suggested that the difference between these two meanings of "intuition", Kant's and the particular sense of 'intuition that' that we are discussing, deriving from "*intuitus*", was created by translating Kant's "*Anschauung*" into English (and French) as "intuition". But what Kant referred to as "*Anschauung*" in the *Critique of Pure Reason*, he sometimes parenthetically called "intuitus" and also referred to exclusively as *intuitus* in his earlier *Inaugural Dissertation*, written in Latin. (See for example Sect. 10.) Thus, in using the term "*Anschauung*", he was merely translating the Latin into German: no new meaning was created by our translation of "*Anschauung*"; it was already there in *his* own use of the term "*intuitus*." An interesting question, which I won't attempt to answer here, is why Kant adapted the term *intuitus* in the way he did.

[17]Thus, the intuitive truths in this sense are the a priori truths (i.e. the 'first principles') in the original sense of that term.

the connection between subject and predicate is mediated by *construction*. The demonstration of the proposition begins with the 'construction of the concept' S. Thus, to take one of Kant's examples, to demonstrate that the interior angles of a triangle equal two right angles, we first construct the concept 'triangle'. We then construct some auxiliary lines, and then compute the equality of the sums of two sets of angles, using the Postulate "All right angles are equal" and the Common Notions "Equals added to (subtracted from) equals are equal". The construction of a concept is according to a rule, which Kant calls the *schema* of the concept. In the case of geometric concepts these are or at least include the rules of construction given by Euclid's 'to construct' postulates, Postulates 1–3 and 5. Of course, these rules are rules to construct objects from given objects. For example, given three points A, B, C, we can construct the three lines joining them and thereby, assuming that they are non-collinear, construct the triangle ABC. About constructing the concept Kant writes

> For the construction of the concept ... a *non-empirical* intuition is required, which consequently, as intuition, is an *individual* object, but that must nevertheless, as the construction of the concept (of a general representation), express in the representation universal validity for all possible intuitions that belong under the same concept.

The nature of these 'non-empirical intuitions' remains one of the main issues in the study of Kant's critical philosophy. When, in the *Discipline of Pure Reason in Dogmatic Use* (B 741–2), he actually gives the above example of the proof that the interior angles of a triangle equal two right angles, he speaks of constructing an empirical figure or one in imagination, where in the former case (at least) one abstracts from everything we do not intend to be part of the figure.

Kant had in fact very little to say specifically about arithmetic, in the *Critique of Pure Reason* or elsewhere; and what he did say is subject to different readings. He identifies *number* as the schema of magnitude, including both quantity and geometric magnitude. (In the latter case, he has in mind the fundamental role of number in measurement, i.e. in defining ratios in Book V of Euclid's *Elements*.[18]) From his discussion of it in the *Schematism*, number seems to be identified with the rule of representing something in intuition as a finite sequence of objects, and so a particular number, say 5, is the property of a representation of an object in intuition as a sequence of 5 things. Presumably, reasoning about numbers begins with 'constructing' one or more in pure inner intuition (time). But if, in analogy with the case of geometry (see the passage quoted above), the construction is to be 'of the concept of a general representation', then reasoning would seem to begin with 'constructions of *arbitrary* numbers. The development of this conception might indeed lead to a theory of number founded on the principle of reasoning by induction (see Tait

[18] If A, B and C, D are pairs of like magnitudes, then $A : B \leq C : D$ if and only if for all positive numbers m and n, $mB \leq nA$ implies $mC \leq nD$.

1981)[19]; and this is what I meant by suggesting that, at the end of the day, in spite of the difference between Kant's and Poincaré's use of "intuition", they are essentially in agreement about the foundations of arithmetic. But it is certainly a stretch to think that Kant anticipated such a development or even had a clear idea of arithmetic as opposed to algebra.[20]

13. In name at least, Kant plays a significant role in the ultimate response by Hilbert and Bernays to Poincaré's charge of circularity against Hilbert's earlier approach to proof theory. The role of finitary reasoning in Hilbert's program, as it developed in the 1920s, was this: In order to be assured that the axioms, say of first- or second-order number theory, indeed do define a structure, we must prove them consistent. In non-trivial cases, such a proof would itself seem to involve non-trivial mathematics. If the mathematics involved in the consistency proof were itself founded on a system of axioms, we would be in a circle: Poincaré's circle. Therefore, **a different conception of mathematics** needs to be invoked in founding the methods used in consistency proofs. These methods must themselves be immune to the demand for consistency proof.

For this, Hilbert and Bernays went back to an older conception of mathematics, which is indeed Kantian, according to which mathematics is construction and computation—a conception which, if one didn't look too closely at least, worked quite well in Kant's time and indeed so long as $\epsilon - \delta$ arguments—i.e. logic—could be successfully hidden behind the use of infinitesimals. Of course, $\epsilon - \delta$ arguments were there early in Greek mathematics, in applications of the method of exhaustion. Moreover, Newton was explicitly aware that the use of infinitesimals was just a shorthand and had to be backed up with an $\epsilon - \delta$ arguments. (Leibniz may not have thought that the elimination of infinitesimals was necessary, but he explicitly believed that it was always possible.) However in Kant's time in the eighteenth century, the *calculus* truly reigned. If one looks at Euler's books on function theory, *Introduction to the Analysis of the Infinite* or the *Foundations of Differential Calculus*, after a very brief indication of the justification for using infinitesimals in the Preface, the text looks to be entirely logic-free calculation.

But of course for Hilbert and Bernays, it wasn't all of mathematics that needs to be founded in this way on computation and construction. And this was the appeal of their conception over the severely restrictive view of Kronecker, that *all* of mathematics must be finitist, or the less restrictive view of Brouwer but whose intuitionism would nevertheless still reject much of the analysis developed in the nineteenth century. For Hilbert and Bernays, only the discrete mathematics that is involved in the consistency proofs needed to be founded on

---

[19]One would have, on Kant's behalf, to admit, given the construction of a number $f(X)$ from the arbitrary number X, the iteration $f^Y(X)$ of this construction along the arbitrary number Y.

[20]In his discussion of mathematical reasoning in contrast with philosophical reasoning in the *Discipline of Pure Reason*, he speaks of geometric reasoning and algebraic reasoning but indicates no awareness of the special character of reasoning about the natural numbers.

this 'Kantian' conception. Once consistency of the formal axiom system was established, the full range of methods coded in it would be available.

So in their quest for a foundation of proof theory, Hilbert and Bernays did indeed turn to Kant, at least in the sense that they returned to a conception of mathematics that was prevalent in Kant's times and that was, indeed, embraced by Kant. But their claims to Kant's authority go beyond that: Hilbert wrote in "On the Infinite" (1926)

> Kant already taught [...] that mathematics has at its disposal a content secured independently of all logic and hence can never be provided with a foundation by means of logic alone; .... Rather, as a condition for the use of logical inferences and the performance of logical operations, something must already be given to our faculty of representation, certain extralogical concrete objects that are intuitively present as immediate experience prior to all thought. If logical inference is to be reliable, it must be possible to survey these objects completely in all their parts, and the fact that they occur, that they differ from one another, and that they follow each other, or are concatenated, is immediately given intuitively, together with the objects, as something that neither can be reduced to anything else nor requires reduction. This is the basic philosophical position that I consider requisite for mathematics and, in general, for all scientific thinking, understanding, and communication. And in mathematics, in particular, what we consider is the concrete signs themselves, whose shape, according to the conception we have adopted, is immediately clear and recognizable.

Bernays, in "The philosophy of mathematics and Hilbert's proof theory" (Bernays 1930–1931) endorses what he takes to be

> Kant's fundamental idea that mathematical knowledge and also the successful application of logical inference rests on intuitive knowledge

while distinguishing this from the "particular form that Kant gave to this idea in his theory of space and time" and he sketches a theory of such intuitive knowledge in terms of his notion of 'formal abstraction' and a 'formal object'.

But what has been abandoned, in addition to Kant's particular views about space and time, is his Schematism, the idea that the mind is equipped with rules that govern the application of concepts, i.e. our reasoning about formal objects—our computations or constructions. In Hilbert (1926) we are given only a negative injunction that is essentially Kronecker's principle, that the concepts we use should be algorithmic—so, for example, we must reject infinitary quantification in general. But nothing is said about where reasoning about these objects is to begin. Bernays (1930–1931, Part II, Sect. 1) argues that definition and proof by recursion are valid on this finitist conception when we take the natural numbers to be the signs (formal objects) $||\ldots|$; but his argument for this is in essence simply the usual one, based not on the particular nature of the individual formal objects (the particular nature of 0 and the successor operation $n \mapsto n|$), but on the way that they are generated—by iterating the successor operation finitely often. (See Sect. 8 above.) But *the concept of a formal object does not contain this notion of finite iteration.* The gap in Bernays' argument becomes most evident when he acknowledges our 'empirical limitations', the fact that arithmetic concerns numbers such as $10^{10^{20}}$

which are unlikely to occur any way in physical reality. He writes "But intuitive abstraction is not constrained by such limits on the possibility of realization. For the limits are accidental from the formal standpoint. Formal abstraction finds no earlier place, so to speak, to make a principled distinction than at the difference between finite and infinite." But our projection from the number-signs that can be perceived by us, much less empirically realized, to those that cannot is mediated by the concept of finite iteration ("We go on-and-on like that"); and it is *this* concept that is the essence of arithmetic. As Poincaré put it: "these definitions, whatever they may be, do not enter into the course of the reasoning."

14. Kronecker's principle, stated above, allows us to introduce a function only when its definition yields an algorithm for computing its values. But as we know, the question of whether or not the definition of the function actually yields such an algorithm is itself in general a nontrivial arithmetic problem whose solution may depend upon what methods of proof we are willing to admit. On what basis do we accept that the algorithm works—that the definition is legitimate? We might agree that what *can* be proved *should* be proved; but obviously proof has to start somewhere. So, unless we abandon the idea of absolute proof in arithmetic, there must be some principles of arithmetic reasoning that are immune to the demand that we prove legitimacy. These must be the principles that follow from the very conception of the natural numbers and are, as I argued in my paper "Finitism" (Tait 1981), precisely the principle of definition and proof by induction.

References

Bernays, P. 1930–1931. Die Philosophie der Mathematik und die Hilbertsche Beweistheorie. *Blätter für deutsche Philosophie* 4: 326–367. Reprinted in Bernays (1976). A translation by P. Mancosu appears in Mancosu (1998), 234–265.

Bernays, P. 1976. *Abhandlungen zur philosophie der mathematik.* Darmstadt: Wissenschaftliche Buchgesellschaft.

Cantor, G. 1883a. *Grundlagen einer allgemeinen Mannigfaltigheitslehre. Ein mathematisch-philosophischer Versuch in der Lehre des Unendlichen.* Leipzig: Teubner. A separate printing of Cantor (1883b), with a subtitloe, preface and some footnotes added. A translation *Foundations of a general theory of manifolds: a mathmatico-philosophical investigation into the theory of the infinite* by W. Ewald is in (Ewald, 1996, 639–920).

Cantor, G. 1883b. Über unendliche, lineare Punktmannigfaltigkeiten, 5. *Mathematische Annalen* 21: 545–586. In Cantor (1932).

Cantor, G. 1932. In *Gesammelte Abhandlungen mathematischen und philosophischen Inhalts*, ed. E. Zermelo. Berlin: Springer.

Curry, H. 1940. A formalization of recursive arithmetic. *American Journal of Mathematics* 63(1941): 263–282.

Dedekind, R. 1872. *Stetigkeit und irrationale Zahlen.* Braunschweig: Vieweg. In Dedekind (1932). Republished in 1969 by Vieweg and translated in Dedekind (1963).

Dedekind, R. 1888. *Was sind und was sollen die Zahlen?* Braunschweig: Vieweg. In Dedekind (1932). Republished in 1969 by Vieweg and translated in Dedekind (1963).

Dedekind, R. 1890. Letter to Keferstein. Translated in van Heijenoort (1967), 99–103. Cambridge: Harvard University Press.
Dedekind, R. 1932. In *Gesammelte Werke,* vol. 3, ed. R. Fricke, E. Noether, and O. Ore. Braunschweig: Vieweg.
Dedekind, R. 1963. *Essays on the theory of numbers*. New York: Dover. English translation by W.W. Berman of Dedekind (1872) and Dedekind (1888).
Dugac, P. 1976. *Richard Dedekind et les fondements des mathematiques (avec de nombeux textes inédits).* Paris: Librairie Philosophique J. Vrin.
Ewald, W. (ed.). 1996. *From Kant to Hilbert: a source book in the foundations of mathematics.* Oxford: Oxford University Press. Two volumes.
Frege, G. 1879. *Begriffsschrift, eine der arithmetischen nachgebildete Formalsprache des reinen Denkens*. Halle: L. Nebert.
Frege, G. 1893. *Grundgesetze der Arithmetik: Begriffsschriftlich abgeleitet, Band I,* Jena: H. Pohle. Reprinted in 1962 along with Frege (1903) by Hildesheim: Georg Olms.
Frege, G. 1903. *Grundgesetze der Arithmetik: Begriffsschriftlich abgeleitet, Band II.* Jena: H. Pohle.
Gödel, K. 1958. Über eine bisher noch nicht benützte Erweiterung des finiten Standpunktes. *Dialectica* 12: 280–287. Reprinted with an Englsh translation in (Gödel, 1990, 240–252). Gödel (1972) is a revised version.
Gödel, K. 1972. On an extension of finitary mathematics which has not yet been used. In *Collected works*, vol. II Gödel (1990), 271–280. Revised version of Gödel (1958).
Gödel, K. 1990. *Collected works*, vol. II. Oxford: Oxford University Press.
Goldfarb, W. 1988. Poincaré against the logicists. In *History and philosophy of modern mathematics: minnesota studies in the philosophy of science*, vol. XI, ed. W. Aspray and P. Kitcher, 61–81. Minneapolis: The University of Minnesota Press.
Goodstein, R. 1945. Function theory in an axiom-free equation calculus. *Proceedings of the London Mathematical Society* 48: 401–34.
Goodstein, R. 1957. *Recursive number theory*. Amsterdam: North-Holland.
Grassmann, H. 1904. *Gesammelte mathematische und physikalische Werke*, vol. 2. Leipzig: Druck und Verlag von B.G. Teubner.
Hilbert, D. 1905. Über die Grundlagen der Logik und der Arithmetik. In *Verhandlungen des Dritten Internationalen Mathematiker-Kongress*. Leipzig: Teubner.
Hilbert, D. 1918. Axiomatisches denken. *Mathematishe Annalen* 78: 405–15. Reprinted in (Hilbert, 1932-9325, vol. 3, 1105–1115). Translated by W. Ewald in (Ewald, 1996, vol. 2).
Hilbert, D. 1922. Neubegründung der Mathematik: Erste Mitteilung. *Abhandlungen aus dem Seminar der Hamburgischen Universität* 1: 157–177. English translation in (Mancosu, 1998, 198–214) and (Ewald, 1996, 1115–1134).
Hilbert, D. 1923. Die logischen Grundlagen der Mathematik. *Mathematische Annalen* 88: 151–165. English translation in (Ewald, 1996, 1134–1148).
Hilbert, D. 1926. Über das Unendliche. *Mathematische Annalen* 95: 161–90. Translated by Stefan Bauer-Mengelberg in *From Frege to Gödel: a source book in mathematical logic*, 367–92.
Hilbert, D. 1932–9325. *Gesammelte Abhandlungen*. Bedrlin: Springer. 3 volumes.
Hilbert, D. 2011. In *David Hilbert's Lectures on the Foundations of Arithmetic and Logic 1917–1933*, ed. M. Hallett, W. Ewald, W. Sieg and U. Majer. Berlin: Springer.
Kronecker, L. 1881. Grundzüge einer arithmetischen Theorie der algebraischen Grössen. In *Leopold Kronecker's Werke*, vol. 2, ed. K. Hensel, 236–387 New York: Chelsea.
Kronecker, L. 1886. Über einige Anwendungen der Modulsysteme auf elementare algebraische Fragen. In *Leopold Kronecker's Werke*, vol. 3, ed. K. Hensel, 147–208. New York: Chelsea.
Kronecker, L. 1887. ber den zahlbegriff. In *Leopold Kronecker's Werke*, ed. K. Hensel, 251–274, New York: Chelsea.
Mancosu, P. (ed.). 1998. *From Brouwer to Hilbert: The Debate on the Foundations of Mathematics in the 1920's*. Oxford: Oxford University Press.
Mancosu, P. 1999. Between Russell and Hilbert: Behmann on the foundations of mathematics. *Bulletin of Symbolic Logic* 5: 303–330.

Martin-Löf, P. 1973. An intuitionistic theory of types: predicative part. In *Logic colloqium '73*, ed. H. E. Rose and J. C. Shepherdson. Amsterdam: North-Holland.

Martin-Löf, P. 1998. An intuitionistic theory of types. In *Twenty-five years of constructive type theory*, ed. G. Sambin and J. Smith, 221–244. Oxford: Oxford University Press.

Netto, E. 1901. *Lehrbuch der Combinatorik*. Leipzig: Verlag von B.G. Teubner.

Poincaré, H. 1894. Sur la Nature du Raisonnement mathématique. *Revue de métaphysique et de morale* 2: 371–84. Translation by George Bruce Halsted in Ewald (1996), vol. 2, 972–982.

Poincaré, H. 1900. Du rôle de l'intuition et de la logique en mathématiques. In *Compte rendu du Deuxiéme congrès international des mathématiciens tenu à Paris du 6 au 12 août 1900*, 210–22. Pais: Gauthier-Villars. Translation by George Bruce Halsted, reprinted Ewald (1996), vol. 2, 1021–1038.

Poincaré, H. 1905. Les mathématiques et la logique. *Revue de métaphysique et de morale* 13: 815–35. Translation by George Bruce Halsted in Ewald (1996), vol. 2, 1021–1038.

Poincaré, H. 1906a. Les mathématiques et la logique. *Revue de métaphysique et de morale* 14: 17–34. Translation by George Bruce Halsted in Ewald (1996), vol. 2, 1038–1052.

Poincaré, H. 1906b. Les mathématiques et la logique. *Revue de métaphysique et de morale* 14: 294–317. Translation by George Bruce Halsted in Ewald (1996), vol. 2, 1052–1071.

Poincaré, H. 1906c. A propos de la logistique. *Revue de métaphysique et de morale* 14: 866–868.

Poincaré, H. 1909. Le llogique de l'infin i. *Revue de métaphysique et de morale* 17: 461–82. Translation by George Bruce Halsted in Ewald (1996), vol. 2, 1038–1052.

Robinson, J. 1947. Primitive recursive functions. *Bulletin of the American Mathematical Society* 53: 925–942.

Sieg, W. 1999a. Hilbert's programs: 1917–1922. *Bulletin of Symbolic Logic* 5: 1–44.

Skolem, T. 1923. Einige Bemerkungen zur axiomatischen Begründung der Mengenlehre. *Matematikerkongressen in Helsingfors 4–7 Juli 1922, Den femte skandinaviske matematikerkongressen, Redogörelse*, 217–232. Helsingfors: Akademiska Bokhandeln.

Skolem, T. 1947. The development of recursive arithmetic. In *Copenhagen: Proceedings of the Tenth Congress of Scandinavian Mathematicians*, 1–16. Reprinted in Skolem (1970), 499–514. Copenhagen.

Skolem, T. 1956. A version of the proof of equivalence between complete induction and the uniqueness of primitive recursions. *Kongelige Norske Videnskabsselskabs Forhandlinger* XXIX: 10–15.

Skolem, T. 1970. In *Selected works in logic*, ed. J.E. Fenstad. Oslo: Universitetsforlaget.

Spector, C. 1962. Provably recursive functionals of analysis: a consistency proof of analysis by an extension of the principles formulated in current intuitionistc mathematics. In *Recursive function theory, proceedings of symposia in pure mathematics*, vol. 5, ed. J. Dekker, 1–27. Providence: American Mathematical Society.

Tait, W. 1981. Finitism. *Journal of Philosophy* 78: 524–556.

Tait, W. 2000. Cantor's *Grundlagen* and the paradoxes of set theory, Between Logic and Intuition: Essays in honor of Charles Parsons. (ed. G. Sher and R. Tieszen). Cambridge: Cambridge Univeristy Press, 269–290. Reprinted in Tait (2005b), 252–275.

Tait, W. 2005a. Proof-theoretic semantics for classical mathematics. In *Proof-theoretic semantics for classical mathematics*, ed. R. Kahle and P. Schroeder-Heister. Special edition of Synthese. Synthese 148(3): 603–622.

Tait, W. 2005b. *The provenance of pure reason: essays in the philosophy of mathematics and its history*. Oxford: Oxford University Press.

Tait, W. 2006. Gödel's interpretation of intuitionism. *Philosophia Mathematica* 14: 208–228.

van Heijenoort, J. (ed.). 1967. *From Frege to Gödel: A Source Book in Mathematical Logic*. Cambridge: Harvard University Press.

Weyl, H. 1921. Über die neue Grundlagenkrise der Mathematik. *Mathematische Zeitschrift* 10: 39–79. Translated by P. Mancosu in Mancosu (1998).

Part II
Foundations

Chapter 9
Type Theory and Homotopy

Steve Awodey

9.1 Introduction

The purpose of this informal survey article is to introduce the reader to a new and surprising connection between Logic, Geometry, and Algebra which has recently come to light in the form of an interpretation of the constructive type theory of Per Martin-Löf into homotopy theory and higher-dimensional category theory. This connection was discovered quite recently, and various aspects of it are now under active investigation by several researchers. (See Awodey and Warren 2009; Awodey et al. 2009; Warren 2008; van den Berg and Garner 2010, 2012; Gambino and Garner 2008; Garner 2009b; Lumsdaine 2009; Voevodsky 2006.)

9.1.1 Type Theory

Martin-Löf type theory is a formal system originally intended to provide a rigorous framework for constructive mathematics (Martin-Löf 1975, 1984, 1998). It is an extension of the typed λ-calculus admitting dependent types and terms. Under the Curry-Howard correspondence (Howard 1980), one identifies types with propositions, and terms with proofs; viewed thus, the system is an extension of first-order logic, and it is known to interpret constructive set theory (Aczel 1974). Indeed, Martin-Löf type theory has been used successfully to formalize large parts of constructive mathematics, such as the theory of generalized recursive definitions (Nordström et al. 1990; Martin-Löf 1979). Moreover, it is also employed extensively as a framework for the development of high-level programming languages, in virtue

S. Awodey (✉)
Department of Philosophy, Carnegie Mellon University, Pittsburgh, PA 15213, USA

P. Dybjer et al. (eds.), *Epistemology versus Ontology*, Logic, Epistemology, and the Unity of Science 27, DOI 10.1007/978-94-007-4435-6_9,

of its combination of expressive strength and desirable proof-theoretic properties (Nordström et al. 1990; Streicher 1991).

In addition to simple types $A, B, \ldots$ and their terms $x : A \vdash b(x) : B$, the theory also has dependent types $x : A \vdash B(x)$, which are regarded as indexed families of types. There are simple type forming operations $A \times B$ and $A \rightarrow B$, as well as operations on dependent types, including in particular the sum $\sum_{x:A} B(x)$ and product $\prod_{x:A} B(x)$ types (see the appendix for details). The Curry-Howard interpretation of the operations $A \times B$ and $A \rightarrow B$ is as propositional conjunction and implication, of course; the dependent types $x : A \vdash B(x)$ are predicates, or more generally, relations,

$$x_1 : A_1, \ldots, x_n : A_n \vdash R(x_1, \ldots, x_n),$$

and the sum $\sum$ and product $\prod$ operations are the existential $\exists$ and universal $\forall$ quantifiers, respectively.

It is now natural to further extend the type theory with a primitive equality relation, corresponding to the equality formulas of first-order logic. Specifically, given two terms a, b of the same type A, one can form a new **identity type** $\mathtt{Id}_A(a, b)$, representing the proposition that a and b are equal; a term of this type thus represents a proof of the proposition that a equals b. One therefore has two notions of equality: **propositional equality** is the notion represented by the identity types, and two terms are propositionally equal just if their identity type $\mathtt{Id}_A(a, b)$ is inhabited by a term. By contrast, **definitional equality** is a primitive relation on terms and is not represented by a type; it behaves much like equality between terms in the simply-typed lambda-calculus, or any conventional equational theory.

If the terms a and b are definitionally equal, then (since they can be freely substituted for each other) they are also propositionally equal; but the converse is generally not true in the intensional version of the theory (the rules for identity types are given in the appendix). In the extensional theory, by contrast, the two notions of equality are forced by an additional rule to coincide. As a consequence, the extensional version of the theory is essentially a dependent type theory with an extensional equality relation. As is well-known, however, the price one pays for this simplification is a loss of desirable proof-theoretic properties, such as strong normalization and decidable type checking and equality of terms (Streicher 1991, 1993; Hofmann 1995a).

In the intensional theory, each type A is thus endowed by the identity types $\mathtt{Id}_A(a, b)$ with a non-trivial structure. Indeed, this structure was observed by Hofmann and Streicher in (1995) to satisfy conditions analogous to the familiar laws for groupoids.[1] Specifically, the posited reflexivity of propositional equality

[1] A **groupoid** is like a group, but with a partially-defined composition operation. Precisely, a groupoid can be defined as a category in which every arrow has an inverse. A group is thus a groupoid with only one object. Groupoids arise in topology as generalized fundamental groups, not tied to a choice of basepoint (see below).

produces identity proofs $\mathtt{r}(a) : \mathtt{Id}_A(a, a)$ for any term $a : A$, playing the role of a unit arrow 1_a for a; and when $f : \mathtt{Id}_A(a, b)$ is an identity proof, then (corresponding to the symmetry of identity) there also exists a proof $f^{-1} : \mathtt{Id}_A(b, a)$, to be thought of as the inverse of f; finally, when $f : \mathtt{Id}_A(a, b)$ and $g : \mathtt{Id}_A(b, c)$ are identity proofs, then (corresponding to transitivity) there is a new proof $g \circ f : \mathtt{Id}_A(a, c)$, to be thought of as the composite of f and g. Moreover, this structure on each type A can be shown to satisfy the usual groupoid laws, but significantly, only **up to propositional equality**. We shall return to this point below.

The constructive character, computational tractability, and proof-theoretic clarity of the type theory are owed in part to this rather subtle treatment of equality between terms. Unlike the extensional theory, which is computationally intractable, the intensional theory leads to a system that is both powerful and expressive while retaining its important computational character. The cost of intensionality, however, has long been the resulting difficulty of finding a natural, conventional semantic interpretation. (See Hofmann 1995b, 1997; Cartmell 1986; Dybjer 1996 for previous semantics).

The new approach presented here constructs a bridge from constructive type theory to algebraic topology, exploiting both the axiomatic approach to homotopy of Quillen model categories, as well as the related algebraic methods involving (weak) higher-dimensional groupoids. This at once provides **two** new domains of interpretation for type theory. In doing so, it also permits logical methods to be combined with the traditional algebraic and topological approaches to homotopy theory, opening up a range of possible new applications of type theory in homotopy and higher-dimensional algebra. It also allows the importation into homotopy theory of computational tools based on the type theory, such as the computer proof assistants Coq and Agda (cf. Théry et al. 2006).

9.1.2 Homotopy Theory

In homotopy theory one is concerned with spaces and continuous mappings up to homotopy; a **homotopy** between continuous maps $f, g \colon X \to Y$ is a continuous map $\vartheta \colon X \times [0, 1] \to Y$ satisfying $\vartheta(x, 0) = f(x)$ and $\vartheta(x, 1) = g(x)$. Such a homotopy ϑ can be thought of as a "continuous deformation" of f into g. Two spaces are said to be homotopy-equivalent if there are continuous maps going back and forth, the composites of which are homotopical to the respective identity mappings. Such spaces may be thought of as differing only by a continuous deformation. Algebraic invariants, such as homology or the fundamental group, are homotopy-invariant, in that any spaces that are homotopy-equivalent must have the same invariants.

When we consider a space X, a distinguished point $p \in X$, and the paths in X beginning and ending at p, and identify such paths up to homotopy, the result is the **fundamental group** $\pi(X, p)$ of the space at the point. Pursuing an idea of Grothendieck's (1983), modern homotopy theory generalizes this classical construction in several directions: first, we remove the dependence on the basepoint p by considering the **fundamental groupoid** $\pi(X)$, consisting of all points and all paths up to homotopy. Next, rather than identifying homotopic paths, we can consider the homotopies between paths as distinct, new objects of a higher dimension (just as the paths themselves are homotopies between points). Continuing in this way, we obtain a structure consisting of the points of X, the paths in X, the homotopies between paths, the homotopies between these homotopies, and so on. The resulting structure $\pi_\infty(X)$ is called the **fundamental weak ∞-groupoid of** X. Such higher-dimensional algebraic structures now play a central role in homotopy theory (see e.g. Kapranov and Voevodsky 1991); they capture much more of the homotopical information of a space than does the fundamental group $\pi(X, p)$, or the groupoid $\pi(X) = \pi_1(X)$, which is a quotient of $\pi_\infty(X)$ by collapsing the higher homotopies. As discussed in Sect. 9.2.4 below, it has recently been shown that such higher-dimensional groupoids also arise naturally in intensional type theory.

Another central concept in modern homotopy theory is that of a **Quillen model structure**, which captures axiomatically some of the essential features of homotopy of topological spaces, enabling one to "do homotopy" in different mathematical settings, and to express the fact that two settings carry the same homotopical information. Quillen (1967) introduced model categories as an abstract framework for homotopy theory which would apply to a wide range of mathematical settings. Such a structure consists of the specification of three classes of maps (the fibrations, weak equivalences, and cofibrations) satisfying certain conditions typical of the leading topological examples. The resulting framework of axiomatic homotopy theory allows the development of the main lines of classical homotopy theory (fundamental groups, homotopies of maps, strong and weak equivalence, homotopy limits, etc.) independently of any one specific setting. Thus, for instance, it is also applicable not only in spaces and simplicial sets, but also in new settings, as in the work of Voevodsky on the homotopy theory of schemes (Morel and Voevodsky 1999), or that of Joyal (2002, in prep) and Lurie (2009) on quasicategories. In the work under consideration here (Sect. 9.2.3), it is shown that Martin-Löf type theory can be interpreted in any model category. This allows the use of type theory to reason formally and systematically about homotopy theory.

9.2 The Homotopy Interpretation

9.2.1 Background

Among recent treatments of the **extensional** type theory are the two papers (Moerdijk and Palmgren 2000, 2002) by Moerdijk and Palmgren from 2000 and 2002. The authors also announced a projected third paper devoted to the intensional theory, which never appeared. Their intention was presumably to make use of higher categories and, perhaps, Quillen model categories. No preliminary results were stated, but see (Palmgren 2003).

In 2006, Vladimir Voevodsky gave a series of lectures at Stanford University entitled "Homotopy lambda-calculus", in which an interpretation of intensional type theory into simplicial sets was proposed (see Voevodsky 2006). At the same time, and independently, the author and his doctoral student Michael Warren established the interpretation of intensional type theory in Quillen model categories, following a suggestion of Moerdijk.

All of these approaches derive from the pioneering work of Hoffmann and Streicher (1995), which we now summarize.

9.2.2 Groupoid Semantics

A model of type theory is *extensional* if the following reflection rule is satisfied:

$$\frac{p : \mathtt{Id}_A(a,b)}{a = b : A}\ \text{Id-reflection}$$

I.e., the identity type $\mathtt{Id}_A(a,b)$ in extensional models captures no more information than whether or not the terms a and b are definitionally equal. Although type checking is decidable in the intensional theory, it fails to be so in the extensional theory obtained by adding Id-reflection as a rule governing identity types. This fact is the principal motivation for studying intensional rather than extensional type theories (cf. Streicher 1991 for a discussion of the difference between the intensional and extensional forms of the theory). A good notion of a model for the extensional theory is due to Seely (1984), who showed that one can interpret type dependency in locally cartesian closed categories in a very natural way. (There are certain coherence issues, prompting a later refinement by Hofmann (1997), but this need not concern us here.) Of course, intensional type theory can also be interpreted this way, but then the interpretation of the identity types necessarily becomes trivial in the above sense

The first natural, non-trivial semantics for intensional type theory was developed by Hoffmann and Streicher (1995) using **groupoids**. The category of groupoids is not locally cartesian closed (Palmgren 2003), and the model employs certain fibrations (equivalently, groupoid-valued functors) to model type dependency. Intuitively, the identity type over a groupoid G is interpreted as the groupoid $G^{\rightarrow}$ of arrows in G, so that an identity proof $f : \mathtt{Id}_A(a,b)$ becomes an arrow $f : a \to b$ in G. The interpretation no longer validates extensionality, since there can be different elements a, b related by non-identity arrows $f : a \to b$. Indeed, there may be many different such arrows $f, g : a \rightrightarrows b$; however—unlike in the type theory—these cannot in turn be non-trivially related by identity terms of higher type $\vartheta : \mathtt{Id}_{\mathtt{Id}_A}(f, g)$, since a (conventional) groupoid has no such higher-dimensional structure. Thus the groupoid semantics validates a certain truncation principle, stating that all higher identity types are trivial—a form of extensionality one dimension up. In particular, the groupoid laws for the identity types are strictly satisfied in these models, rather than holding only up to propositional equality.

This situation suggests the use of the higher-dimensional analogues of groupoids, as arising in homotopy theory, in order to provide models admitting non-trivial higher identity types. Such higher groupoids occur naturally as the (higher) fundamental groupoids of spaces (as discussed above). A step in this direction was made by Garner (2009b), who uses a 2-dimensional notion of fibration to model intensional type theory in a higher-dimensional category, and shows that when various truncation axioms are added, the resulting theory is sound and complete with respect to this semantics. In his dissertation (Warren 2008), Warren showed that strict, infinite-dimensional groupoids also give rise to a model, which validates no such additional truncation axioms (see also Warren 2010). Such models do, however, satisfy type-theoretically unprovable strictness conditions such as the associativity of composition. It seems clear that one will ultimately need to use *weak* infinite dimensional groupoids in order to faithfully model the full intensional type theory (see Sect. 9.2.4 below).

9.2.3 Homotopical Models of Type Theory

Groupoids and their homomorphisms arise in homotopy theory as a "model" (i.e. a representation) of topological spaces with homotopy classes of continuous maps. There are other models as well, such as simplicial sets. The idea of a Quillen model category (cf. Quillen 1967; Bousfield 1977) is to axiomatize the common features of these different models of homotopy, allowing one to develop the theory in an abstract general setting, and to compare different particular settings. An object of a Quillen model category can be regarded as an abstract "space" and a morphism as an abstract "continuous map". Since there is an abstract notion of "path space" in any model category, one can define "homotopy" and related notions.

This axiomatic framework also provides a convenient way of specifying a general semantics for intensional type theory, not tied to a particular choice of groupoids, 2-groupoids, ∞-groupoids, simplicial sets, etc., or even spaces themselves. The basic result in this connection states that it is possible to model the intensional type theory in any Quillen model category (Awodey and Warren 2009) (see also Warren 2008); indeed, one requires only the simpler notion of a "weak factorization system" (for which see below). The idea is that a type is interpreted as an abstract "space" X and a term $x : X \vdash a(x) : A$ as a continuous function $a\colon X \to A$. Thus e.g. a closed term $a : A$ is a point a of A, an identity term $p : \mathtt{Id}_A(a,b)$ is then a path $p : a \rightsquigarrow b$ in A (a homotopy between points!). A "higher" identity term $\vartheta : \mathtt{Id}_{\mathtt{Id}_A(a,b)}(p,q)$ is a homotopy between the paths p and q, and so on for even higher identity terms and higher homotopies. In this interpretation, one uses abstract "fibrations" to interpret dependent types, and abstract "path spaces" to model identity types, recovering the groupoid model and its relatives as special cases.

In (Gambino and Garner 2008) it was then shown that the type theory itself carries a natural homotopical structure (again, a weak factorization system), so that the theory is not only sound, but also (essentially) logically complete with respect to such abstract homotopical semantics. While some "coherence" issues regarding the strictness of the interpretation remain to be worked out (see Warren 2008, as well as van den Berg and Garner 2012), together these results clearly establish not only the viability of the homotopical interpretation as a semantics for type theory, but also the possibility of using type theory to reason in Quillen model categories. That is to say, they suggest that intensional type theory can be seen as a "logic of homotopy theory".

In order to describe the interpretation in somewhat more detail, we first recall a few standard definitions. In any category $\mathcal{C}$, given maps $f : A \to B$ and $g : C \to D$, we write $f \pitchfork g$ to indicate that f has the *left-lifting property* (LLP) with respect to g: for any commutative square

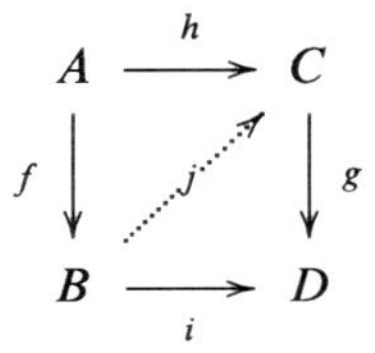

there exists a diagonal map $j : B \to C$ such that $j \circ f = h$ and $g \circ j = i$. If M is any collection of maps in $\mathcal{C}$, we denote by ${}^{\pitchfork}\mathsf{M}$ the collection of maps in $\mathcal{C}$ having the LLP with respect to all maps in M. The collection of maps $\mathsf{M}^{\pitchfork}$ is defined similarly. A *weak factorization system* (L, R) in a category $\mathcal{C}$ consists of two collections L (the "left-class") and R (the "right-class") of maps in $\mathcal{C}$ such that:

(1) Every map $f : A \to B$ has a factorization as $f = p \circ i$, where $i \in \mathsf{L}$ and $p \in \mathsf{R}$.

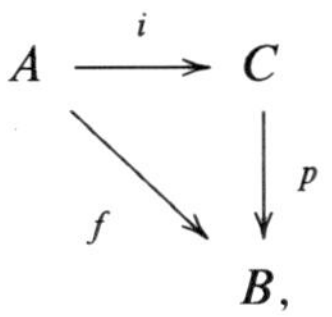

(2) $\mathsf{L} = {}^{\pitchfork}\mathsf{R}$ and $\mathsf{L}^{\pitchfork} = \mathsf{R}$.

A *(closed) model category* (Quillen 1967) is a bicomplete category $\mathcal{C}$ equipped with subcategories F (fibrations), C (cofibrations) and W (weak equivalences), satisfying the following two conditions: (1) Given any maps $g \circ f = h$, if any two of f, g, h are weak equivalences, then so is the third; (2) both $(\mathsf{C}, \mathsf{F} \cap \mathsf{W})$ and $(\mathsf{C} \cap \mathsf{W}, \mathsf{F})$ are weak factorization systems. A map f in a model category is a *trivial cofibration* if it is both a cofibration and a weak equivalence. Dually, a *trivial fibration* is a map which is both a fibration and a weak equivalence. An object A is said to be *fibrant* if the canonical map $A \to 1$ is a fibration. Dually, A is *cofibrant* if $0 \to A$ is a cofibration.

Examples of model categories include the following:

(1) The category Top of topological spaces, with fibrations the Serre fibrations, weak equivalences the weak homotopy equivalences, and cofibrations those maps which have the LLP with respect to trivial fibrations. The cofibrant objects in this model structure are the retracts of CW-complexes, spaces constructed by attaching cells.
(2) The category SSet of simplicial sets, with cofibrations the monomorphisms, fibrations the Kan fibrations, and weak equivalences the weak homotopy equivalences. The fibrant objects for this model structure are the Kan complexes.
(3) The category Gpd of (small) groupoids, with cofibrations the homomorphisms that are injective on objects, fibrations the Grothendieck fibrations, and weak equivalences the categorical equivalences. Here all objects are both fibrant and cofibrant.

See e.g. Dwyer and Spalinski (1995) and Hovey (1999) for further examples and details.

Finally, recall that in any model category $\mathcal{C}$, a *(very good) path object* A^I for an object A consists of a factorization

$$\begin{array}{ccc} A & \xrightarrow{r} & A^I \\ & \searrow^{\Delta} & \downarrow{\scriptstyle p} \\ & & A \times A, \end{array} \tag{9.1}$$

of the diagonal map $\Delta \colon A \to A \times A$ as a trivial cofibration r followed by a fibration p (see Hovey 1999). Paradigm examples of path objects are given by exponentiation by a suitable "unit interval" I in either Gpd or, when the object A is a Kan complex, in SSet. In e.g. the former case, G^I is just the "arrow groupoid" $G^{\rightarrow}$, consisting of all arrows in the groupoid G. Path objects always exist, but are not uniquely determined. In many examples, however, they can be chosen functorially.

We can now describe the homotopy interpretation of type theory more precisely. Whereas the idea of the Curry-Howard correspondence is often summarized by the slogan "Propositions as Types", the idea underlying the homotopy interpretation is instead "Fibrations as Types". In classical topology, and in most model categories, a fibration $p \colon E \to X$ can be thought of as a family of objects E_x varying continuously in a parameter $x \in X$. (The path-lifting property of a topological fibration describes how to get from one fiber $E_x = p^{-1}(x)$ to another E_y along a path $f \colon x \rightsquigarrow y$). This notion gives the interpretation of type dependency.

Specifically, assume that $\mathcal{C}$ is a finitely complete category with (at least) a weak factorization system (L, R). Because most interesting examples arise from model categories, we refer to maps in L as trivial cofibrations and those in R as fibrations. A judgement $\vdash A : \mathtt{type}$ is then interpreted as a fibrant object A of $\mathcal{C}$. Similarly, a dependent type $x : A \vdash B(x) : \texttt{type}$ is interpreted as a fibration $p: B \to A$. Terms $x : A \vdash b(x) : B(x)$ in context are interpreted as sections $b: A \to B$ of $p: B \to A$, i.e. $p \circ b = 1_A$. Thinking of fibrant objects as types and fibrations as dependent types, the natural interpretation of the identity type $\texttt{Id}_A(a, b)$ should then be as the *fibration of paths* in A from a to b, so that the type $x, y : A \vdash \texttt{Id}_A(x, y)$ should be the "fibration of all paths in A". That is, it should be a path object for A.

Theorem 9.1 (Awodey and Warren 2009). *Let $\mathcal{C}$ be a finitely complete category with a weak factorization system and a functorial choice of stable path objects A^I: i.e., given any fibration $A \to X$ and any map $f: Y \to X$, the evident comparison map is an isomorphism,*

$$f^*(A^I) \cong f^*(A)^I.$$

Then $\mathcal{C}$ is a model of Martin-Löf type theory with identity types (up to coherence, to be discussed below).

The proof exhibits the close connection between type theory and axiomatic reasoning in this setting: We verify the rules for the identity types (see the Appendix). Given a fibrant object A, the judgement $x, y : A \vdash \texttt{Id}_A(x, y)$ is interpreted as the path object fibration $p: A^I \to A \times A$, see (9.2.3). Because p is then a fibration, the formation rule

$$x, y : A \vdash \texttt{Id}_A(x, y) : \texttt{type}$$

is satisfied. Similarly, the introduction rule

$$x : A \vdash \texttt{r}(x) : \texttt{Id}_A(x, x)$$

is valid because the interpretation $r: A \to A^I$ is a section of p over $\Delta: A \to A \times A$. For the elimination and conversion rules, assume that the following premisses are given

$$x : A, y : A, z : \texttt{Id}_A(x, y) \vdash D(x, y, z) : \texttt{type},$$
$$x : A \vdash d(x) : D(x, x, \texttt{r}(x)).$$

We have, therefore, a fibration $q: D \to A^I$ together with a map $d: A \to D$ such that $q \circ d = r$. These data yields the following (outer) commutative square:

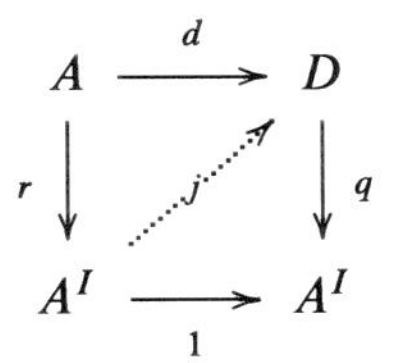

Because q is a fibration and r is, by definition, a trivial cofibration, there exists a diagonal filler j, which we choose as the interpretation of the term:

$$x, y : A, z : \mathtt{Id}_A(x, y) \vdash \mathtt{J}(d, x, y, z) : D(x, y, z).$$

Commutativity of the bottom triangle ensures that j validates the elimination rule, and commutativity of the top triangle is the required conversion rule:

$$x : A \vdash \mathtt{J}(d, x, x, \mathtt{r}(x)) = d(x) : D(x, x, \mathtt{r}(x)).$$

Examples of categories satisfying the hypotheses of this theorem include groupoids, simplicial sets, and many simplicial model categories (Quillen 1967) (including, e.g., simplicial sheaves and presheaves). There is a question of selecting the diagonal fillers j as interpretations of the $\mathtt{J}$-terms in a "coherent way", i.e. respecting substitutions of terms for variables. Some solutions to this problem are discussed in (Awodey and Warren 2009; Warren 2008). One neat solution is provided by the notion of a "natural" weak factorization system, which can be constructed in many model categories that are cofibrantly generated; see (Garner 2007, 2009a). The recent work of Riehl (2011) on "algebraic" Quillen model structures is related. A systematic investigation of the issue of coherence, along with several examples of coherent models derived from homotopy theory, can be found in the recent work (van den Berg and Garner 2012) of van den Berg and Garner.

9.2.4 Higher Algebraic Structures

Given the essential soundness and completeness of type theory with respect to the homotopical interpretation we may further ask, how *expressive* is the logical system as a language for homotopy theory? From this point of view, we think of the types in the intensional theory as spaces, the terms of the type A as the points of the "space" A, the identity type $\mathtt{Id}_A(a, b)$ as the collection of paths from a to b, and the higher identities as homotopies between paths, homotopies between homotopies of paths, etc. We can then ask what homotopically relevant facts, properties, and structures are logically expressible. The topological fact that paths and homotopies do not form a groupoid, but only a groupoid up to homotopy, is of course reminiscent of the logical fact that the identity types only satisfy the groupoid laws up to propositional equality. This apparent *analogy* between homotopy theory and type theory can now be made precise, and indeed can be recognized as one and the same fact, resting entirely on the homotopical interpretation of the logic. The fundamental weak ω-groupoid of a space is namely a construction entirely within the logical system—it belongs, as it were, to the logic of homotopy theory, as we now proceed to explain.

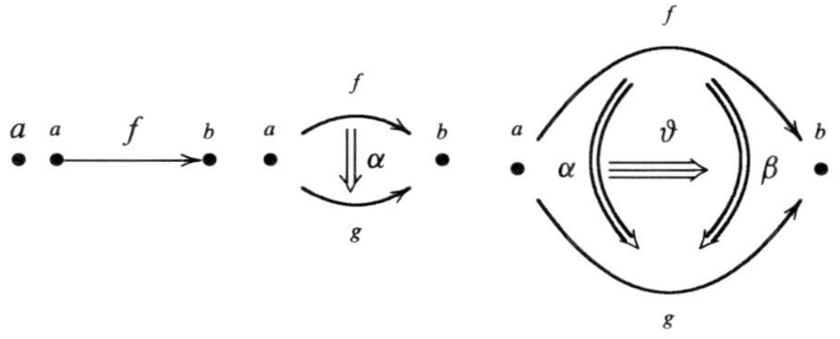

Fig. 9.1 Some cells in dimensions 0–3

9.2.4.1 Weak ω-Groupoids

It has recently been shown by Lumsdaine (2009) and, independently, van den Berg and Garner (2006, 2010), that the tower of identity types over any fixed base type A in the type theory bears an infinite dimensional algebraic structure of exactly the kind arising in homotopy theory, called a weak ω-groupoid (there are several notions of weak ω-groupoid in the literature: Kapranov and Voevodsky 1991; Leinster 2002; Cheng 2007; Brown 1987).

In somewhat more detail, in the globular approach to higher groupoids (Leinster 2004; Batanin 1998), a weak ω-groupoid has objects ("0-cells"), arrows ("1-cells") between objects, 2-cells between 1-cells, and so on, with various composition operations and laws depending on the kind of groupoid in question (strict or weak, n- or ω-, etc.). We first require the notion of a globular set, which may be thought of as an "infinite-dimensional" graph. Specifically, a *globular set* (Batanin 1998; Street 2000) is a presheaf on the category $\mathbb{G}$ generated by arrows

$$0 \underset{t}{\overset{s}{\rightrightarrows}} 1 \underset{t}{\overset{s}{\rightrightarrows}} 2 \rightrightarrows \ldots$$

subject to the equations $ss = ts$, $st = tt$. More concretely, a globular set $A_\bullet$ has a set A_n of "n-cells" for each $n \in \mathbb{N}$, and each $(n+1)$-cell x has parallel source and target n-cells $s(x)$, $t(x)$ (Fig. 9.1). (Cells x, y of dimension > 0 are *parallel* if $s(x) = s(y)$ and $t(x) = t(y)$; all 0-cells are considered parallel.)

For example, given a type A in a type theory $\mathbb{T}$, the terms of types

$$A, \mathtt{Id}_A, \mathtt{Id}_{\mathtt{Id}_A}, \ldots,$$

together with the evident indexing projections, e.g. $s(p) = a$ and $t(p) = b$ for $p : \mathtt{Id}_A(a, b)$, form a globular set $\hat{A}$.

A strict ω-groupoid is an infinite-dimensional groupoid satisfying, in all dimensions, associativity, unit, and inverse laws given by equations between certain cells. Such a groupoid has an underlying globular set consisting of cells of each dimension, and any globular set $A_\bullet$ generates a free strict ω-groupoid $F(A_\bullet)$—just as any set generates a free group, and any graph, a free groupoid. The cells of $F(A_\bullet)$ are free (strictly associative) pastings-together of cells from $A_\bullet$ and their formal duals, including degenerate pastings from the identity cells of $F(A_\bullet)$. In a *strict* ω-groupoid, cells can be composed along a common boundary in any lower dimension, and the composition satisfies various associativity, unit, and interchange laws, captured by the generalized associativity law: each labelled pasting diagram has a unique composite.

In a *weak* ω-groupoid, by contrast, we do not expect strict associativity, and so we may have multiple composition maps for each shape of pasting diagram; but we do demand that these composites agree *up to cells of the next dimension*, and that these associativity cells satisfy coherence laws of their own, and so on.

Now, this is exactly the situation we find in intensional type theory. For instance, even in constructing a term witnessing the transitivity of identity, one finds that there is no single canonical candidate. Specifically, as a composition for the pasting diagram

$$\cdot \longrightarrow \cdot \longrightarrow \cdot$$

or more concretely, a term c such that

$$x, y, z : X, p : \mathtt{Id}(x, y), q : \mathtt{Id}(y, z) \vdash c(q, p) : \mathtt{Id}(x, z),$$

there are the two equally natural terms c_l, c_r obtained by applying ($\mathtt{Id}$-ELIM) to p and q respectively. These are not definitionally equal, but are propositionally equal, i.e. equal up to a 2-cell, for there is a term e with

$$x, y, z : X, p : \mathtt{Id}(x, y), q : \mathtt{Id}(y, z) \vdash e(q, p) : \mathtt{Id}(c_l(q, p), c_r(q, p)).$$

Indeed, we have the following:

Theorem 9.2 (Lumsdaine 2009; van den Berg and Garner 2010). *Let A be any type in a system $\mathbb{T}$ of intensional Martin-Löf type theory. Then the globular set $\hat{A}$ of terms of type*

$$A, \mathtt{Id}_A, \mathtt{Id}_{\mathtt{Id}_A}, \ldots$$

carries a natural weak ω-groupoid structure.

It is now quite natural to ask what special properties this particular ω-groupoid has in virtue of its type-theoretic construction. In light of related syntactic constructions of other types of free algebras, a reasonable conjecture might be that it is in some sense a **free** weak ω-groupoid generated by syntactic primitive data, and up to a suitable notion of equivalence. We return to this question below.

9.2.4.2 Weak n-Groupoids

A further step in exploring the connection between type theory and homotopy is to investigate the relationship between type theoretic "truncation" (i.e. higher-dimensional extensionality principles) and topological "truncation" of the higher fundamental groups. Spaces for which the homotopy type is already completely determined by the fundamental groupoid are called **homotopy 1-types**, or simply 1-types (Baues 1995). More generally, one has n-types, which are thought of as spaces which have no homotopical information above dimension n. One of the goals of homotopy theory is to obtain good models of homotopy n-types. For example, the category of groupoids is Quillen equivalent to the category of 1-types; in this precise sense, groupoids are said to model homotopy 1-types. A famous conjecture of Grothendieck's is that (arbitrary) homotopy types are modeled by weak ∞-groupoids (see e.g. Batanin 1998 for a precise statement).

Recent work (Awodey et al. 2009) by Pieter Hofstra, Michael Warren and the author has shown that the 1-truncation of the intensional theory, arrived at by adding the analogue of the Id-reflection rule for all terms of identity type, generates a category of structured graphs with a Quillen model structure, and this model category is Quillen equivalent to that of groupoids. In a precise sense, the truncated system of 1-dimensional type theory thus models the homotopy 1-types.

In a bit more detail, for every globular set $A_\bullet$ one can define a certain system of type theory $\mathbb{T}(A_\bullet)$ over a single basic type A, the basic terms of which are the elements of the various A_n, typed as terms of the corresponding identity types over A: so for each $a \in A_0$ there is a new basic term "a" or type A, and for each $b \in A_1$ is a new basic term of type $\mathtt{Id}_A(\text{“}s(a)\text{”}, \text{“}t(a)\text{”})$, where $s, t \colon A_1 \rightrightarrows A_0$ are the source and target maps, at dimension 1, of $A_\bullet$, and so on. Since we know from the result of Lumsdaine et al. (Lumsdaine 2009, van den Berg and Garner 2010), just reviewed, that for any type X, the underlying globular set of terms of the various identity types $X, \mathtt{Id}_X, \mathtt{Id}_{\mathtt{Id}_X}, \ldots$ gives rise to a weak ω-groupoid, we can infer that in particular the globular set of terms over the ground type A_0 in the theory $\mathbb{T}(A_\bullet)$ form such a groupoid, **generated type-theoretically** from the arbitrary globular set $A_\bullet$. Let us call this weak ω-groupoid $G_\omega(A_\bullet)$, the **type-theoretically free** weak ω-groupoid generated by $A_\bullet$. This construction is investigated in depth in Awodey et al. (2009), where certain groupoids of this kind are termed **Martin-Löf complexes** (technically, these are the algebras for the globular monad just described).

It is clearly of interest to investigate the relationship between this type-theoretic construction of higher groupoids and both the algebraically free higher groupoids, on the one hand, and the higher groupoids arising from spaces as fundamental groupoids, on the other. As a first step, one can consider the 1-dimensional truncation of the above construction, and the resulting (1-) groupoid $G_1(A_\bullet)$. For that case, the following result relating $G_1(A_\bullet)$ to the usual, algebraically free groupoid is established in the work cited:

Theorem 9.3 (Awodey et al. 2009). *The type-theoretically free groupoid is equivalent to the algebraically free groupoid.*

Furthermore, it is shown that the 1-truncated Martin-Löf complexes admit a Quillen model structure equivalent to that of (1-) groupoids. The following then results from known facts from homotopy theory:

Theorem 9.4 (Awodey et al. 2009). *The 1-truncated Martin-Löf complexes classify homotopy 1-types.*

Obviously, one could now proceed to higher groupoids and the corresponding type theories truncated at higher dimensions.

9.3 Conclusion: The Logic of Homotopy

The application of logic in geometry and topology via categorical algebra has a precedent in the development of topos theory. Invented by Grothendieck as an abstract framework for sheaf cohomology, the notion of a topos was soon discovered to have a logical interpretation, admitting the use of logical methods into topology (see e.g. Joyal and Tierney 1984 for just one of many examples). Equally important was the resulting flow of geometric and topological ideas and methods into logic, e.g. sheaf-theoretic independence proofs, topological semantics for many non-classical systems, and an abstract treatment of realizability (see the encyclopedic work Johnstone 2003).

An important and lively research program in current homotopy theory is the pursuit (again following Grothendieck 1983) of a general concept of "stack," subsuming sheaves of homotopy types, higher groupoids, quasi-categories, and the like. Two important works in this area have just appeared (Lurie, *Higher Topos Theory* 2009; Joyal, *Theory of Quasi-Categories* in prep). It may be said, somewhat roughly, that the notion of a "higher-dimensional topos" is to homotopy what that of a topos is to topology (as in Joyal and Tierney 1991). This concept also has a clear categorical-algebraic component via Grothendieck's "homotopy hypothesis", which states that n-groupoids are combinatorial models for homotopy n-types, and ∞-groupoids are models for arbitrary homotopy types of spaces. Still missing from the recent development of higher-dimensional toposes, however, is a logical aspect analogous to that of (1-dimensional) topos theory. The research surveyed here suggests that such a logic is already available in intensional type theory. The homotopy interpretation of Martin-Löf type theory into Quillen model categories, and the related results on type-theoretic constructions of higher groupoids, are analogous to the basic results interpreting *extensional* type theory and higher-order logic in (1-) toposes. They clearly indicate that the logic of higher toposes—i.e., the logic of homotopy—is, rather remarkably, a form of intensional type theory.

A.1 Appendix A. Rules of Type Theory

This appendix recalls (some of) the rules of intensional Martin-Löf type theory. See Martin-Löf (1984), Nordström et al. (1990), and Jacobs (1999) for detailed presentations.

Judgement forms. There are four basic forms of judgement:

$$A : \mathtt{type} \qquad a : A$$
$$a = b : A \qquad A = B : \mathtt{type}$$

Each form can occur also with free variables: e.g. if A is a type, then

$$x : A \vdash B(x) : \mathtt{type}$$

is called a *dependent type*, regarded as an A-indexed family of types. The part $x : A$ to the left of the turnstile $\vdash$ is called the *context* of the judgement. More generally, a list of variable declarations $x_1 : A_1, x_2 : A_2, \ldots, x_n : A_n$ is a context whenever the judgements $A_1 : \mathtt{type}$ and

$$x_1 : A_1, \ldots, x_m : A_m \vdash A_{m+1} : \mathtt{type}$$

are derivable, for $1 \leq m < n$. Given such a context Γ, the judgement $\Gamma \vdash A : \mathtt{type}$ means that A is a type (in context Γ), while $\Gamma \vdash a : A$ indicates that a is a term of type A (in context Γ); the equality judgements have their usual meaning.

Formation rules. Given an A-indexed family of types $B(x)$, the dependent sum $\sum_{x:A} . B(x)$ and product $\prod_{x:A} . B(x)$ can be formed. The identity type introduces a new dependent type $\mathtt{Id}_A$ for any type A.

$$\frac{x : A \;\vdash\; B(x) : \mathtt{type}}{\prod_{x:A} B(x) : \mathtt{type}} \; \textstyle\prod \text{ formation}$$

$$\frac{x : A \;\vdash\; B(x) : \mathtt{type}}{\sum_{x:A} B(x) : \mathtt{type}} \; \textstyle\sum \text{ formation}$$

$$\frac{A : \mathtt{type}}{x : A, y : A \;\vdash\; \mathtt{Id}_A(x, y) : \mathtt{type}} \; \mathtt{Id} \text{ formation}$$

Under the Curry-Howard correspondence, sums correspond to existential quantifiers, products to universal quantifiers, and identity types to equations. The behavior of each of these types is specified by introduction, elimination and conversion rules.

Rules for dependent products.

$$\frac{x : A \vdash f(x) : B(x)}{\lambda x.f(x) : \prod_{x:A} B(x)} \prod \text{introduction}$$

$$\frac{a : A \qquad f : \prod_{x:A} B(x)}{\mathtt{app}(f, a) : B(a).} \prod \text{elimination}$$

$$\frac{a : A \qquad x : A \vdash f(x) : B(x)}{\mathtt{app}(\lambda x.f(x), a) = f(a) : B(a)} \prod \text{conversion}$$

The introduction rule states that for every family of terms $f(x) : B(x)$ there is a term $\lambda x.f(x)$ of type $\prod_{x:A} B(x)$. The elimination rule corresponds to the application of a term f of the indexed product to $a : A$. Finally, the conversion rule for states that the application term $\mathtt{app}(-, a)$ behaves correctly when applied to a term of the form $\lambda x.f(x)$.

Rules for dependent sums.

$$\frac{a : A \qquad b : B(a)}{\langle a, b\rangle : \sum_{x:A} B(x)} \sum \text{introduction}$$

$$\frac{p : \sum_{x:A} B(x) \qquad x : A, y : B(x) \vdash c(x, y) : C(\langle x, y\rangle)}{\sigma(c, p) : C(p)} \sum \text{elimination}$$

$$\frac{a : A \qquad b : B(a) \qquad x : A, y : B(x) \vdash c(x, y) : C(\langle x, y\rangle)}{\sigma(c, \langle a, b\rangle) = c(a, b) : C(\langle a, b\rangle)} \sum \text{conversion}$$

The variables $x : A, y : B(a)$ are bound in the notation $\sigma(c, p)$.

Note that when A and B are types in the same context, the usual product $A \times B$ and function $A \to B$ types from the simply typed λ-calculus are recovered as $\sum_{x:A} B$ and $\prod_{x:A} B$, respectively.

Rules for identity types.

$$\frac{a : A}{\mathtt{r}(a) : \mathtt{Id}_A(a, a)} \mathtt{Id} \text{ introduction}$$

$$\frac{\begin{array}{cc} & x : A, y : A, z : \mathtt{Id}_A(x, y) \vdash B(x, y, z) : \mathtt{type} \\ c : \mathtt{Id}_A(a, b) & x : A \vdash d(x) : B\big(x, x, \mathtt{r}(x)\big) \end{array}}{\mathtt{J}(d, a, b, c) : B(a, b, c)} \mathtt{Id} \text{ elimination}$$

$$\frac{a : A}{\mathtt{J}\big(d, a, a, \mathtt{r}(a)\big) = d(a) : B\big(a, a, \mathtt{r}(a)\big)} \; \mathtt{Id} \text{ conversion}$$

The introduction rule provides a witness $\mathtt{r}(a)$ that a is identical to itself, called the *reflexivity term*. The distinctive elimination rule can be recognized as a form of Leibniz's law. The variable $x : A$ is bound in the notation $\mathtt{J}(d, a, b, c)$.

Acknowledgements Thanks to Pieter Hofstra, Peter Lumsdaine, and Michael Warren for their contributions to this article, and to Per Martin-Löf and Erik Palmgren for supporting this work over many years.

References

Aczel, P. 1974. The strength of Martin-Löf's type theory with one universe. In *Proceedings of the symposium on mathematical logic*, Oulu, ed. S. Miettinen and J.J. Vaananen, 1–32.

Awodey, S., and M.A. Warren. 2009. Homotopy theoretic models of identity types. *Mathematical Proceedings of the Cambridge Philosophical Society* 146: 45–55.

Awodey, S., P. Hofstra, and M.A. Warren. 2009. Martin-Löf complexes. Submitted, on the arXiv as `arXiv:0906.4521`.

Batanin, M.A. 1998. Monoidal globular categories as a natural environment for the theory of weak n-categories. *Advances in Mathematics* 136(1): 39–103.

Baues, H.-J. 1995. Homotopy types. In *Handbook of algebraic topology*, ed. I.M. James, 1–72. Amsterdam: North-Holland.

Bousfield, A.K. 1977. Constructions of factorization systems in categories. *Journal of Pure and Applied Algebra* 9: 207–220.

Brown, R. 1987. From groups to groupoids. *Bulletin of the London Mathematical Society* 19: 113–134.

Cartmell, J. 1986. Generalised algebraic theories and contextual categories. *Annals of Pure and Applied Logic* 32(3): 209–243.

Cheng, E. 2007. An ω-category with all duals is an ω-groupoid. *Applied Categorical Structures* 15(4): 439–453.

Dwyer, W.G., and J. Spalinski. 1995. Homotopy theories and model categories. In *Handbook of algebraic topology*, ed. I.M. James, 73–126. Amsterdam: North-Holland.

Dybjer, P. 1996. Internal type theory. In *Proceedings of the BRA TYPES workshop*, Torino, June 1995. Lecture notes in computer science, vol. 1158. Berlin: Springer.

Gambino, N., and R. Garner. 2008. The identity type weak factorisation system. *Theoretical Computer Science* 409(3): 94–109.

Garner, R. 2007. Cofibrantly generated natural weak factorisation systems. On the arXiv as `math.CT/0702290`.

Garner, R. 2009a. Understanding the small object argument. *Applied Categorical Structures* 17(3): 247–285.

Garner, R. 2009b. Two-dimensional models of type theory. *Mathematical Structures in Computer Science* 19(4): 687–736.

Grothendieck, A. 1983. Pursuing stacks. Unpublished letter to Quillen,

Hofmann, M. 1995a. Extensional concepts in intensional type theory. Ph.D. thesis, University of Edinburgh.

Hofmann, M. 1995b. On the interpretation of type theory in locally cartesian closed categories. In *Computer science logic 1994*, ed. J. Tiuryn and Leszek Pacholski, 427–441. Berlin/New York: Springer.

Hofmann, M. 1997. Syntax and semantics of dependent types. In *Semantics and logics of computation*, Publications of the Newton Institute. ed. P. Dybjer and A.M. Pitts 79–130. Cambridge: Cambridge University Press.
Hofmann, M., and T. Streicher. 1995. The groupoid interpretation of type theory. In *Twenty-five years of constructive type theory*. Oxford logic guides, vol. 36, ed. G. Sambin and J. Smith, 83–111. Oxford: Oxford University Press.
Hovey, M. 1999. *Model categories*, Mathematical surveys and monographs, vol. 63. Providence: American Mathematical Society.
Howard, W.A. 1980. The formulae-as-types notion of construction. In *To H. B. Curry: Essays on combinatory logic, lambda Calculus and formalism*, ed. J.P. Seldin and J.R. Hindley, 479–490. London: Academic Press.
Jacobs, B. 1999. *Categorical logic and type theory*. Amsterdam: North-Holland Publishing Co.
Johnstone, P.T. 2003. *Sketches of an elephant*, vol. 2. Oxford: Oxford University Press.
Joyal, A. *The theory of quasi-categories*. In preparation.
Joyal, A. 2002. Quasi-categories and Kan complexes. *Journal of Pure and Applied Algebra* 175: 207–222.
Joyal, A., and M. Tierney. 1984. *An extension of the galois theory of grothendieck*, Memoirs of the American Mathematical Society, vol. 51. Providence: American Mathematical Society.
Joyal, A., and M. Tierney. 1991. Strong stacks and classifying spaces. In *Category theory (Como, 1990)*, Lecture notes in mathematics, vol. 1488, 213–236. Berlin: Springer.
Kapranov, M.M., and V.A. Voevodsky. 1991. ∞-groupoids and homotopy types. *Cahiers de Topologie et Géometrie Différentielle Catégoriques* 32(1): 29–46.
Leinster, T. 2002. A survey of definitions of n-category. *Theory and Applications of Categories* 10: 1–70 (electronic).
Leinster, T. 2004. *Higher operads, higher categories*, London mathematical society lecture note series, vol. 298. Cambridge: Cambridge University Press.
Lumsdaine, P.L. 2009. Weak ω-categories from intensional type theory. In *Typed Lambda-calculus and its applications*. 172–187. Brasilia, Brazil: Springer-Verlag, Berlin, Heidelberg.
Lurie, J. 2009. *Higher topos theory*. Princeton: Princeton University Press.
Martin-Löf, P. 1975. An intuitionistic theory of types: Predicative part. In *Logic Colloquium 73*, ed. H.E. Rose and J.C. Shepherdson, 73–118. Amsterdam: North-Holland.
Martin-Löf, P. 1979. Constructive mathematics and computer programming. In *Proceedings of the 6th international congress for logic, methodology and philosophy of science*. Amsterdam: North-Holland.
Martin-Löf, P. 1984. *Intuitionistic type theory*. Napoli: Bibliopolis.
Martin-Löf, P. 1998. An intuitionistic theory of types. In *Twenty-five years of constructive type theory*, Oxford logic guides, vol. 36, ed. G. Sambin and J. Smith, 127–172. Oxford: Oxford University Press. This paper was originally a 1972 preprint from the Department of Mathematics at the University of Stockholm.
Moerdijk, I., and E. Palmgren. 2000. Wellfounded trees in categories. *Annals of Pure and Applied Logic* 104: 189–218.
Moerdijk, I., and E. Palmgren. 2002. Type theories, toposes and constructive set theory: Predicative aspects of AST. *Annals of Pure and Applied Logic* 114: 155–201.
Morel, F., and V. Voevodsky. 1999. A^1-homotopy theory of schemes. *Publications Mathématiques de l'I.H.E.S.* 90: 45–143.
Nordström, B., K. Petersson, and J.M. Smith. 1990. *Programming in Martin-Löf's type theory. An introduction*. Oxford: Oxford University Press.
Palmgren, E. 2003. Groupoids and local cartesian closure. Department of Mathematics Technical Report 2003:21, Uppsala University.
Quillen, D. 1967. *Homotopical algebra*, Lecture notes in mathematics, vol. 43. Berlin/Heidelberg: Springer.
Riehl, E. 2011. Algebraic model structures. *New York Journal of Mathematics* 17: 173–231.
Seely, R.A.G. 1984. Locally cartesian closed categories and type theory. *Mathematical Proceedings of the Cambridge Philosophical Society* 95: 33–48.

Street, R. 2000. The petit topos of globular sets. *Journal of Pure and Applied Algebra* 154: 299–315.

Streicher, T. 1991. *Semantics of type theory*, Progress in theoretical computer science. Basel: Birkhauser.

Streicher, T. 1993. *Investigations into intensional type theory*. Habilitationsschrift, Ludwig-Maximilians-Universität München.

Théry, L., P. Letouzey, and G. Gonthier. 2006. Coq. In *The seventeen provers of the world*, Lecture notes in computer Science, ed. F. Wiedijk, 28–35. Berlin/Heidelberg: Springer.

van den Berg, B. 2006. Types as weak ω-categories. Lecture delivered in Uppsala, and unpublished notes.

van den Berg, B., and R. Garner. 2010. Types are weak ω-groupoids. *Proceedings of the London Mathematical Society* 102: 370–394

van den Berg, B., and R. Garner. 2012. Topological and simplicial models of identity types. *ACM Transactions on Computational Logic* 13(1): 1–44.

Voevodsky, V. 2006. A very short note on the homotopy λ-calculus. Unpublished note.

Warren, M.A. 2008. Homotopy theoretic aspects of constructive type theory. Ph.D. thesis, Carnegie Mellon University.

Warren, M. 2010. The strict omega-groupoid interpretation of type theory. Forthcoming in *Models, logics and higher-dimensional categories: A tribute to the work of Mihály Makkai*. Providence: American Mathematical Society

Chapter 10
A Computational Interpretation of Forcing in Type Theory

Thierry Coquand and Guilhem Jaber

10.1 Introduction

In a previous work (Coquand and Jaber 2010), we considered intuitionistic type theory with a type of natural numbers N and a type of Booleans N_2. The type $C = N \rightarrow N_2$ represents *Cantor space*, the space of functions from natural numbers to Booleans, and it has a natural topology, with basic compact open subsets defined by a finite set of conditions of the form $f\ n_i = b_i$ about a function $f : C$. We have shown (Coquand and Jaber 2010) that any definable functional $F : C \rightarrow N_2$ is *uniformly continuous*. This means that we can find a partition of Cantor space in a finite number of conditions $p_1, \ldots, p_n$ with corresponding Boolean values $b_1, \ldots, b_n$ such that $F\ f = b_i$ whenever f satisfies the condition p_i. The argument in Coquand and Jaber (2010) is constructive, and thus can be seen *implicitly* an algorithm which computes an uniform modulus of continuity. We explicitate here a possible algorithm, which given such a functional F, produces a covering $p_1, \ldots, p_n$ and a list $b_1, \ldots, b_n$. To simplify the presentation, we limit ourselves to a type system which is an extension of Gödel system T (Gödel 1990) with a type of Booleans. This computation can be readily expressed in a functional programming language, here Haskell, using the notion of monads (Wadler 1992).

We briefly outline the paper. We first recall the syntax for terms and conditions. We then give a simple operational semantics corresponding to forcing. We prove the termination of this evaluation, and give an algorithm to compute the modulus of continuity of a given functional. These computation combines in a non trivial

T. Coquand (✉)
Chalmers tekniska högskola, Data-och informationsteknik, 412 96 Göteborg, Sweden
e-mail: coquand@chalmers.se

G. Jaber
Département Informatique – École des Mines de Nantes 4, rue Alfred Kastler, 44307 Nantes, France
e-mail: guilhem.jaber@mines-nantes.fr

P. Dybjer et al. (eds.), *Epistemology versus Ontology*, Logic, Epistemology, and the Unity of Science 27, DOI 10.1007/978-94-007-4435-6_10,

way realizability and Beth models, and we end by commenting on this point, following Goodman (1978). A first appendix presents a representation in the programming language Haskell and a second appendix explains how we can give computational sense to universal quantification over Cantor space by iterating this forcing construction.

10.2 Terms, Types and Conditions

10.2.1 Terms

The terms of Type Theory are untyped λ-calculus extended with constants, and with the following syntax.

$$t, u ::= x \mid \lambda x.t \mid t\ t \mid \mathsf{natrec}(t,t) \mid \mathsf{boolrec}(t,t) \mid S(t) \mid 0 \mid 1$$

We consider terms up to α-conversion. Besides β-reduction, natrec and boolrec have the reduction rules

$$\mathsf{natrec}(a,g)\ 0 \to a \qquad \mathsf{natrec}(a,g)\ S(n) \to g\ a\ (\mathsf{natrec}(a,g)\ n)$$

and

$$\mathsf{boolrec}(a_0,a_1)\ 0 \to a_0 \qquad \mathsf{boolrec}(a_0,a_1)\ 1 \to a_1$$

This forms an extension of β-reduction which still has the Church-Rosser property (Martin-Löf 1998), sometimes called β, ι-reduction (Barendregt 1997).

If k is a natural number, we write $\overline{k}$ the term $S^k(0)$.

10.2.2 Typing Rules

The basic types are N, for natural numbers, and N_k, for finite types with k elements. If A, B are types then so is $A \to B$. The typing judgements are of the form $\Gamma \vdash t{:}A$, where Γ is a context $x_1{:}A_1, \ldots, x_n{:}A_n$ (with $x_i \neq x_j$ for $i \neq j$).

The typing rules are as follows.

$$\frac{(x{:}A) \in \Gamma}{\Gamma \vdash x{:}A} \qquad \frac{\Gamma, x{:}A \vdash t : B}{\Gamma \vdash \lambda x.t : A \to B} \qquad \frac{\Gamma \vdash v : A \to B \quad \Gamma \vdash u{:}A}{\Gamma \vdash v\ u : B}$$

$$\frac{}{\Gamma \vdash 0 : N} \qquad \frac{\Gamma \vdash t : N}{\Gamma \vdash S(t) : N} \qquad \frac{\Gamma \vdash a : B \quad \Gamma \vdash g : N \to B \to B}{\Gamma \vdash \mathsf{natrec}(a,g) : N \to B}$$

$$\frac{}{\Gamma \vdash 0 : N_2} \qquad \frac{}{\Gamma \vdash 1 : N_2} \qquad \frac{\Gamma \vdash a_0 : B \quad \Gamma \vdash a_1 : B}{\Gamma \vdash \mathsf{boolrec}(a_0,a_1) : N_2 \to B}$$

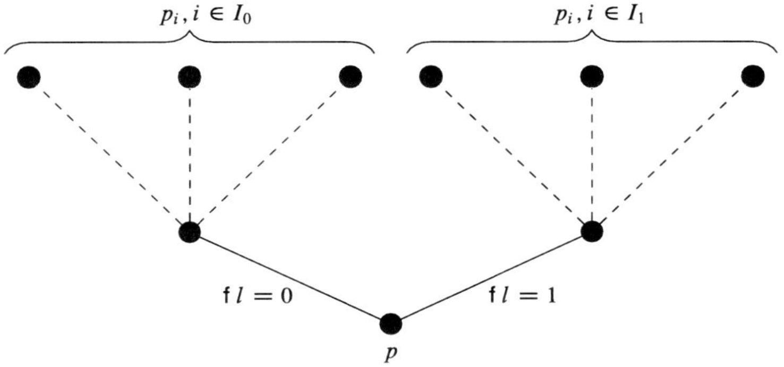

Fig. 10.1 An example of condition

10.2.3 Conditions

The conditions $p, q, \ldots$ represent finite amount of information about the infinite object we want to describe. Since we want to force the addition of a Cohen real, the conditions are finite sub-graphs of function from natural numbers to Booleans. Thus the conditions can be represented as a finite list of equations

$$\mathsf{f}\, n_1 = b_1 \quad \ldots \quad \mathsf{f}\, n_k = b_k$$

where $n_1, \ldots, n_k$ are distinct natural numbers and $b_1, \ldots, b_k$ Booleans. The *domain* $\mathsf{dom}(p)$ of this condition p is the finite set $n_1, \ldots, n_k$. We write $q \leq p$ if the condition q extends the condition p. One can think of a condition p as a compact open subset C_p of Cantor space C, which is the space of functions from natural numbers to the discrete space of Booleans, with the product topology. A condition p represents also some finite amount of information about a generic element of Cantor space. If p and q are compatible conditions, we can consider $pq = qp$, by taking the union of the conditions p and q. We clearly have $C_q \subseteq C_p$ if $q \leq p$ and $C_{pq} = C_p \cap C_q$ if p and q are compatible. Any condition p can be considered to be the product of elementary conditions $\mathsf{f}\, n = b$. If n is not in the domain of p then the two conditions $p(\mathsf{f}\, n = 0)$ and $p(\mathsf{f}\, n = 1)$ form an *elementary partition* of p. By iterating this construction, we obtain the general notion of *partition* $p_1, \ldots, p_l$ of a condition p (this includes as well the trivial partition p of p.) In general a non trivial partition $p_i,\ i \in I$ of p is built from one partition $p_i,\ i \in I_0$ of $p(\mathsf{f}\, l = 0)$ and one partition $p_i,\ i \in I_1$ of $p(\mathsf{f}\, l = 1)$ for some l not in the domain of p (Fig. 10.1).

10.2.4 Generic Function

We extend the syntax of terms with a new function symbol f. To each condition p we associate the reduction relation $\rightarrow_p$ which extends β, ι reduction with the rule

$\mathsf{f}\,\overline{n} \rightarrow_p b$ whenever $\mathsf{f}\ n = b$ is in p. This extension still satisfies the Church-Rosser property, by the usual Martin-Löf/Tait argument (as presented for instance in Martin-Löf 1998). We define then $t =_p u$ to mean that t and u have a common reduct for $\rightarrow_p$.

10.3 Computational Interpretation of Forcing

10.3.1 Operational Semantics

In ordinary type theory, the computation is described by a rewriting relation $t \rightarrow t'$ between terms. Here the computation deals with a pair pt of a condition p (which can be thought as a state) and a term t. Furthermore the computation (process) may open during the computation independent computations, and the computation step is a relation $pt \rightarrow \alpha$ between pt and a formal sum $\alpha = \Sigma p_i t_i$ where $p_1, \ldots, p_n$ is a partition of p. The definition is the following.

$$\frac{pt \rightarrow \Sigma p_i t_i}{p(t\ u) \rightarrow \Sigma p_i (t_i\ u)} \qquad \frac{}{p((\lambda x.t)\ u) \rightarrow pt[x/u]}$$

$$\frac{}{p(\mathsf{boolrec}(t_0, t_1)\ 0) \rightarrow pt_0} \qquad \frac{}{p(\mathsf{boolrec}(t_0, t_1)\ 1) \rightarrow pt_1}$$

$$\frac{}{p(\mathsf{natrec}(t_0, t_1)\ 0) \rightarrow pt_0} \qquad \frac{}{p(\mathsf{natrec}(t_0, t_1)\ S(t)) \rightarrow p(t_1\ t\ (\mathsf{natrec}(t_0, t_1)\ t))}$$

$$\frac{pt \rightarrow \Sigma p_i t_i}{p(\mathsf{natrec}(t_0, t_1)\ t) \rightarrow \Sigma p_i (\mathsf{natrec}(t_0, t_1)\ t_i)}$$

$$\frac{pt \rightarrow \Sigma p_i t_i}{p(\mathsf{boolrec}(t_0, t_1)\ t) \rightarrow \Sigma p_i (\mathsf{boolrec}(t_0, t_1)\ t_i)}$$

The remaining crucial rules are that $p(\mathsf{f}\ \overline{k}) \rightarrow pb$ if $\mathsf{f}\ k = b$ is in p and otherwise $p(\mathsf{f}\ \overline{k}) \rightarrow p_0 0 + p_1 1$ with $p_i = p(\mathsf{f}\ k = i)$. Finally, we have $p(\mathsf{f}\ S^n(t)) \rightarrow \Sigma p_i (\mathsf{f}\ S^n(t_i))$ whenever $pt \rightarrow \Sigma p_i t_i$.

We can then define the computation of the normal form (for ground types):

$$\frac{}{p0 \Rightarrow p0} \qquad \frac{}{p1 \Rightarrow p1} \qquad \frac{pt \Rightarrow \Sigma p_i \overline{k_i}}{pS(t) \Rightarrow \Sigma p_i \overline{1 + k_1}} \qquad \frac{pt \rightarrow \Sigma p_i t_i \quad p_i t_i \Rightarrow \alpha_i}{pt \Rightarrow \Sigma \alpha_i}$$

Lemma 10.1. *If $pt \rightarrow \Sigma p_i t_i$ or $pt \Rightarrow \Sigma p_i t_i$ then (p_i) is a partition of p and $t \rightarrow^*_{p_i} t_i$.*

If $\alpha = \Sigma p_i t_i$ is a formal sum, with (p_i) partition of p, and $q \leq p$ we can define $q\alpha = \Sigma (qp_i) t_i$ where we limit the sum to the p_i compatible with q.

Lemma 10.2. *If $pt \rightarrow \alpha$ and $q \leq p$ then $qt \rightarrow q\alpha$. If $pt \Rightarrow \alpha$ and $q \leq p$ then $qt \Rightarrow q\alpha$.*

10.3.2 Computability Predicate

We define $p \Vdash \varphi_N(t)$ inductively

- $p \Vdash \varphi_N(0)$
- $p \Vdash \varphi_N(S(t))$ if $p \Vdash \varphi_N(t)$
- $p \Vdash \varphi_N(t)$ if $pt \to \Sigma p_i t_i$ with $p_i \Vdash \varphi_N(t_i)$ for all i

This is equivalent to the fact that we have a relation $t \Rightarrow \Sigma p_i \overline{k_i}$. Similarly $p \Vdash \varphi_{N_2}(t)$ is defined by the clauses

- $p \Vdash \varphi_{N_2}(0)$
- $p \Vdash \varphi_{N_2}(1)$
- $p \Vdash \varphi_{N_2}(t)$ if $pt \to \Sigma p_i t_i$ with $p_i \Vdash \varphi_{N_2}(t_i)$ for all i

and this is equivalent to the fact that $pt \Rightarrow \Sigma p_i v_i$ with $v_i = 0$ or $v_i = 1$ for all i. Finally, $p \Vdash \varphi_{A \to B}(t)$ means that $q \leq p$ and $q \Vdash \varphi_A(u)$ implies $q \Vdash \varphi_B(t\ u)$.

$p \Vdash \varphi_A(t)$ can be read as "p forces that t is computable at type A". In the case $A = N$ or $A = N_2$ this means that we have $pt \Rightarrow \alpha$ for some α, i.e. that the computation of pt terminates.

Lemma 10.3. *If $p \Vdash \varphi_A(t)$ and $q \leq p$ then $q \Vdash \varphi_A(t)$.*

Proof. This is direct if A is a function type and follows from Lemma 10.2 in the case $A = N$ or $A = N_2$.

Lemma 10.4. *If $pt \to \Sigma p_i t_i$ and $p_i \Vdash \varphi_A(t_i)$ for all i then $p \Vdash \varphi_A(t)$.*

Proof. This is clear if $A = N$ or $A = N_2$. If $A = A_1 \to A_2$ and $pt \to \Sigma p_i t_i$ and $p_i \Vdash \varphi_A(t_i)$ for all i and if $q \leq p$ then we have $qt \to \Sigma(qp_i)t_i$ by Lemma 10.2. If $q \Vdash \varphi_{A_1}(u)$ we have $q(t\ u) \to \Sigma(qp_i)(t_i\ u)$ and $qp_i \Vdash \varphi_{A_2}(t_i\ u)$. By induction we have $q \Vdash \varphi_{A_2}(t\ u)$ as desired.

Lemma 10.5. *If $p \Vdash \varphi_A(t_0)$ and $p \Vdash \varphi_{N \to A \to A}(t_1)$ then $p \Vdash \varphi_{N \to A}(\mathsf{natrec}(t_0, t_1))$. Similarly, if $p \Vdash \varphi_A(t_0)$ and $p \Vdash \varphi_A(t_1)$ then $p \Vdash \varphi_{N_2 \to A}(\mathsf{boolrec}(t_0, t_1))$.*

Proof. This follows from Lemma 10.4.

Lemma 10.6. *The generic function is computable, i.e. $p \Vdash \varphi_{N \to N_2}(\mathsf{f})$ for all p.*

Proof. We assume $p \Vdash \varphi_N(t)$ and we prove $p \Vdash \varphi_{N_2}(\mathsf{f}\ t)$. We have $pt \Rightarrow \Sigma p_i \overline{k_i}$ and, using Lemma 10.4, we are reduced to prove that $p \Vdash \varphi_{N_2}(\mathsf{f}\ \overline{k})$, which is direct, by case if k is in the domain of p or not.

Theorem 10.1. *If $x_1{:}A_1, \ldots, x_n{:}A_n \vdash t{:}A$ and $p \Vdash \varphi_{A_1}(t_1), \ldots, p \Vdash \varphi_{A_n}(t_n)$ then we have $p \Vdash \varphi_A(t[x_1/t_1, \ldots, x_n/t_n])$. In particular, if $\vdash t{:}A$ then $p \Vdash \varphi_A(t)$ for all p.*

Proof. By induction on the proof of $x_1{:}A_1, \ldots, x_n{:}A_n \vdash t{:}A$ using Lemmas 10.5 and 10.6.

If we have $\vdash F : C \rightarrow N_2$ it is possible to use this result and compute a modulus of uniform continuity for F as follows. Using Theorem 10.1 and Lemma 10.6, we have $\Vdash \varphi_{N_2}(F\ h)$. Hence we have $F\ h \Rightarrow \Sigma p_i v_i$ with $v_i = 0$ or $v_i = 1$ for all i, and p_i is a partition of Cantor space. By Lemma 10.1, we have $F\ h \rightarrow^*_{p_i} v_i$. We can see the modulus of continuity of F as the greatest k such that a condition of the form $\mathsf{f}\ k = b$ appears in one of the p_i.

10.3.3 Baire Space

Our argument can be adapted to the case of Baire space $N \rightarrow N$ instead of Cantor space. The generic function f is now of type $N \rightarrow N$ and an elementary condition is of the form $\mathsf{f}\ n = m$, where n and m are natural numbers. The partitions are not finite objects anymore but well-founded trees. The inductive definition of partition is the following: the condition p itself is a (trivial) partition of p, and if n is not in the domain of p, and for each m we have a partition P_m of $p(\mathsf{f}\ n = m)$, then the union of all P_m is a partition of p. Similarly the formal sums $\Sigma p_i t_i$ are now indexed by well-founded trees: we have the formal sum pt over p, and if n is not in the domain of p, and for each m we have a formal sum σ_m over $p(\mathsf{f}\ n = m)$, then the formal sum $\Sigma_m \sigma_m$ is a formal sum over p. The operational semantics have the same rules, except that $p(\mathsf{f}\ \overline{k}) \rightarrow p\overline{l}$ if $\mathsf{f}\ k = l$ is in p and $p(\mathsf{f}\ \overline{k}) \rightarrow \Sigma p_n \overline{n}$ with $p_n = p(\mathsf{f}\ k = n)$ otherwise. Whenever $\vdash F : (N \rightarrow N) \rightarrow N$ it is possible in this way to associate to F a well-founded tree (a *bar* on Baire space) with natural numbers at each leaves, by computing $F\ \mathsf{f}$. This gives a strong form of the continuity of definable functionals on Baire space.[1]

10.4 Conclusion

In the reference (Goodman 1978), Goodman compares recursive realizability and Kripke/Beth models as follows. Recursive realizability "emphasizes the active aspect of constructive mathematics… However, Kleene's notion has the weakness that it disregards that aspect of constructive mathematics which concern epistemological change.… Precisely that aspect of constructive mathematics which Kleene's notion neglects is emphasized by Kripke's semantics for intuitionistic logic.… However, Kripke's notion makes it appear that the constructive mathematician is a passive observer of a structure which gradually reveals itself. What is lacking is the emphasis on the mathematician as active which Kleene's notion provides."

[1]This result is stated for instance in the reference Bishop (1970). In this reference, Bishop argues that an appropriate approach to Brouwer's theory of choice sequence is to express them as part of the metatheory of a system similar to Gödel System T.

He then presents a combination of realizability and Kripke semantics. We think that our work illustrates these remarks in a simple and concrete framework. Usual computation rules in type theory, with a rewriting relation on terms, don't involve "epistemological change". In our framework, the condition p represents a state of knowledge. While in usual Kripke/Beth semantics, these states of knowledge are independent of the computations, they are here needed in the computation, and the computation may create new states of knowledge.

A.1 Appendix 1: Representation in Haskell

The operational semantics given in the previous section has a natural representation in the programming language Haskell, using the notion of monad (Wadler 1992). Written in this way, the program is quite close to an ordinary evaluation program for Gödel system T by head reduction. The monad we use is a composition of the list monad (for nondeterminism) and of the state monad (Wadler 1992).

```
type Name = String

data Exp =
   Zero | One | Succ Exp | App Exp Exp | Natrec Exp Exp
  | Boolrec Exp Exp | Lam Name Exp | Var Name | Gen

-- closed substitution

subst :: Exp -> Name -> Exp -> Exp

subst t x e = case t of
 Var y -> if x == y then e else t
 Lam y t1 -> if x == y then t else Lam y (subst t1 x e)
 App t1 t2 -> App (subst t1 x e) (subst t2 x e)
 Natrec t1 t2 -> Natrec (subst t1 x e) (subst t2 x e)
 Boolrec t1 t2 -> Boolrec (subst t1 x e) (subst t2 x e)
 Succ t1 -> Succ (subst t1 x e)
 _ -> t

type Cond = [(Int,Exp)]        -- uses only Zero or One

newtype M a = M (Cond -> [(Cond,a)])

app :: M a -> Cond -> [(Cond,a)]
app (M f) p = f p

instance Monad M where
 return x = M (\p -> [(p,x)])
 l >>= k = M (\p -> concat (map (\(p,a) -> app (k a) p)
                                   (app l p)))
```

```
-- split determines if the condition p contains the value in k,
-- and otherwise forks between the two possibilities

split :: Int -> M Exp

split k = M (\ p ->  case lookup k p of
                     Just b -> [(p,b)]
                     Nothing -> [((k,Zero):p,Zero),
                                 ((k,One):p,One)])

-- gen k e computes e before applying it to split

gen :: Int -> Exp -> M Exp

gen k Zero = split k
gen k (Succ e) = gen (k+1) e
gen k e = do e' <- step e
             gen k e'

-- step implements the reduction

step :: Exp -> M Exp

step (App (Lam x t) u) = return (subst t x u)
step (App (Natrec t0 t1) Zero) = return t0
step (App (Natrec t0 t1) (Succ t)) =
  return (App (App t1 t) (App (Natrec t0 t1) t))
step (App (Boolrec t0 t1) Zero) = return t0
step (App (Boolrec t0 t1) One) = return t1
step (App (Natrec t0 t1) t) =
  do t' <- step t
     return (App (Natrec t0 t1) t')
step (App (Boolrec t0 t1) t) =
  do t' <- step t
     return (App (Boolrec t0 t1) t')
step (App Gen u) = gen 0 u
step (App t u) = do t' <- step t
                    return (App t' u)
step t = error("step " ++ show t)

-- app (eval t) [] outputs a covering of
-- Cantor space if t is of type N2

eval :: Exp -> M Exp

eval Zero = return Zero
eval One = return One
eval t = do t' <- step t
            eval t'
```

A.2 Appendix 2: Quantification on Cantor Space

A.2.1 New Conditions

We explain how one can use this operational interpretation of forcing to give a new computational interpretation of an universal quantification $\forall : (C \to N_2) \to N_2$ on Cantor space. There are already computational interpretations (Escardo 2007; Simpson 1998), using a general recursive program.[2] The interpretation we suggest relies on iterating the previous construction and introducing infinitely generic functions $\mathsf{f}_0, \mathsf{f}_1, \ldots$ It is reminiscent of iterated forcing in set theory, and of the interpretation of choice sequences in intuitionism (Troelstra and van Dalen 1988).

The first step is to extend the notion of condition. So far, a condition p represents a compact open subset of Cantor space. We can in the same way consider conditions $r, s, \ldots$ which represent compact open subsets of the product space $C^{\mathbb{N}}$. The elementary conditions are now of the form $\mathsf{f}_l\ k = i$, given an information about the generic function f_l, and a condition r is a finite product of compatible elementary conditions. The set of conditions P is the union of the sets P_n of condition containing only $\mathsf{f}_l\ k = i$ with $l < n$. The conditions we need $p, q, \ldots$ are pairs $p = (r, n)$, with r in P_n. Such a condition represents a compact open subset X of C^n. We define $(s, m) \leq (r, n)$ to mean $n \leq m$ and $s \leq r$. To summarize, each condition $p = (r, n)$ represents a finite amount of information about a finite number of generic functions, and to refine this condition we can either add new informations, or add a new generic function. (Intuitively, the conditions represent compact open subsets of a "variable" space.)

The reduction relations $pt \to \alpha$, $pt \Rightarrow \alpha$ are as before, with $p = (n, r)$, and t a term which may contain $\mathsf{f}_0, \ldots, \mathsf{f}_{n-1}$ and α is now a formal sum $\Sigma p_i t_i$ where $p_i = (n, r_i)$ and (r_i) is a partition of r.

A.2.2 Universal Quantification as Projection

An element r of P_n represents a compact open subset X of C^n. A formal sum of Booleans $\alpha = \Sigma p_i v_i$ with $p_i = (n, r_i)$ and r_i partition of r represents a continuous function f_α from X to the discrete space N_2.

We define the conjunction operation on formal sums of Booleans $\alpha \wedge \beta$ as

$$(\Sigma p_i v_i) \wedge (\Sigma q_j w_j) = \Sigma p_i q_j (v_i \wedge w_j)$$

in such a way that we have $f_{\alpha \wedge \beta} = f_\alpha \wedge f_\beta$.

[2]The termination of this program relies on classical logic and the fact that definable functionals are continuous.

If r is a condition in P_{n+1}, we can write $r = r's$ with r' in P_n and s a product of conditions of the form $\mathsf{f}_n\ k = i$. The condition $(n+1, r)$ can thus be thought as representing a product $X \times Y$, with $X \subseteq C^n$ corresponding to the condition (n, r') and Y corresponding to s. If we consider a partition (r_i) of r in P_{n+1}, the formal sum $\alpha = \Sigma p_i v_i$, with $p_i = (n+1, r_i)$ represents a continuous function $f_\alpha : X \times Y \to N_2$. We are going to define the formal sum $\mathsf{p}(\alpha) = \Sigma(n, s_j)w_j$ which represents the function $f_{\mathsf{p}(\alpha)} : X \to N_2$ such that $f_{\mathsf{p}(\alpha)}(x) = 1$ iff $f_\alpha(x, y) = 1$ for all y in Y.

This definition is by induction on the fact that (r_i) is a partition of r. If (r_i) is the unit partition then we take $\mathsf{p}((n+1, r)v) = (n, r')v$. If it is a partition formed of a partition $(r_i,\ i \in I_0)$ of $r(\mathsf{f}_l\ k = 0)$ and a partition $(r_i,\ i \in I_1)$ of $r(\mathsf{f}_l\ k = 1)$, we can consider by induction

$$\beta_0 = \mathsf{p}(\Sigma_{i \in I_0} p_i v_i) \qquad \beta_1 = \mathsf{p}(\Sigma_{i \in I_1} p_i v_i)$$

If $l = n$, we define $\mathsf{p}(\alpha) = \beta_0 \wedge \beta_1$ and if $l < n$, we define $\mathsf{p}(\alpha) = \beta_0 + \beta_1$.

A.2.3 Computation Rules

The only new reduction rule is the following

$$\frac{(n+1, r)(F\ \mathsf{f}_n) \Rightarrow \alpha}{(n, r)(\forall\ F) \to \mathsf{p}(\alpha)}$$

The intuition is that we want to compute $\forall\ F$ and we know that F mentions only the generic functions $\mathsf{f}_0, \ \ldots, \ \mathsf{f}_{n-1}$, satisfying the condition r. We compute then $F\ \mathsf{f}_n$, where f_n is "fresh" for F, and from the result of this computation we can compute $\forall\ F$ using the function p.

The computability relation $p \Vdash \varphi_A(t)$ is defined as before, for $p = (n, r)$ and t a term which may contain $\mathsf{f}_0, \ldots, \mathsf{f}_{n-1}$.

Lemma 10.7. *All constant f_l are computable, i.e. $(n, 1) \Vdash \varphi_C(\mathsf{f}_l)$ if $l < n$. The constant $\forall$ is computable, i.e. $\Vdash \varphi_{C \to N_2}(\forall)$.*

Proof. The proof that f_l is computable is the same as the proof of Lemma 10.6.

If we have $(n, r) \Vdash \varphi_{C \to N_2}(F)$ we show that $(n, r) \Vdash \varphi_{N_2}(\forall\ F)$. For this it is enough to show that $(n+1, r) \Vdash \varphi_{N_2}(F\ \mathsf{f}_n)$, which follows from $(n, r) \Vdash \varphi_{C \to N_2}(F)$ and $(n+1, r) \Vdash \varphi_C(\mathsf{f}_n)$.

References

Barendregt, H. 1997. The impact of the lambda calculus. *Bulletin of Symbolic Logic* 3: 181–215.
Bishop, E. 1970. Mathematics as a numerical language. In *Intuitionism and proof theory*, ed. A. Kino, J. Myhill, and R.E. Vesley. Amsterdam: North-Holland.

Escardo, M. 2007. Infinite sets that admit fast exhaustive search. In *LICS 2007*, Wroclaw, 443–452.
Coquand, Th., and G. Jaber. 2010. A note on forcing in type theory. to appear in *Fundamenta Informatica* 100: 43–52.
Gödel, K. 1990. On a hitherto unexploited extension of the finitary standpoint. In *Collected Works*, vol. II. Publications 1938–1974. New York: Oxford University Press.
Goodman, N. 1978. Relativised realizability in intuitionistic arithmetic at all finite types. *Journal of Symbolic Logic* 43: 23–44.
Martin-Löf, P. 1998. An intuitionistic theory of types In *Twenty-five years of type theory*, ed. G. Sambin and J. Smith. New York: Oxford University Press (reprinted version of an unpublished report from 1972).
Simpson, A. 1998. Lazy functional algorithms for exact real functionals. In *Mathematical foundations of computer science 1998*, Lecture notes in computer science, vol. 1450, ed. L. Brim, J. Gruska, and J. Zlatuška 456–464. Berlin: Springer.
Troelstra, A.S., and D. van Dalen. 1988. *Constructivism in mathematics*, vol. II. Amsterdam: North-Holland.
Wadler, Ph. 1992. The essence of functional programming. In *Conference record of the nineteeth annual symposium of principle of programming languages*, New Mexico.

Chapter 11
Program Testing and the Meaning Explanations of Intuitionistic Type Theory

Peter Dybjer

11.1 Introduction

Consider the following often cited remark by Knuth (1977):

> Beware of bugs in the above code; I have only proved it correct, not tried it.

How come? If you have proved your program correct, you should be certain that it works! However, several things can go wrong:

- The formal specification may fail to capture the intended behaviour of the program.
- The formal representation of the program in the logical system may fail to capture what actually happens when you run the program.
- The proof can be wrong. This easily happens with manual proofs, but even mechanically assisted proofs can be wrong. The logical principles may not be implemented correctly.

What does this have to do with the foundations of mathematics? There is yet another possibility:

- The logical principles employed may themselves be wrong! Maybe the logical system is inconsistent.

Can you "test" a logical law? Does the following make sense?

> Beware of bugs in the above proof; I have only followed inference rules, not run it.

However, you cannot in general test a proof in the same way as you can test a program. For example, how would you "run" a proof in Zermelo-Fraenkel set theory?

P. Dybjer (✉)
Department of Computer Science and Engineering, Chalmers University of Technology, Rännvägen 6, 412 96 Göteborg, Sweden
e-mail: peterd@chalmers.se

P. Dybjer et al. (eds.), *Epistemology versus Ontology*, Logic, Epistemology, and the Unity of Science 27, DOI 10.1007/978-94-007-4435-6_11,

On the other hand, in Martin-Löf's *intuitionistic type theory* you can run proofs. In this theory the basic unit is that of a *judgement*. There are four forms: A *type*, $A = A'$, $a \in A$, $a = a' \in A$. Each of these can be hypothetical, that is, depend on a context $x_1 \in A_1, \ldots, x_n \in A_n$. I shall argue that the following makes sense:

> When you've made your judgement evident to yourself, then you'd better run it, to make sure it's valid!

I shall explain what I mean by this, and let me immediately say that I do not mean running the type-checking algorithm used by proof assistants based on intuitionistic type theory! Instead I will base the discussion on the computation of closed expressions to canonical form, which underlies Martin-Löf's meaning explanations from the paper *Constructive Mathematics and Computer Programming* (Martin-Löf 1982). We shall see how the testing point of view provides a way to reformulate the meaning explanations using vocabulary from programming rather than from philosophy and logic. In other words, we look at the meaning explanations from the point of view of the computer programmer (or better the computer user) rather than from the point of view of the constructive mathematician.

Originally, the testing point of view that we explore in this paper was not meant to provide alternative meaning explanations of intuitionistic type theory, only an alternative *presentation* of these meaning explanations. Nevertheless, it seems that my interpretation of both hypothetical judgements and of type equality differs from Martin-Löf's (1982, 1984). A consequence is that I only provide meaning to identity types $I(A, a, b)$ if the equality on A is a decidable, but not otherwise, for example, when A is a function type $N \to N$.

Another aim has been to pave the way for meaning explanations for other systems than Martin-Löf type theory. If the essence of the meaning explanations is that they explain how to test judgements, then we can ask ourselves whether we can write testing manuals for other logical systems than intuitionistic type theory. As an example, we shall discuss how to test the judgements of Coquand and Huet's Calculus of Constructions (Coquand and Huet 1988), an impredicative intuitionistic type theory.

In the future I hope to provide testing interpretations of systems including coinductive types and partial types and functions. It would also be interesting to rephrase insights about the computational content of classical logic (Coquand 1995) as Martin-Löf style meaning explanations. A research program with this aim is Hayashi's *Limit Computable Mathematics*: "To test formalization of proofs by experiments (animation) via Gold's limiting recursive functions" (Hayashi 2007). Already in the 1980s Hayashi pioneered the idea of *proof animation*, whereby he utilized the Curry-Howard isomorphism to test formal proofs in his system PX (Hayashi and Nakano 1989) in much the same way as one tests computer programs (Hayashi et al. 2002). Limit Computable Mathematics is the extension of this idea to classical logic. Hayashi's ideas have been an important source of inspiration for the present work.

11.1.1 Plan of the Paper

In Sect. 11.2 we review the meaning explanations for intuitionistic type theory without committing ourselves to either a *pre-mathematical* or a *meta-mathematical* interpretation. In Sect. 11.3 we recall a meta-mathematical realizability interpretation following Allen (1987a). The reader who is familiar with the meaning explanations can skip Sects. 11.2 and 11.3 and go straight to Sect. 11.4 which is the principal part of the paper. There we propose a pre-mathematical testing interpretation involving evaluation of terms and generation of inputs. In Sect. 11.5 we briefly discuss the possibility of a pre-mathematical testing interpretation for the Calculus of Constructions.

11.2 Meaning Explanations

11.2.1 History

Martin-Löf's meaning explanations for intuitionistic type theory were first presented in 1979 in the paper *Constructive mathematics and computer programming* (Martin-Löf 1982). These ideas were elaborated on in the book *Intuitionistic Type Theory* (Martin-Löf 1984). In both works meaning explanations are used to justify an extensional polymorphic version of intuitionistic type theory. Another useful reference for the philosophical basis of meaning explanations is *On the meaning of the logical constants and the justification of the logical laws* (Martin-Löf 1996) from 1983, although this is concerned with meaning explanations for ordinary intuitionistic predicate logic (without formal proof objects) rather than for intuitionistic type theory.

The meaning explanations are also refered to as *direct semantics*, *intuitive semantics*, *informal semantics*, *standard semantics*, or the *syntactico-semantical* approach to meaning theory. Semantics is here understood *pre-mathematically* rather than *meta-mathematically*, as is clear from the following quotation from the first paragraph of *Intuitionistic Type Theory*.

> Mathematical logic and the relation between logic and mathematics have been interpreted in at least three different ways:
>
> 1. Mathematical logic as symbolic logic, or logic using mathematical symbolism;
> 2. Mathematical logic as foundations (or philosophy) of mathematics;
> 3. Mathematical logic as logic studied by mathematical methods, as a branch of mathematics.
>
> We shall here mainly be interested in mathematical logic in the second sense. What we shall do is also mathematical logic in the first sense, but certainly not in the third.

In the present paper we shall also do mathematical logic in the third sense, when we interpret the meaning explanations meta-mathematically in Sect. 11.3.

Hilbert's original use of the word meta-mathematics assumed that only finitistic methods were used on the meta-level. However, in Sect. 11.3 we assume that general mathematical (classical set-theoretic) methods are available, although we could argue informally that we only use parts of set theory which are constructively valid. In Sect. 11.4, we shall interpret the meaning explanations in the second sense, that is, pre-mathematically, which means that we restrict ourselves to everyday concepts relating to programming: running a program, observing its result, generating input, etc.

The meaning explanations were a profound contribution to the semantics of intuitionistic type theory (and to intuitionism in general) when they were first presented in 1979. Prior versions of intuitionistic type theory did not come with such meaning explanations, but were justified by normalization proofs. Consistency then follows from the Church-Rosser property. The first version, *A theory of types* (Martin-Löf 1971), which had an axiom that there is a type of all types, was actually inconsistent, although it was proved to have the normalization property. The problem was that the meta-theory of the normalization proof was itself inconsistent. This problem was rectified in *An intuitionistic theory of types* from 1972 (Martin-Löf 1998), where the type of all types was replaced by a type of *small* types (a universe), and a correct normalization proof was provided. The first published version of type theory *An intuitionistic theory of types: predicative part* from 1973 (Martin-Löf 1975) had an infinite sequence of universes and contained a proof of normalization by an intuitionistic, but meta-mathematical, model construction.

11.2.2 Terms and Computation Rules

We shall now give an informal account of Martin-Löf's meaning explanations for the fragment of his intuitionistic type theory where the only types are natural numbers N, identity types $I(A, a, b)$, cartesian products of families of types $\Pi(A, B)$, and a universe of small types U. These type formers suffice to illustrate the main points. The discussion of other type formers is analogous.

We shall stay rather close to the accounts given in *Constructive Mathematics and Computer Programming* (Martin-Löf 1982) and *Intuitionistic Type Theory* (Martin-Löf 1984), although there will be three (relatively minor) differences:

- We will present type theory based on a theory of expressions (as proposed in the preface of *Intuitionistic Type Theory*, Martin-Löf 1984, see also Nordström et al. 1989). This choice is inessential; similar meaning explanations can be provided for both earlier and later versions of intuitionistic type theory, and can also be extended to deal with other types if details are modified suitably.
- Another inessential difference is that we will present both computation rules and meaning explanations using inference rule notation, although such notation is

not employed in (Martin-Löf 1982, 1984). Later on we will discuss in detail how to read these inference rules both meta-mathematically (Sect. 11.3) and pre-mathematically (Sect. 11.4).

- Martin-Löf (1982, 1984) actually considers an extensional type equality: two canonical types are equal iff they have equal objects and equal object equality. However, in our testing semantics it will be easier to justify an intensional type equality, where two canonical types are equal only if they begin with the same type constructor. Both extensional and intensional type equality were previously discussed by Allen (1987a,b).

The *theory of expressions* is nothing but the simply typed lambda calculus with one base type ι. The types are called "arities" to distinguish them from the types of type theory which will be introduced later. Abstraction in the theory of expressions is written $(x)a$, and application is written $f(a)$. Furthermore, we add constants for the type formers, and for the canonical and non-canonical term constructors. The arities of the constants in our fragment of intuitionistic type theory are

$$
\begin{aligned}
N &: \iota \\
0 &: \iota \\
s &: \iota \to \iota \\
R &: \iota \to \iota \to (\iota \to \iota \to \iota) \to \iota \\
I &: \iota \to \iota \to \iota \to \iota \\
r &: \iota \\
J &: \iota \to \iota \to \iota \\
\Pi &: \iota \to (\iota \to \iota) \to \iota \\
\lambda &: (\iota \to \iota) \to \iota \\
Ap &: \iota \to \iota \to \iota \\
U &: \iota
\end{aligned}
$$

We abbreviate $f(a_1, \ldots, .a_n) = f(a_1) \cdots (a_n)$. Moreover, $(\lambda x)a = \lambda((x)a)$ and $(\Pi x \in A)B = \Pi(A, (x)B)$.

Universes are formulated à la Russell (1984) and there is a common syntactic category of types and terms.

Judgements are interpreted in terms of the relation $a \Rightarrow v$ between *closed* terms of base arity, meaning "a has canonical form v". Canonical forms are *lazy*:

$$v ::= \Pi(a, a)\,\lambda(a) \mid 0 \mid s(a) \mid I(a, a, a) \mid r \mid U \mid$$

where a ranges over arbitrary, not necessarily canonical, terms. The canonical form relation is given by the following computation rules:

$$\frac{c \Rightarrow 0 \quad d \Rightarrow v}{R(c,d,e) \Rightarrow v} \qquad \frac{c \Rightarrow s(a) \quad e(a, R(a,d,e)) \Rightarrow v}{R(c,d,e) \Rightarrow v}$$

$$\frac{c \Rightarrow r \quad d \Rightarrow v}{J(c,d) \Rightarrow v}$$

$$\frac{c \Rightarrow \lambda(b) \quad b(a) \Rightarrow v}{Ap(c,a) \Rightarrow v}$$

in addition to the rule

$$v \Rightarrow v$$

stating that a canonical term has itself as value.

11.2.3 The Meaning of Judgement Forms

The general principle is that if A is a type, then it has a canonical type as value:

$$\frac{A \Rightarrow C(a_1, \ldots, a_m) \quad \cdots}{A \; type}$$

where C is an m-place type constructor and $\cdots$ stand for additional requirements on $a_1, \ldots .a_m$, which are part of the definition of what it means to be a type.

The types A and A' are equal if their canonical forms begin with the same type constructor:

$$\frac{A \Rightarrow C(a_1, \ldots, a_m) \quad A' \Rightarrow C(a'_1, \ldots, a'_m) \quad \cdots}{A = A'}$$

where $\cdots$ stand for additional requirements on $a_1, a'_1, \ldots, a_m, a'_m$,. (Note again that this an intensional notion of type equality which differs from the extensional type equality in Martin-Löf's (1982) meaning explanations.)

Furthermore, a is an element of A provided

$$\frac{A \Rightarrow C(a_1, \ldots, a_m) \quad a \Rightarrow c(b_1, \ldots, b_n) \quad \cdots}{a \in A}$$

where c is an n-place term constructor for C, and where $\cdots$ stand for additional requirements on $a_1, \ldots, a_m, b_1, \ldots, b_n$.

The elements a and a' are equal elements of A provided

$$\frac{A \Rightarrow C(a_1, \ldots, a_m) \quad a \Rightarrow c(b_1, \ldots, b_n) \quad a' \Rightarrow c(b'_1, \ldots, b'_n) \quad \cdots}{a = a' \in A}$$

and where $\cdots$ stand for additional requirements on $a_1, \ldots, a_m, b_1, b'_1 \ldots, b_n, b'_n$.

11.2.4 General Schema for Type Formers

Martin-Löf does not elaborate on what is allowed as $\cdots$ in the schematic rules above. The schema for inductive-recursive definitions (Dybjer 2000; Dybjer and Setzer 1999, 2006) provides a general form which covers most type formers existing in the literature. Beyond that, Setzer has proposed several proof-theoretically stronger types: a Mahlo universe (Setzer 2000) and in unpublished work an autonomous Mahlo universe and a Π_3-reflecting universe. We hope that the present article helps explaining what is involved in claiming that these large types are constructively valid, and why their justification is *predicative*, although we do not discuss their testing semantics explicitly.

In this article, however, we restrict ourselves to a few crucial instances of canonical types: N, $I(A, a, b)$, $\Pi(A, B)$, and U.

11.2.5 Natural Numbers

$$\frac{A \Rightarrow N}{A\ type}$$

$$\frac{A \Rightarrow N \quad A' \Rightarrow N}{A = A'}$$

$$\frac{A \Rightarrow N \quad a \Rightarrow 0}{a \in A} \qquad \frac{A \Rightarrow N \quad a \Rightarrow s(b) \quad b \in N}{a \in A}$$

$$\frac{A \Rightarrow N \quad a \Rightarrow 0 \quad a' \Rightarrow 0}{a = a' \in A} \qquad \frac{A \Rightarrow N \quad a \Rightarrow s(b) \quad a' \Rightarrow s(b') \quad b = b' \in N}{a = a' \in A}$$

11.2.6 Identity Types

$$\frac{A \Rightarrow I(B, b, b') \quad b \in B \quad b' \in B}{A\ type}$$

$$\frac{A \Rightarrow I(B, b, c) \quad A \Rightarrow I(B', b', c') \quad B = B' \quad b = b' \in B \quad c = c' \in B}{A = A'}$$

$$\frac{A \Rightarrow I(B, b, b') \quad a \Rightarrow r \quad b = b' \in B}{a \in A}$$

$$\frac{A \Rightarrow I(B, b, b') \quad a \Rightarrow r \quad a' \Rightarrow r \quad b = b' \in B}{a = a' \in A}$$

11.2.7 Remarks on the Meaning of Identity Types

Note that the judgement $A = A'$ presupposes the judgements A, A' $type$, since the former judgement can be valid only if the latter are. Similarly, $a \in A$ presupposes A $type$, and $a = a' \in A$ presupposes $a, a' \in A$, and A $type$. We do not write out presupposed judgements in the rules.

Note that $r \in I(B, b, b')$ iff $b = b' \in B$. Hence these meaning explanations justify the rule

$$\frac{\Gamma \vdash c \in I(B, b, b')}{\Gamma \vdash b = b' \in B}$$

which is present in extensional type theory, but not in intensional type theory (Martin-Löf 1975, 1998; Nordström et al. 1989).

Note that we can form the type $I(B, b, b')$ for any type B in intuitionistic type theory (Martin-Löf 1975, 1982, 1984), and that this construction will be validated by the meta-mathematical model construction (realizability) in the next section. However, this rule will not be validated by our pre-mathematical testing semantics for general B, only for B with decidable equality. (This does not mean that we cannot define a type expressing the extensional equality of two functions, only that this type is not primitive.)

11.2.8 Function Types

$$\frac{A \Rightarrow \Pi(B, C) \quad y \in B \vdash C(y)\ type}{A\ type}$$

$$\frac{A \Rightarrow \Pi(B, C) \quad A' \Rightarrow \Pi(B', C') \quad B = B' \quad y \in B \vdash C(y) = C'(y)}{A = A'}$$

$$\frac{A \Rightarrow \Pi(B, C) \quad a \Rightarrow \lambda(c) \quad y \in B \vdash c(y) \in C(y)}{a \in A}$$

$$\frac{A \Rightarrow \Pi(B, C) \quad a \Rightarrow \lambda(c) \quad a' \Rightarrow \lambda(c') \quad y \in B \vdash c(y) = c'(y) \in C(y)}{a = a' \in A}$$

Note that some premises are hypothetical judgements. For example, $y \in B \vdash C(y)\ type$ means that $C(y)$ is a type under the assumption that $y \in B$. At this point we shall not discuss the meaning of hypothetical judgements further. The reader is referred to Sect. 11.3 for a meta-mathematical interpretation and to Sect. 11.4 for a pre-mathematical interpretation. Note that we have again omitted presupposed judgements, and that if $x \in A$ is an assumption in a valid hypothetical judgement, then A $type$ is also valid and can be presupposed.

11.2.9 The universe (à la Russell) of Small Types

$$\frac{A \Rightarrow U}{A\ type}$$

$$\frac{A \Rightarrow U \quad A' \Rightarrow U}{A = A'}$$

$$\frac{A \Rightarrow U \quad a \Rightarrow N}{a \in A} \qquad \frac{A \Rightarrow U \quad a \Rightarrow I(b,c,d) \quad b \in U \quad c,d \in b}{a \in A}$$

$$\frac{A \Rightarrow U \quad a \Rightarrow \Pi(b,c) \quad b \in U \quad y \in b \vdash c(y) \in U}{a \in A}$$

$$\frac{A \Rightarrow U \quad a \Rightarrow N \quad a' \Rightarrow N}{a = a' \in A}$$

$$\frac{A \Rightarrow U \quad a \Rightarrow I(b,c,d) \quad a' \Rightarrow I(b',c',d') \quad b = b' \in U \quad c = c' \in b \quad d = d' \in b}{a = a' \in A}$$

$$\frac{A \Rightarrow U \quad a \Rightarrow \Pi(b,c) \quad a' \Rightarrow \Pi(b',c') \quad b = b' \in U \quad y \in b \vdash c(y) = c'(y) \in U}{a = a' \in A}$$

11.2.10 Other Inductive and Inductive-Recursive Types

Analogous rules can be given for the remaining type formers of intuitionistic type theory (Martin-Löf 1982, 1984). Even more, analogous rules can be given for inductive types and inductive-recursive types (Dybjer 2000; Dybjer and Setzer 1999). It would be interesting to discuss the rules for inductive and inductive-recursive families (Dybjer 2000; Dybjer and Setzer 2006) from the testing point of view, but this is outside the scope of the present paper.

11.3 Meta-mathematical Reading: Realizability

We shall now interpret the meaning explanations meta-mathematically, using set theory as meta-language and essentially following Allen (1987a,b). It is sometimes said that the meaning explanations are nothing but a realizability interpretation (in the sense of Kleene), but this is fundamentally misleading. Realizability provides a meta-mathematical and not a pre-mathematical interpretaion! Nevertheless, it helps us to be precise and to understand the details involved in the meaning explanations.

Realizability interpretations of intuitionistic type theory go back to Aczel (1977b, 1980), Beeson (1982), and Smith (1984). Note that Aczel (1980), Beeson (1982), and Allen (1987a,b) are "semantic" interpretations where set theory is the meta-language, whereas Aczel (1977b) and Smith (1984) provide syntactic translations into versions of intuitionistic predicate logic. Allen's version is closest to Martin-Löf's meaning explanations; it is a rather direct mathematical interpretation of the relevant passages in Martin-Löf (1982, 1984).

In the meta-mathematical interpretation, the set of terms and the arity relation are defined inductively (as usual for the simply typed lambda calculus). Furthermore, the canonical terms is an inductively defined subset of the set of terms and the canonical form relation is an inductively defined relation between terms and canonical terms of arity ι.

The question is then how to give a meta-mathematical interpretation of the rules providing the meaning of the judgements? A first try would be to let them inductively generate four relations on terms – one for each form of judgement. However, this would not yield a positive inductive definition, since the $\in$-relation appears negatively in the rules for Π and U. To turn these rules into a meaningful set-theoretic definition is the main problem which was solved in different ways by Aczel, Beeson, and Allen.

Allen uses the fact that it suffices to interpret the judgement $a = a' \in A$; the interpretation of the other judgement can then also be derived, like in other work on partial equivalence relation (per) models. He defines a relation $\mathcal{E}l \subseteq \Lambda \times \mathcal{P}(\Lambda^2)$, such that $a = a' \in A$ is valid in the model iff there is an $\mathcal{R}$ such that $(A, \mathcal{R}) \in \mathcal{E}l$ and $(a, a') \in \mathcal{R}$. In other words, $\mathcal{E}l$ is the graph of the function which maps a type A to its notion of element equality, a partial equivalence relation on Λ. With this method the rules can be understood as the rules of a positive inductive definition of $\mathcal{E}l$.

Allen follows Martin-Löf (1982) and interprets type equality extensionally: two types are equal iff they have the same canonical elements and the same equal canonical elements. However, he points out that if we want to interpret a theory with intensional type equality then we should instead inductively generate $\mathcal{E}l \subseteq \Lambda^2 \times \mathcal{P}(\Lambda^2)$, but does not present the details. However, since our testing interpretation in the next section naturally gives rise to intensional type equality we shall sketch its realizability counterpart here.

The $\mathcal{E}l$-relation interprets an auxiliary heterogenous equality judgement:

- $A \in a = a' \in A'$ is valid in the model iff there is an $\mathcal{R}$ such that $((A, A'), \mathcal{R}) \in \mathcal{E}l$ and $(a, a') \in \mathcal{R}$.

$\mathcal{E}l$ is the graph of a function which maps equal types to the per they denote. The domain of this function is a per $\sim$ which will interpret type equality:

- $A = A'$ is valid in the model iff here is an $\mathcal{R}$ such that $((A, A'), \mathcal{R}) \in \mathcal{E}l$.

The other judgement forms are interpreted as follows:

- $A\ type$ is interpreted as $A = A$;
- $a = a' \in A$ is interpreted as $A \in a = a' \in A$;
- $a \in A$ is interpreted as $a = a \in A$;

The interpretation of a hypothetical judgement is

- $x_1 \in A_1, \dots, x_n \in A_n \vdash \mathcal{J}$ is interpreted as whenever $a_1 \in A_1, \dots, a_n \in A_n$ then $\mathcal{J}[a_1/x_1, \dots, a_n/x_n]$.

We will use Aczel's rule sets (Aczel 1977a) to present the inductive definition of $\mathcal{E}l$. In this setting an inductive definition is given by a set of rules, where a *rule* on a set V is a pair $\frac{X}{x}$, such that $X \subseteq V$ and $x \in V$. A set $A \subseteq V$ is closed under this rule iff $X \subseteq A$ implies $x \in A$. A is said to be inductively generated by a set of rules Φ iff it is the least set closed under all rules in Φ. (Aczel's notion of inductive definition given by rule sets is equivalent to the notion given by monotone operators.)

To interpret N we inductively generate the per $\mathcal{N}$ of equal natural numbers. The rule set is

$$\left\{ \frac{\emptyset}{(a,a')} \mid a \Rightarrow 0, a' \Rightarrow 0 \right\} \cup \left\{ \frac{\{(b,b')\}}{(a,a')} \mid a \Rightarrow s(b), a' \Rightarrow s(b') \right\}$$

To interpret Π-types, we define the cartesian product of a doubly indexed famiy of pers. Let $\mathcal{Q}$ be a per, and $\mathcal{R}(y, y')$ be a per for y, y' such that $y\mathcal{Q}y'$. Then

$$\prod(\mathcal{Q}, \mathcal{R}) = \{(a,a') | a \Rightarrow \lambda(c), a' \Rightarrow \lambda(c'), (\forall y\mathcal{Q}y')c(y)\mathcal{R}(y,y')c'(y')\}$$

To interpret I-types we let $\mathcal{Q}$ be a per, and define the per of identity proofs:

$$\mathcal{I}(\mathcal{Q}, b, c) = \{(a,a') | a \Rightarrow r, a' \Rightarrow r, b\mathcal{Q}c\}$$

To interpret the type of small types U we first inductively generate the relation $\mathcal{T} \subseteq \Lambda^2 \times \mathcal{P}(\Lambda^2)$, the graph of the function mapping two equal *small* types to the per they denote.

$$\left\{ \frac{\emptyset}{((A,A'),\mathcal{N})} | A \Rightarrow N, A' \Rightarrow N \right\}$$

$$\cup$$

$$\left\{ \frac{\{((B,B'),\mathcal{Q})\} \cup \{((C(y), C'(y'), \mathcal{R}(y,y')) | y\mathcal{Q}y'\}}{((A,A'), \prod(\mathcal{Q},\mathcal{R}))} | \right.$$
$$\left. A \Rightarrow \Pi(B,C), A' \Rightarrow \Pi(B',C') \right\}$$

$$\cup$$

$$\left\{ \frac{\{((B,B'),\mathcal{Q})\}}{((A,A'),\mathcal{I}(\mathcal{Q},b,c))} | A \Rightarrow I(B,b,c), A' \Rightarrow I(B',b',c'), b\mathcal{Q}b', c\mathcal{Q}c' \right\}$$

We then interpret U as the per $\mathcal{U}$ of equal small types, that is, the domain of $\mathcal{T}$:

$$\mathcal{U} = \{(A,A') | \exists \mathcal{R}.(A,A')\mathcal{T}\mathcal{R}\}$$

Finally we inductively generate $\mathcal{E}l \subseteq \Lambda^2 \times \mathcal{P}(\Lambda^2)$, the graph of the function mapping two equal possibly *large* types to the per they denote. The rules for $\mathcal{E}l$ is the union of the rule set for $\mathcal{T}$ and the rule set

$$\left\{ \frac{\emptyset}{((A, A'), \mathcal{U})} | A \Rightarrow U, A' \Rightarrow U \right\}$$

As explained above we can interpret all the judgement forms of intuitionistic type theory using $\mathcal{E}l$, and also show that all the inference rules of this theory are valid. However, we refer to Allen for the details and limit ourselves to validate the rule of N-elimination:

$$\frac{C(x)\ type \qquad a \in N \qquad d \in C(0) \qquad x \in N, y \in C(x) \vdash e(x, y) \in C(s(x))}{R(a, d, e) \in C(a)}$$

We have here omitted the context Γ which is common to the premises and the conclusion.

To prove the validity of this rule we do induction on $\mathcal{N}$, the interpretation of N. We know that if $a \in N$ is valid, then either $a \Rightarrow 0$ or $a \Rightarrow s(b)$ where $b \in N$ is valid. If $a \Rightarrow 0$ then $R(a, d, e) \Rightarrow v$ iff $d \Rightarrow v$. Thus $d \in C(0)$ implies $R(a, d, e) \in C(a)$, since $\in$ is invariant under evaluation: if $a \Rightarrow v$ and $a' \Rightarrow v'$ then $a \in C(a')$ iff $v \in C(v')$ (a lemma to be proved). The validation of the case where $a \Rightarrow s(b)$ is similar.

11.4 Pre-mathematical Reading: A Testing Manual for Intuitionistic Type Theory

The meta-mathematical reading of the meaning explanations does not have foundational significance. To prove that mathematical induction (N-elimination) is valid in the model we rely on (something more complex than) the principle of mathematical induction in the meta-language. Similarly, to justify the correctness of proof-theoretically strong extensions of Martin-Löf type theory, such as Setzer's Mahlo universe (Setzer 2000), autonomous Mahlo universe, and Π_3-reflecting universe, we need analogous notions in the meta-language: Mahlo cardinals, autonomous Mahlo cardinals, and Π_3-reflecting cardinals.

Instead we shall understand Martin-Löf's meaning explanations as a manual for testing the validity of judgements. For example, to test the rule of N-elimination, we test the primitive recursion combinator $R(c, d, e)$ for $c = 0, c = s(0), c = s(s(0)), \ldots$, and for arbitrary base case d and arbitrary step case e which satisfy the assumptions of the rule. Note that, such tests are necessarily incomplete since there are infinitely many possible inputs. As in inductive inference in philosophy of science, a judgement is a conjecture which can be corroborated by a successful test or refuted by an unsuccessful test, see Popper (1963). A justification of the rule of induction is a justification of our belief that all tests of the primitive recursion combinator will succeed. Such justifications are of course fallible, although the risk that the rule of N-elimination is incorrect might seem slim. However, if we remember that it is only meaningful to test an *implementation* of the primitive recursion combinator, we realize that there is a real danger of falsification.

Similarly, to give meaning explanations for Setzer's Mahlo universe, autonomous Mahlo universe, and Π_3-reflecting universe involves explaining how judgements containing them are to be tested. To justify the rules for these universes we must explain why there is reason to believe that all such tests will succeed. Here the risk of falsification is greater, although these universes are defined in such a way that it should be possible to "see" that they pass all tests. This is achieved by explaining how they are "built up from below" or how some measure decreases during computation over them. Again, such explanations are made by humans and fallible. We cannot say that these rules are constructively valid in an absolute sense, only that we believe them to be constructively valid in the sense that we believe that they will pass all tests. And when we say that they are "predicatively valid" we mean that we can provide a justification which suggests a "well-founded" structure. Again, such a justification is fallible.

The program testing point of view emphasizes that meaning explanations are about what *really* happens! It is a process in space-time. An expression is something static (in space). Computation is something dynamic (in time). The formal proposition $a \Rightarrow v$ is a static mathematical representation of the *real* fact that a will turn into v a little later after some computation is done. This relates to Martin-Löf's term "syntactico-semantical". Semantics is what happens during execution. The meaning is the extension which is gradually unfolded as time goes by.

Coming back to the remark in the beginning of the introduction: the above discussion of validity and testing is rather obvious in the context of software testing. It is something hardly worth saying. But when we present the same idea in the context of intuitionistic type theory it nevertheless seems surprising.

11.4.1 Is Mathematics an Empirical Science?

The relevance of a similar perspective has recently been argued by Miquel in his essay *The experimental effectiveness of mathematical proof* (Miquel 2010, p. 38):

> We can thus argue (against Popper) that mathematics fulfills the demarcation criterion that makes mathematics an empirical science. The only specificity of mathematics is that the universal empirical hypothesis underlying mathematics is (almost) never stated explicitly.

11.4.2 Pre-mathematical Understanding of Terms and Computation Rules

We now read the grammar and typing rules as concrete prescriptions for generating well-formed terms of type theory. Contrast this to the meta-mathematical interpretation in terms of the mathematical concept of an inductive definition. Similarly, the computation rules should be read as concrete prescriptions for how to evaluate a term to canonical form. Here this is a prescription for manual evaluation of a term, and hopefully this prescription is sufficiently precise to reproduce the same value each time. Alternatively, we can replace this prescription by an implementation, a real machine which can perform the evaluation.

11.4.3 Testing Categorical Judgements

We shall now read the rules of the meaning explanations, which in the meta-mathematical interpretation were made to inductively generate the realizability interpretation, as a *testing manual*, a manual for *falsification of conjectures*, or *bug-finding*. A tester only needs to push a button "execute program" and inspect the results. He or she is only a "user" who does not need to know logic or programming.

We first show how to test the *categorical* judgements, that is, judgements without hypotheses.

11.4.4 How to Test A Type?

The relevant rules are

$$\frac{A \Rightarrow N}{A\ type} \qquad \frac{A \Rightarrow I(B,b,c) \quad b,c \in B}{A\ type}$$

$$\frac{A \Rightarrow \Pi(B,C) \quad y \in B \vdash C(y)\ type}{A\ type} \qquad \frac{A \Rightarrow U}{A\ type}$$

As before, we do not write out presupposed assumptions. For example, the assumption that $B\ type$ in the second and third rules is omitted. This is reflected in the testing manual: we do not need to test presupposed assumptions, since their validity is a consequence of a successful test of another assumption.

To test $A\ type$ we always begin by evaluating A to canonical form, where the outermost form is a constructor.

- If it has canonical form N, then the test is successful.
- If it has canonical form $I(B,b,c)$, then test $b \in B$ and $c \in B$.

- If it has canonical form $\Pi(B,C)$, then test $B\ type$ and $y \in B \vdash C(y)\ type$. The latter is a hypothetical judgement the testing of which will be explained below.
- If it has canonical form U, then the test is successful.
- If it has a canonical form which does not begin with a type constructor then the test fails.

If A has no canonical form, then the judgement is not valid either. In this case we will wait forever – at no stage will we observe a canonical form. Nevertheless, if we observe the intermediate stages of the computation of A we may for example detect an infinite loop, and thus conclude that it will never terminate. However, this requires higher level reasoning than the simple observations of canonical forms.

11.4.5 How to Test $A = A'$?

The relevant rules are

$$\frac{A \Rightarrow N \quad A' \Rightarrow N}{A = A'}$$

$$\frac{A \Rightarrow I(B,b,c) \quad A \Rightarrow I(B',b',c') \quad B = B' \quad b = b' \in B \quad c = c' \in B}{A = A'}$$

$$\frac{A \Rightarrow \Pi(B,C) \quad A' \Rightarrow \Pi(B',C') \quad B = B' \quad y \in B \vdash C(y) = C'(y)}{A = A'}$$

$$\frac{A \Rightarrow U \quad A' \Rightarrow U}{A = A'}$$

To test $A = A'$ we always begin by evaluating A and A' to canonical form, where the outermost form is a constructor.

- If both have canonical form N, then the test is successful.
- If $A \Rightarrow I(B,b,c)$ and $A' \Rightarrow I(B',b',c')$, then first test $B = B'$ and if successful test $b = b' \in B$ and then $c = c' \in B$.
- If $A \Rightarrow \Pi(B,C)$ and $A' \Rightarrow \Pi(B',C')$, then test $B = B'$ and $y \in B \vdash C(y) = C'(y)$. The latter is a hypothetical judgement the testing of which will be explained below.
- If both have canonical form U, then the test is successful.
- If A and A' have canonical forms with different constructors or if one of them does not begin with a type constructor, then the test fails.

If neither A nor A' has canonical form, then the test fails, see the discussion of non-termination above.

11.4.6 How to Test $a \in A$?

The relevant rules are

$$\frac{A \Rightarrow N \quad a \Rightarrow 0}{a \in A} \qquad \frac{A \Rightarrow N \quad a \Rightarrow s(b) \quad b \in N}{a \in A}$$

$$\frac{A \Rightarrow I(B,b,b') \quad a \Rightarrow r \quad b = b' \in B}{a \in A}$$

$$\frac{A \Rightarrow U \quad a \Rightarrow N}{a \in A} \qquad \frac{A \Rightarrow U \quad a \Rightarrow I(b,c,d) \quad b \in U \quad c,d \in b}{a \in A}$$

$$\frac{A \Rightarrow U \quad a \Rightarrow \Pi(b,c) \quad b \in U \quad y \in b \vdash c(y) \in U}{a \in A}$$

The testing manual states that we begin by running both A and a! We need both canonical forms in order to know which rule applies. The further premises of that rule tell us what further tests we need to perform:

- If $A \Rightarrow N$ and $a \Rightarrow 0$, then the test is successful.
- If $A \Rightarrow N$ and $a \Rightarrow s(b)$, then test whether $b \in N$.
- If $A \Rightarrow I(B,b,b')$ and $a \Rightarrow r$, then test whether $b = b' \in B$
- If $A \Rightarrow U$ and $a \Rightarrow N$, then the test is successful.
- If $A \Rightarrow U$ and $a \Rightarrow I(b,c,d)$, then test whether $b \in U$ and $c,d \in b$.
- If $A \Rightarrow U$ and $a \Rightarrow \Pi(b,c)$, then test whether $b \in U$ and $y \in b \vdash c(y) \in U$.
- If A and a have non-matching canonical forms, that is, combinations of canonical forms for which there is no rule, then the test fails.

If neither A nor a has canonical form, then the test fails, see the discussion of non-termination above.

11.4.7 How to Test $a = a' \in A$?

We leave it to the reader to write the testing manual for this judgement form.

11.4.8 Testing Hypothetical Judgements and Function Types

Let us now turn to hypothetical judgements:

$$\Gamma \vdash \mathcal{J}$$

For example, how do we read the rule for Π-types

$$\frac{A \Rightarrow \Pi(B, C) \qquad a \Rightarrow \lambda(c) \qquad x \in B \vdash c(x) \in C(x)}{a \in A}$$

as a rule in our testing manual? What action should we take to test

$$x \in B \vdash c(x) \in C(x)?$$

In *Constructive Mathematics and Computer Programming* (Martin-Löf 1982) and *Intuitionistic Type Theory* (Martin-Löf 1984) the meaning of this hypothetical judgement is that $c(b) \in C(b)$ provided $b \in B$ (and also that $c(b) = c(b') \in C(b)$ provided $b = b' \in B$). However, this is not a satisfactory answer when we look at it from the testing point of view, because we must ask ourselves how we obtained $b \in B$. What if it came from a malicious hacker?

Instead we had better manufacture our own tests! To this end the rules for the judgement form $a \in A$ will be given a second reading: how to generate input to hypothetical tests! This is a point which to my knowledge has not previously been discussed in the context of Martin-Löf's meaning explanations. On the other hand, input generation is an important aspect of software testing, as in the testing tools *QuickCheck* (Claessen and Hughes 2000) and *SmallCheck* (Runciman et al. 2008) for the functional programming language Haskell.

11.4.9 Input Generation

The relevant rules are those which define the meaning of the judgement $a \in A$. We display those rules again but in order to suggest that they should now be read as rules for instantiating variables, we replace the letters $a, b, \ldots$, by $x, y, \ldots$. For this reason we also reorder the premises of some of the rules.

$$\frac{A \Rightarrow N \quad x \Rightarrow 0}{x \in A} \qquad \frac{A \Rightarrow N \quad x \Rightarrow s(y) \quad y \in N}{x \in A}$$

$$\frac{A \Rightarrow I(B, b, b') \quad b = b' \in B \quad x \Rightarrow r}{x \in A}$$

$$\frac{A \Rightarrow \Pi(B, C) \quad x \Rightarrow \lambda(z) \quad y \in B \vdash z(y) \in C(y)}{x \in A}$$

$$\frac{A \Rightarrow U \quad x \Rightarrow N}{x \in A} \qquad \frac{A \Rightarrow U \quad x \Rightarrow I(y, z, z') \quad y \in U \quad z, z' \in y}{x \in A}$$

$$\frac{A \Rightarrow U \quad x \Rightarrow \Pi(y, z) \quad y \in U \quad t \in b \vdash z(t) \in U}{x \in A}$$

If we want to generate $x \in A$, we begin by computing the canonical form of A:

- If it has canonical form N, then either generate $x = 0$ or generate $x = s(y)$ and then generate $y \in N$.
- If it has canonical form $I(B, b, b')$, then we need to test whether $b = b' \in B$. If so generate $x = r$. If not, there is no element to generate. Note that this requires us to *decide* in finite time whether $b = b' \in B$. (This test presupposes the test of $B\ type$, if this fails, then the whole judgement, in which $x \in A$ is a hypothesis, fails.)
- If it has canonical form $\Pi(B, C)$, then generate $x = \lambda(z)$ where z is a function such that $y \in B \vdash z(y) \in C(y)$. However, we do not generate function terms z but input-output pairs, as we shall explain below.
- If it has canonical form U, then either generate $x = N$, or generate $x = I(y, z, z')$ and then generate $y \in U, z, z' \in y$, or generate $x = \Pi(y, z)$, and then generate $y \in U$ and $t \in y \vdash z(t) \in U$. As above, we do not generate function terms z, but input-output pairs.

A question arises: should we fully instantiated a variable x to a closed expression before computing the expression which it is part of, or should we have a lazy instantiation procedure, where we compute with open expressions, and only instantiate x when it blocks further computation, for example, when we get an expression $R(x, d, e)$ which cannot be computed further unless we know the canonical form of x. As we shall see it will be important to evaluate open expressions, when we generate functional input.

11.4.10 Functional Input Generation

To simplify the discussion, let us consider the generation of numerical functions only. How do we read the rule

$$\frac{x \Rightarrow \lambda(z) \quad y \in N \vdash z(y) \in N}{x \in N \to N}$$

as a rule for generating x? It states that we generate $x = \lambda(z)$ and then generate z so that $z(y) \in N$ for $y \in N$. Now, it would be wrong to try to read $y \in N \vdash z(y) \in N$ as syntactic derivability in some formal system for Martin-Löf type theory. Instead we want testing to be "local", that is, not depending on the formal system as a whole. We want the "semantic" notion, not a syntactic one!

This observation is relevant to the question of the impredicativity of types of functionals. A functional $g \in (N \to N) \to N$, is an expression such that $g(f) \in N$ for all $f \in N \to N$, including those f which contain g. This may seem circular if we take a syntactic point of view (generating function terms in a given type system), but not the semantic view (generating input-output pairs) in the sense to follow. It will follow that the justification of types of functionals is predicative in the testing interpretation.

When generating input-output pairs, it is useful to recall the lessons of domain theory (the continuity principle) and game semantics. What we need to do is to generate as many pairs (m, n) with $m, n \in N$ as needed! Consider for example testing

$$x \in N \to N \vdash b(x) \in N$$

- We begin testing $b(x) \in N$ without knowing x. This means we try to compute the canonical form of $b(x)$.
- At some stage the computation may get stuck because it does not know x. For example, $R(Ap(x,0), d, e)$ needs to know the canonical form of $Ap(x, 0)$.
- So we generate an input-output pair $(0, y)$ for the function x, where the output $y \in N$ will be generated as described above. Now the computation can go on, until we either arrive at the canonical form or get stuck again.
- Etc.

We will not here provide a complete description of the generation of function input and leave it to future work on game semantics for dependent types.

11.4.11 Are Function Identity Types Meaningful?

As we saw above we do not provide testing semantics for function identity types. We are led to restrict the formation rule for the types $I(B, b, b')$ to cases where the judgement $b = b' \in B$ is decidable (under the assumption that $b, b' \in B$). This is the case if $B = N$ or if B is another algebraic data types, but it is also the case $B = I(N, c, c')$, for example. Compare Hedberg's coherence theorem (Hedberg 1998), which is valid only for decidable identity types.

Although we do not justify types such as $I(N \to N, f, g)$ as a primitive identity type, we can of course still define it as $(\Pi x \in N)I(N, Ap(f, x), Ap(g, x)$, expressing the extensional equality of $f, g \in N \to N$. This type can be tested by following the manual for testing Π-types and I-types over the type N.

11.5 Impredicative Type Theory

We have explained how the meaning explanations for intuitionistic type theory are about testing judgements by running programs interactively, that is, by generating input, and observing results. We repeat the process for several inputs. In order to test higher order functions we need to consider repeated interactive testing in the style of game semantics.

Can we provide similar meaning explanations for other theories? That is, can we describe the meaning of the judgements of the theory by running programs interactively, generating input, observing results, and thereby corroborating or falsifying results? Is this what constructive validity is all about?

In this section we shall ask ourselves what such meaning explanations for impredicative type theory would be like. We would like to write testing manuals for impredicative type theories such as

- System F of Girard (1971)?
- The Calculus of Constructions of Coquand and Huet (1988)?
- The Calculus of Inductive Constructions of Coquand and Paulin (1993), the theory of the Coq-system (Bertot and Castéran 2004)?

These systems have to my knowledge only been analyzed meta-mathematically. For example, Girard (1971) showed that System F is strongly normalizing. In this proof he relied on impredicative aspects of his meta-language: classical set theory.

We shall now ask ourselves what a pre-mathematical testing semantics would be like for these systems? Note that they have real users, especially the last one (the Coq-users). What do these users expect when they "run" their programs?

Let us here consider Coquand and Huet's pure Calculus of Constructions (Coquand and Huet 1988). This theory has an impredicative universe U (usually called *Prop*, the universe of impredicative propositions). It is impredicative since we can form the cartesian product of a family of small types indexed by an arbitrary (not necessary small) type:

$$\frac{A\ type \quad x \in A \vdash B \in U}{(\Pi x \in A)B \in U}$$

In particular we can put $A = U$ and define a new element of U by quantification over all elements in U including the one we are defining.

Contrast this to the type of small types in Martin-Löf type theory. This is predicative: we can only form a family of small types indexed by an arbitrary small type:

$$\frac{A \in U \quad x \in A \vdash B \in U}{(\Pi x \in A)B \in U}$$

Unlike in Martin-Löf type theory,we can thus define the type of natural numbers as the type of Church numerals in the Calculus of Construction as follows:

$$N = (\Pi X \in U)X \to (X \to X) \to X \in U$$

We can also define the type of identity proofs over an arbitrary type (Leibniz equality):

$$I(A, a, b) = (\Pi X \in A \to U)X(a) \to X(b) \in U$$

Since they can be encoded, the pure Calculus of Constructions does not have primitive types N and $I(A, a, b)$. This coding can be done for a very general class of inductive types. However, some aspects of this encoding are unsatisfactory. As a consequence Coquand and Paulin decided that primitive inductive types and families should be added. The resulting theory is the Calculus of Inductive Constructions, the core of the Coq system (Bertot and Castéran 2004).

11.5.1 A Testing Manual for the Calculus of Constructions Based on the Evaluation of Closed Expressions

As we saw above, there is great similarity between Martin-Löf type theory and the Calculus of Constructions, except that the latter

- Only has types U and $(\Pi x \in B)C$ and no primitive data types $N, I(A, a, b), \ldots$;
- Has U which is closed under impredicative Π.

Let us first try to make a minimal modification of the testing manual for Martin-Löf type theory to accommodate the types of the impredicative universe U. The difference appears in the test for elements of U. The only rule is:

$$\frac{A \Rightarrow U \quad a \Rightarrow (\Pi x \in B)C \quad B\ type \quad x \in B \vdash C \in U}{a \in A}$$

Hence, there is no base case! It is not clear how the testing procedure would ever stop.

11.5.2 Testing Based on Normalization of Open Expressions

Martin-Löf type theory is a functional programming language. A user will run a program in much the same way as a user of an ordinary lazy functional language such as Haskell, that is, the user will evaluate *closed* expressions to canonical form. (When we base Martin-Löf type theory on the theory expression this corresponds to evaluating closed expressions of arity ι to canonical form.)

However, a user of the Calculus of Constructions will instead evaluate programs to full normal form, for example when computing with Church numerals. To obtain this full normal form we need to evaluate expressions under λ, that is, we need to evaluate *open* expressions. We will therefore consider a testing manual for the Calculus of Constructions based on the evaluation of open expressions.

11.5.3 Evaluation of Open Expressions in Martin-Löf Type Theory

Incidentally, Martin-Löf has recently considered introducing meaning explanations based on the evaluation of open expressions for intensional intuitionistic type theory.

The abstract of his talk *Evaluation of open expressions* (given at the *Symposium on Programming, Types, and Languages* in Gothenburg in March 2009) states his aim as follows:

> The informal, or intuitive, semantics of type theory makes it evident that closed expressions of ground type evaluate to head normal form, whereas metamathematics, ..., is currently needed to show that expressions which are open or of higher type can be reduced to normal form. The question to be discussed is: Would it be possible to modify the informal semantics in such a way that it becomes evident that all expressions, also those that are open or of higher type, can be reduced to full normal form?

The aim is to provide meaning explanations for intensional type theory which do not validate the rule of equality reflection, and which match the type-checking algorithm for intensional type theory.

11.5.4 Testing Manual for the Calculus of Constructions

The terms are

$$a ::= U \mid (\Pi x \in a)a \mid (\lambda x)a \mid a(a) \mid x$$

Note that this is the syntax of the ordinary untyped lambda calculus extended with U and Π. We do not employ the theory of expressions here. These terms will be evaluated to open weak head normal forms:

$$v ::= U \mid (\Pi x \in a)a \mid (\lambda x)a \mid x(a, \ldots, a)$$

where we use the abbreviation $x(a_1, \ldots, a_n) = x(a_1) \cdots (a_n)$. Note that a in the above productions of values range over arbitrary terms.

There is only one computation rule

$$\frac{c \Rightarrow (\lambda x)b \quad b[a/x] \Rightarrow v}{c(a) \Rightarrow v}$$

in addition to the rule that a canonical term has itself as value

$$v \Rightarrow v$$

To test the hypothetical judgement $\Gamma \vdash A\ type$ we evaluate the open expression A. We need to consider three possibilities:

$$\frac{A \Rightarrow (\Pi y \in B)C \quad \Gamma \vdash B\ type \quad \Gamma, y \in B \vdash C\ type}{\Gamma \vdash A\ type} \qquad \frac{A \Rightarrow U}{\Gamma \vdash A\ type}$$

$$\frac{A \Rightarrow x(b_1, \ldots, b_n) \qquad \Gamma \vdash b_1 \in B_1 \quad \cdots \quad \Gamma \vdash b_n \in B_n}{\Gamma \vdash A\ type}$$

In the last rule the type of x in Γ is $(\Pi \in B_1) \cdots (\Pi y_n \in B_n)$.

To test the hypothetical judgement $\Gamma \vdash a \in A$ we evaluate the open expressions a and A:

$$\frac{A \Rightarrow U \qquad a \Rightarrow (\Pi y \in B)C \qquad \Gamma \vdash B\ type \qquad \Gamma, y \in B \vdash C \in U}{\Gamma \vdash a \in A}$$

$$\frac{A \Rightarrow (\Pi y \in B)C \quad a \Rightarrow (\lambda y)c \quad \Gamma, y \in B \vdash c \in C}{\Gamma \vdash a \in A}$$

$$\frac{a \Rightarrow x(b_1, \ldots, b_n)\ \Gamma \vdash b_1 \in B_1\ \cdots\ \Gamma \vdash b_n \in B_n\ \Gamma \vdash C[b_1/y_1, \ldots, b_n/y_n] = A}{\Gamma \vdash a \in A}$$

In the last rule the type of x in Γ is $(\Pi \in B_1) \cdots (\Pi y_n \in B_n)$.

We leave it to the reader to write the testing manual for the hypothetical judgements $\Gamma \vdash A = A'$ and $\Gamma \vdash a = a' \in A$.

Note that there is no need for input generation. The meaning of hypothetical judgements is defined directly in terms of evaluation to open canonical forms. Note that this testing manual is nothing but the type-checking algorithm for the Calculus of Constructions! The correctness of the Calculus of Constructions thus simply amounts to the termination with appropriate canonical forms of the type-checking algorithm. There is circumstantial evidence that this algorithm is correct since it has been corroborated many times. However, there is not yet a "well-founded" explanation of why this is the case. Nevertheless, many, but not all, researchers express certainty that it will indeed always terminate correctly.

11.6 Conclusion

To conclude, I will discuss some important distinctions.

11.6.1 Pre-mathematical Versus Meta-mathematical Semantics

I have emphasized the difference between on the one hand (formal) meta-mathematical semantics, understood as the translation into another mathematical language (usually set theory) the meaningfulness of which is taken for granted, and on the other hand, pre-mathematical semantics, where the meaning is explained in terms of real world interactions with and observations of expressions and their computations. This difference is crucial, although I would not like to claim that it is unproblematic. Formality may not be a requirement of meta-mathematics in the usual sense, see for example the discussion by Kleene (1952).

I believe pre-mathematical and meta-mathematical aspects are complementary. Meta-mathematical semantics can provide further insight into the pre-mathematical situation. But meta-mathematical semantics may also introduce extraneous issues which are not relevant for the pre-mathematical understanding. For example, it is not clear that the meta-mathematical work of turning the meaning explanations into a positive inductive definition in set theory has much foundational significance.

11.6.2 Game Semantics Versus Realizability Semantics

The meta-mathematical semantics in Sect. 11.3 does not match the pre-mathematical semantics in Sect. 11.4. This is because the realizability semantics is not sufficiently fine-grained to adequately capture the details of the testing manual in Sect. 11.4. It would be interesting to provide a meta-mathematical treatment of the testing manual which captures more such details, perhaps in the style of game semantics.

11.6.3 Input Generation Versus Output Computation

Martin-Löf's meaning explanations for intuitionistic type theory are based on the evaluation of expressions to canonical form, that is, they are based on the computation of output. One of the key ideas of this paper is that a dual discussion about methods for input generation is needed for meaning explanations of hypothetical judgement.

11.6.4 Judgement Versus Proposition

It is important to note that we test *judgements* and not *propositions* in type theory.

11.6.5 Molecular Versus Holistic View of Meaning

Another important facet of the present notion of meaning is that it is *molecular* in the sense that the meaning of a judgement only depends on the meaning of its constituent parts. This is to be contrasted to the *holistic* view, where meaning depends on the whole formal system which the judgement is part of. Consistency is an example of a holistic property.

11.6.6 Validated by Testing Versus Made Evident by Thinking

Prior discussions of meaning explanations have focussed on their capacity to make the rules of inference *evident*. However, to be evident is a subjective notion:

something is evident to *somebody*. Here we emphasize that judgements can also be validated by testing, and that this is an objective notion. The result of the test does not depend on who executed the test. For further discussion of *epistemological* vs *ontological* aspects of truth and proof in intuitionistic logic the reader is referred to Prawitz' article in this volume (Prawitz 2012).

11.6.7 Primary School Computation of Closed Expressions Versus Secondary School Computation of Open Expressions

Martin-Löf style meaning explanations are based on the computation of *closed* expressions. This is after all the primary notion of computation; normalization of open expressions is a secondary notion of computation, although it may be relevant for meaning explanations of impredicative type theory.

11.6.8 Meaning as Testing and Meaning as Use

The notion of a user occupies a central stage in the testing view of meaning explanations. It seems to fit well into Wittgenstein's "meaning-as-use" paradigm which has been further developed by Dummett (1973) and Prawitz (1977) in connection with the philosophical basis of intuitionism.

11.6.9 Some Historical Notes

This is not the place to trace the historical origin of the idea that testing is fundamental for the meaning of logic. I would just like to mention that Gödel's Dialectica interpretation (Gödel 1958) is a proof as programs interpretation where the correctness is justified in terms of tests. However, note that the Dialectica interpretation differs from Kleene realizability. The game interpretation of logic goes back to Lorenzen (1978). It seems that Lorenzen's dialogical semantics for intuitionistic predicate logic would be closely related to a potential game interpretation of intuitionistic type theory.

Acknowledgements I would like to express my deep gratitude to Per Martin-Löf for his profound ideas and for many discussions and much help over the years. The present paper owes a lot to these discussionson, for example, on the nature of the meaning explanations, on the distinction between the pre-mathematical and the meta-mathematical, and on the meaning of induction.

I would also like to thank Erik Palmgren and an anonymous referee for useful feedback on a preliminary version of this paper. The paper is based on a talk given several times. I am grateful for interesting comments by many people who attended these talks, for example, Andreas Abel, Peter Aczel, Pierre Clairambault, Thierry Coquand, Peter Hancock, Bengt Nordström, Andrew Pitts, Gordon Plotkin, Michael Rathjen, Dag Prawitz, Peter Schroeder-Heister, Helmut Schwichtenberg, Anton Setzer, and Sören Stenlund.

References

Aczel, P. 1977a. An introduction to inductive definitions. In *Handbook of mathematical logic*, ed. J. Barwise, 739–782. Amsterdam: North-Holland.

Aczel, P. 1977b. The strength of Martin-Löf's type theory with one universe. In *Proceedings of the symposium on mathematical logic* (Oulu 1974), ed. S. Miettinen and J. Väänanen, 1–32. Report No 2 of Department of Philosophy, University of Helsinki.

Aczel, P. 1980. Frege structures and the notions of proposition, truth, and set. In *The Kleene symposium*, ed. J. Barwise, H.J. Keisler, and K. Kunen, 31–59. Amsterdam: North-Holland.

Allen, S.F. 1987a A non-type-theoretic definition of Martin-Löf's types. In *Proceedings of second IEEE symposium on logic in computer Science*, New York, 215–224.

Allen, S.F. 1987b. A non-type-theoretic semantics for type-theoretic language. Ph.D. thesis, Cornell University.

Beeson, M. 1982. Recursive models for constructive set theories. *Annals of Mathematical Logic* 23: 127–178.

Bertot, Y., and P. Castéran. 2004. *Interactive theorem proving and program development Coq'Art: The calculus of inductive constructions*, Texts in theoretical computer science. An EATCS series. Berlin/New York: Springer.

Claessen, K., and J. Hughes. 2000. QuickCheck: A lightweight tool for random testing of Haskell programs. In *Proceedings of the ACM SIGPLAN international conference on functional programming*, Montreal, ACM SIGPLAN notices, vol. 35.9, 268–279. ACM Press.

Coquand, T. 1995. A semantics of evidence for classical arithmetic. *Journal of Symbolic Logic* 60(1): 325–337.

Coquand, T., and G. Huet. 1988. The calculus of constructions. *Information and Computation* 76: 95–120.

Dummett, M. 1973. The philosophical basis of intuitionistic logic. In *Logic colloquium '73*, ed. H.E. Rose and J.C. Shepherdson, 5–40. Amsterdam: North Holland.

Dybjer, P. 2000. A general formulation of simultaneous inductive–recursive definitions in type theory. *Journal of Symbolic Logic* 65(2): 525–549.

Dybjer, P., and A. Setzer. 1999. A finite axiomatization of inductive–recursive definitions. In *Typed lambda calculi and applications*, Lecture notes in computer science, vol. 1581, ed. J.-Y. Girard, 129–146. Berlin/London: Springer.

Dybjer, P., and A. Setzer. 2006. Indexed induction–recursion. *Journal of Logic and Algebraic Programming* 66: 1–49.

Girard, J.-Y. 1971. Une extension de l'interpretation de Gödel à l'analyse, et son application à l'èlimination des coupures dans l'analyse et la théorie des types. In *Proceedings 2nd Scandinavian logic symposium*, ed. J.E. Fenstad, 63–92. Amsterdam: North Holland.

Gödel, K. 1958. Über eine bisher noch nicht benützte Erweiterung des finites Standpunktes. *Dialectica* 12: 280–287.

Hayashi, S. 2007. PA/LCM home page. http://www.shayashi.jp/PALCM/

Hayashi, S., and H. Nakano. 1989. *PX: A computational logic*. Cambridge: MIT.

Hayashi, S., R. Sumitomo, and K. Shii. 2002. Towards the animation of proofs – Testing proofs by examples. *Theoretical Computer Science* 272(1–2): 177–195.

Hedberg, M. 1998. A coherence theorem for Martin-Löf's type theory. *Journal of Functional Programming* 8(4): 413–436.

Kleene, S.C. 1952. *Introduction to meta-mathematics*. Amsterdam: North-Holland.

Knuth, D. 1977. Notes on the van Emde Boas construction of priority deques: An instructive use of recursion. Memo sent to Peter van Emde Boas, see http://www-cs-faculty.stanford.edu/~uno/faq.html.

Lorenzen, P., and K. Lorenz. 1978. *Dialogische Logik*. Darmstadt: Wissenschaftliche Buchgesellschaft.

Martin-Löf, P. 1971. A theory of types. Preprint. Stockholm: Stockholm University.

Martin-Löf, P. 1975. An intuitionistic theory of types: Predicative part. In *Logic colloquium '73*, ed. H.E. Rose and J.C. Shepherdson, 73–118. Amsterdam: North Holland.

Martin-Löf, P. 1982. Constructive mathematics and computer programming. In *Logic, methodology and philosophy of science, VI, 1979*, ed. L.J. Cohen et al., 153–175. Amsterdam: North-Holland.

Martin-Löf, P. 1984. *Intuitionistic type theory: Notes by Giovanni Sambin of a series of lectures given in Padua, June 1980*. Napoli: Bibliopolis.

Martin-Löf, P. 1996. On the meaning of the logical constants and the justifications of the logical laws. *Nordic Journal of Philosophical Logic* 1(1): 11–60.

Martin-Löf, P. 1998. An intuitionistic theory of types. In *Twenty-five years of constructive type theory*, ed. G. Sambin and J. Smith. New York: Oxford University Press. Reprinted version of an unpublished report from 1972.

Miquel, A. 2010. The reasonable effectiveness of mathematical proof. In *Anachronismes logiques*, ed. M. Quatrini and S. Tronon. Logique, Langage, Sciences, Philosophie. Publications de la Sorbonne. To appear.

Nordström, B., K. Petersson, and J.M. Smith. 1989. *Programming in Martin-Löf's type theory – An introduction*. New York: Oxford University Press.

Paulin-Mohring, C. 1993. Inductive definitions in the system Coq – Rules and properties. In *Proceedings typed λ-calculus and applications*, Lecture notes in computer science, 328–245. Berlin: Springer.

Popper, K. 1963. *Conjectures and refutations: The growth of scientific knowledge*. London: Routledge.

Prawitz, D. 2012. Truth and proof in intuitionism, in Epistemology versus Ontology (eds. P. Dybjer, S. Lindström, E. Palmgren, and G. Sundholm) in Logic, Epistemology, and the Unity of Science 27. This volume.

Prawitz, D. 1977. Meaning and proofs: On the conflict between classical and intuitionistic logic. *Theoria* 43: 11–40.

Runciman, C., M. Naylor, and F. Lindblad. 2008. SmallCheck and Lazy SmallCheck – Automatic exhaustive testing for small values. In *Proceedings of the first ACM SIGPLAN symposium on Haskell*, Victoria, 37–48,

Setzer, A. 2000. Extending Martin-Löf type theory by one Mahlo universe. *Archive for Mathematical Logic* 39(3): 155–181.

Smith, J. 1984. An interpretation of Martin-Löf's type theory in a type-free theory of propositions. *Journal of Symbolic Logic* 49(3): 730–753.

Chapter 12
Normativity in Logic

Jean-Yves Girard

À Per Martin-Löf, incomparable défricheur.

Foundational questions are cognitive: "What can we know?", "How do we know?", "What are our preconceptions?". Thus, the open problems in algorithmic complexity, which address the efficiency of computation, are foundational, although far from the stereotyped problem of consistency. Foundations can—and must—question everything... including questions, the only limit to this interrogation being efficiency. And the rigid departure between syntax and semantics—which is only appropriate in "usual" situations—is the first dogma that foundations should put into question.

Consider natural numbers: several systems, yet only one model, the *standard* $\mathbb{N}$. This extreme poverty (incompleteness) of the semantic universe is a by-product of *normativity*. Indeed, the question "what is standard?" is booby-trapped: it induces a meaningless dichotomy between standard, normal, integers and non-standard, abnormal ones. There is no way to escape this *aporia* while sticking to the rigid distinction between syntax and semantics, where subject and object—both clearly individuated—relate according to a fixed protocol. This normativity is *ready-made*, i.e., hidden and external: it goes without saying, moreover it proceeds from the sky. By making normative requirements explicit, by internalising them as parts of the "semantic", objective, universe, we produce new "models" for natural numbers. In some sense, normativity appears as a mobile curtain separating the object from the subject; by the way, this would not be the only mobile curtain of logic, think of the departure between sets and proper classes.

This supposes a *constructive* viewpoint, i.e., an emphasis on the construction of natural numbers, with eventually a radical change of framework: the replacement

J.-Y. Girard (✉)
Institut de Mathématiques de Luminy, 13288 Marseille Cedex 9, France
e-mail: girard@iml.univ-mrs.fr

P. Dybjer et al. (eds.), *Epistemology versus Ontology*, Logic, Epistemology, and the Unity of Science 27, DOI 10.1007/978-94-007-4435-6_12,

of combinatorics with operator algebras. This is rather natural, since quantum physics—which radically puts into question the departure subject/object—dwells in those spaces. *Geometry of Interaction* (GoI) yields, for each $n \in \mathbb{N}$, infinitely many isomorphic *representations* N_n, none of them more "standard" than the others. In order to avoid *interference*, this intrinsic isomorphy must be bridled, whence normativity. Syntactical devices (formal system, typed calculus, complexity class) should therefore correspond to various ways of taming the isomorphy classes of integers, thus inducing a sort of normativity—no longer absolute, ready-made.

Normativity is usually what goes without saying. Take, for instance, the handling of *variables*: outside their range of operationality, i.e., when bound, variables are up to renaming (isomorphism). This discipline avoids accidental coincidences, i.e., interference.[1] Due to the inherent rigidity of syntax, there is no alternative normativity: this explains why the literature on bound variables is so afflictive. In GoI, the various N_n are the same "up to renaming", i.e., up to isomorphy: what will eventually be recognised as a variable, i.e., excluded from the interaction, depends upon the possible interactions, i.e., upon the context. In this way, GoI proposes a sort of "not-yet-frozen syntax".

We shall implement these ideas in the framework of algorithmic complexity, and produce "logspace integers", corresponding to logspace computations. The "model" thus constructed operates a departure of its own between isomorphic objects, some becoming standard, the others non-standard. Besides these relative non-standard integers do exist truly non-standard ones: those in charge of normativity.

The main technical references for this paper are Girard (2007, 2011). The ideas sketched here will be developed in a forthcoming paper devoted to *transcendental syntax*. Thanks to Damiano Mazza for his careful reading and to Paulin Jacobé de Naurois for his expertise on LOGSPACE.

12.1 One Reality, Several Systems

The completeness theorem says that syntax refers to semantics. However, when dealing with natural numbers, *incompleteness* prevails: anybody acquainted with formal logic may easily name half a dozen formal arithmetics, but only one model for them, $\mathbb{N}$. Incompleteness thus means that, although the difference between remarkable systems can be accounted for by models, such models are by no means remarkable, there are but *ad hoc* doohickeys. This concrete incompleteness—the absence of any *convincing* interpretation—is a common drawback of all logico-computational approaches to natural numbers: formal arithmetics, typed calculi, complexity classes.

[1] This point was never questioned, not even by the most outrageous AI-oriented "logics".

12.1.1 *Formal Systems, Typed Calculi and Complexity Classes*

12.1.1.1 Formal Systems

There is no standard axiomatisation of natural numbers: besides the widely advertised Peano arithmetic **PA**, coexist the alternative (and somewhat more flexible) $\mathbf{PA}_2$ (second order Peano arithmetic) or **ZF** (Zermelo-Fraenkel set theory); not to speak of various subsystems of the former—e.g., weak arithmetics—introduced for proof-theoretic reasons. None of these systems can claim to be "the" system, since, by Gödel's incompleteness, there is always a true arithmetical formula not provable in it. Incompleteness can indeed be turned into a (rather empty) machinery providing fresh axiomatisations: if **T** is a sound system of arithmetic, then $\mathbf{T} + Con(\mathbf{T})$ is still sound, but distinct from **T**.

Incompleteness offers no *explanation* for this plethora of systems whose meaning remains unclear. According to the book, two classical systems are distinct when separated by a model; but the only models distinguishing **PA** from $\mathbf{PA} + Con(\mathbf{PA})$ are non-standard ones satisfying $\neg Con(\mathbf{PA})$. Such crazy models—obtained through a completion of $\mathbf{PA} + \neg Con(\mathbf{PA})$—are nothing but an illegible rewriting of the second incompleteness theorem: the difference between **PA** and $\mathbf{PA} + Con(\mathbf{PA})$ accounts for the model, not the other way around! If these systems can only be separated through non-standard models, it is not because one of them is fishy: what is fishy here is the very notion of model!

12.1.1.2 Typed Calculi

Let us adopt a constructive viewpoint: mathematical objects are no longer given to us, since we *construct* them. We can thus imagine a way out from the *aporia* leading to models and non-standard integers: focusing on functions of natural numbers, i.e., on the various ways of constructing them. Formal systems are replaced with typed calculi, e.g., Martin-Löf's theory of types. These calculi enable one to define computable functions from $\mathbb{N}$ to $\{0, 1\}$ and Cantor's diagonalisation—incompleteness *ante litteram*—ensures that none of them is complete: yet another empty machine providing fresh systems.

The reference universe for the constructive approach is category theory. Two systems can be distinguished by the choice of their *morphisms*. This is undoubtedly a progress w.r.t. models: categories, morphisms etc. are usually effective and meaningful, in sharp contrast to non-standard integers—the meaningless and non-effective scions of model theory. However, putting the burden on morphisms is *reculer pour mieux sauter*: there is no reasonably natural notion of morphism from $\mathbb{N}$ to $\{0, 1\}$ that will account for the choice of such and such functional system. For instance, the functions of that type definable in system **F** have but one characterisation... that of coming from system **F**, barely a manageable criterion. This comes from the fact that—up to minor details—a morphism from $\mathbb{N}$ to $\{0, 1\}$

always translates as a plain subset of $\mathbb{N}$. There are presumably "not enough" integers, but where to find them? Or perhaps something essential—some missing structure—has not been taken into account.

12.1.1.3 Complexity Classes

The two previous approaches are basically equivalent: typed calculi present the effective side of formal systems. Thus, a provably terminating algorithm of $\mathbf{PA}_2$ can be represented in system **F**; conversely, a function of system **F** is provably terminating in $\mathbf{PA}_2$. If we turn our attention towards extremely weak formal systems, typed calculi are bound to represent complexity classes: this is the case for *light logics* like **LLL** or **ELL**, two variants of linear logic corresponding to polytime and elementary complexities.

The question is, so to speak, a refinement of the previous one: how can we manage to "force" the morphisms from $\mathbb{N}$ to $\{0, 1\}$ to be polytime or logspace? Where to find the "missing integers" or the "missing structure"? Are there polytime, logspace, integers?

12.1.2 Alternative Integers

We cannot content ourselves with the usual *credo* saying that systems live their own formalist life, thus reducing a system to a list of theorems—a list mostly out of reach, thanks to undecidability. It is reasonable to require that distinct axiomatisations of natural numbers describe distinct "realities", i.e., distinct "notions" of natural numbers.

First observe that not all formal systems are of interest: typically, *unsound* systems such as $\mathbf{PA} + \neg\, Con(\mathbf{PA})$ should not be accounted for. The same applies to most sound systems, typically $\mathbf{PA} + Con(\mathbf{PA})$, more a "PhD system" than one in which one would like to formalise arithmetic. Summing up, we are not supposed to explain "all" systems: it will be enough to explain a few *meaningful* ones. The same holds for the constructive, functional, aspect of the problem: not all typed systems are of interest. In the same way, one should not seek a systematic account of complexity classes: some of them may be just "PhD classes".

The general pattern is obviously that of *alternative integers*: by this, I mean integers "besides" the usual ones. I avoided the adjective "non-standard" which would corner us to the rut of non-standard models: non-standard integers are non-effective, moreover each belong in a model of its own, outside of which it makes no sense. Even being aware of this basic misconception, the idea of an alternative integer is booby-trapped. Besides usual, *natural*, numbers, there should be abnormal individuals: this distinction normal/abnormal is the result of a hidden *normativity*. Logic is usually unaware of this normativity, more precisely tries to hide it behind general considerations that one is in right to question at the foundational level.

We shall therefore try to expose the hidden normativity of logic, with the secret hope that there is nothing like a standard, optimal, normativity; it will turn out that the same number—say 3—has infinitely many isomorphic *representations*; w.r.t. an evaluative context, two isomorphic representations may behave differently. The role of normativity is to restrict the choice of possible representations so as to make the evaluation "objective", independent of the representation. Abnormality thus becomes *relative* to the evaluating context. This is not the end of the story: the internalisation of normative constraints produces "watchdogs" which behave like integers w.r.t. evaluations, but are clearly of a different nature.

12.1.3 A Misfire: Ordinals

Among the traditional explanations for the diversity of formal arithmetics, one should mention the use of ordinal numbers, typically Gentzen's assignment of the ordinal ϵ_0 to **PA**. Ordinals are normative devices, whose role is to forbid *draws* (see *infra*) in the game-theoretic interpretation of proofs. The good point is that normativity is made explicit; unfortunately, this normativity remains external. The approach is indeed a disappointment:

1. There is no conceptual background for ordinal assignments. The München School produced, in its day, a correspondence between a long list of not-too-meaningful subsystems of $\mathbf{PA}_2$ and a list of not-too-understandable denumerable ordinals. The limited interest of both list was not compensated for by some enlightening explanation as to the *nature* of their relation.
2. The technique works for systems "not too far" from **PA**. It is completely inadequate for the full $\mathbf{PA}_2$—no understandable ordinal can be found—and for very weak complexity sensitive systems.
3. The stereotyped relation between ordinals and formal systems is modelled on the following: if f, from $\mathbb{N}$ to $\mathbb{N}$, is a provably terminating recursive function of **PA**, then $f \leq \phi_\alpha$ for some $\alpha < \epsilon_0$, where ϕ_α is a hierarchy of recursive functions. This style of relation fails for complexity-sensitive systems: complexity issues usually deal with functions from $\mathbb{N}$ to $\{0, 1\}$, a priori bounded by the constant 1!

12.2 On Normativity

12.2.1 Ready-made Normativity

The ready-made conception of logic (associated with the names of Frege and Tarski) supposes that everything has its predetermined place. In this *essentialist* world, the language (syntax) refers to the reality (semantics). However, the same reality—$\mathbb{N}$—is handled by distinct syntaxes which can only be distinguished by "non-realities", i.e., non-standard models.

Back to the failed promises of categories, observe that the word "morphism" refers to the form, i.e., the essence. But, what is a morphism from $\mathbb{N}$ to $\{0, 1\}$, besides a plain function, what is it supposed to preserve, comply with? Since there is no answer to that question, there only remain discretionary definitions of the kind: "The morphisms from $\mathbb{N}$ to itself are polytime functions" or "The morphisms... are those functions definable in system **F**".

12.2.2 A Priori vs. A Posteriori

A priori normativity constructs objects according to rules, like a construction kit. Typically, in a typed λ-calculus such as system **F**, the constructors must preserve the types, i.e., obey to the law.

An example of normativity a posteriori is given by the pure λ-calculus. λ-terms are defined and interact (through application and normalisation) independently of any logical commitment. Now, we can decide to regroup certain λ-terms into sets called *types* and define logical operations between such types. For the same type of system **F**, we thus get two approaches:

1. The typed (system **F**) pattern: $\Lambda X \lambda x^X x$ *is of* type $\forall X(X \Rightarrow X)$.
2. The untyped pattern: $\lambda x x$ *belongs to* the type $\forall X(X \Rightarrow X)$.

The two approaches are related: every typed λ-term of system **F** yields, if we forget everything pertaining to types (here ΛX and the type superscript ^{X}) an untyped term belonging to the same type. In certain cases, the converse is true: for instance, any closed and normal pure λ-term *in* the type **nat** $:= \forall X((X \Rightarrow X) \Rightarrow (X \Rightarrow X))$ comes from a typed λ-term *of* the same type in **F**. This establishes, for those types, a *completeness* of the former approach (normativity a priori) w.r.t. the latter (normativity a posteriori).

Normativity a posteriori yields more objects: they may belong to a type without being constructible according to the book—i.e., predetermined analytic patterns. This approach is therefore more promising.

However, everything depends upon the *quality* of the interpretation. In spite of its robustness, λ-calculus is not suited for our purpose: we just observed that **nat**, the type of integers, has no "stowaways". The limitation of this—by other standards, excellent—system is presumably due to the fact that it is syntactic a priori. What still remains of the disputable departure syntax/semantics is the setting apart of a combinatorial world—that of language. How can we seriously question natural numbers when $\mathbb{N}$ plays, under the disguise of syntax, such a prominent role?

12.2.3 Games

The interpretation of logic by games, initiated by Gentzen, is subject to two approaches:

12.2.3.1 Game Semantics

A prenex formula $A := \forall m \exists n \forall p \ldots$ with k alternated quantifiers can be described as a *game*: A is true when player `I` has a winning strategy. In this game of finite duration k, winning strategies are Skolem functions: being non-effective, they cannot be considered as proofs. By allowing "remorse", i.e., replays corresponding to the rule of contraction, Gentzen was able to give an effective version of the same game: winning strategies become sorts of infinite proofs in a game of duration[2] ϵ_0. Peano arithmetic can be seen as a construction kit yielding winning strategies—the ones induced by proofs.

Although technically correct, this approach is very reductive and somehow misses the point. The first hint is the remark that, of the two partners, `I`, which tries to prove A, "plays syntax", whereas `II`, which tries to confute A, "plays semantics". Since this approach encompasses the familiar departure syntax/semantics, the expression "game semantics"—which insists upon an obsolete opposition—is, at least, misleading. Indeed, since there is no essential difference between `I` and `II`, syntax and semantics are no longer separated by a Great Wall: nobody forbids `I` from "playing semantics", `II` from "playing syntax", not to speak of intermediate or joint possibilities—for instance, both playing semantics.

12.2.3.2 Norm as a Game

Indeed, the idea of a game is so rich that normativity itself can be the thing at stake! To sum up, game semantics reduces the debate to the question "Is this true?", an *evaluative* query; whereas the alternative approach developed, e.g., in *ludics* poses the more general question "Is this appropriate?", a *deontic* query which encompasses the evaluative questioning about truth.

Indeed, "Is this true?" supposes at least that we know the question at stake; foundationally speaking, we cannot escape the most basic questioning "What is the question?". I am not playing on words, gilding the (meta) lily, and, by the way, questioning the question has deep technical implications. Typically, in pure λ-calculus, we can see a type as a question whose answer lies in its inhabitants: thus, λxx can be seen as an answer to the questions of the form $A \Rightarrow A$, and also $\forall X(X \Rightarrow X)$. This "anteriority" of the answer over the question prompts the issue of *subtyping, intersection types*, which might as well be called subquestioning, intersection of questions...

In terms of games, this questioning about questions poses the problem of the *rule of the game*; indeed, specifying a rule is the same thing as choosing the question. Now, proposing a game "without rules" as a solution seems inadequate, since a

[2] Ordinal "duration" is topsy-turvied: moves are labelled by *decreasing* ordinals: this ensures termination.

game without rules is still a game—with a lax rule, so lax that it makes it of little interest! Everything clarifies if we take into account the possibility of a *draw*.

1. In a game-with-a-rule (game semantics), there is no draw: one of the players wins. Typically, ordinals such as ϵ_0 ensure this absence of draw.
2. In a game-without-a-rule (ludics), there can be draws: typically an infinite play, but also a too long (i.e., infinite) delay before a move. The idea is to forbid draws, thus forcing the players into a mutual discipline. This discipline is a sort of rule of the game, no longer proceeding from the sky: it is a by-product of interaction. The contention between players is therefore not primarily about truth (winning) but about norm (agreeing): what is *appropriate*?

Let us illustrate this point by an example (loosely) inspired from ludics: the treatment of the formula $\exists m \in X\ \forall n \in Y\ m \leq n\ (X, Y \subset \mathbb{N})$. In game semantics, a strategy for $\mathtt{I}$ is of the form $\sigma = \{m_0\} \times Y$ for some $m_0 \in X$, a strategy for $\mathtt{II}$ is a function τ from X to Y; $\sigma \cap \tau$ is a singleton $\{(m, n)\}$: if $m \leq n$, $\mathtt{I}$ wins, otherwise $\mathtt{II}$ wins. Now, let us introduce *designs* (which are to strategies what pure λ-terms are to typed ones): a design is any subset of $\mathbb{N} \times \mathbb{N}$. If σ, τ are designs for $\mathtt{I}$, $\mathtt{II}$, then $\sigma \cap \tau$ need no longer be a singleton, in which case the result of the "play" is a draw. To the game $A := \exists m \in X\ \forall n \in Y\ m \leq n$, let us associate the following sets of designs:

- $A_{\mathtt{I}}$: all designs of the form $\sigma = \{m_0\} \times Y'$, with $m_0 \in X$, $Y' \supset Y$.
- $A_{\mathtt{II}}$: all designs τ such that $\tau \cap (X \times \mathbb{N})$ is a function from X to Y.

It is easy to verify that $\sigma \in A_{\mathtt{I}}$ iff for all $\tau \in A_{\mathtt{II}}$, $\sigma \cap \tau$ is a singleton, i.e., if $\sharp(\sigma \cap \tau) = 1$; in the same way, $\tau \in A_{\mathtt{II}}$ iff for all $\sigma \in A_{\mathtt{I}}$, $\sharp(\sigma \cap \tau) = 1$. In other terms, $A_{\mathtt{I}}$ and $A_{\mathtt{II}}$ refer to each other, and not to some external *rule of the game*. Indeed, the role of the "rule for $\mathtt{I}$" is played by the designs $\tau \in A_{\mathtt{II}}$, and vice versa, via the constraint "no draw". The replacement of strategies with designs has thus the following consequences:

Internalisation: the rule of the game no longer proceeds from the sky.
Subtyping: if $B := \exists m \in X'\ \forall n \in Y'\ m \leq n$, with $X \subset X'$, $Y' \subset Y$, then $A_{\mathtt{I}} \subset B_{\mathtt{I}}$ and $B_{\mathtt{II}} \subset A_{\mathtt{II}}$.
Incarnation: strategies are definable as the *minimal* designs of $A_{\mathtt{I}}$ and $A_{\mathtt{II}}$.

12.2.4 Negation

Player $\mathtt{I}$ tries to prove A, player $\mathtt{II}$ tries to refute, i.e., to *negate*, A. This is why the logical explanation of the most basic operation on games—the swapping $\mathtt{I}/\mathtt{II}$—is precisely negation. Since swapping is involutive, intuitionistic negation does not fit into this pattern; classical negation neither, because structural rules, especially contraction, are not self-dual.[3] This explanation can only be carried out in the

---

[3] Witness Gentzen's entangled "cross-cuts".

framework of *linear logic*, where linear negation quite represents the swapping `I/II`. Summing up, we see that negation does not merely *refute*, it *forbids*!

We observed in Sect. 12.2.2 that the restriction to combinatorial methods (this applies to syntax, but also to various games, including ludics) is booby-trapped: how can we put $\mathbb{N}$ into question while presupposing it? This is the reason for a drastic change of paradigm, *Geometry of Interaction* (GoI). By replacing combinatorics with operator algebras (matrix algebras and their generalisation: von Neumann algebras), we put ourself in a more *constructive* situation w.r.t. integers; this is mainly due to the *non-commutativity* at work in those structures.

12.3 Integers in GoI

12.3.1 An Example: The Number 4

In the absence of contraction,[4] proofs can be represented by plain matrices. Thus, the proof of $X \Rightarrow X, X \Rightarrow X, X \Rightarrow X, X \Rightarrow X \vdash X \Rightarrow X$ (corresponding to the function $f, g, h, k \rightsquigarrow k \circ h \circ g \circ f$) is interpreted by:

$$L_4 := \begin{bmatrix} 0 & 0 & 0 & 0 & 0 & 0 & 0 & 0 & 1 & 0 \\ 0 & 0 & 1 & 0 & 0 & 0 & 0 & 0 & 0 & 0 \\ 0 & 1 & 0 & 0 & 0 & 0 & 0 & 0 & 0 & 0 \\ 0 & 0 & 0 & 0 & 1 & 0 & 0 & 0 & 0 & 0 \\ 0 & 0 & 0 & 1 & 0 & 0 & 0 & 0 & 0 & 0 \\ 0 & 0 & 0 & 0 & 0 & 0 & 1 & 0 & 0 & 0 \\ 0 & 0 & 0 & 0 & 0 & 1 & 0 & 0 & 0 & 0 \\ 0 & 0 & 0 & 0 & 0 & 0 & 0 & 0 & 0 & 1 \\ 1 & 0 & 0 & 0 & 0 & 0 & 0 & 0 & 0 & 0 \\ 0 & 0 & 0 & 0 & 0 & 0 & 0 & 1 & 0 & 0 \end{bmatrix} \tag{12.1}$$

L_4 is indexed by the ten occurrences $X_1, \ldots, X_{10}$ of X; $L_{ij} = 1$ iff there is an identity axiom—i.e., a *link*—between X_i and X_j, i.e., when one of $(i, j), (j, i)$ belongs to $\{(1, 9), (2, 3), (4, 5), (6, 7), (8, 10)\}$.

The contraction rule—fingernail of infinity in this finite world—enables one to replace the four left occurrences of $X \Rightarrow X$ with a single one, thus yielding $X \Rightarrow X \vdash X \Rightarrow X$, hence $\vdash \forall X((X \Rightarrow X) \Rightarrow (X \Rightarrow X))$. This proof is Curry-Howard isomorphic to the integer 4 of system **F**: if f is of type $A \Rightarrow A$, then $4\{A\}(f) = f \circ f \circ f \circ f$. In order to perform the contraction, we use the indices 1, 2, 3, 4 to distinguish the four contracted occurrences, 0 being used for the occurrence of $X \Rightarrow X$ on the right. L_4 can be

[4] I.e., in multiplicative-additive linear logic **MALL**.

transformed into a 20 × 20 matrix (indexed by $\{1, \ldots, 4\} \times \{0, \ldots, 4\}$): the indices $1, \ldots, 10$ respectively become: $1.1, 2.1, 1.2, 2.2, 1.3, 2.3, 1.4, 2.4, 3.0, 4.0$; the absent $1.0, 2.0, 3.1, 3.2, 3.3, 3.4, 4.1, 4.2, 4.3, 4.4$ induce blank lines and columns (zeros). This inflated L_4 can be written as the 4 × 4 matrix:

$$M_4 := \begin{bmatrix} 0 & v & u & 0 \\ v^* & 0 & 0 & w \\ u^* & 0 & 0 & 0 \\ 0 & w^* & 0 & 0 \end{bmatrix} \tag{12.2}$$

whose entries are in turn 5 × 5 matrices:

$$u := \begin{bmatrix} 0&0&0&0&0 \\ 1&0&0&0&0 \\ 0&0&0&0&0 \\ 0&0&0&0&0 \\ 0&0&0&0&0 \end{bmatrix} \quad v := \begin{bmatrix} 0&0&0&0&0 \\ 0&0&0&0&0 \\ 0&1&0&0&0 \\ 0&0&1&0&0 \\ 0&0&0&1&0 \end{bmatrix} \quad w := \begin{bmatrix} 0&0&0&0&1 \\ 0&0&0&0&0 \\ 0&0&0&0&0 \\ 0&0&0&0&0 \\ 0&0&0&0&0 \end{bmatrix} \tag{12.3}$$

Let $\pi_0, \ldots, \pi_4$ be the orthoprojections:

$$\pi_0 = \begin{bmatrix} 1&0&0&0&0 \\ 0&0&0&0&0 \\ 0&0&0&0&0 \\ 0&0&0&0&0 \\ 0&0&0&0&0 \end{bmatrix} \quad \pi_1 = \begin{bmatrix} 0&0&0&0&0 \\ 0&1&0&0&0 \\ 0&0&0&0&0 \\ 0&0&0&0&0 \\ 0&0&0&0&0 \end{bmatrix} \quad \pi_2 = \begin{bmatrix} 0&0&0&0&0 \\ 0&0&0&0&0 \\ 0&0&1&0&0 \\ 0&0&0&0&0 \\ 0&0&0&0&0 \end{bmatrix} \ldots \tag{12.4}$$

then $uu^* = \pi_1, u^*u = ww^* = \pi_0, w^*w = \pi_4, vv^* = \pi_2 + \pi_3 + \pi_4, v^*v = \pi_1 + \pi_2 + \pi_3$.

The *partial isometries* u, v, w are such that:

$$\begin{aligned} u\pi_0 &= \pi_1 u & u\pi_i &= 0 \quad (i \neq 0) \\ v\pi_i &= \pi_{i+1} v \quad (i = 1, 2, 3) & v\pi_i &= 0 \quad (i = 0, 4) \\ w\pi_4 &= \pi_0 w & w\pi_i &= 0 \quad (i \neq 4) \end{aligned}$$

So to speak, u, v, w organise a "round-trip" $\pi_0 \ldots \pi_4$ in $5 = 4 + 1$ steps.

12.3.2 Representations

We can, more generally, interpret the number n by a 4 × 4 matrix M_n whose entries are $n + 1 \times n + 1$ matrices:

$$M_0 = \begin{bmatrix} 0 & 0 & 0 & 0 \\ 0 & 0 & 0 & 0 \\ 0 & 0 & 0 & 1 \\ 0 & 0 & 1 & 0 \end{bmatrix} \qquad M_n = \begin{bmatrix} 0 & v_n & u_n & 0 \\ v_n^* & 0 & 0 & w_n^* \\ u_n^* & 0 & 0 & 0 \\ 0 & w_n & 0 & 0 \end{bmatrix} \tag{12.5}$$

We would like to encode all integers within the same $\mathcal{M}_4(\mathcal{H})$; for this we need an algebra $\mathcal{H}$ together with embeddings $\mathcal{M}_{n+1}(\mathbb{C}) \overset{\phi_{n+1}}{\rightarrowtail} \mathcal{H}$. An obvious choice for $\mathcal{H}$ is a type $\mathbf{II}_1$ von Neumann algebra, the *hyperfinite factor*, in which usual matrix algebras embed (in a non unique way). If $a_n = \phi_{n+1}(u_n), b_n = \phi_{n+1}(v_n), c_n = \phi_{n+1}(w_n)$, define $N_n \in \mathcal{M}_4(\mathcal{H})$:

$$N_0 = \begin{bmatrix} 0 & 0 & 0 & 0 \\ 0 & 0 & 0 & 0 \\ 0 & 0 & 0 & I \\ 0 & 0 & I & 0 \end{bmatrix} \qquad N_n = \begin{bmatrix} 0 & b_n & a_n & 0 \\ b_n^* & 0 & 0 & c_n^* \\ a_n^* & 0 & 0 & 0 \\ 0 & c_n & 0 & 0 \end{bmatrix} \tag{12.6}$$

i.e., $N_n := \mathcal{M}_4(\phi_{n+1})(M_n)$; $a_n + b_n + c_n$ is a sort of circular permutation of $n+1$ projections $\pi_{n.0}, \ldots, \pi_{n.n}$ such that $\pi_{n.0} + \ldots + \pi_{n.n} = I$. More precisely:

$$a_n = \pi_{n.1}\, a_n\, \pi_{n.0} \tag{12.7}$$

$$b_n = \pi_{n.2}\, b_n\, \pi_{n.1} + \pi_{n.3}\, b_n\, \pi_{n.2} + \ldots + \pi_{n.n}\, b_n\, \pi_{n.n-1} \tag{12.8}$$

$$c_n = a_n^*(b_n^*)^{n-1} \qquad \text{hence} \tag{12.9}$$

$$c_n = \pi_{n.0}\, c_n\, \pi_{n.n} \tag{12.10}$$

Thus the $\pi_{n,i}$ can in turn be recovered as:

$$\pi_{n.0} := c_n\, c_n^* \tag{12.11}$$

$$\pi_{n.1} := a_n\, a_n^* \tag{12.12}$$

$$\pi_{n.i+1} := b_n\, \pi_{n.i}\, b_n^* \qquad (1 \leq i < n) \tag{12.13}$$

If $n \in \mathbb{N}$, a matrix $N_n \in \mathcal{M}_4(\mathcal{H})$ of the type (12.7) and (in case $n \neq 0$) enjoying (12.7–12.9) is called a *representation* of n. This is the only sensible definition: since there is no standard embedding $\mathcal{M}_{n+1}(\mathbb{C}) \overset{\phi_{n+1}}{\rightarrowtail} \mathcal{H}$, the entries a_n, b_n, c_n can only be characterised up to isomorphism.

12.3.3 Measurement as a Determinant

For $n \neq 0$, there is a continuum of representations N_n of n, all of them isomorphic: if N, N' are two representations of the same $n \in \mathbb{N}$, there is a unitary $u \in \mathcal{H}$

such that $N' = \mathcal{M}_4(u)(N')$ (if u_0 is any partial isometry from $\pi_{n.0}$ to $\pi'_{n.0}$, define $u_1 := a'_n u_0 a^*_n$, $u_{i+1} := b'_n u_i b^*_n$ $(1 \leq i < n)$ and let $u := u_0 + \ldots + u_n$).

Thus, among all representations of n, none is "more standard" than the others. From an uncouth logicist standpoint, we gave but another definition of natural numbers—the length of a round-trip replacing the cardinality of a set—, nothing really exciting! Now remember that operator algebras were designed to cope with quantum physics, in particular with the process of *observation* and the possible *interference* with the object observed. In other terms, when dealing with an integer, the *measurement* may depend upon the representation in a rather intricate way: this is non-commutativity. One should not consider this possibility as irrelevant to a our discussion:

- The logic tradition, from Frege to category theory, consider objects up to isomorphism; this supposes that the *form*—what isomorphisms preserve—is given in advance. This *essentialism*, which can be advocated in other contexts, is foundationally suspect.
- Complexity theory is about the *difficulty* of computation; it is a theory without operating concepts, reduced to a phenomenology of Turing machines. It is thus legitimate to seek an *explanation* of complexity classes through the neighbouring idea of *observation*.

In Geometry of Interaction, the operator $A \in \mathcal{K}$ is "observed" by an operator $B \in \mathcal{K}$—the process being, like in games, symmetrical. The output of this observation, the *measurement*, is the real number:

$$\ll A \mid B \gg := \det(I - AB) \qquad (= \det(I - BA)) \tag{12.14}$$

Here we must say something about the scalar $\det(\cdot)$ in a type $\mathbf{II}_1$ von Neumann algebra (like $\mathcal{H}$ or the isomorphic $\mathcal{M}_4(\mathcal{H})$). The idea is that the determinant should be invariant under the embeddings $\mathcal{M}_n(\mathbb{C}) \overset{\phi_n}{\rightarrowtail} \mathcal{H}$: $\det(\phi_n(M)) = \det(M)$ for $M \in \mathcal{M}_n(\mathbb{C})$. To make the long story short, this rests upon the invariance under the $\phi_{n.k} : \mathcal{M}_n(\mathbb{C}) \mapsto \mathcal{M}_{nk}(\mathbb{C})$ which replace each entry with a diagonal $k \times k$-matrix: $\det(\phi_{n.k}(M)) = \det(M)$; this does not hold for the usual determinant $\mathrm{Det}(\cdot)$ of $\mathcal{M}_n(\mathbb{C})$ which must be "normalised" as $\det(u) := |\mathrm{Det}(u)|^{\frac{1}{n}}$: the exponent fixes the problems of dimension, and the absolute value accounts for the impossibility of defining $z^{\frac{1}{n}}$ for all $z \in \mathbb{C}$. Remembering that a vN algebra of type $\mathbf{II}_1$ admits a trace; in $\mathcal{H}$, the trace extends the normalised trace $\mathrm{tr}(u) := \frac{1}{n}\mathrm{Tr}(u)$ of $\mathcal{M}_n(\mathbb{C})$.

$$-\log\det(1 - u) = \mathrm{tr}(u) + \frac{\mathrm{tr}(u^2)}{2} + \frac{\mathrm{tr}(u^3)}{3} + \cdots \tag{12.15}$$

expresses the determinant when u is hermitian (or a product AB of hermitians) and the spectral radius of u is < 1, e.g., when $\|u\| < 1$.

12.3.4 Interference

We shall thus "observe" our represented integers by means of matrices $\Phi, \Psi, \ldots \in \mathcal{M}_4(\mathcal{H})$. Then the following question arises: when observing N_n are we observing n or a specific representation? Indeed, although two representions N_n, N_n' of the same n are isomorphic, the measurements $\ll N_n \mid \Phi \gg$ and $\ll N_n' \mid \Phi \gg$ need not be the same; indeed $\ll \Phi \mid N_n \gg$ may be rather unpredictable, and certain measurements should be discarded as "meaningless". This is similar to the usual requirement about bound variables: when combining formulas, make sure that their bound variables are distinct. The act of discarding a measurement is strongly normative and should not be pushed under the carpet. The point about GoI is that we get the objects, so to speak, with their bound variables, although there is no clear renaming technique like in logic. *Objectivity of measurement*:

If N_n, N_n' are representations of $n \in \mathbb{N}$, if Φ is an observation, then
$$\ll \Phi \mid N_n \gg = \ll \Phi \mid N_n' \gg .$$

is a normative requirement, calling for restrictions upon the shape of representations and their observations.

12.3.5 An Example: Commutation

Consider the "observation":

$$\Phi := \begin{bmatrix} 0 & v & 0 & 0 \\ v^* & 0 & 0 & 0 \\ 0 & 0 & u & 0 \\ 0 & 0 & 0 & w \end{bmatrix} \tag{12.16}$$

Proposition 12.1. *If u, v, w commute with the entries a_n, b_n, c_n of N_n, then*

$$\ll \Phi \mid N_n \gg = (\det(I - v^n w v^{*n} u))^{\frac{1}{2n+2}} \tag{12.17}$$

Proof. we restrict to the particular case where (12.15) converges. Observe that $\mathrm{tr}(N_n \Phi)^p = 0$ when p is not a multiple of $2n+2$, hence:

$$-\log\det(1 - N_n\Phi) = \frac{\mathrm{tr}((N_n\Phi)^{2n+2})}{2n+2} + \frac{\mathrm{tr}((N_n\Phi)^{4n+4})}{4n+4} + \frac{\mathrm{tr}((N_n\Phi)^{6n+6})}{6n+6} + \ldots \quad \text{and}$$

$$(N_n\Phi)^{(2n+2)k} = \begin{bmatrix} A_{n.k} & 0 & 0 & 0 \\ 0 & B_{n.k} & 0 & 0 \\ 0 & 0 & C_{n.k} & 0 \\ 0 & 0 & 0 & D_{n.k} \end{bmatrix} \tag{12.18}$$

with:

$$A_{n.k} = \pi_{n.1}\, v^{*n-1}u(v^n wv^{*n}u)^{k-1}v^n wv^* + \ldots + \pi_{n.n-1}\, v^* u(v^n wv^{*n}u)^{k-1}v^n wv^{*n-1}$$

$$B_{n.k} = \pi_{n.n}\, v^{n-1}wv^{*n}u(v^n wv^{*n}u)^{k-1}v + \ldots + \pi_{n.2}\, vwv^{*n}u(v^n wv^{*n}u)^{k-1}v^{n-1}$$

$$C_{n.k} = \pi_{n.1}\, (v^n wv^{*n}u)^k$$

$$D_{n.k} = \pi_{n.0}\, (v^{*n}uv^n w)^k$$

$\mathrm{tr}(C_{n.k}) = \mathrm{tr}(D_{n.k}) = \frac{\mathrm{tr}((v^n wv^{*n}u)^k)}{n+1}$, $\mathrm{tr}(A_{n.k}) = \mathrm{tr}(B_{n.k}) = n \cdot \mathrm{tr}(C_{n.k})$ yield $\mathrm{tr}((N_n\Phi)^{(2n+2)k}) = \frac{\mathrm{tr}(A_{n.k}+B_{n.k}+C_{n.k}+D_{n.k})}{4} = \frac{\mathrm{tr}((v^n wv^{*n}u)^k)}{2}$, hence (12.17). □

Proposition 12.1 establishes objectivity of measurement under the hypothesis that the coefficients u, v, w of the observation do commute with the entries a_n, b_n, c_n of the representation: metaphorically, the "bound variables" of N_n and Φ do not interfere. However, this remedy cures the disease by killing the patient: if u, v, w must commute with all a_n, b_n, c_n, they must be scalars, and one hardly sees how to express any non-trivial algorithm in this way! We can loosen the situation by assuming a certain amount of commutation a priori: this is the *dialectal* (or idiomatic) maintenance of GoI, at work in Girard (2011). We thus modify the definition of observations:

Definition 12.1 (Observations). A *dialect* $\mathcal{D}$ is a matrix space $\mathcal{M}_k(\mathbb{C})$; an *observation* (of dialect $\mathcal{D}$) is an element of $\mathcal{M}_4(\mathcal{H}) \otimes \mathcal{D}$, i.e., a 4×4 matrix with entries in $\mathcal{H} \otimes \mathcal{D} = \mathcal{M}_k(\mathcal{H})$. The output of the observation of N_n by Φ is the *measurement*:

$$\ll \Phi \mid N_n \gg := \det(I - \Phi(N_n \otimes I_k)) \tag{12.19}$$

Even in this relaxed setting, commutation remains too drastic, i.e., leaves very few interesting observations.

12.3.6 Normativity by Subalgebras

The translation of logical rules done in Girard (2011)—especially of the exponentials et work in the encoding of Dedekind integers—induces a restriction on the observations together with a co-restriction on the "objects", i.e., on the representations. This joint restriction ensures *objectivity of measurement* without assuming commutation.

Without entering into something as normative as the logical maintenance of exponentials, we can directly seek joint restrictions of the pairs observation/ representation. The simplest idea is that of a restriction to specific subalgebras: we shall seek pairs $(\mathcal{I}, \mathcal{O})$ of subalgebras of $\mathcal{H}$ such that the restriction $N_n \in \mathcal{M}_4(\mathcal{I})$, $\Phi \in \mathcal{M}_4(\mathcal{O}) \otimes \mathcal{D}$ ($\mathcal{D}$ arbitrary), ensures objectivity:

$$\forall N_n, N_n' \in \mathcal{M}_4(\mathcal{I})\ \forall \Phi \in \mathcal{M}_4(\mathcal{O}) \otimes \mathcal{D} \qquad \ll \Phi \mid N_n \gg = \ll \Phi \mid N_n' \gg \tag{12.20}$$

Let us call such a pair $(\mathcal{I}, \mathcal{O})$ a *normative pair*. Among normative pairs, $(\mathcal{H}, \mathbb{C} \cdot I)$: if the entries of Φ belong to the dialect space $\mathcal{D}$, then we are in—rather, isomorphic to—the situation of Proposition 12.1: the objectivity of measurement is ensured for *all* representations of natural numbers. More generally, if $\mathcal{I}, \mathcal{O}$ are such that, whenever $u \in \mathcal{I}, v \in \mathcal{O}$, then $uv = vu$, then (12.20) holds.

12.4 Logspace Integers

If we seek a non-trivial (i.e., non-commuting) normative pair, then the most natural example is given by the crossed product to be defined below; the big surprise is that this restriction corresponds to logspace computation!

12.4.1 A Normative Pair

Consider[5] the infinite tensor power $\mathcal{K} := \bigotimes_{n>0} \mathcal{H}'$ of ω copies of some $\mathcal{H}'$ isomorphic to $\mathcal{H}$. For future reference, we note $\mathcal{H}_i$ the subalgebra of $\mathcal{K}$ consisting of the $I \otimes \ldots \otimes u \otimes I \otimes \ldots$ (the $\bigotimes u_n$ s.t. $u_n = I$ for $n \neq i$). The group $\mathfrak{S}$ of (finite) permutations of $\mathbb{N}$ operates on $\mathcal{K}$ by $\sigma(\bigotimes u_n) := \bigotimes u_{\sigma(n)}$. The crossed product $\mathcal{K} \rtimes \mathfrak{S}$ *internalises* $\mathfrak{S}$, the action of σ becoming an inner automorphism:

$$\sigma \cdot \bigotimes u_n = \left(\bigotimes u_{\sigma(n)}\right) \cdot \sigma \tag{12.21}$$

Let $\mathcal{H} := \mathcal{K} \rtimes \mathfrak{S}$. Among the remarkable subalgebras of $\mathcal{H}$: the $\mathcal{H}_i$, $\mathcal{K}$ and the algebra $\mathcal{S}$ generated by $\mathfrak{S}$. These subalgebras ar all isomorphic to $\mathcal{H}$, the *unique* hyperfinite factor of type $\mathbf{II}_1$.

Proposition 12.2. *Any automorphism θ_1 of $\mathcal{H}_1$ can be uniquely extended into an automorphism θ of $\mathcal{H}$ which is the identity on $\mathcal{S}$.*

Proof. if $\theta_1(u \otimes I \otimes \ldots) = \vartheta(u) \otimes I \otimes \ldots$, define $\theta(\sigma \cdot \bigotimes_n u_n) := \sigma \cdot \bigotimes_n \vartheta(u_n)$.

Corollary 12.1. *$(\mathcal{H}_1, \mathcal{S})$ is a normative pair.*

Proof. assume that $N_n, N'_n \in \mathcal{M}_4(\mathcal{H}_1)$; then $N'_n = \mathcal{M}_4(\theta_1)(N_n) = \mathcal{M}_4(\theta)(N_n)$ for some automorphism θ_1 of $\mathcal{H}_1$. If $\Phi \in \mathcal{M}_4(\mathcal{S}) \otimes \mathcal{M}_k(\mathbb{C})$, then:

$$\begin{aligned}\det(I - \Phi \cdot (\mathcal{M}_4(\theta_1)(N_n) \otimes I_k)) &= \det(I - \mathcal{M}_{4k}(\theta)(\Phi \cdot (N_n \otimes I_k))) \\ &= \det(I - \Phi \cdot (N_n \otimes I_k))\end{aligned}$$

since the determinant is invariant under the isomrphism $\mathcal{M}_{4k}(\theta)$. □

---

[5]This section requires some familiarity with vN algebras, especially with crossed products, see, e.g., Kadison and Ringrose (1986).

12.4.2 Logspace Operators

We now turn our attention towards computation.

Definition 12.2 (Logspace operators). A *logspace operator* is any $\Phi \in \mathcal{M}_4(\mathcal{S}) \otimes \mathcal{D}$, where $\mathcal{D} = \mathcal{M}_k(\mathbb{C})$ is a matrix algebra, such that the entries of the $4k \times 4k$ matrix Φ are finite linear combinations $\sum \lambda_i s_i$ of elements $s_i \in \mathfrak{S}$ with positive coefficients $\lambda_i > 0$.

Φ being a logspace operator, consider the set:

$$[\Phi] := \{n \in \mathbb{N}\ ;\ \forall N_n \in \mathcal{M}_4(\mathcal{H}_1) \qquad \ll \Phi \mid N_n \gg = 1\} \tag{12.22}$$

Theorem 12.1 (Logspace integers). *The set* $[\Phi]$*, as a set of tallies (see Remark 12.1,* infra*) is in* NL *(non-deterministic logspace).*

Proof. let us compute $\Phi(N_n \otimes I_k)$ and its iterates. The elements of $\mathfrak{S}$ occurring in the entries of Φ generate a finite subgroup $\mathfrak{S}_\Phi$; let $N \in \mathbb{N}$ be such that $\sigma(i) = i$ for all $\sigma \in \mathfrak{S}_\Phi$ and $i \geq N$. We can, w.l.o.g., place ourself in $\bigotimes_{1,\dots,N} \mathcal{H} \rtimes \mathfrak{S}[1,\dots,N]$. Since $(\mathcal{H}_1, \mathcal{S})$ is a normative pair, we can replace the entries of N_n with $n+1 \times n+1$ matrices whose entries are $0, 1$; in particular, $\mathcal{H}$ is replaced with $\mathcal{M}_{n+1}(\mathbb{C})$. Our computation eventually takes place in $\mathcal{M}_4(\mathbb{C}) \otimes \mathcal{M}_k(\mathbb{C}) \otimes (\mathcal{M}_{n+1}(\mathbb{C}) \otimes \dots \otimes \mathcal{M}_{n+1}(\mathbb{C})) \rtimes \mathfrak{S}[1,\dots,N]$. Φ and $N_n \otimes I_k$ have thus been reduced to finite-dimensional operators, on a space of dimension $4k(n+1)^N \cdot N!$ whose canonical base can be written

$$\{(ai(j_1,\dots,j_N);\sigma)\ ;\ 1 \leq a \leq 4,\ 1 \leq i \leq k,\ j_1 \dots, j_N \leq n,\ \sigma \in \mathfrak{S}[1,\dots,N]\}$$

- $\Phi((ai(j_1,\dots,j_N);\sigma))$ is a sum: if τ "occurs" in the entry $\Phi_{a'i',ai}$, then $(a'i'(j_{\tau(1)},\dots,j_{\tau(N)});\tau\sigma)$ occurs in $\Phi(ai(j_1,\dots,j_N);\sigma)$ with the same multiplicity.
- $(N_n \otimes I_k)((ai(j_1,\dots,j_N);\sigma)) = 0$ if the entries $N_{a',j';a,j_1}$ are all null. Otherwise, let $N_{a',j';a,j_1}$ be the only nonzero entry of N_n of this form; then $(N_n \otimes I_k)((ai(j_1,\dots,j_N);\sigma)) = (a'i(j'_1, j_2,\dots,j_N);\sigma)$.

$\det(I - \Phi(N_n \otimes I_k)) = 1$ iff $\Phi(N_n \otimes I_k)$ is nilpotent. This is the same as saying that the iterates $(\Phi(N_n \otimes I_k))^p((ai(j_1,\dots,j_N);\iota))$ are all null for sufficiently great p (e.g., $p = 4k(n+1)^N$), ι denoting the identity permutation. Now, Φ being fixed, it is plain that the process of iteration yielding the $(\Phi(N_n \otimes I_k))^p((ai(j_1,\dots,j_N);\iota))$ is logspace: indeed it takes place in a universe of size $s(n) = 4k(n+1)^N \cdot N!$ whose elements can be written with approximately $\log(n).N + \log(4kN!)$ digits. Indeed, non-deterministic logspace: when computing $\Phi((ai(j_1,\dots,j_N);\sigma))$, several choices $\tau \in \Phi_{a'i',ai}$ are available. Nilpotency is therefore coNL, which is the same as NL. □

Remark 12.1. The theorem should be stated for binaries, see Sect. 12.4.4 for the exact relation with NL. Theorem 12.1 is only a prototype which relies on the dumb tally representation of natural numbers; since, as a binary, the tally n encodes the number 2^n, Theorem 12.1 indeed says that $2^{[\Phi]} := \{2^n\ ;\ n \in [\Phi]\}$ is in NL.

The choice of the coefficients λ_i in the entries $\sum \lambda_i s_i$ is irrelevant, as long as they stay positive. In particular, they can be chosen small enough to ensure $\|\Phi\| \leq 1$, an essential requirement of GoI. One can also require them to be rational: this may simplify technical issues.

12.4.3 Normative vs. Non-standard

Normativity occurs because we are specifically interested in measurements, i.e., in observations; and, as logicians, in general properties of observations: "What can we observe?", "Is this style of observation more efficient that one?"...

At the level of the objects "observed", normativity induces a departure standard/ non-standard. We must distinguish between two forms of non-standardness, *relative* and *absolute*.

12.4.3.1 Relative Non-standardness

W.r.t. normativity by the subalgebras $\mathcal{H}_1, \mathcal{S}$, the integers—rather their representations—$N_n \in \mathcal{M}_4(\mathcal{H}_1)$ are standard. This means that $N_n, N'_n \in \mathcal{M}_4(\mathcal{H}_1)$ cannot be distinguished by observations: they are, so to speak, the same "up to bound variables". The other representations should be styled "non-standard"; they are, however, plainly isomorphic to standard integers and their "non-standardness" is only relative to our observational normativity.

Non-standard integers yield additional objects to which the observation may be applied. Due to interference, the measurement thus obtained may be completely meaningless. But, this need not be always the case: for instance, $(\mathcal{H}_2, \mathcal{S})$ is also a normative pair, hence the non-standard integers $N_n \in \mathcal{M}_4(\mathcal{H}_2)$ yield consistent "alternative" measurements: the example is a bit too simple, since the same result can be achieved by applying $\tau_{12}\Phi\tau_{12}$—where τ_{12} is the transposition $1 \leftrightarrows 2$—to "standard" N_n.

12.4.3.2 Absolute Non-standardness

Normativity by subalgebras is external: both the object (the natural number "observed") and the subject (the "observation" Φ) are coerced into algebras of their own, so as to inhibit unwanted interferences.

An internal normativity is much more satisfactory; we can internalise normativity on both sides—numbers and observations—the case of observations being the most interesting one. So how can we ensure that $\Phi \in \mathcal{M}_4(\mathcal{S}) \otimes \mathcal{D}$? In a second step, how can we ensure that the entries of Φ are finite linear combinations $\sum \lambda_i s_i$ of elements of $\mathfrak{S}$, with[6] $\lambda_i \in \mathbb{Q}^+$?

The answer lies in the introduction of additional objects, the "watchdogs of normativity". These objects are *formal* linear combinations $\mathfrak{a} = \bigoplus \lambda_i \bullet A_i$ of operators. The measurement $\ll A \mid B \gg$ of (12.14) is generalised into:

$$\ll \bigoplus \lambda_i \bullet A_i \mid \bigoplus \mu_j \bullet B_j \gg := \prod_{ij} \ll A_i \mid B_j \gg^{\lambda_i \mu_j} \tag{12.23}$$

In Girard (2011), internal normativity takes the form $\ll \mathfrak{a} \mid \mathfrak{b} \gg \neq 0, 1$. Assuming that $\ll N_0 \mid \Phi \gg \neq 0, 1$, most normative queries can be expressed under the form:

$$\ll N_0 \oplus \lambda \bullet A \ominus \lambda \bullet B \mid \Phi \gg \neq 0, 1 \tag{12.24}$$

For instance, if (12.24) holds for all $\lambda \in \mathbb{R}$, then $\ll A \mid \Phi \gg = \ll B \mid \Phi \gg$. Taking $A \in \mathcal{M}_4(\mathcal{K})$, $B := \theta(A)$, where θ_1, θ are as in Proposition 12.2, we can thus express the constraint $\Phi \in \mathcal{M}_4(\mathcal{S}) \otimes \mathcal{D}$. The constraint $N_n \in \mathcal{M}_4(\mathcal{H}_1)$ can in turn be recovered from $\ll N_n \mid \Phi \gg = \ll N_n \mid \Psi \gg$, for any observations Φ and $\Psi := \mathcal{M}_{4k}(\sigma)(\Phi)$, where σ is any element of $\mathcal{S}$ such that $\sigma(1) = 1$.

Since non "integer-like", the proper linear combinations $\mathfrak{a} = \bigoplus \lambda_i \bullet A_i$ in charge of the law, are intrinsically non-standard. The question is to determine whether or not they can be of some use, i.e., if the measurements $\ll \mathfrak{a} \mid \Phi \gg$ are meaningful. The question extends, of course, to those $\mathfrak{a}$ in charge of other "laws" that Φ may or may not break.

12.4.4 Logspace Binaries

In view of our concern for complexity issues, the tallies just discussed must be replaced with the set $\mathbb{S}$ of lists of 0, 1, which is in bijection with $\mathbb{N}^*$: the list s encodes the binary number $1s$. The empty sequence therefore encodes 1; the maps $s \rightsquigarrow s0$ and $s \rightsquigarrow s1$ respectively encode the functions $n \rightsquigarrow 2n$ and $n \rightsquigarrow 2n+1$. These binaries can be typed in system **F** by
bin $:= \forall X((X \Rightarrow X) \Rightarrow ((X \Rightarrow X) \Rightarrow (X \Rightarrow X)))$, that GoI handles by means of 6×6 matrices (instead of the 4×4 matrices used for **nat**).

[6]The replacement $\mathbb{R}^+ \rightsquigarrow \mathbb{Q}^+$ ensures that there are only denumerably many entries.

To a list s of 0, 1, we associate representations:

$$B_{<>} = \begin{bmatrix} 0&0&0&0&0&0 \\ 0&0&0&0&0&0 \\ 0&0&0&0&0&0 \\ 0&0&0&0&0&0 \\ 0&0&0&0&0&I \\ 0&0&0&0&I&0 \end{bmatrix} \qquad B_s = \begin{bmatrix} 0 & c_s & 0 & e_s & a_s & 0 \\ c_s^* & 0 & d_s^* & 0 & 0 & g_s^* \\ 0 & d_s & 0 & f_s & b_s & 0 \\ e_s^* & 0 & f_s^* & 0 & 0 & h_s^* \\ a_s^* & 0 & b_s^* & 0 & 0 & 0 \\ 0 & g_s & 0 & h_s & 0 & 0 \end{bmatrix} \tag{12.25}$$

If $s =< s_1, \dots, s_n >$ ($s_i \in \{0,1\}$), then the entries $a_s, \dots, h_s$ are partial isometries. Indeed, consider the sets:

$$a_s := \{0\},\ b_s = \emptyset \quad \text{if } s_1 = 0 \qquad b_s := \{0\},\ a_s = \emptyset \quad \text{if } s_1 = 1$$

$$c_s := \{i \neq 0, n\ ;\ s_i = s_{i+1} = 0\} \qquad d_s := \{i \neq 0, n\ ;\ s_i = 0, s_{i+1} = 1\}$$

$$e_s := \{i \neq 0, n\ ;\ s_i = 1, s_{i+1} = 0\} \qquad f_s := \{i \neq 0, n\ ;\ s_i = s_{i+1} = 1\}$$

$$g_s := \{n\},\ h_s = \emptyset \quad \text{if } s_n = 0 \qquad h_s := \{n\},\ g_s = \emptyset \quad \text{if } s_n = 1$$

The entries $a_s, \dots, h_s$ are characterised by the existence of projections $\pi_{s.0}, \dots, \pi_{s.n}$ such that $I = \pi_{s.0} + \dots + \pi_{s.n}$ and:

$$a_s = \sum_{i \in a_s} \pi_{s.i+1}\, a_s\, \pi_{s.i} \qquad b_s = \sum_{i \in b_s} \pi_{s.i+1}\, b_s\, \pi_{s.i}$$

$$c_s = \sum_{i \in c_s} \pi_{s.i+1}\, c_s\, \pi_{s.i} \qquad d_s = \sum_{i \in d_s} \pi_{s.i+1}\, d_s\, \pi_{s.i}$$

$$e_s = \sum_{i \in e_s} \pi_{s.i+1}\, e_s\, \pi_{s.i} \qquad f_s = \sum_{i \in f_s} \pi_{s.i+1}\, f_s\, \pi_{s.i}$$

$$g_s = \sum_{i \in g_s} \pi_{s.i+1}\, g_s\, \pi_{s.i} \qquad h_s = \sum_{i \in h_s} \pi_{s.i+1}\, h_s\, \pi_{s.i}$$

$$\pi_{s,0} = (g_s + h_s)(c_s + d_s + e_s + f_s)^{n-1}(a_s + b_s)$$

From which we can define the notions of *representation* of s. A pair $(\mathcal{I}, \mathcal{O})$ of subalgebras of $\mathcal{H}$ ensuring *objectivity*:

$$\forall B_s, B_s' \in \mathcal{M}_6(\mathcal{I})\ \forall \Phi \in \mathcal{M}_6(\mathcal{S}) \otimes \mathcal{D} \qquad \ll \Phi \mid B_s \gg = \ll \Phi \mid B_s' \gg \tag{12.26}$$

is called a *normative pair*. Again, the typical normative pair is $(\mathcal{H}_1, \mathcal{S})$.

Definition 12.3 (Logspace operators). A *logspace operator* is any $\Phi \in \mathcal{M}_6(\mathcal{S}) \otimes \mathcal{D}$, where $\mathcal{D} = \mathcal{M}_k(\mathbb{C})$ is a matrix algebra such that the entries $\Phi_{a.p,b.q}$ Φ (as a $6k \times 6k$ matrix) are finite linear combinations $\sum \lambda_i s_i$ of elements $s_i \in \mathfrak{S}$ with $\lambda_i > 0$.

Φ being a normative operator, consider the set:

$$[\Phi] := \{s \in \mathbb{S}\ ;\ \forall B_s \in \mathcal{M}_6(\mathcal{H}_1) \quad \ll \Phi \mid B_s \gg = 0\} \tag{12.27}$$

Then we get the following (immediate) analogue of Theorem 12.1.

Theorem 12.2 (Logspace integers). *The set* $[\Phi]$ *is in* NL *(non-deterministic logspace).*

Conversely, consider a non-deterministic logspace algorithm F, applying to binaries. F makes use of N "fingers" simultaneously visiting a binary $s \in \mathbb{S}$ of length n, with locations $\sharp 1, \ldots, \sharp n$ occupied by the digits 0 or 1, and an additional location, the *origin* $\sharp 0$; depending on the *configuration* $(i_1, \ldots, i_N; a)$, i.e., the data (0, 1, `origin`) simultaneously read by the N fingers and the current state (represented by $a \in A$, A finite), one can prompt certain *transitions*, which combine three actions: a change of state, a rearranging of the fingers and a move of the "thumb" (finger $\sharp 1$) forwards or backwards (next or previous location, the origin standing after $\sharp n$ and before $\sharp 1$); let $f_1, \ldots, f_r$ be the possible transitions, so that $F = \{f_1, \ldots, f_r\}$. We say that *s is accepted by F* when F, acting on s, has no loops. This definition implies that certain configurations may prompt no transition at all; otherwise, due to the finiteness of the configuration space, the algorithm must loop.

The computation will be encoded in the algebra $\mathcal{M}_{6^N}(\mathcal{H}) \otimes \mathcal{M}_A(\mathbb{C})$, which can be written: $\mathcal{M}_6(\mathcal{H}) \otimes \mathcal{D}$, with $\mathcal{D} := \mathcal{M}_{6^{N-1} \times A}(\mathbb{C})$. In order to encode F by an operator, it will be enough to encode—in a faithful way—each transition $f_1, \ldots, f_r \in F$ by adequate operators $\phi_1, \ldots, \phi_r \in \mathcal{M}_6(\mathcal{S}) \otimes \mathcal{D}$ and define $\Phi := \mu_1\phi_1 + \ldots + \mu_r\phi_r$, with $\mu_1, \ldots, \mu_r > 0$.

The execution of F applied to s will therefore be represented by the iterates $((B_s \otimes I_{\mathcal{D}})\Phi)^p$ $(p \in \mathbb{N})$, which are linear combinations of "monomials", i.e., alternated products $B_s f B_s g B_s \ldots B_s h$ of transitions $f, g, \ldots, h$ and B_s (indeed, $B_s \otimes I_{\mathcal{D}}$). Each monomial is a partial isometry whose final projection is of the form $(m_{i_1 i_1} \otimes \ldots \otimes m_{i_N i_N} \otimes m_{aa}) \cdot (\pi_{s.q_1} \otimes \ldots \otimes \pi_{s.q_N})$, where n is the length of s, $i_1, \ldots, i_N \in \{1, \ldots, 6\}$, $q_1, \ldots, q_N \in \{0, \ldots, n\}$ and $\pi_{s.0}, \ldots, \pi_{s.n}$ are the projections associated with B_s; such a projection is the product of $(i_1, \ldots, i_N; a) := m_{i_1 i_1} \otimes \ldots \otimes m_{i_N i_N} \otimes m_{aa}$, representing the current configuration and $\pi_{s.q_1} \otimes \ldots \otimes \pi_{s.q_N}$ representing the simultaneous location $(q_1, \ldots, q_N)$ of the N digits $q_1, \ldots, q_N$; let us abbreviate it as $((i_1, q_1), \ldots, (i_N, q_N); a)$. Obviously:

if $i = 1, 2$	then $q_i \in c_s \cup d_s$
if $i = 3, 4$	then $q_i \in e_s \cup f_s$
if $i = 5, 6$	then $q_i = 0$

The integers $i_1, \ldots, i_N \in \{1, \ldots, 6\}$ encode the data possibly read by the fingers, respectively $0, 0, 1, 1,$ `begin`, `end`. This encoding is redundant: each of the basic data "digit 0", "digit 1", "origin" gets two possible encodings, respectively $\{1, 2\}, \{3, 4\}, \{5, 6\}$. Indeed, our representations do combine a "forward trip", leading from $\{2, 4, 5\}$ to $\{1, 3, 6\}$ and an adjoint "backward trip", leading from $\{1, 3, 6\}$ to $\{2, 4, 5\}$. This duality of encoding is thus of a dynamic nature.

To the transition f, prompted by the configuration $(i_1, \ldots, i_N; a)$ we can associate $\phi \in \mathcal{M}_6(\mathcal{S}) \otimes \mathcal{D}$:

$$\phi := (m_{ki_1} \otimes m_{i_{\sigma(2)}i_2} \otimes \ldots \otimes m_{i_{\sigma(N)}i_N} \otimes m_{ba}) \cdot \sigma \tag{12.28}$$

where $\sigma \in \mathfrak{S}(\{1, \ldots, N\}) \subset \mathfrak{S}$ is the reordering of the fingers, which induces an operator of $\mathcal{S}$, still noted σ, $b \in A$ is the next state. k is defined as "$i_{\sigma(1)}$ up to a change of direction"; in other terms, $k \in \{2, 4, 5\}$ if the thumb "moves forward", $k \in \{1, 3, 6\}$ if the thumb "moves backward" and $\{i_{\sigma(1)}, k\}$ is included in one of the sets $\{1, 2\}, \{3, 4\}, \{5, 6\}$.

$B_s \cdot \phi \cdot ((i_1, q_1), \ldots, (i_N, q_N); a)$ is a partial isometry with final projection $\nu = ((i_k, q_{\sigma(1)\pm 1}), (i_{\sigma(2)}, q_{\sigma(2)}) \ldots, (i_{\sigma(N)}, q_{\sigma(N)}); b)$, where $(k, q_{\sigma(1)\pm 1})$ is the next location of the thumb: in case of a forward move $\nu = ((i_k, q_{\sigma(1)+1}), \ldots)$ with $k \in \{1, 3, 6\}$, in case of a backward move $\nu = ((i_k, q_{\sigma(1)-1}), \ldots)$ with $k \in \{2, 4, 5\}$.

We just proved (or rather sketched):

Theorem 12.3. *If $X \subset \mathbb{S}$ is in* NL, *then $X = [\Phi]$ for a logspace operator Φ.*

NON SI NON LA

References

Girard, J.-Y. 2006–2007. *Le point aveugle, tome 1: vers la perfection, tome 2: vers l'imperfection.* Paris: Visions des Sciences. Hermann, 296pp. +299pp.

Girard, J.-Y. 2011. Geometry of interaction V : logic in the hyperfinite factor. *Theoretical Computer Science* 412: 1860–1883. *Girard's Festschrift, eds. Ehrhard, Faggian and Laurent.*

Kadison, R. V., and J. R. Ringrose. 1986. *Fundamentals of the theory of operator algebras.* Pure and applied mathematics, vol. II. Orlando: Academic, 32887.

Chapter 13
Constructivist Versus Structuralist Foundations

Erik Palmgren

13.1 Introduction

The mathematical philosophies of constructivism and structuralism may at first appear to be at odds with each other. The emphasis on direct construction and lack of a full-fledged abstract set-theoretic or type-theoretic language in early approaches seemed to preclude a structuralist view of mathematics in constructivism. One may in particular note the restriction in type level of what Brouwer called sets or species: sets of (choice) sequences were the upper limit of sets considered. The latter deficit was certainly remedied in the 1970s with the introduction of Martin-Löf type theory and Aczel–Myhill set theory, which have all the expected abstraction powers of Zermelo-Fraenkel set theory, though not its proof strength.

Errett Bishop's book *Foundations of Constructive Analysis* from 1967 however contains a chapter on set theory. This set theory, apart from being informal, is quite unlike any of the theories of Zermelo–Fraenkel or Gödel–Bernays, which are derived from the iterative concept of set.

> "A set is not an entity which has an ideal existence: a set exists only when it has been defined. To define a set we prescribe, at least implicitly, what we (the constructing intelligence) must do in order to construct an element of the set, and what we must do to show that two elements are equal." (Bishop 1967, p. 2)

We find a similar explanation of what a set is also in the type theory of Martin-Löf (1984). Both explanations are aligned to Cantor's early explanation of sets from 1882 in the respect that they mention conditions for equality of elements explicitly. See Tait (2000) for a discussion. Bishop (1967, p. 74) emphasizes that two elements may not be compared unless they belong to some common set. This indicates a type-theoretic attitude to the foundations. Bishop's version of set theory has, despite its

E. Palmgren (✉)
Department of Mathematics, Stockholm University, 106 91 Stockholm, Sweden
e-mail: palmgren@math.su.se

P. Dybjer et al. (eds.), *Epistemology versus Ontology*, Logic, Epistemology, and the Unity of Science 27, DOI 10.1007/978-94-007-4435-6_13,

constructiveness, a more abstract character than e.g. ZF set theory in that it does not concern coding issues for basic mathematical objects. It defines a subset of a set X to be a pair (A, i_A) where $i_A : A \longrightarrow X$ is function so that $a = b$ if, and only if, $i(a) = i(b)$. An element $x \in X$ is a member of the subset if $x = i_A(a)$ for some $a \in A$. That the subset (A, i_A) is included in another subset (B, i_B) of X is defined by requirement that there is a function $f : A \longrightarrow B$ so that $i_A = i_B \circ f$, i.e. that the diagram

$$\begin{array}{ccc} A & \xrightarrow{\quad f \quad} & B \\ & {\scriptstyle i_A} \searrow \quad \swarrow {\scriptstyle i_B} & \\ & X & \end{array} \tag{13.1}$$

commutes. The subsets are equal in case f is a bijection. Unions and intersection are only defined when the involved sets are subsets of the same underlying set. These and other features of Bishop's set theory are remarkably reminiscent of Lawvere's *Elementary Theory of the Category of Sets* (ETCS) introduced in 1964. ETCS is obtained by singling out category-theoretic universal properties of various set construction in such a way that they become invariant under isomorphism; see McLarty (2004) and the introduction McLarty (2005) to Lawvere (2005), the full version of the 1964 paper. This invariance is of course fundamental for a *structuralist foundation*. ETCS is an elementary theory in the sense that it uses classical first order logic as a basis, and make no special assumption on existence of second order or higher order objects. The theory is equivalent to the axioms of a well-pointed topos with the axiom of choice (McLarty 2004; Mac Lane 1998).

Bishop (1970a, Bishop, 1970b. Compiling mathematics into algol, unpublished text for a seminar) considered various versions of Gödel's system T as a possible foundation for his set theory. At the basis of the interpretation is a system of computable functions and functionals, which in effect are the core operations of certain modern programming languages. Full-fledged systems suitable for the formalization of constructive mathematics in the style of Bishop emerged later with the constructive type theory of Martin-Löf (1975) and the constructive set theories CST (Myhill 1975) and CZF (Aczel 1978). Of these, the type-theoretic system is the more fundamental from a constructive semantical point of view, since it describes explicitly how the computation of functions are carried out. Indeed, the mentioned set-theoretic system, CZF, can be justified on the grounds of Martin-Löf's type theory (MLTT) as shown by Aczel (1978) by a model construction. In MLTT the explanation of when elements of a set (type) are equal halts at the level of definitional equality. There are no quotient constructions, so it is customary to consider a type together with an equivalence relation, as a set-like object, a so-called *setoid*. This gives two possible conceptions of constructive sets based on the formal theories CZF and MLTT, namely iterative sets (sets as trees) and setoids respectively.

In this paper, whose technical part is a shorter edition of Palmgren (2012), we present a constructive version of ETCS, called CETCS. This theory is obtained abstracting on category-theoretic properties of CZF sets and of setoids in a universe in MLTT. A first requirement on CETCS is of course that we use intuitionistic first order logic instead of the customary classical logic. CETCS has however the property that by adding the law of excluded middle and the axiom of choice (AC), we get a theory equivalent to ETCS. Furthermore the theories of Aczel–Myhill and Martin-Löf are (generalized) predicative, so that general separation and power set principles are not valid. Thus a constructive ETCS cannot be based on the Lawvere-Tierney elementary theory of toposes. In Moerdijk and Palmgren (2000, 2002) a notion of predicative topos was introduced taking the setoids of MLTT with a hierarchy of universes as a standard model. Other variants of predicative toposes have been introduced and studied (van den Berg 2005); see also Maietti (2005) and Awodey and Warren (2005). A drawback of the category of setoids, as opposed categories of sets, is that there is no canonical choice of pullbacks (Sect. 13.6, Hofmann 1994). This makes the formulation of some axioms a bit less concise, but also more general. The idea of avoiding choice in category theory is advocated by Makkai (1996).

We emphasize that ETCS does not deal with the set-class distinction or replacement axioms. ETCS with replacement has however been considered (Osius 1974; McLarty 2004). A constructive treatment of the set-class distinction was given by Joyal and Moerdijk (1995) by the introduction of notion of small map. Predicatively acceptable versions of this were developed in Moerdijk and Palmgren (2002) and van den Berg and Moerdijk (2008).

An outline of the paper is a follows: In Sect. 13.2 a standard first-order logic definition of categories is given. The axioms of ETCS and CETCS are presented in parallel and compared in Sect. 13.3. In Sect. 13.4 some elementary set-theoretic consequence are drawn from CETCS, which indicates its usefulness for Bishop style constructive mathematics. It is shown that CETCS together with the axiom of choice and classical logic gives the original ETCS. The relation of CETCS to standard category theory notions is given in Sect. 13.5, but the full details are presented in Palmgren (2012). The final section contains some reflections on category theory as a constructivist structuralist foundation.

13.2 Relations in Elementary Categories

We shall take care to formulate all the axioms so that they maybe easily cast in many sorted first-order (intuitionistic) logic. Following the notation of Mac Lane (1998), a category $\mathcal{C}$ is specified by an algebraic signature consisting of three collections $\mathcal{C}_0, \mathcal{C}_1, \mathcal{C}_2$ (for objects, mappings (or arrows), composable mappings) and six functions $\mathsf{id} : \mathcal{C}_0 \longrightarrow \mathcal{C}_1$, $\mathsf{dom}, \mathsf{cod} : \mathcal{C}_1 \longrightarrow \mathcal{C}_0$, $\mathsf{comp} : \mathcal{C}_2 \longrightarrow \mathcal{C}_1$, $\mathsf{fst}, \mathsf{snd} : \mathcal{C}_2 \longrightarrow \mathcal{C}_1$. The intention is that dom gives the domain of the mapping while cod gives its codomain. The collection $\mathcal{C}_2$ is supposed to consist of composable mappings

$$\cdot \xrightarrow{f} \cdot \xrightarrow{g} \cdot$$

and fst gives the first of these mappings while snd gives the second mapping. Then comp is the composition operation.

We introduce abbreviations: for mappings f, g, h write $h \equiv g \circ f$ for $(\exists u \in \mathcal{C}_2)[\mathsf{fst}(u) = f \,\&\, \mathsf{snd}(u) = g \,\&\, \mathsf{comp}(u) = h]$, that is, the diagram

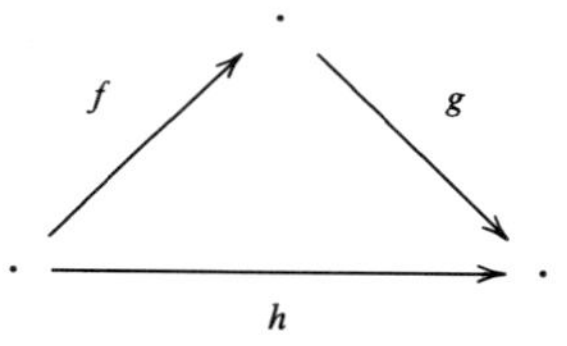

is composable and commutes. Take $f : a \longrightarrow b$ and $a \xrightarrow{f} b$ to be abbreviations for the conjunction $\mathsf{dom}\, f = a \,\&\, \mathsf{cod}\, f = b$. We shall often omit $\circ$ and write $h \equiv gf$ for $h \equiv g \circ f$. Moreover $\equiv$ is often replaced by $=$ when there is no danger of confusion.

The axioms for a category may then easily be expressed as first-order formulas in this signature, for details, see for instance Palmgren (2012).

13.2.1 Subobjects

We may define the notion of an n-ary relation in any category. Recall that a mapping $f : A \longrightarrow B$ is *monic* or *is a mono* if for any mappings $h, k : U \longrightarrow A$ with $fh = fk$ it holds that $h = k$. We write in this case $f : A \rightarrowtail B$. This notion can be generalized to several mappings. A sequence of mappings $r_1 : R \longrightarrow X_1, \ldots, r_n : R \longrightarrow X_n$ are *jointly monic,* if for any $f, g : U \longrightarrow R$

$$r_1 f = r_1 g, \ldots, r_n f = r_n g \Longrightarrow f = g.$$

In this case we write $(r_1, \ldots, r_n) : R \rightarrowtail (X_1, \ldots, X_n)$. We regard this as an *$n$-ary relation between the objects $X_1, \ldots, X_n$*. In particular, a *binary relation between X_1 and X_2* is a pair of mappings $r_1 : R \longrightarrow X_1$ and $r_2 : R \longrightarrow X_2$ which are jointly monic. Another particular case is: if the category has a terminal object $\mathbf{1}$, a 0-ary relation $() : R \rightarrowtail ()$ means that the unique map $R \longrightarrow \mathbf{1}$ is a mono.

Consider a category $\mathcal{C}$ with a terminal object $\mathbf{1}$. An *element* of an object A of $\mathcal{C}$ is a mapping $x : \mathbf{1} \longrightarrow A$. For a monic $m : M \longrightarrow X$ and element x of X write $x \,\epsilon\, m$ if $(\exists a : \mathbf{1} \longrightarrow M) ma = x$. We say that *$x$ is a member of m*. More generally, if $(m_1, \ldots, m_n) : M \rightarrowtail (X_1, \ldots, X_n)$ and $(x_1, \ldots, x_n) : \mathbf{1} \rightarrowtail (X_1, \ldots, X_n)$ we write $(x_1, \ldots, x_n) \,\epsilon\, (m_1, \ldots, m_n)$ if there is $a : \mathbf{1} \longrightarrow M$ so that $m_i a = x_i$ for all $i = 1, \ldots, n$.

To simplify notation we often write $x \in X$ and $(x_1, \ldots, x_n) \in (X_1, \ldots, X_n)$ for $x : \mathbf{1} \longrightarrow X$ and $(x_1, \ldots, x_n) : \mathbf{1} \longrightarrow (X_1, \ldots, X_n)$, respectively. Note the difference between the signs $\in$ (elementhood) and ϵ (membership).

We shall be interested in categories where there is no canonical construction for products, but where it is merely assumed that they exist. Recall that an *n-ary product diagram* in a category is a sequence of mappings $X \xrightarrow[p_i]{} X_i$ $(i = 1, \ldots, n)$ so that for any sequence of mappings $C \xrightarrow[p_i]{} X_i$ $(i = 1, \ldots, n)$ there is a unique $h : C \longrightarrow X$ such that $f_i \equiv hp_i$ for all $i = 1, \ldots, n$. We write

$$h \equiv \langle f_1, \ldots, f_n \rangle_{\bar{p}}$$

when $f_i \equiv hp_i$ for all $i = 1, \ldots, n$, where $\bar{p} = p_1, \ldots, p_n$. It is convenient to drop the subscripts $\bar{p}$ when the product diagrams are obvious from the context.

Proposition 13.1. *Suppose that* $X \xrightarrow[p_i]{} X_i$ $(i = 1, \ldots, n)$ *is a product diagram. If* $(r_1, \ldots, r_n) : R \longrightarrow (X_1, \ldots, X_n)$, $r' : R \longrightarrow X$ *and* $r' \equiv \langle r_1, \ldots, r_n \rangle_{\bar{p}}$, *then* r' *is monic iff* $(r_1, \ldots, r_n)$ *are jointly monic. Moreover, for* $(x_1, \ldots, x_n) \in (X_1, \ldots, X_n)$, $x' \in X$ *with* $x' \equiv \langle x_1, \ldots, x_n \rangle_{\bar{p}}$, *we have*

$$x' \epsilon\, r' \Longleftrightarrow (x_1, \ldots, x_n) \epsilon\, (r_1, \ldots, r_n). \qquad \square$$

A binary relation $f = (\xi, \upsilon) : R \rightarrowtail (X, Y)$ is a *partial function* in case ξ is mono. It is a *total function* in case ξ is iso. A relation

$$f = (\xi_1, \ldots, \xi_n, \upsilon) : R \rightarrowtail (X_1, \ldots, X_n, Y)$$

is a *partial function of n variables* if $(\xi_1, \ldots, \xi_n) : R \rightarrowtail (X_1, \ldots, X_n)$. We write

$$f : (X_1, \ldots, X_n) \rightharpoonup Y.$$

It is *total function of n variables* if $R \xrightarrow{\xi} X_i$ $(i = 1, \ldots, n)$ is a product diagram. We write

$$f : (X_1, \ldots, X_n) \longrightarrow Y.$$

For $x_1 \in X_1, \ldots, x_n \in X_n$ and $y \in Y$ we write

$$f(x_1, \ldots, x_n) \equiv y$$

in case $(x_1, \ldots, x_n, y) \epsilon\, f$.

13.3 Axioms of ETCS and CETCS

Lawvere's theory ETCS (Lawvere 2005) has eight axioms: (L1) finite roots exist, (L2) the exponential of any pair of objects exist, (L3) there is a Dedekind-Peano object, (L4) the terminal object is separating, (L5) axiom of choice, (L6) every object not isomorphic to an initial object contains an element, (L7) Each element of a sum is a member of one of its injections, (L8) there is an object with more than one element.

We present a constructive version of ETCS, called CETCS, and some extensions, by laying down axioms for a category $\mathcal{C}$. (It should be evident that the following axioms may be formulated in first-order logic in a language with $\mathcal{C}_0, \mathcal{C}_1, \mathcal{C}_2$ as sorts and the function symbols $\mathsf{id}, \mathsf{dom}, \mathsf{cod}, \mathsf{comp}, \mathsf{fst}, \mathsf{snd}$ as indicated in Sect. 13.2.)

Lawvere's (L1) says that the category is *bicartesian,* i.e. both cartesian and cocartesian.

Recall that $\mathcal{C}$ is *cartesian* if the conditions (C1)–(C3) are satisfied:

(C1) There is a terminal object $\mathbf{1}$ in $\mathcal{C}$.

(C2) Binary products exist: For any pair of objects A and B there exists an object P and two mappings

$$A \xleftarrow{p} P \xrightarrow{q} B$$

which are such that if $A \xleftarrow{f} X \xrightarrow{g} B$ then there exists a unique $h : X \longrightarrow P$ so that $ph \equiv f$ and $qh \equiv g$.

(C3) Equalizers exist: For any parallel pair of mappings $A \underset{g}{\overset{f}{\rightrightarrows}} B$ there exists a mapping $e : E \longrightarrow A$ so that $fe \equiv ge$ and such that whenever $h : X \longrightarrow A$ satisfies $fh \equiv gh$ then there exists a unique $k : X \longrightarrow E$ with $ek \equiv h$.

A category $\mathcal{C}$ is *cocartesian* if it satisfies (D1)–(D3), which are the categorical duals of (C1)–(C3).

(D1) There is an initial object $\mathbf{0}$ in $\mathcal{C}$.

(D2) Binary sums exist: For any pair of objects A, B there is a diagram

$$A \xrightarrow{i} S \xleftarrow{j} B \tag{13.2}$$

such that if $A \xrightarrow{f} T \xleftarrow{g} B$ then there is a unique $h : S \longrightarrow T$ with $hi \equiv f$ and $hj \equiv g$.

(D3) Coequalizers exist: For any parallel pair of mappings $A \underset{g}{\overset{f}{\rightrightarrows}} B$ there exists a mapping $q : B \longrightarrow Q$ so that $qf \equiv qg$ and such that whenever $h : B \longrightarrow Y$ satisfies $hf \equiv hg$ then there exists a unique $k : Q \longrightarrow E$ with $kq \equiv h$.

The axiom (L2) of ETCS says together with (L1) that the category is cartesian closed. Instead, we take for an axiom the following (Π) which, together with cartesianess and axiom (G) below, states that the category is locally cartesian closed. (This axiom is a theorem of ETCS.)

(Π) Dependent products exist: For any mappings $Y \xrightarrow{g} X \xrightarrow{f} I$ there exists a commutative diagram

$$\begin{array}{ccccc} Y & \xleftarrow{\text{ev}} & P & \xrightarrow{\pi_1} & F \\ & \searrow^{g} & \downarrow{\scriptstyle \pi_2} & & \downarrow{\scriptstyle \varphi} \\ & & X & \xrightarrow[f]{} & I \end{array} \tag{13.3}$$

where the square is a pullback, and which is such that for any element $i \in I$ and any partial function $\psi = (\xi, \upsilon) : R \rightharpoondown (X, Y)$ such that

(a) For all $(x, y) \in (X, Y)$, $(x, y) \epsilon \psi$ implies $gy \equiv x$ and $fx \equiv i$,
(b) If $fx \equiv i$, then there is $y \in Y$ with $(x, y) \epsilon \psi$,

then there is a unique $s \in F$ so that $\varphi s = i$ and for all $(x, y) \in (X, Y)$,

$$(s, x, y) \epsilon \alpha \iff (x, y) \epsilon \psi. \tag{13.4}$$

Here $\alpha = (\pi_1, \pi_2, \text{ev}) : P \rightarrowtail (F, X, Y)$.

A diagram (13.3) satisfying these properties is called a *universal dependent product diagram* or shortly a *universal Π-diagram* for $Y \xrightarrow{g} X \xrightarrow{f} I$.

The third axiom (L3) of ETCS says, in now common terminology, that there exists a *natural numbers object* (NNO). A category $\mathcal{C}$ has an NNO if there is a sequence of mappings (the NNO) $\mathbf{1} \xrightarrow{0} N \xrightarrow{s} N$ so that for any other sequence of mappings $\mathbf{1} \xrightarrow{b} A \xrightarrow{h} A$ there is a unique $f : N \longrightarrow A$ with $f0 \equiv b$ and $fS \equiv hf$.

Axiom (L4) states in modern terminology that $\mathbf{1}$ is a separating object, i.e. as in Proposition 13.2. We consider instead a stronger axiom (G) which is a theorem of ETCS. A mapping $f : A \longrightarrow B$ of $\mathcal{C}$ is *onto* if for any $y \in B$ there exists an $x \in A$ so that $y \equiv fx$. Our axiom is

(G) Any mapping which is both onto and mono, is an isomorphism.

The fifth axiom (L5) of ETCS states the axiom of choice in peculiar way; see Sect. 5.2. A more standard way is to first define an object P of $\mathcal{C}$ to be a *choice object*, if for any onto $f : A \longrightarrow P$ there is a $g : P \longrightarrow A$ with $fg = \mathsf{id}_P$. The *axiom of choice* (AC) says that every object is a choice object. This is a far too strong assumption in a constructive setting. There is a constructively acceptable weakening which accords well with Bishop's distinction of operations and functions, the *presentation axiom* (Aczel 1978):

(PA) For any object A there is an onto mapping $P \twoheadrightarrow A$ where P is a choice object.

Axiom (L6) of ETCS says in contrapositive form: if an object has no elements then it is an initial object. We take instead

(I) The object $\mathbf{0}$ has no elements.

This together with (G) implies (L6).

The Axiom (L7) of ETCS is *each element of a sum is a member of one of its injections.* We adopt this axiom unaltered but call it the *disjunction principle (DP)* as it connects sums to disjunctions:

(DP) In a sum diagram $A \xrightarrow{i} S \xleftarrow{j} B$: for any $z \in S$, $z \epsilon i$ or $z \epsilon j$.

The final axiom (L8) of ETCS states that there exists object with at least two elements. We state this as

(NT, Non-triviality) For any sum diagram $\mathbf{1} \xrightarrow{x} S \xleftarrow{y} \mathbf{1}$ it holds that $x \neq y$.

There are two further axioms that we shall consider, which are in fact theorems of ETCS.

(Fct) Factorization. Any mapping f can be factored as $f \equiv i\,e$ where i is mono and e is onto.

(Eff) All equivalence relations are effective. For each equivalence relation $(r_1, r_2) : R \rightarrowtail (X, X)$ there is some mapping $e : X \twoheadrightarrow E$ so that

$$(x_1, x_2) \,\epsilon\, (r_1, r_2) \iff e x_1 \equiv e x_2$$

for all $(x_1, x_2) \in (X, X)$.

In summary, the theory CETCS consists of the axioms (C1–C3), (D1–D3), (Π), (G), (PA), (I), (DP), (NT), (Fct) and (Eff). We observe that it is a finitely axiomatized theory just as ETCS.

Remark 13.1. Note that it is not assumed that the (co)products or (co)equalizers are given as functions of their data. The axiom (G) is in the terminology of Johnstone (2002) that $\mathbf{1}$ *generates* $\mathcal{C}$. It entails that one can "reason using elements" as the two following results exemplify. This gives a substantial simplification of the internal logic.

Proposition 13.2. *Let $\mathcal{C}$ be a cartesian category which satisfies (G). Then*

(a) For any pair of mappings $f, g : A \longrightarrow B$, $f = g$ whenever $(\forall x \in A)(fx = gx)$.
(b) A mapping $f : A \longrightarrow B$ is monic if and only if $(\forall x, y \in A)(fx = fy \Rightarrow x = y)$.

Proof. (b) follows easily from (a). To prove the non-trivial direction of (a): assume that $(\forall x \in A)(fx = gx)$. Construct an equalizer $E \xrightarrow{e} A \underset{g}{\overset{f}{\rightrightarrows}} B$ of f and g. Then e is monic. By the assumption and the equalizing property it is also easy to see its onto. Hence by (G) e is an isomorphism. Since $fe = ge$ we get $f = g$. □

Define an element-wise inclusion relation for monos $m : M \longrightarrow X$ and $n : N \longrightarrow X$

$$m \dot{\subseteq} n \Longleftrightarrow_{\text{def}} (\forall x \in X)(x \,\epsilon\, m \Rightarrow x \,\epsilon\, n)$$

The standard inclusion relation in a category is given by $m \leq n \Longleftrightarrow_{\text{def}} (\exists f : M \longrightarrow N)(m = nf)$. Compare diagram (13.1). Their correspondence is given by:

Proposition 13.3. *Let $\mathcal{C}$ be a cartesian category which satisfies (G). Then for all monos $m : M \longrightarrow X$ and $n : N \longrightarrow X$,*

$$m \dot{\subseteq} n \Longleftrightarrow m \leq n.$$

Proof. ($\Leftarrow$) This is straightforward in any category with a terminal object. ($\Rightarrow$) Suppose that $m : M \rightarrowtail X$ and $n : N \rightarrowtail X$ satisfies $m \dot{\subseteq} n$. Form a pullback square

$$\begin{array}{ccc} P & \xrightarrow{p} & N \\ {\scriptstyle q}\downarrow & & \downarrow{\scriptstyle n} \\ M & \xrightarrow[m]{} & X \end{array}$$

To prove $m \leq n$ it is evidently enough to show that q is an isomorphism. Now q is the pullback of a mono, so it is a mono as well. By (G) it is sufficient to show that q is onto. Let $y \in M$. Thus $my \,\epsilon\, m$ and by assumption also $my \,\epsilon\, n$. There is thus $t \in N$ with $my = nt$. Hence by the pullback square there is a unique $u \in P$ so that $qu = y$ and $pu = t$. In particular, this shows that q is onto. □

Functions as a graphs and as morphisms can be characterized as follows.

Proposition 13.4. *Let $\mathcal{C}$ be a cartesian category which satisfies (G). Let $r = (r_1, r_2) : R \longrightarrow (X, Y)$ be a relation. Then*

(a) r is a partial function if and only if

$$(\forall x \in X)(\forall y, z \in Y)[(x, y) \,\epsilon\, r \;\&\; (x, z) \,\epsilon\, r \Rightarrow y = z]. \tag{13.5}$$

(b) r is a total function if and only if

$$(\forall x \in X)(\exists! y \in Y)(x, y) \,\epsilon\, r. \tag{13.6}$$

(c) (Unique Choice) *If* $(\forall x \in X)(\exists! y \in Y)(x, y) \,\epsilon\, r$, *then there is* $f : X \longrightarrow Y$ *with*

$$(\forall x \in X)(x, fx) \,\epsilon\, r.$$

Proof. (a): by definition r is a partial function if and only if r_1 is mono. By Proposition 13.2, r is thus a partial function precisely when

$$(\forall s, t \in R)[r_1 s = r_1 t \Rightarrow s = t].$$

This is easily seen to be equivalent to (13.5). (b, $\Rightarrow$): Suppose r is a total function. Then r_1 is iso. For $x \in X$, we have $(x, y) \,\epsilon\, r$ with $y = r_2 r_1^{-1} x$. By (a) it follows that y is unique.

(b, $\Leftarrow$): Suppose (13.6) holds. By (a) r_1 is mono. For each $x \in X$ there is some $t \in R$ and $y = r_2 t$ so that $(x, y) \,\epsilon\, r$. Thus r_1 is onto, and by (G) r_1 is iso.

(c): This is clear from (b, $\Leftarrow$) since then r_1 is invertible, and we may take $f = r_2 r_1^{-1}$: for $x \in X$, $x = r_1 r_1^{-1} x$ and $fx = r_2 r_1^{-1} x$ so $(x, fx) \,\epsilon\, r$. □

13.4 Basic Set-theoretic Consequences

We mention some easy consequences of the axioms.

Proposition 13.5 (Quotient sets). *Suppose that the bicartesian category $\mathcal{C}$ satisfies (G). For any equivalence relation* $r =_{\text{def}} (r_1, r_2) : R \rightarrowtail (X, X)$ *there exists a mapping* $q : X \longrightarrow Q$ *so that for all* $(x_1, x_2) \in (X, X)$

$$(x_1, x_2) \,\epsilon\, r \Longrightarrow qx_1 = qx_2 \tag{13.7}$$

and if $f : X \longrightarrow Y$ *is any mapping with*

$$(x_1, x_2) \,\epsilon\, r \Longrightarrow fx_1 = fx_2. \tag{13.8}$$

then there exists a unique $h : Q \longrightarrow Y$ *with* $hq = f$.

In case the category also satisfies (Eff) it follows that (13.7) is an equivalence.

Proof. Construct a coequalizer diagram

$$R \underset{r_2}{\overset{r_1}{\rightrightarrows}} X \overset{q}{\longrightarrow} Q.$$

Since the diagram commutes, the implication (13.7) holds. Let $f : X \longrightarrow Y$ be any mapping satisfying the implication (13.8). Thus for any $t \in R$, $fr_1t = fr_2t$. Thus by Proposition 13.2 (a) we have $fr_1 = fr_2$ and since q is a coequalizer, there is a unique $h : Q \longrightarrow Y$ with $hq = f$.

From Axiom (Eff) that there is some $e : X \longrightarrow E$ such that

$$(x_1, x_2) \mathrel{\epsilon} r \iff ex_1 = ex_2 \tag{13.9}$$

for all $(x_1, x_2) \in (X, X)$. Thus $er_1 = er_2$. Let $e' : Q \longrightarrow E$ be the unique mapping so that $e'q = e$. Thus if $qx_1 = qx_2$, it follows that $ex_1 = ex_2$ and hence $(x_1, x_2) \mathrel{\epsilon} r$ by (13.9). □

Proposition 13.6 (Induction). *Assume that C is a cartesian category which satisfies (G) and (NNO). Let $r : R \rightarrowtail N$. Suppose that $0 \mathrel{\epsilon} r$ and that for each $n \in N$, $n \mathrel{\epsilon} r$ implies $Sn \mathrel{\epsilon} r$. Then for all $n \in N$, $n \mathrel{\epsilon} r$.*

Proposition 13.7 (Exponential objects). *Assume that C is a cartesian category that satisfies (G) and (Π). Then for any objects X and Y there is an object E and a total function $e : (E, X) \longrightarrow Y$ such that for every morphism $f : X \longrightarrow Y$ there is a unique $s \in E$ such that for $x \in X$ and $y \in Y$:*

$$e(s, x) \equiv y \iff f \circ x \equiv y$$

Theorem 13.1 (Dependent choices). *Assume that C is a cartesian category that satisfies CETCS. Then for any object X, any total relation $r = (r_1, r_2) : R \rightarrowtail (X, X)$ and any $x \in X$ there is a morphism $f : N \longrightarrow X$ with $f0 = x$ and for all $n \in N$*

$$(fn, f \circ Sn) \mathrel{\epsilon} r. \tag{13.10}$$

Proof (Sketch). Take a projective cover $p : P \longrightarrow X$ of X. Since r is total, we have thus for each $u \in P$ some $v \in P$ with $(pu, pv) \mathrel{\epsilon} r$. As P is a choice object, there is a morphism $g : P \longrightarrow P$ with $(pu, pgu) \mathrel{\epsilon} r$ for all $u \in P$. Let $x \in P$. Then $p \circ w \equiv x$ for some $w \in P$. Now $\mathbf{1} \overset{0}{\longrightarrow} N \overset{S}{\longrightarrow} N$ is a natural numbers object, so there is $h : N \longrightarrow P$ with $h0 = w$ and $hS = gh$. Now it is easy to check by induction that $f =_{\text{def}} ph$ satisfies (13.10). □

13.4.1 Constructing New Relations

We review some of the possibilities to construct relations in a category satisfying CETCS. For proofs and sharper results see Palmgren (2012).

On any object X the identity mapping gives a universally true relation $\mathsf{t}_X = \mathsf{id}_X : X \twoheadrightarrow X$, i.e. for all $x \in X$

$$x \,\epsilon\, \mathsf{t}_X .$$

The unique mapping from the initial object $\mathsf{f}_X : \mathbf{0} \longrightarrow X$ gives an universally false relation, i.e. for all $x \in X$,

$$\neg(x \,\epsilon\, \mathsf{f}_X).$$

If $E \xrightarrow{e} X \overset{g}{\underset{h}{\rightrightarrows}} Y$ is an equalizer diagram, then for $x \in X$

$$x \,\epsilon\, e \Longleftrightarrow gx = hx.$$

Relations can be combined using the logical operations ($\wedge, \vee, \Rightarrow$) and quantifiers ($\forall$,$\exists$) in the following way:

Let $r = (r_1, \ldots, r_n) : R \rightarrowtail (X_1, \ldots, X_n)$ and $s = (s_1, \ldots, s_n) : S \rightarrowtail (X_1, \ldots, X_n)$. Then exists $(r \wedge s), (r \vee s), (r \Rightarrow s) : R \rightarrowtail (X_1, \ldots, X_n)$ so that for all $x = (x_1, \ldots, x_n) \in (X_1, \ldots, X_n)$

(a) $x \,\epsilon\, (r \wedge s)$ if and only if $x \,\epsilon\, r$ and $x \,\epsilon\, s$,
(b) $x \,\epsilon\, (r \vee s)$ if and only if $x \,\epsilon\, r$ or $x \,\epsilon\, s$,
(c) $x \,\epsilon\, (r \Rightarrow s)$ if and only if $x \,\epsilon\, r$ implies $x \,\epsilon\, s$,

Moreover, if $m : M \rightarrowtail (X_1, \ldots, X_n, Y)$ then there is $\forall(m) : A \rightarrowtail (X_1, \ldots, X_n)$ and $\exists(m) : E \rightarrowtail (X_1, \ldots, X_n)$ so that for all $x = (x_1, \ldots, x_n) \in (X_1, \ldots, X_n)$

(d) $x \,\epsilon\, \forall(m)$ if and only if for all $y \in Y$, $(x_1, \ldots, x_n, y) \,\epsilon\, m$,
(e) $x \,\epsilon\, \exists(m)$ if and only if for some $y \in Y$, $(x_1, \ldots, x_n, y) \,\epsilon\, m$.

13.4.2 Decidable Relations and Classical Logic

Let $\mathcal{C}$ be a CETCS category. Construct a two element set using the sum axiom $\mathbf{1} \xrightarrow{f} \mathbf{2} \xleftarrow{t} \mathbf{1}$. If $r : P \rightarrowtail X$ is decidable, i.e. for all $x \in X$,

$$x \,\epsilon\, r \text{ or } \neg x \,\epsilon\, r,$$

then we can construct $\chi_r : X \longrightarrow \mathbf{2}$ so that for all $x \in X$

$$x \,\epsilon\, r \wedge \chi_r(x) = t \text{ or } (\neg x \,\epsilon\, r) \wedge \chi_r(x) = f,$$

It follows that χ_r is the unique map $X \longrightarrow \mathbf{2}$ such that $x \in r$ iff $\chi_r(x) = t$. Thus $\mathbf{1} \xrightarrow{t} \mathbf{2}$ classifies decidable relations. In case we take the axioms of CETCS with classical logic every relation is decidable, and hence $\mathbf{1} \xrightarrow{t} \mathbf{2}$ is a full subobject classifier for the category. In this case $\mathcal{C}$ is a topos.

The Lawvere's choice axiom (L5) states: If $f : A \longrightarrow B$ is mapping and A contains at least one element, then there is a mapping $g : B \longrightarrow A$ so that $fgf = f$.

Theorem 13.2. *In CETCS with classical logic (AC) and (L5) are equivalent.*

Corollary 13.1. *ETCS and CETCS +PEM + AC have the same theorems.*

13.5 Relation to Standard Categorical Formulations

In standard category-theoretic terms (Johnstone, 2002) various combinations of the CETCS axioms can be characterized by the following theorems. Proofs appear in Palmgren (2012).

Theorem 13.3. *Let $\mathcal{C}$ be a cartesian category satisfying (G). Then:*

(i) $\mathcal{C}$ satisfies (Fct) if, and only if, $\mathcal{C}$ is a regular category where the terminal object is projective.
(ii) $\mathcal{C}$ is locally cartesian closed if and only if $\mathcal{C}$ satisfies the axiom (Π).

Theorem 13.4. *Let $\mathcal{C}$ be a category. Then $\mathcal{C}$ satisfies CETCS if, and only if, $\mathcal{C}$ is a locally cartesian closed pretopos such that*

(i) it has NNO,
(ii) its terminal object is projective and generates $\mathcal{C}$,
(iii) $\mathbf{0} \not\cong \mathbf{1}$,
(iv) it satisfies the disjunction property,
(v) it has enough projectives.

13.6 Reflections on Constructivist Structuralist Foundations

There is a debate whether category theory is an adequate or natural foundation for mathematics, with opponents like Feferman and Hellman; see McLarty (2004, 2005). An elementary textbook with a category theory as foundation is available (Lawvere and Rosebrugh 2003). Does it (secretly) require classical ZF set theory as a foundation or motivation? Studies like Joyal and Moerdijk (1995), Moerdijk and Palmgren (2002) and van den Berg and Moerdijk (2009) shows that it is not tied to classical foundations. The presented theory CETCS is a small and easy example of this. The question whether category theory is a natural foundation for mathematics can be answered as follows:

- It describes well the practice of certain disciplines of mathematics.
- It is less clear that it explains the foundations as simple as possible.

It is seems to be a general experience among practioneers that a category-theoretic framework tends to favour constructive modes of reasoning. Why is this so?

- Many categorical constructions can be expressed in fragments of logic (regular logic or geometric logic). For these fragments uses of the principle of excluded middle can always be eliminated (Barr's Theorem).
- Fundamental constructions are defined by universal properties, which tends to give unique and canonical constructions.
- Fundamental constructions are made to work in sheaves over a topological space, which is akin to Kripke models, and thus forces use of intutionistic logic.

Acknowledgements The main results of this article were obtained while the author was a fellow of the Swedish Collegium for Advanced Study, January–June 2009. Many thanks go to the Collegium and its principal Professor Björn Wittrock for the opportunity to work in this most stimulating research environment, and for the challenging task to give a talk on philosophy of mathematics to researchers in sociology and history.

References

Aczel, P. 1978. The type theoretic interpretation of constructive set theory. *Logic Colloquium '77 (Proc. Conf., Wroclaw, 1977)*. Studies in logic and the foundations of mathematics, vol. 96, 55–66. Amsterdam/New York: North-Holland.

Awodey, S., and M.A. Warren. 2005/2006. Predicative algebraic set theory. *Theory and Applications of Categories* 15(1): 1–39.

Bishop, E. 1967. *Foundations of constructive analysis*. New York: McGraw-Hill.

Bishop, E. 1970a. Mathematics as a numerical language. In *Intuitionism and proof theory*, 53–71. Amsterdam: North-Holland.

Hofmann, M. 1994. On the interpretation of type theory in locally cartesian closed categories. In *Proceedings of the Computer Science Logic '94, Kazimierz, Poland*, Lecture notes in computer science, vol. 933, ed. J. Tiuryn and L Pacholski. New York: Springer.

Johnstone, P.T. 2002. *Sketches of an elephant: a Topos theory compendium*, vols. 1, 2. New York: Oxford University Press.

Joyal, A., and I. Moerdijk. 1995. *Algebraic set theory*. London: Cambridge University Press.

Lawvere, F.W. 2005. An elementary theory of the category of sets (long version). *Theory and Applications of Categories* 11: 7–35.

Lawvere, F.W., and R. Rosebrugh. 2003. *Sets for mathematics*. New York: Cambridge University Press.

Mac Lane, S. 1998. *Categories for the working mathematician*, 2nd ed. New York: Springer.

Maietti, M.E. 2005. Modular correspondence between dependent type theories and categories including pretopoi and topoi. *Mathematical Structures for Computer Science* 15(6): 1089–1149.

Makkai, M. 1996. Avoiding the axiom of choice in general category theory. *Journal of Pure and Applied Algebra* 108: 109–173.

Martin-Löf, P. 1975. An intuitionistic theory of types: predicative part. In *Logic Colloquium '73*, ed. H.E. Rose and J. Shepherdson. Amsterdam: North-Holland.

Martin-Löf, P. 1984. Intuitionistic type theory. *Notes by Giovanni Sambin*. Naples: Bibliopolis.
McLarty, C. 2004. Exploring categorical structuralism. *Philosophy of Mathematics* 12: 37–53.
McLarty, C. 2005. ETCS and philosophy of mathematics. *Theory and Applications of Categories* 11: 2–4.
Moerdijk, I., and E. Palmgren. 2000. Well-founded trees in categories. *Annals of Pure and Applied Logic* 104: 189–218.
Moerdijk, I., and E. Palmgren. 2002. Type theories, toposes and constructive set theory: predicative aspects of AST. *Annals of Pure and Applied Logic* 114: 155–201.
Myhill, J. 1975. Constructive set theory. *Journal of Symbolic Logic* 40(3): 347–382.
Osius, G. 1974. Categorical set theory: a characterization of the category of sets. *Journal of Pure and Applied Algebra* 4: 79–119.
Palmgren, E. 2012. Constructivist and structuralist foundations: Bishop's and Lawvere's theories of sets. *Annals of Pure and Applied Logic*, In press.
Tait, W.W. 2000. Cantor's Grundlagen and the paradoxes of set theory. In *Between logic and intuition: essays in honor of Charles Parsons*, ed. G. Sher and R. Tieszen, 269–290. Cambridge: Cambridge University Press.
van den Berg, B. 2005. Inductive types and exact completions. *Annals of Pure and Applied Logic* 134: 95–121.
van den Berg, B., and I. Moerdijk. 2008. Aspects of predicative algebraic set theory I: exact completion. *Annals of Pure and Applied Logic* 156: 123–159.
van den Berg, B., and I. Moerdijk. 2009. A unified approach to algebraic set theory. Logic Colloquium 2006, 18–37, Lecture Notes in Logic., 32, Assoc. Symbol. Logic, Chicago, IL.

Chapter 14
Machine Translation and Type Theory

Aarne Ranta

To Per Martin-Löf.

14.1 Introduction

Machine translation was one of the first applications envisaged for digital computers. The need came from the U.S. military: computers would help intelligence by automatically translating Russian documents to English. This enterprise was encouraged by the success of cryptography during the Second World War. Russian was seen as an encrypted form of English, and translation was a matter of cracking the code.

The main ideas were summarized by Weaver in 1947 (Hutchins 2000). He proposed several approaches, but the most influential one was the use of the noisy channel model developed by Shannon (1948). In practice, most of the early work had to do with word-to-word translation—how to store large dictionaries and perform efficient lookup with the machines of the time. But Weaver and Shannon also envisaged the generalization of this model to *n-grams* of words. The rationale was that single words are often ambiguous and have many alternative translations, but they can be disambiguated in a context.

For example, *even* in English has French translations such as *même* (adverb), *égal, plat, pair* (adjectives).; In the 2-gram *even number*, *pair* is the likely translation, whereas *not even* might become *même pas*, as in *he does not even smile*. However, in the sentence *7 is not even*, the adjective *pair* is again the right choice, which might be detected if 4-grams are considered, and so on.

A. Ranta (✉)
Department of Computer Science and Engineering, Chalmers University of Technology and University of Gothenburg, Gothenburg, Sweden
e-mail: aarne@chalmers.se

P. Dybjer et al. (eds.), *Epistemology versus Ontology*, Logic, Epistemology, and the Unity of Science 27, DOI 10.1007/978-94-007-4435-6_14,

It is easy to find counterexamples to n-gram based translation for any fixed n. And even for relatively small values of n, the method has scalability problems. A natural language may have millions of words, if all word forms are counted separately, as customary in approaches that don't use grammars. The number of n-grams is exponential in n, and *sparse data* becomes a problem. If the n-gram model is based on a corpus of text, there is no hope to find all relevant n-grams. Thus the actual translation usually has to recur to smoothing with smaller n-grams, which results in lower quality.

The problems with statistical n-gram based translation were of course known to their developers. They did not expect the translation to give more than approximations, or *raw translations*, which were to be improved by human post-processing. Alternative methods were studied in parallel. Thus Bar-Hillel soon published the idea to use *categorial grammars* for translation (Bar-Hillel 1953). This was an elaboration of the idea of Ajdukiewicz (1935) to use simple type theory to formalize the rules of natural language. The advantage of grammar rules compared to n-grams is that they can easily cope with arbitrarily long sequences. Thus for instance, in *this number was always believed but never proven to be even*, there are eight words between *number* and *even*, but these words are related by grammar, which is easy to describe by Bar-Hillel's model. More generally, since a grammar can cover an unlimited number of sentences, it can overcome the problem of sparse data.

Bar-Hillel devised an algorithm for what was later to be known as context-free parsing. Much of his work was not possible to implement on the computers of the time and remained theoretical, although seminal. But his most famous contribution to the field was his eventual rejection of the whole enterprise of machine translation (Bar-Hillel 1964). He showed with some simple examples that the translation problem may require unlimited intelligence and universal knowledge. His examples used the word *pen*, which can mean either a writing utensil or a play area for children. In most languages, these two senses of *pen* have two different words. Now, the two sentences *the pen is in the box* and *the box is in the pen* probably use *pen* in these two different senses. But how do we know? The knowledge does not come from grammar, but from our familiarity with the sizes of objects in the world. Bar-Hillel concluded that fully automatic high-quality translation is impossible, not only in foreseeable future but in principle, because there is no hope to formalize all this knowledge.

After Bar-Hillel's paper, the ALPAC report (Automatic Language Processing Advisory Committee) was published in 1966 (Pierce et al. 1966). It presented an evaluation of the results obtained with all the massive funding given to machine translation in the post-war era. Its drastic conclusion was that the investment was wasted—that machine translation had not delivered anything useful. The consequence of the ALPAC report was that machine translation funding in the U.S. was withdrawn and the projects laid down.

Interestingly, the main argument in the ALPAC report was not that machine translation was bound to be unreliable, but that it was too expensive. It was not denied that machine translation *can* give good results, but the problem was that, to achieve these results, so much work was needed that manual translation was cheaper.

The explanation pointed out was that machine translation had been pursued as pure engineering task in an *ad hoc* manner—that one should first have done scientific groundwork and described the languages before attacking the complex engineering problem of machine translation. And indeed, what replaced machine translation as the occupation of many groups and individuals was the field of *computational linguistics*, which now emerged with more modest and realistic goals than machine translation.

In the 1970s and 1980s, most of computational linguistics focused on creating mathematical language models and collecting data, rather than building ambitious machine translation systems. Many of the exceptions had to do with more modest goals than full-scale machine translation: thus *domain-specific translation systems* were able to build accurate models for limited areas such as weather reports (Chandioux 1976). On such limited areas, one can often avoid the problem with word-sense ambiguity; for instance, if the area is elementary arithmetic, then *even* can be assumed to mean divisibility by 2. More generally, the semantics of a limited domain can be formalized in such a way that meaning preservation becomes a rigorously defined concept.

Another idea that emerged after the ALPAC report was to make translation *interactive*. In 1979, Kay wrote an influential paper (eventually published as Kay 1997) that showed with more examples in the spirit of Bar-Hillel that proper translation needs human judgement. Many of Kay's examples had to do with the interpretation of pronouns. For instance, *il* in French can be translated by both *he* and *it* in English. Thus *il est possible que...* becomes *it is possible that...*, whereas *il est convaincu que...* is more likely to become *he is convinced that....* Translating the pronoun is, just like lexical disambiguation in Bar-Hillel's case, the matter of an unpredictable amount of information about the context of use and the world.

We will return to the problem of translating pronouns in Sect. 14.4. Kay's main suggestion was that, since there is no automatic solution, translation tools should make human intervention seamless—must more seamless than post-processing. This idea indeed led to practicals tool widely used by professional translators—basically text editors in which translation is performed manually. Such tools are equipped with a *translation memory*—a repository of sentences and phrases translated earlier, reducing the need of translating the same phrase more than once.

The post-ALPAC pessimism around machine translation was finally removed at the end of the 1980s, in the project funded by IBM and led by Jelinek (Brown et al. 1990). The "IBM approach" was another try with Shannon's noisy channel model and n-grams, but now with the benefit of more powerful computers, adequate bilingual training data (from the Canadian parliament), and experience from statistical speech recognition likewise conducted by Jelinek a decade earlier (Jelinek 2009). A mathematician and engineer himself, Jelinek was disappointed by the over-sophisticated and complex models of linguists, which required so much human effort and gave so little concrete results. The IBM approach created with relatively little effort a system that performed fully automatic translation of unlimited text.

The quality reached by the IBM approach was not perfect, but it was continuously improved, and methods were developed for building translation systems with minimal human intervention. The most prominent achievement of this approach is Google's on-line translation system (`translate.google.com`), which at the time of writing covers 57 languages. Google translate is *robust* in the sense that it produces a result for any input. Moreover, it has been created with a minimal human effort, in particular, with no or little linguistic knowledge about the languages involved. Instead of linguistics, it uses techniques such as *phrase alignment* (Och and Ney 2004), which extract combinations of words ("phrases") and their translations from parallel texts. Actually, it has become a rule of the game that no linguistic, language-specific knowledge may be used, but translation systems must be built by language-neutral algorithms. This can require a lot of data, because alignment can be a many-to-many relation and work on a long distance. The following picture shows a set of alignments between English, Finnish, French, and German.

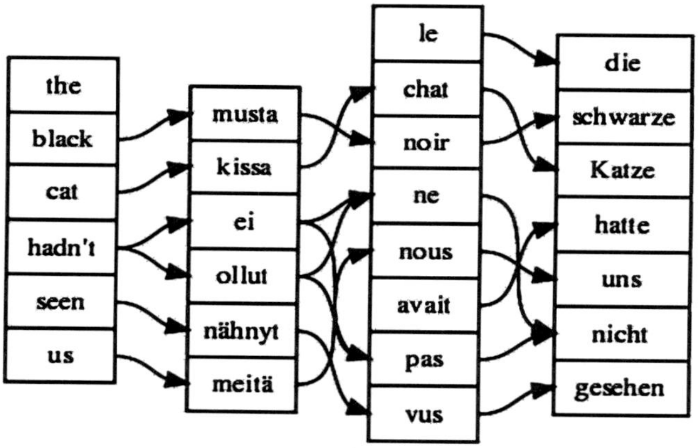

One of Google's obvious strengths in machine translation is that the company has access to so much text, perhaps more than any other organization in the world. In this field, "there is no data like more data". Hence, if anyone is able to solve problems by introducing more data, then Google should be. Nevertheless, the results are not satisfactory in all respects. Google translate is useful in giving quick rough translations, and impressive in its speed of increasing the number of languages. But its quality cannot be trusted, and it is doubtful whether it can ever become fully reliable. This is due to the general reasons given by Bar-Hillel and Kay, but also to the specific limitations of purely statistical language models. To remove the latter kind of problems, much of the current research in machine translation targets *hybrid models*, which combine statistical processing with linguistic information (Lopez 2008 gives an excellent survey of such methods).

14.2 Models of Translation

The history of machine translation shows a back and forth movement between statistical and ruled-based methods. In its purest form, statistical translation is Shannon's noisy channel decoding based on *n*-gram probabilities. Rule-based translation, on the other hand, applies manually written translation functions and performs disambiguation by deep semantic analysis. This model is actually the same as is used in *compilers* for programming languages.

A compiler is a translator from a source language, such as C or Java, to a target language, such as the Intel x86 machine language or JVM (Java Virtual Machine). In early times, compilers were implemented as *transducers* directly converting source code to target code. This was accomplished by means of *semantic actions* attached to the grammar rules of the source language; the mathematical model of this was given in the *attribute grammars* of Knuth (1968). For example, the rule for addition expressions could look as follows:

```
Exp ::= Exp "+" Exp
  { compile $1 ; compile $3 ; emit (ADD (type $1 $3)) }
```

This rule implements infix expressions with the operator `+`. It compiles the first operand `$1` (emitting whatever code belongs to it), recognizes the + sign (which would be denoted by `$2`), compiles the second operand `$3`, and finally emits the `ADD` instruction computed from the types of the operands by the attribute `type`. This attribute makes it possible to use an *overloaded* addition operator in the source language, at the same time as the machine language typically has separate addition instructions for integers, floating point numbers, etc. Notice that the resolution of overloading is similar to word sense disambiguation in natural language translators: the source language can have words such as *even* in English, which require an inspection of the operands (e.g. the nound modified) to decide about the correct translation in a target language like French.

Semantic actions in transducing compilers can be very complex, since they simultaneously define several different operations, such as type checking and code generation in the example above. Furthermore, one-pass transduction poses serious constraints on the source language, which has to be closer to machine languages and thus less natural for humans than modern high-level programming languages are. For instance, mutual recursion is difficult to deal with in one-pass compilers.

Modern compilers thus favour several passes, most of which operate on an *abstract syntax*, which is an intermediate representation between the source and the target language (Appel 1998). The abstract syntax can be formalized as a system of datatypes, where the data objects are abstract syntax trees. The first phase of the compiler is *parsing*, which converts the source code string into an abstract syntax tree. The last phase is *linearization*, which converts abstract syntax into target code. Between these phases, several operations of code analysis and optimization can be performed to manipulate the abstract syntax tree. For instance, GCC (the GNU Compiler Collection, Stallman 2001) can make dozens of passes before emitting the target code.

In addition to modularizing the compiler, the use of an abstract syntax makes it language-neutral: it can be applied to new source and target languages by just changing the parsing and linearization components. The hard work (semantic analysis and optimizations) is performed on the abstract syntax level. Thus GCC, which was originally created for translating C into Motorola 68020, currently supports several source and target languages.

The two compilation methods discussed above have counterparts in the translation of natural language. The transduction model corresponds to *transfer*, i.e. translation functions defined separately for each pair of languages. The abstract syntax model corresponds to *interlingua*. Just like in compilers, the interlingua is an abstract representation of meaning, and translation is performed by meaning-preserving mappings between the interlingua and the languages involved. Thus the translation from English to French is the composition of first translating English to the interlingua and then the interlingua to French.

The advantages of the interlingua approach are the same in machine translation as in compilers. A well-designed, semantically grounded interlingua is an excellent platform for the analysis of the source language, and tasks such as word sense disambiguation and anaphora resolution. It is also useful when selecting the most natural expressions in the target language—an operation similar to *optimizations* in the case of compilers. Another advantage is similar to the multi-source multi-target compilers: work is saved, both in the semantic operations (which are language-independent) and in the number of translation functions. An interlingual system involving n languages needs just $2n$ functions: from each language to the interlingua and back. If separate transfer functions were used for each pair of languages, $n(n-1)$ functions would be needed for n languages.

The transfer/interlingua distinction is orthogonal to the statistical/rule-based distinction. In both types of translation, it is the interlingua approach that scales up into highly multilingual systems. Thus Google translate uses an interlingua for most of its 57×56 language pairs. This interlingua is English (as confirmed by Franz Och in personal communication).

From the semantic point of view, English (or any natural language), might sound like a strange interlingua, because it is ambiguous and destroys distinctions found in other languages. To take a typical example, the distinction between singular and plural *you* disappears in English. Consequently, the translation between, for instance, Swedish and French is not guaranteed to preserve this distinction. Swedish *jag älskar dig* ("I love you" (singular/familiar)) and *jag älskar er* ("I love you" (plural/polite)) are currently both translated as *je t'aime* ("I love you" (singular/familiar)), although the plural/polite form should be translated *je vous aime*.

Nevertheless, English is probably the best choice for training statistical translation models, because there is much more data available for Swedish and English in parallel and for French and English in parallel than for Swedish and French in parallel. There is, moreover, a compelling reason for using a natural language as an interlingua: there simply is no formal language capable of expressing everything that can be expressed in natural languages. This requirement for an interlingua was formulated already by Descartes in 1629, when he proposed a universal language that would support translation:

> [the universal language must] establish an order among all thoughts that can enter in the human spirit, in the same way as there is a natural order among numbers, and as one can learn in one day the names of all numbers up to infinity and write them in an unknown language, even though they are an infinity of different words...
>
> The invention of this language depends on the true philosophy; for it is impossible otherwise to denumerate all thoughts of men and order them, or even distinguish them into clear and simple ones...
>
> (Descartes, letter to Mersenne 1629)

The need of such precision, of a "true philosophy", is demonstrated by the examples of Bar-Hillel and Kay: if the interlingua were to determine how to express the meaning of the source in the target language, it has to be unambiguous and make all the required distinctions. Now, as centuries of philosophers and logicians have in vain been looking for such a formalism, shouldn't we admit that it is just an unrealistic dream?

A natural idea is to use logic and type theory when building an interlingua. An early proposal to this effect was made by Curry (1961). It was applied at a larger scale in the Rosetta system (Rosetta 1994) at Philips. Rosetta was based on the grammar and logical semantics of Montague (1974). They were generalized from English to a multilingual grammar in a way that contained many ingredients of the method discussed in Sect. 14.6.

14.3 A Framework for Translation

The previous discussion has identified two distinctions within machine translation:

- Statistical vs. rule-based
- Transfer vs. interlingua

We will now propose a framework for translation, which is rule-based and uses an interlingua. But we will later show how this model can be extended with statistical components and transfer. We will also meet the main challenge of the ALPAC report and show that the framework is economically viable and useful.

The framework has the same structure as multi-source multi-target compilers: a translator consists of an abstract syntax together with mappings to and from concrete languages. The concrete languages can be varied *ad libitum*; the technique should apply to all natural languages. But it can also deal with formal languages, in tasks such as translating between predicate logic and English.

While having the same structure as GCC, a translation framework must be more general, so that it can deal with different subject matters and not only with computer programs. Thus it must have a more expressive abstract syntax than GCC. It might seem that we would need the power of a universal interlingua—but fortunately we don't. Instead, we apply the idea of a *logical framework* (LF, Harper et al. 1993), originally designed to be a *framework for defining logics*, as a *framework for defining interlinguas*. Then we can define *domain-specific* interlinguas, corresponding to semantically coherent and formalizable domains. To coin a slogan, *the Rosetta stone is not a monolith but a boulder field.*

Logical frameworks were born in the constructivist tradition, which abandoned the idea of a monolithic foundation of mathematics. Instead of reducing all mathematics into one formal theory, such as axiomatic set theory, a logical framework makes it possible to define separate theories for different parts of mathematics. With the expressive power of *dependent types*, this extends to the possibility to define new systems of inference rules, that is, new logics. The framework itself doesn't determine a logic, but provides an infrastructure with a common notation, algorithms for proof checking and proof search, and a generic user interface. Ever since the early times of LEGO (Luo and Pollack 1992), Coq (Dowek et al. 1993), and ALF (Magnusson 1994), logical frameworks have provided an economical way to implement logics and experiment with them. Due to the infrastructure provided by the framework, the implementation of a new logic boils down to writing a set of declarative definitions.

The logical frameworks LEGO and ALF were based on the constructive type theory of Martin-Löf (Martin-Löf 1984; Nordström et al. 1990). Constructive type theory has also proven usable for meaning representation in natural languages (Ranta 1994). The type checking and proof search machinery provided by a logical framework gives tools for the kind of semantic analysis needed in machine translation. What is missing are the parsing and linearization functions for the natural languages themselves. To fulfill this need, the *Grammatical Framework*, GF (Ranta 2004, 2011), was developed. GF is an extension of a logical framework with a component called *concrete syntax*.

If LF is a framework for defining logics, GF is a framework for defining *multilingual grammars*. A multilingual grammar is a pair

$$< \mathcal{A}, \{\mathcal{C}_1, \ldots, \mathcal{C}_n\} >$$

where $\mathcal{A}$ is an abstract syntax (a logic in the sense of LF) and $\mathcal{C}_1, \ldots, \mathcal{C}_n$ are concrete syntaxes. A concrete syntax is a mapping between the abstract syntax trees of $\mathcal{A}$ and the strings in some string language, such as English, French, Java, or JVM.

As a first example of multilingual grammars in GF, consider the translation of addition expressions. The abstract syntax defines a *function* (`fun`), and each concrete syntax defines a *linearization* (`lin`). The following grammar covers Java, JVM, English, and French.

```
fun EPlus : Exp -> Exp -> Exp
lin EPlus x y = x ++ "+" ++ y
lin EPlus x y = x ++ y ++ "iadd"
lin EPlus x y = "the sum of" ++ x ++ "and" ++ y
lin EPlus x y = "la somme de" ++ x ++ "et de" ++ y
```

The `lin` rules of GF are *reversible mappings*: they can be used both for the linearization of trees into strings and for the parsing of strings into trees. How to do linearization is obvious: just think of the `lin` rules as clauses in the definition of a recursive function, where the variables `x` and `y` stand for the linearizations of the arguments. The parsing direction is more tricky and can be stated as a search problem. A general solution was found by Ljunglöf (2004), who moreover showed that the parsing complexity in GF is polynomial.

The above example is a valid GF grammar, but it is oversimplified in many ways. First we might notice that the abstract syntax `fun` rule doesn't indicate the type of the expression (integer, float, etc). The JVM rule is, in an arbitrary way, directed to integer addition (`iadd`) only. But this problem can be solved by making `Exp` into a *dependent type*, which takes the object language type (type `Typ` in this grammar) as its argument. Then we can write

```
fun EPlus : (t : Typ) -> Exp t -> Exp t -> Exp t
```

to force the operands and the value to be of the same type, and

```
lin EPlus t x y = x ++ y ++ add t
```

to select the proper JVM instruction `add t` as a function of the type `t`. (Precisely how the `add` function is defined in GF is omitted here.) In the other three languages, the type argument is *suppressed*. For instance,

```
lin EPlus _ x y = x ++ "+" ++ y
```

We use the wildcard _ for arguments that are suppressed, that is, don't appear on the right of the equality sign. Now assume the following rules for numeric literals and program variables, with linearizations in Java:

```
fun EInt : Int -> Exp TInt
lin EInt i = i
fun EVar : (t : Typ) -> Var t -> Exp t
lin EVar _ v = v
```

In JVM, the type of the variable has to be known by the instruction that loads the values of variables from memory:

```
lin EVar t v = load t v
```

As Java suppresses the type arguments of `EPlus` and `EVar`, the expression `2 + x` is initially parsed by introducing *metavariables*:

```
EPlus ?1 (EInt 2) (EVar ?2 x)
```

If `x` is an integer variable, well-known algorithms for type checking and constraint solving, similar to those used in ALF (Magnusson 1994), now manage to instantiate the metavariables:

```
EPlus TInt (EInt 2) (EVar TInt x)
```

From this syntax tree, we can generate the JVM code

```
iconst_2
iload_0
iadd
```

which uses the `i` (integer) variants of the `load` and `add` instructions. (It moreover maps the variable `x` to the memory address 0, but we omit the details about how this is done.)

14.4 Types and Disambiguation

Although the translation of `2 + x` to JVM is elementary, it illustrates some fundamental aspects of machine translation:

- Word order can vary from one language to another (here: infix in Java, postfix in JVM)
- One language may suppress information that another language needs (here: the type of the addition operator)
- Suppressed information can be restored by semantic analysis (here: type checking and metavariable solving)

A natural language example with the same features is *anaphora resolution*, that is, the interpretation of pronouns. Consider the following examples (from Hutchins and Somers 1992):

the monkey ate the banana because it was hungry
the monkey ate the banana because it was ripe
the monkey ate the banana because it was tea-time

The focal point is the pronoun *it*. The proper translation into German is different in each of the three sentences. In the first one, *it* refers to the monkey (*der Affe*), and becomes the masculine *er*. In the second one, *it* refers to the banana (*die Banane*), and becomes the feminine *sie*. In the third one, *it* is the formal, impersonal subject, translated by the neutrum pronoun *es*.

What is the algorithm for translating *it* in the three described ways? As is clear from the explanations given to each translation, it has to do with the *reference* of the pronoun, not just its syntactic form. It also has to do with the *type* of applicability of the adjective that it predicated of the pronoun. The outline of the algorithm presented in Ranta (1994) is the following:

1. Analyse the context of the pronoun to collect all possible referents with their types, thus forming the *referent space* $\{r_1 : R_1, \ldots, r_n : R_n\}$ of objects given in the context.
2. Analyse the occurrence of the pronoun and collect all types $\{T_1, \ldots, T_m\}$ that an object may have in that position.
3. Consider the set of those elements $r_i : R_i$ whose type R_i matches some of the types T_j.

   (a) If the set is singleton $\{r_i : R_i\}$, then r_i is the referent and its type is R_i.
   (b) If the set is empty, then report an anaphora resolution error (or widen the referent space).
   (c) If the set has many elements, then ask the user to disambiguate interactively (or look for more constraints in the context).

This algorithm does a half of the job—it finds the referent of the pronoun with its type. The other half is to generate the translation. But this part is easy once the referent and its type are found, because pronouns can be given the abstract syntax

```
fun Pron : (t : Typ) -> Ref t -> Exp t
```

and the German concrete syntax chooses the proper word as a function of the gender of the type,

```
lin Pron t _ = case (gender t) of {
  Masc  => "er" ;
  Fem   => "sie" ;
  Neutr => "es"
  }
```

(showing only the nominative forms for simplicity). When the parser encounters the English pronoun *it*, the initial abstract syntax tree is

```
Pron ?1 ?2
```

But as soon as the resolution algorithm has found a value for `?1`, the translation can be performed.

The above algorithm is a sketch, as it uses undefined concepts and leaves alternatives open. First, we need to know how to "analyse the context". We use *context* in the technical sense of type theory: the sequence of variables with their types that are in scope. This context is maintained by the type checker when it analyses the syntax tree. The *possible referents*, then, are a closure of the context under simple operations such as the projections p and q of Sigma types. Ranta (1994) explains in more detail how different constructs of natural language contribute to the context.

Secondly, what are the types that "an object may have at the position" where the pronoun occurs? This can be defined by considering the wider syntax tree around the pronoun. In general, there can be many such trees because natural language is syntactically and lexically ambiguous. These trees have the form $t_i(x)$, where x is the slot for the pronoun. The types $T_i : \{T_1, \ldots, T_m\}$ are thus all the types that x can have in all the trees $t_i(x)$. In addition, the pronoun of the type T_i must be the one actually being resolved (i.e. have the same gender); this is the only condition referring to concrete syntax in the algorithm, which otherwise works on the abstract syntax level. (Notice that we have here assumed that the number of types is finite; if this doesn't hold, the problem may become undecidable.)

The third concept left undefined is *type matching*. The baseline is equality: $R_i = T_j$. But this can be extended by using techniques such as *coercive subtyping* (Luo and Callaghan 1999). Johannisson (2005) and Angelov and Enache (2010) show how to do this in GF. Considering the first of the examples above, the predicate *hungry* might be defined as a propositional function over animals,

```
fun Hungry : Exp Animal -> Prop
```

The monkey, on the other hand, might be introduced as a referent of type `Monkey`,

```
r : Ref Monkey
```

But a coercion

```
c : Exp Monkey -> Exp Animal
```

will establish

```
c (Pron Monkey r)
```

as a possible argument of `Hungry`. Notice that we use the coercion on the `Exp` level rather than `Ref`, so that the gender of the subtype is preserved.

Fourthly, what does it mean to *widen the search space* when no referent is found? One way is to widen the class of the operations under which the referent space is closed. Subtyping coercions can be seen as an instance of this, but any functions may have to be considered. This shows that anaphora resolution can be as hard as *proof search* in general. Another way is to widen the context that generates the search space. For instance, if the referent cannot be found in the same sentence as the pronoun, earlier sentences may have to be taken into account. One reason for the undecidability found by Bar-Hillel (1964) and Kay (1997) is that the search space may be infinite.

Fifthly, what does it mean to *look for more constraints* when the referent is not unique? One possibility is to shrink the search space. For instance, if the context has been widened by earlier sentences, priority can still be given to referents found in later ones. Another possibility is to use probabilities. Consider, for instance,

the monkey ate the banana because it was so sweet
the monkey ate the banana because it had fallen from the tree

The above algorithm may construct both the monkey and the banana as possible arguments, but one of them may be more likely, and this can be defined by using *probabilistic GF grammars* (see Sect. 14.8 below). However, in the end maybe none of the referents comes out as the clear winner, or the translation task may be so critical that no guesses are tolerated. It is in these cases that the system may need user input and hence be interactive. In a good translation system, interactive disambiguation should be smooth and intuitive. The systems should, for instance, not display types or abstract syntax trees, but rather pose simple questions in natural language, in a manner similar to how a human would do: *do you mean the monkey or the banana?*

In its full generality, type-theoretical anaphora resolution is undecidable—just as the arguments of Kay (1997) suggest. The translation of large documents may preclude the use of interaction. But the algorithm can still be seen as the *specification* of what anaphora resolution should ideally do. Practical approximations can be created by omitting too complex proof search or far-away parts of the context.

Also statistical models can give approximations of anaphora resolution: since all words in *it was hungry* fit into one 3-gram, it may well happen that a model "knows" that *it* is related to *hungry* and guesses the translation of the pronoun right.

14.5 User Interaction

As fully reliable translation cannot be fully automatic, user interfaces are an essential part of machine translation. Post-processing bad machine translation is hardly a sufficient form of interaction; one of the conclusions of the ALPAC report (Pierce et al. 1966) was that translators found it slow and unpleasant, and would have preferred manual translation from the beginning. In grammar-based translation, grammars that are accurate enough for translation can hardly be complete. Hence their users easily end up in situations where the input is not recognized, and the response from the system is "syntax error". While this is accepted in compilers, where the grammars can be learnt from manuals, it is hardly acceptable in parsers of natural language, where grammars are theoretical constructs not known by native speakers.

The model that has been applied in the user interfaces of GF is that of a *syntax editor* (Teitelbaum and Reps 1981; Donzeau-Gouge et al. 1975). Syntax editors have been a standard interface for logical frameworks, where they are backed by a rich metatheory of editing actions (Magnusson 1994; Norell 2007). The main idea of a syntax editor is that the user is manipulating abstract syntax trees and not texts; the text is just a special *view* of the tree, produced by linearization. One advantage of syntax editors is that they avoid the problem of parsing. An early application of this was the WYSIWYM system ("What You See Is What You Mean", Power and Scott 1998), which replaced translation by *multilingual generation*. The user of WYSIWYM would directly construct an abstract representation, from which translations in different languages were generated automatically.

Pure syntax editing can however be heavy and slow, and it requires the awareness of an abstract, sometimes complex, structure. When comparing parsers and syntax editors, Welsh et al. (1991) ended up recommending *pluralistic editors*, which combine parsing and syntax-based editing. Now, since parsing has been supported for GF grammars from the beginning, the syntax editors built for GF have been pluralistic (Khegai et al. 2003). Their basic functionality is the stepwise construction of syntax trees by *refinements*, which are selections of constructors that build trees in a top-down fashion. For example, the construction of an arithmetic expression can have as its intermediate state the tree

```
EPlus ?1 ?2
```

In this state, a refinement is expected for the first metavariable, `?1`. This refinement can be selected from a *menu*, which contains all constants and variables whose value type is possible for `?1`. In this case, the menu might contain the constructors `EInt`, `EVar`, and `EPlus`. If `EPlus` is selected, the next state is

```
EPlus (EPlus ?11 ?12) ?2
```

However, as the editor is pluralistic, it also accepts an expression written in a concrete syntax as a refinement. Refining `?1` by `the sum of x and 5` would thus result in

```
EPlus (EPlus (EVar x) (EInt 5)) ?2
```

compressing five refinement steps into one short string. Since the editor continuously type-checks the tree, it of course makes sure that the variable x is actually available in context and has a correct type.

The disadvantage of parsing compared with syntax editing is that syntax errors are possible. Fortunately, a pluralistic editor can solve this issue by *incremental parsing*—a process in which the input is analysed word by word, and the set of possible next words is computed after each word. Thus a user may start typing

```
every number is _
```

and get a list of suggestions: *divisible*, *equal*, *even*, *not*, *odd*, *prime*, etc. Every suggestion is guaranteed to lead, eventually, to a correct sentence and thereby an abstract syntax tree. If the list of suggestions is long, it can be narrowed down by typing the beginning of a word: with

```
every number is e_
```

only words beginning with an *e* are suggested. If the incremental parsing algorithm is efficient and the interface well implemented, its usage can be as fast as the input of free text. It can even be faster, because typos are excluded and unique word choices can be auto-completed. Angelov (2009) defines the incremental parsing algorithm actually used in GF. A later version of the algorithm (used in Angelov and Enache 2010) integrates dependent type checking and variable binding analysis to narrow down the suggestions to semantically correct ones.

For most users, incremental parsing is the method of choice when source text is created in the first place. However, syntax editing can still be useful in later edits of a text. Consider the following business letter written in French:

> *Chère Madame X, j'ai l'honneur de vous informer que vous avez été promue chargée de projet.*

("Dear Mrs X, I have the honour to inform you that you have been promoted to a project manager"). If Mrs X declines and the letter is sent to Mr Y instead, just changing the recipient will result in

> ***Chère** Monsieur Y, j'ai l'honneur de vous informer que vous avez été **promue chargée** de projet.*

The boldface parts of the letter are now grammatically incorrect, since they are in feminine forms, in agreement with *Madame X*; embarrassingly, they may disclose to *Monsieur Y* that he was not the first choice for the position. But this embarrassment can be avoided if the letter is constructed in a syntax editor and the abstract syntax tree is saved. If the tree has the form

```
Letter (Dear (Mrs X)) (Honour (Promote ProjectManager))
```

then the one-place change of `Mrs X` to `Mr Y` results in

```
Letter (Dear (Mr Y)) (Honour (Promote ProjectManager))
```

Now the linearization knows how to inflect the relevant parts in agreement with the new recipient, which results in the letter

> ***Cher*** *Monsieur Y, j'ai l'honneur de vous informer que vous avez été* ***promu chargé*** *de projet.*

Boldface is here used for marking the parts that have changed as a result of agreement and are now correct.

At the end of the previous section, we identified disambiguation as a critical part of interactive systems. Parsing user input may lead to several trees from which only the user is able to choose. The simplest way to display the alternatives is to show the abstract syntax trees, but this is hardly user-friendly, and we want the translation system to be usable without awareness of the abstract syntax. An alternative is to use a *disambiguation grammar*—a grammar that is similar to that of the source language, but contains supplementary information that makes it unambiguous.

Consider, for example, the disambiguation needed in the example with *the donkey ate the banana*. Since no humans are involved, the English grammar for pronouns could be simply

```
lin Pron _ _ = "it"
```

A disambiguation grammar for translation into German should make at least the type explicit.

```
lin Pron t _ = "it" ++ "(" ++ "the" ++ t ++ ")"
```

Hence, the question posed to the user when translating *it had fallen from the tree* would display the menu items: *it (the donkey)* and *it (the banana)*. Full disambiguation would of course also show the referent:

```
lin Pron t r = "it" ++ "(" ++ "the" ++ t ++ r ++ ")"
```

As illustrated by the MOLTO Phrasebook (Détrez et al. 2012), disambiguation grammars can be constructed with minor additions to the original, ambiguous grammars.

14.6 Variations in Concrete Syntax

How is it possible for different languages to share an abstract syntax? We have mainly considered two ways of achieving this: in different concrete syntaxes,

- *Words* can be different;
- The *order* of words can be different.

However, more freedom is needed to enable an abstract syntax really to abstract away from language-dependent facts. Fortunately, two more things have proven to be enough to achieve this:

– *Parameters*: words and phrases can have different inflectional forms and features;
– *Discontinuity*: the translation of a word can consist of separate parts.

Let us first consider the parameters. English verbs (with the exception of *be*) have five forms, exemplified by *write, writes, wrote, written, writing*. German verbs have at least 20 finite forms, and moreover dozens of adjectival forms of the participles. French verbs have 51 forms, Latin verbs a couple of hundreds, Finnish verbs several thousands depending on how one counts. The grammar of each language has to define precisely how these forms are created for each verb and how they are used in sentences. Yet we want to have, in the abstract syntax, a common category of verbs, and common rules (i.e. functions) for combining verbs with their subjects and objects.

Here is an example: a function that forms a sentence (S) by combining a two-place verb (V2) with a subject and an object. The subject and the object are *noun phrases* (NP), such as pronouns (*she*), proper names (*Mary*), or nouns with determiners (*the banana*). The abstract syntax thus has three categories and one function:

```
cat S ; V2 ; NP
fun PredV2 : V2 -> NP -> NP -> S
```

To achieve a complete description in little space, let us restrict the grammar to present indicative sentences. The simplest possible concrete syntaxes are

```
lin PredV2 v s o = s ++ v ++ o
lin PredV2 v s o = s ++ o ++ v
```

and four other permutations; thus there is no problem to treat so-called SVO and SOV languages with the same abstract syntax. Swedish is almost as simple as this, since the verb has just one form for the present indicative. However, pronouns have separate nominative and accusative forms, used for the subject and the object, respectively. Thus for Swedish, we have to change the *linearization type* of noun phrases from strings to *tables*, which assign a string to each of the cases nominative and accusative. We write

```
lincat NP = Case => Str
```

to say that noun phrases are linearized to case-to-string tables. We write

```
param Case = Nom | Acc
```

to define the *parameter type* of cases in Swedish. And finally, we write

```
lin PredV2 v s o = s ! Nom ++ v ++ o ! Acc
```

to linearize subject-verb-object sentences in Swedish. The *selection* operator `!` is used for retrieving values from tables. The tables themselves are given with a special expression form, as shown in the rule for the pronoun *she*:

```
fun She : NP
lin She = table {Nom => "hon" ; Acc => "henne"}
```

German is more complex than Swedish in three ways: noun phrases have four cases instead of two; verbs have five forms instead of one; and the object can have different cases depending on the verb. For instance, *lieben* ("love") takes its object in the accusative, but *folgen* ("follow") in the dative. We need a more complex system of parameters and linearization types:

```
param Case = Nom | Acc | Dat | Gen
param Number = Sg | Pl
param Person = Per1 | Per2 | Per3
lincat NP = {s: Case => Str; n: Number; p: Person}
lincat V2 = {s: Number => Person => Str; c: Case}
```

These linearization types use yet another data structure provided by GF: *records*. A record is a collection of objects of possibly different types. Thus the German NP has a case-to-string table, a number, and a person. These objects can be retrieved from the record by the *projection* operator . (dot). Now we can write a German linearization rule that selects the correct forms of the subject, the verb, and the object:

```
lin PredV2 v s o= s.s! Nom++v.s! s.n! s.p++o.s! v.c
```

Even though this rule is more complex than the Swedish rule, the abstract syntax is still the same. To see how much they differ, consider the language fragment generated from two verbs and three pronouns. The abstract syntax trees are given by the expression

```
PredV2 (Love|Follow) (I|You|She) (I|You|She)
```

where | marks alternatives. Thus there are 18 different trees (which include questionable ones such as *I love me*). The following picture shows finite automata representing the concrete syntaxes, Swedish on the left and German on the right. The German automaton is more complex. One can notice, for instance, that the number of different words (word forms) is almost twice the number of word forms in Swedish (14 vs. 8).

The higher the number of word forms in a language, the less probable is the occurrence of each word in a corpus. This is a problem for statistical string-based language models. For instance, a system may fail to find some less common German verb forms at all. A remedy to this is to introduce grammatical knowledge into the system by analysing the words into their dictionary forms and morphological description tags. Thus for instance

```
du folgst mir
```

becomes something like

```
du<+Pron+Nom> folgen<+Verb+Ind+Pres+Sg+2> ich<+Pron+Acc>
```

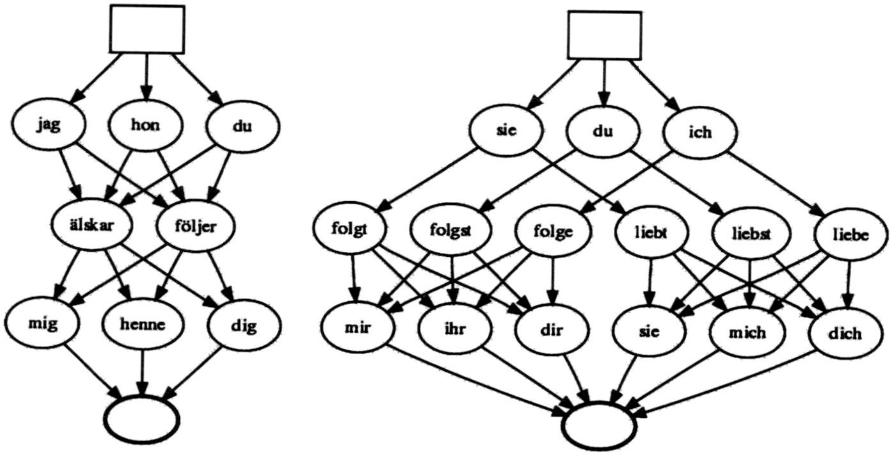

Then an *n*-gram model can be built for the sequences of description tags, which are much more frequent than the verb forms, and the translations of dictionary forms can be defined separately. This idea is known as *factored translation models* and studied in Koehn and Hoang (2007).

A prime example of *discontinuous constituents* are the compound verbs of Germanic languages. Thus the German verb *umbringen* ("kill") consists of the verb *bringen* ("bring") and the particle *um* ("around"). In the sentence

er bringt mich um

("he kills me", word to word "he-brings-me-around"), these parts fit into one and the same 3-gram, whereas in

er bringt meinen besten Freund um

("he kills my best friend") the chances are that a purely statistical system misses the whole point of the sentence.

In GF, discontinuous constituents can be modelled with records. Thus we extend the German linearization type of V2 with a field for the particle:

```
lincat V2 = {s : Number => Person => Str ; p : Str ; c : Case}
```

The predication rule becomes correspondingly

```
lin PredV2 v s o =
  s.s ! Nom ++ v.s ! s.n ! s.p ++ o.s ! v.c ++ v.p
```

The full grammar of subject-verb-object predication is of course much more complex than shown above. In German, we have to take into account the word order variation in main clauses and subordinate sentences. In French, the position of the object is different for pronouns from heavier noun phrases (*je t'aime* "I-you-love" vs. *j'aime Marie* "I-love-Marie"), with subtle agreement differences in compound tenses, and so on. Nevertheless, the two data structures of GF (tables and records) have proven sufficient for the expression of all these rules in a compact and efficient manner. Thus the *GF Resource Grammar Library* (Ranta 2009a) covers a large fragment of syntax of 16 languages using the same abstract syntax for all languages. As we will see in next section, this library plays a key role in the economical production of translation systems.

Interestingly, the GF Resource Grammar Library also demonstrates that the morphology and syntax, despite huge differences in surface variation, has a similar complexity in each language. This is not only reflected in the shared abstract syntax, but also in the size of the source code. The following table shows the code size for the concrete syntaxes of one and the same abstract syntax for four languages in the library, together with relative standard deviations and the maximum-to-minimum ratios. While the "GF source" column shows little variation, the low-level generated code rules varies much more. The "compressed PGF" column gives the size of the binary code generated by the GF compiler to implement parsing and linearization at run time, compressed by bzip2. The "context-free" column gives the number of rules in a context-free grammar generated as a conservative approximation of the PGF. The "words" column gives the total number or words in the PGF, optimized by a restriction to the words reachable from the start category.

Language	GF source	Compressed PGF	Contex-free	Words
English	3,300	49,000	48,000	1,900
German	3,400	67,000	63,000	4,400
French	5,400	84,000	66,000	4,100
Finnish	4,200	103,000	192,000	21,000
Rel. stdev.	0.21	0.26	0.63	0.97
Max/min	1.9	2.1	4.0	11

The numbers of context-free rules and words reflect the complexity of the language on a low abstraction level, whereas the GF code reflects the amount of information needed for defining the language on a high abstraction level. The PGF size also involves an abstraction in the sense of redundancy reduction due to compression and underlying compiler optimizations such as common subexpression elimination.

14.7 Grammar Engineering

The figures at the end of the previous section suggest that implementing a grammar in GF is not significantly harder for a "complex" language like Finnish than for a "simple" language like English. This result holds for grammars that exploit all the abstractions available in GF. If the grammars had to be written in context-free format, the differences would be much larger. If the language models had to be constructed from the occurrences of words, German and French would need more textual data than English, and Finnish would need even more. This is illustrated by the fact that a Google translate from Finnish often returns Finnish word forms untranslated.

Despite the compactness of GF, grammar writing is not easy. It requires a lot of linguistic knowledge, and only a part of this can be found in standard reference books, in particular as regards syntax. These difficulties may have been a major obstacle to the popularity of grammar-based domain-specific precision-oriented translation systems. While good quality can be reached with the use of grammars on limited domains, writing these grammars is time-consuming. Moreover, it requires the joint competences of a linguist and a *domain expert*, for instance, a mathematician when building a translation system for mathematical texts, or a car engineer when translating car maintenance manuals.

When the first multilingual GF grammars were built for a few domains (mathematics, tourist phrasebooks, restaurant database queries, medical drug descriptions), it soon become obvious that the same linguistic problems arose over and over again. This was the main reason for starting to build the GF Resource Grammar Library (cf. Ranta 2009b). Now that the library is complete for a large fragment of language, it is clearly the most important factor in enhancing the productivity in building translation systems. The library has a high-level API (*Application Programmer's Interface*), which corresponds to an abstract syntax and hides the details like word order, inflection, agreement, and discontinuities. The API is, moreover, language-independent, so that the grammar code for one language can also be used for other languages covered by the library.

To take an example of the use of the library, consider a concept such as *x knows y*, a two-place relation between persons. This might be needed in a translation system for social fora, and defined by the abstract syntax predicate

```
fun Know : Person -> Person -> Fact
```

The concrete syntax can be defined by means of the resource grammar by using NP (noun phrase) as the linearization type of persons and Cl (clause) as the linearization type of facts. The library API displays a function

```
mkCl : NP -> V2 -> NP -> Cl
```

for building a clause from a subject, a two-place verb, and an object. The library moreover provides language-specific lexica of irregular verbs. For instance, the English library has a constant

```
know_V : V
```

There is also a functions for making a verb (V) into a transitive two-place verb (V2),

```
mkV2 : V -> V2
```

Now we can define

```
lin Know x y = mkCl x (mkV2 know_V) y
```

This rule produces all the variation that can occur in a clause. Some of the variations are due to agreement to the subject:

```
Know I She   --> I know her
Know He She --> he knows her
```

but much more is needed when the clause is put into a wider context, such as negation, questions, and tenses:

```
mkS negative_Pol (Know I She)              --> I don't know her
mkQS (Know I She)                          --> do I know her
mkS past_Tense (Know I She)                --> I knew her
mkS fut_Tense anter_Ant (Know I She) --> I will have known her
```

In all these examples, we have just wrapped the clause `Know I She` with resource grammar functions to produce the correct linearizations.

The same API works for German and French, by just changing the verb:

```
lin Know x y = mkCl x (mkV2 kennen_V) y
lin Know x y = mkCl x (mkV2 connaÃőtre_V) y
```

In addition to the variations listed above, German displays word order variations:

```
mkS if_Subj (mkS (Know She I)) (mkS (Know I She)) -->
  wenn sie mich kennt, kenne ich sie
```

("if she knows me I know her"). In French, clitic variations are produced:

```
Know I Marie --> je connais Marie
Know Marie I --> Marie me connait
```

In none of these cases does the user of the library need to know about word order, clitics, inflection, or agreement—she just has to decide what the verb is, select the subject and the object, and perhaps the tense or some other context where the clause is to be used.

A futher effectivization of grammar writing is provided by the use of *functors*. A functor, or a *parametrized module*, is a program module that depends on some undefined constants and can be instantiated by defining these constants. The previous example suggests a functor definition of the predicate *know*:

```
lin Know x y = mkCl x know_V2 y
```

where `know_V2` is an undefined constant. Each of the three languages define it separately:

```
know_V2 = mkV2 know_V
know_V2 = mkV2 kennen_V
know_V2 = mkV2 connaï£¡tre_V
```

Technically, also `mkCl` is an undefined constant. Its definition for each language is given in the GF Resource Grammar Library. In general, functors use two kinds of constants:

- Syntactic constants defined in the library
- Lexical constants defined by the programmer

This way of using functors has become standard in GF projects. Grammarians that use the library thus have to write three kinds of modules:

- Abstract syntax, to define the semantics of a new domain
- Concrete syntax functor, to implement the first language on a new domain
- Domain lexicon, to implement a new language in an old domain

The first two tasks demand domain expertise and knowledge about GF and the Resource Grammar Library, but no detailed linguistic knowledge. The third task demands little more than a native speaker's knowledge of the target language and the terminology of the domain.

The Resource Grammar Library thus minimizes the knowledge requirements for building translation systems: the programmer gives the words, and the grammar governing the words comes from the library. This technique is similar to how human speakers of a foreign language learn new words. If I know the grammar of German, but I don't know how to express a concept such as *x intersects y* in geometry, I can ask someone how to translate this very example. Then I can use my knowledge of German grammar to generalize the translation to more complex cases such as *wenn x nicht y schnitte, würde y auch nicht x schneiden* ("if x didn't intersect y, y wouldn't intersect x either").

The translation of the sentence *x intersects y* as *x schneidet y* can actually be interpreted as the linearization rule

```
lin Intersect x y = mkCl x (mkV2 schneiden_V) y
```

if the sentence *x schneidet y* can be parsed in the German resource grammar as a tree of type Cl. This suggests a method for *example-based grammar writing*, where the library is not used explicitly but via examples, which in this case could be given in a format such as

```
lin Intersect x y = parse Cl "x schneidet y"
```

The crucial piece of information is here the German string. One way to obtain this string is by giving the English string to a human translator, who thus doesn't need any knowledge of GF; she may need to know that the context is that of geometry, to resolve potential word sense ambiguities.

By example-based grammar writing, building a translation system that can deal with an unlimited number of documents boils down to translating just one document, which contains representative examples of all concepts in the domain. This is probably one of the most efficient ways to use human labour in machine translation.

An extreme form of example-based grammar writing is to give the translation examples to a statistical machine translator system, such as Google translate. In fact, properly trained statistical translators are quite reliable with sentences that are short and typical (i.e. frequent n-grams for a small n). Thus, for instance, Google translate gets *x intersects y* right in German. Since the produced GF rule covers all variation due to agreement, tense, and word order, it expands to 288 context-free rules. Most of these derived combinations are not translated correctly in Google translate—but this doesn't matter, since we now have a grammar-based system. In analogy to what we concluded about human labour, providing translations for example-based grammar writing might be one of the most reliable ways of using statistical models in machine translation.

14.8 Transfer and Paraphrasing

We have presented GF as a framework for interlingua-based translation systems. The need of transfer functions (i.e. functions mapping source language trees to target language trees) is less common than in some other grammar formalisms, because the abstract syntax trees of GF can maintain a considerable distance to the concrete trees.

In traditional systems, transfer is used whenever there is a *structural change* between the source and the target language. A typical example is *my name is Bond.* In the English sentence, the subject is *my name.* In the German equivalent, *ich heiße Bond*, the subject is *ich* ("I"), and a special verb *heißen* ("have name") is used. The French translation is *je m'appelle Bond*, literally "I call myself Bond".

In GF, translation is performed via the abstract syntax of a semantic grammar, rather than the syntactic resource grammar. If the abstract syntax has a predicate

```
fun Named : Person -> Name -> Fact
```

it is not a problem to pick different syntactic structures as linearizations,

```
fun Named x y = mkCl (possessive x (mkN "name")) (mkNP y)
fun Named x y = mkCl x (mkV2 heißen_V) (mkNP y)
fun Named x y = mkCl x (mkV2 (reflV appeler_V)) (mkNP y)
```

(Notice that GF provides overloading for sets of functions that have different types, here `mkCl`.) Thus in the run-time translation, no transfer is needed—just parsing and linearization via the interlingua tree. The use of different syntactic structures (as defined by the resource grammar) in linearization rules has the effect of *compile-time transfer*. It eliminates the need of run-time transfer and hence maintains the simple interlingua-based translation model of GF.

The only restriction that GF's interlingua model poses to translation is *compositionality*. More precisely, all linearization rules in GF are compositional, which means that the linearization of every tree must be defined as a function of the *linearizations* of its immediate subtrees (and not of the trees themselves). Using the notation of t^* for the linearization of a tree t and f^* for the linearization function of a function f, compositionality means that

$$(f\ t_1\ \dots\ t_n)^* \ = \ f^*\ t_1^*\ \dots\ t_n^*$$

If linearizations were just strings (as in context-free grammars), it would be impossible to maintain compositionality even in simple translation tasks. But the use of records and tables in GF makes it maintainable in most cases, and it requires some effort to find counterexamples.

One counterexample to compositinal translation is suggested by the *my name is* example above. In German, the predicate has the bearer of the name as its subject, and the subject can be shared by *verb phrase coordination*:

ich heiße Bond und komme aus England

("I have-name Bond and come from England"). The English translation is

my name is Bond and I come from England

But this translation has changed the original's conjunction of predicates (verb phrases) to a conjunction of sentences, because there is no common subject that could be shared. On the abstract syntax tree level, it involves a transfer from a tree of the form

```
PredVP a (ConjVP F G)
```

to a tree of the form

```
ConjS (PredVP a F) (PredVP x G)
```

Also run-time transfer functions can be defined in GF. They are executed on the abstract syntax level between the source text parser and the target text generator. In general, translation from L_1 to L_2 is then a composition of three operations,

$$\text{parse}\, L_1 \implies \text{transfer}\, L_1\, L_2 \implies \text{linearize}\, L_2$$

Interlingual translation is a limiting case, where transfer is the identity mapping. Notice that the role here assigned to transfer is very similar to the operations that *optimizing compilers* such as GCC perform on the intermediate tree language level.

Transfer involves a departure from the interlingual model, destroys reversibility, and may compromise run-time efficiency. Therefore translators written in GF have usually avoided it. However, it can be useful to look at some of the fundamental properties suggested by type theory.

Like many logical frameworks, the abstract syntax part of GF can define a notion of *definitional equality* among syntax trees. For instance, the correspondence between sentence and verb phrase conjunction can be seen as a definitional equality. A reasonable condition for transfer functions is that they must preserve definitional equality. Definitional equality is generated by arbitrary recursive function definitions, which need not be compositional.

The linguistic counterpart of definitional equality is *paraphrasing*. Two concrete syntax expressions are paraphrases, if their syntax trees are definitionally equal. We could say that the *literal translation* of a string is the one obtained by parsing the string and linearizing the resulting tree *t*. If the tree is first converted to a definitionally equal tree *t'*, we obtain a *translation by paraphrase*.

In general, there can be infinitely many trees definitionally equal to a given tree. These trees can be generated by applying equality rules forwards and backwards.

To take the most familiar example from type theory, assume the set of natural numbers defined by the constructors

```
data Zero : Nat
data Succ : Nat -> Nat
```

(the keyword `data` in GF marks a function as a constructor). Then add the constant 1 and the addition operation, with their definitions, written as follows in GF:

```
fun one  : Nat
def one  = Succ Zero

fun plus : Nat -> Nat -> Nat
def plus x Zero = x
def plus x (Succ y)  = Succ (plus x y)
```

The tree `one` now has infinitely many paraphrases, beginning with those obtained in one computation step forward or backward,

```
Succ Zero, plus one Zero, plus (Succ Zero) Zero, ...
```

The *computation distance* of a tree *t'* from a tree *t* is the number of steps needed to obtain *t'* from *t*. A possible condition for translation by transfer is that it should minimize the computation distance. (An alternative measure would be *tree edit distance*, but computation distance has the advantage that the optimal sorting of trees can be straightforwardly generated from the definitions.)

Now, the minimal computation distance is 0, and this cannot always be achieved because there are other constraints. One such constraint is the concrete syntax of a target language can simply lack the construct used in the source language. Another constraint, reasonable in domains that require precision, is *non-ambiguity*: if a string is unambiguous in the source language, its translation should not be ambiguous in the target language. This condition is decidable relative to a fixed grammar, because the GF parser can find all trees corresponding to a string (except in some pathological cases where the number of trees is infinite).

The transfer task can therefore be defined as finding the closest unambiguous paraphrase. A typical example is anaphora. For instance, Finnish has a gender-neutral pronoun, *hän*, corresponding to both *he* and *she*. Therefore a love story written in English often has to paraphrase *he* with *mies* ("the man") and *she* with *nainen* ("the woman"). The optimal translation is more subtle than this: consider a context in which a man and a woman are given, and the sentence

She took his hand.

There are two equally close unambiguous translation of the second sentence,

Hän tarttui miehen käteen. ("She took the man's hand.")
Nainen tarttui hänen käteensä. ("The woman took his hand.")

One does not need to paraphrase both pronouns to remove ambiguity, because the other pronoun can only refer to a different person. If it was the same person, the object position would be rendered as a reflexive.

Yet another ingredient in transfer is *style*. Some constructs used in a source language can be unnatural in the target language, even if they would be possible and unambiguous. For instance, passive constructions with agents are common in English, but are preferably translated by active constructions in Finnish. One way to implement this in GF is to assign *weights* to abstract syntax functions, reflecting their goodness in each target language. The weight of a tree is then the product of the weights of the functions in all of its nodes. Transfer should then find the paraphrase that has the maximal weight in the target language, at the same time as minimizing the computation distance to the literal translation and maintaining non-ambiguity.

One way in which weights can be assigned to abstract syntax functions are via their relative frequencies in some corpus. This leads to the notion of *probabilistic GF grammars* and shows yet another way in which statistical language models can be combined with grammars.

14.9 Specification and Evaluation with Grammars

Translation from one language to another is a function ultimately defined on the level of strings: input a string in the source language, output a string in the target language. But the type `String -> String` is of course too lean as a specification of what translation should do, and must be refined. The standard refinement in the statistical translation community is the *BLEU score* (Papineni et al. 2002), which compares the output of a translating system with some *gold standard* translation produced by a human. The BLEU score is computed by counting the occurrences of words and n-grams, taking into account their order. The best translation is one that matches the gold standard word by word. But also some intuitively bad translations, such as one that matches the original word by word but forgets a negation word, get high scores. And intuitively excellent translations by humans get bad scores, if they don't use the same words as the gold standard.

The problems with the BLEU score are widely acknowledged, but it is still popular because it is automatic. Since the goal of statistical translation systems is often defined as the maximization of the BLEU score, one of its main uses is in the development phase of the systems. A typical system uses several *features* to compute the most likely translation: frequencies from bilingual word alignment, frequencies of n-grams in the source language, etc. Each of these features is given some weight, and the system is tested with several distributions of weights to maximize the BLEU score. While this makes sense, using the BLEU score as a measure of the quality when comparing translation systems is less adequate.

So what alternatives are there for specifying what a translation function should do? The traditional answer is, of course, that the translation should preserve meaning. An implicit presupposition is that it should render grammatically correct output; otherwise it would not count as a target-language expression at all! Fluency and good style (or style that matches the style of the source) are further requirements. For most of these criteria, there is no other evaluation method than human judgement. However, the GF-based method described in this paper suggests a technique for cross-evaluating other systems (such as statistical ones). Assuming that grammaticality is properly defined in GF, GF grammars can be used for assessing the grammaticality of the output from these other systems. The same concerns the preservation of meaning, if we have a GF grammar in which the linearizations and definitional equalities preserve meaning.

A problem in using GF to evaluate statistical translation systems is of course that the coverage of GF is just partial. But it can still be used to test the quality of a statistical translator for those sentences that the GF grammar does cover. The relevant functionality of GF is *multilingual generation*: one can produce a *synthesized corpus*, that is, a set of sentence pairs in a source and a target language, where the sentences in each pair have the same abstract syntax tree. This set can then be used for evaluating a statistical translator, because it gives both the source and the gold standard translations.

Multilingual generation also provides a way to build a statistical translator in the first place. A GF grammar can be used for producing any number of aligned sentences, which can be guaranteed to cover all word forms appearing in the grammar. All sentence pairs can be automatically equipped by correct phrase alignments, which is a good starting point for building the statistical model (Och and Ney 2004). Such an alignment relates to each other the words that have the same smallest spanning subtree in the abstract syntax. The alignments can cross, and they can include many-to-many relations, as shown in the figure in Sect. 14.1, generated from the GF Resource Grammar Library.

The resulting statistical model may perform reasonably for the input covered by the grammar, although never better than the grammar itself. However, the advantage of the statistical model is that it is also able to translate sentences not recognized by the grammar. In this way, the statistical model can be used as a smoothing technique for grammar-based translation. Of course, it is then important to mark clearly which parts of the output are translated by smoothing and which parts come from the grammar. This technique is analogous to the use of GF-generated language models for speech recognition studied by Jonson (2006). The result is a hybrid system where a statistical model is derived from a grammar, and the grammar can in turn have been built as a generalization of a less refined statistical model as described in Sect. 14.7. An experiment with this idea can be found in (Rayner et al. 2011).

14.10 Conclusion

Machine translation was attempted as one of the first applications of digital computers. It was soon realized that fully automatic high-quality translation is impossible. The main conclusion drawn from this was that there is a trade-off between *coverage* and *precision*. The efforts on machine translation are thus roughly divided into open-domain systems aiming at coverage and closed-domain systems aiming at precision. In both kinds of systems, human interaction can be involved.

In open-domain systems, a typical form of interaction is the post-processing of the output to improve its quality. In closed-domain systems, a typical form of interaction is to ask a human to disambiguate. One can also use human interaction to recover from input that is not in the grammar, for instance, to add the translations of unknown words.

The use of type theory in machine translation dates back to the earliest years, when Bar-Hillel applied it to the formal representation of grammar and meaning. GF is a contemporary variant of the idea, providing a notion of multilingual grammars, a framework for applying the method to new domains, a resource grammar library for improving the productivity for new languages, and a set of user interface components (parsing, syntax editing) to help the work of translators.

GF has since 1998 been used for translation on several domains, including mathematics (Hallgren and Ranta 2000; Caprotti 2006), software specifications (Johannisson 2005), and spoken dialogue systems (Bringert et al. 2005; Perera and Ranta 2007). The European MOLTO project (Multilingual On-Line Translation) aims to scale up the methods into larger domains, more languages, easier production and application, and more robustness (i.e. recovery from out-of-the grammar input).

Some of the GF methods presented in this paper have not yet been used in actual translation systems. On the purely type-theoretical side, these include the anaphora resolution algorithm (Sect. 14.4) and the generation of paraphrases via definitional equality (Sect. 14.8). On the hybrid side, the ranking of paraphrases (Sect. 14.8) and the training of statistical translations (Sect. 14.9) have just had their first experiments in the MOLTO project. On the other hand, example-based grammar writing (Sect. 14.7) and disambiguation grammars (Sect. 14.5) are recent ideas both of which have given promising results in the first demonstrator of MOLTO, which is a phrasebook for translating touristic phrases between 14 languages (Détrez et al. 2012).

The first experiences with the Phrasebook and with a translator of mathematical exercises are confirming the basic tenet of MOLTO: that it is possible to build reliable translation systems for limited domains by careful engineering and adequate tools. With almost any language pair and example covered by the grammar, the translation quality is better than the quality produced by general-purpose statistical systems. In fact, it can be better in a crucial way, since grammar-based translation can be guaranteed to be correct by design whereas statistical systems always involve an element of uncertainty.

On the other hand, if we want a system that automatically translates any input, uncertainty cannot be avoided, and statistical methods are the ones that *de facto* yield the best quality in most cases. An interesting exception is translation between closely related languages, such as Swedish and Danish (Tyers and Nordfalk 2009). The lack of bilingual training data can make it impossible to build good statistical systems; at the same time, simple rules (such as replacing words by their equivalents in proper forms) may be sufficient to produce very good quality.

Acknowledgements Per Martin-Löf supervised my PhD thesis and taught me how to think about type theory and language. He also introduced me to the noisy channel model of Shannon. He wondered if statistical models were still considered useful in natural language processing, and they have ever since been a recurrent theme in our discussions. With his unique combination of insights in both statistics and logic, and his accurate knowledge of many languages, Per has continued to be a major resource for my work through the 21 years that have passed since my PhD. When I later started to look closer at statistical methods, I received inspiration and guidance from Joakim Nivre, Lluís Màrquez, and Cristina España. Lauri Carlson has helped me to understand the problems of translation in general. The model described in this paper has received substantial contributions from my own PhD students Peter Ljunglöf, Kristofer Johannisson, Janna Khegai, Markus Forsberg, Björn Bringert, Krasimir Angelov, and Ramona Enache. The insightful comments from an anonymous referee were valuable when preparing the final version of the paper. The research leading to these results has received funding from the European Union's Seventh Framework Programme (FP7/2007–2013) under grant agreement n:o FP7-ICT-247914 (http://www.molto-project.eu).

References

Ajdukiewicz, K. 1935. Die syntaktische Konnexität. *Studia Philosophica* 1: 1–27.

Angelov, K. 2009. Incremental parsing with parallel multiple context-free grammars. In *Proceedings of EACL'09*, Athens.

Angelov, K., and R. Enache. 2010. Typeful ontologies with direct multilingual verbalization. In *CNL 2010, Controlled natural language*, Marettimo Island, ed. N. Fuchs and M. Rosner. New Brunswick: ACL.

Appel, A. 1998. *Modern compiler implementation in ML*. Cambridge/New York: Cambridge University Press.

Bar-Hillel, Y. 1953. A quasi-arithmetical notation for syntactic description. *Language* 29: 27–58.

Bar-Hillel, Y. 1964. *Language and information*. Reading: Addison-Wesley.

Bringert, B., R. Cooper, P. Ljunglöf, and A. Ranta. 2005. Multimodal dialogue system grammars. In *Proceedings of DIALOR'05, ninth workshop on the semantics and pragmatics of dialogue*, Nancy, 53–60.

Brown, P.F., J. Cocke, S.A.D. Pietra, V.J.D. Pietra, F. Jelinek, J.D. Lafferty, R.L. Mercer, and P.S. Roossin. 1990. A statistical approach to machine translation. *Computational Linguistics* 16(2): 76–85.

Caprotti, O. 2006. WebALT! Deliver mathematics everywhere. In *Proceedings of SITE 2006*, Orlando March 20–24. http://webalt.math.helsinki.fi/content/e16/e301/e512/PosterDemoWebALT_eng.pdf.

Chandioux, J. 1976. MÉTÉO: un système opérationnel pour la traduction automatique des bulletins météreologiques destinés au grand public. *META* 21: 127–133.

Curry, H.B. 1961. Some logical aspects of grammatical structure. In *Structure of language and its mathematical aspects: Proceedings of the twelfth symposium in applied mathematics*, ed. R. Jakobson, 56–68. Providence: American Mathematical Society.

Détrez, G., R. Enache, and A. Ranta. 2012. Controlled language for everyday use: the MOLTO phrasebook. In: N. Fuchs and M. Rosner (eds.), CNL 2010 proceedings, Springer LNCS/LNAI 7175: 115–136.

Donzeau-Gouge, V., G. Huet, G. Kahn, B. Lang, and J. J. Levy. 1975. A structure-oriented program editor: A first step towards computer assisted programming. In *International computing symposium (ICS'75)*. Hsinchu: Nat Chiao Tung University.

Dowek, G., A. Felty, H. Herbelin, G. Huet, C. Parent, C. Paulin Mohring, B. Werner, and C. Murthy. 1993. The Coq proof assistant user's guide: version 5.8. Research Report RT-0154, INRIA.

Hallgren, T., and A. Ranta. 2000. An extensible proof text editor. In *LPAR-2000*, Lecture notes in computer science/Lecture notes in artificial intelligence, vol. 1955, ed. M. Parigot and A. Voronkov pp. 70–84. Berlin: Springer. http://www.cse.chalmers.se/~aarne/articles/lpar2000.pdf.

Harper, R., F. Honsell, and G. Plotkin. 1993. A Framework for defining logics. *Journal of the Association for Computing Machinery* 40(1): 143–184.

Hutchins, W.J., and H.L. Somers. 1992. *An introduction to machine translation*. London: Academic.

Hutchins, J. 2000. Early years in machine translation: memoirs and biographies of pioneers. Amsterdam: John Benjamins.

Jelinek, F. 2009. The dawn of statistical ASR and MT. *Computational Linguistics* 35(4): 483–494.

Johannisson, K. 2005. Formal and informal software specifications. Ph.D. thesis, Department of Computing Science, Chalmers University of Technology and Gothenburg University.

Jonson, R. 2006. Generating statistical language models from interpretation grammars in dialogue system. In *Proceedings of EACL06*, Trento.

Kay, M. 1997. The proper place of men and machines in language translation. *Machine Translation* 12(1–2): 3–23.

Khegai, J., B. Nordström, and A. Ranta. 2003. Multilingual syntax editing in GF. In *Intelligent text processing and computational linguistics (CICLing-2003)*, Mexico City, February 2003, Lecture notes in computer science, vol. 2588, ed. A. Gelbukh, 453–464. Springer. http://www.cs.chalmers.se/~aarne/articles/mexico.ps.gz.

Knuth, D. 1968. Semantics of context-free languages. *Mathematical Systems Theory* 2: 127–145.

Koehn, P., and H. Hoang. 2007. Factored translation models. In *EMNLP-CoNLL*, Prague, 868–876. ACL.

Ljunglöf, P. 2004. The expressivity and complexity of grammatical framework. Ph.D. thesis, Department of Computing Science, Chalmers University of Technology and Gothenburg University. http://www.cs.chalmers.se/~peb/pubs/p04-PhD-thesis.pdf.

Lopez, A. 2008. Statistical machine translation. *ACM Computing Surveys* 40(3): 1–49.

Luo, Z., and P. Callaghan (1999). Mathematical vernacular and conceptual well-formedness in mathematical language. In *Logical aspects of computational linguistics (LACL)*, Nancy, Lecture notes in computer science/Lecture notes in artificial intelligence, vol. 1582, ed. A. Lecomte, F. Lamarche, and G. Perrier, 231–250.

Luo, Z., and R. Pollack. 1992. LEGO proof development system. Technical report, University of Edinburgh.

Magnusson, L. 1994. The implementation of ALF – A proof editor based on Martin-Löf's monomorphic type theory with explicit substitution. Ph.D. thesis, Department of Computing Science, Chalmers University of Technology and University of Göteborg.

Martin-Löf, P. 1984. *Intuitionistic type theory*. Napoli: Bibliopolis.

Montague, R. 1974. *Formal philosophy*. New Haven: Yale University Press. Collected papers edited by Richmond Thomason.

Nordström, B., K. Petersson, and J. Smith. 1990. *Programming in Martin-Löf's type theory: An introduction*. Oxford: Clarendon Press.

Norell, U. 2007. Towards a practical programming language based on dependent type theory. Ph.D. thesis, Department of Computer Science and Engineering, Chalmers University of Technology, SE-412 96 Göteborg, Sweden.

Och, F.J., and H. Ney, 2004. The alignment template approach to statistical machine translation. *Computational Linguistics* 30(4): 417–449.

Papineni, K., S. Roukos, T. Ward, and W.-J. Zhu. 2002. BLEU: A method for automatic evaluation of machine translation. In *ACL*, Philadelphia, 311–318.

Perera, N., and A. Ranta (2007). Dialogue system localization with the GF resource grammar library. In *SPEECHGRAM 2007: ACL workshop on grammar-based approaches to spoken language processing*, June 29, 2007, Prague. http://www.cs.chalmers.se/~aarne/articles/perera-ranta.pdf.

Pierce, J.R., and J. B. Carroll et al. 1966. Language and machines – Computers in translation and linguistics. ALPAC report.

Power, R., and D. Scott (1998). Multilingual authoring using feedback texts. In *COLING-ACL 98*, Montreal.

Ranta, A. 1994. *Type theoretical grammar*. Oxford: Oxford University Press.

Ranta, A. 2004. Grammatical framework: A type-theoretical grammar formalism. *The Journal of Functional Programming* 14(2): 145–189. http://www.cse.chalmers.se/~aarne/articles/gf-jfp.pdf.

Ranta, A. 2009a. Grammars as software libraries. In *From semantics to computer science. Essays in honour of Gilles Kahn*, ed. Y. Bertot, G. Huet, J.-J. Lévy, and G. Plotkin, 281–308. Cambridge/New York: Cambridge University Press. http://www.cse.chalmers.se/~aarne/articles/libraries-kahn.pdf.

Ranta, A. 2009b. The GF resource grammar library. *Linguistics in Language Technology* 2: 1–65. http://elanguage.net/journals/index.php/lilt/article/viewFile/214/158.

Ranta, A. 2011. *Grammatical framework: Programming with multilingual grammars*. Stanford: CSLI Publications. ISBN-10: 1-57586-626-9 (Paper), 1-57586-627-7 (Cloth).

Rayner, M., P. Estrella, and P. Bouillon. 2011. Bootstrapping a statistical speech translator from a rule-based one. In *Proceedings of the second international workshop on free/open-source rule-based machine translation*, Barcelona. http://hdl.handle.net/10609/5647.

Rosetta, M.T. 1994. *Compositional translation*. Dordrecht: Kluwer.

Shannon, C. 1948. A mathematical theory of communication. *The Bell System Technical Journal* 27(1): 379–423, 623–656.

Stallman, R. 2001. *Using and porting the GNU compiler collection*. Cambridge: Free Software Foundation.

Teitelbaum, T., and T. Reps. 1981. The Cornell Program Synthesizer: A syntax-directed programming environment. *Communications of the ACM* 24(9): 563–573.

Tyers, F., and J. Nordfalk. 2009. Shallow-transfer rule-based machine translation for Swedish to Danish. In *Proceedings of the first international workshop on free/open-source rule-based machine translation*, Alicante. http://hdl.handle.net/10045/12024.

Welsh, J., B. Broom, and D. Kiong. 1991. A design rationale for a language-based editor. *Software: Practice and Experience* 21: 923–948.

Chapter 15
Constructive Zermelo-Fraenkel Set Theory, Power Set, and the Calculus of Constructions

Michael Rathjen

MSC: 03F50, 03F35

15.1 Introduction

If the power set operation is considered as a definite operation, but the universe of all sets is regarded as an indefinite totality, we are led to systems of set theory having Power Set as an axiom but only Bounded Separation axioms and intuitionistic logic for reasoning about the universe at large. The study of subsystems of **ZF** formulated in intuitionistic logic with Bounded Separation but containing the Power Set axiom was apparently initiated by Pozsgay (1971, 1972) and then pursued more systematically by Tharp (1971), Friedman (1973a), and Wolf (1974). These systems are actually semi-intuitionistic as they contain the law of excluded middle for bounded formulae. Pozsgay had conjectured that his system is as strong as **ZF**, but Tharp and Friedman proved its consistency in **ZF** using a modification of Kleene's method of realizability. Wolf established the equivalence in strength of several related systems.

In the classical context, weak subsystems of **ZF** with Bounded Separation and Power Set have been studied by Thiele (1968), Friedman (1973b) and more recently

*Work on the ideas for this paper started while I was a fellow of SCAS, the Swedish Collegium for Advanced Study, in the period January-June 2009. SCAS provided an exquisite, intellectually inspiring environment for research. I am grateful to Erik Palmgren, Sten Lindström, and the people of SCAS for making this possible. Part of the material is also based upon research supported by the EPSRC of the UK through grant No. EP/G029520/1.

M. Rathjen (✉)
Department of Pure Mathematics, University of Leeds, Leeds LS2 9JT, UK
e-mail: rathjen@maths.leeds.ac.uk

P. Dybjer et al. (eds.), *Epistemology versus Ontology*, Logic, Epistemology, and the Unity of Science 27, DOI 10.1007/978-94-007-4435-6_15,

at great length by Mathias (2001). Mac Lane has singled out and championed a particular fragment of **ZF**, especially in his book *Form and Function* Mac Lane (1992). *Mac Lane Set Theory*, christened **MAC** in Mathias (2001), comprises the axioms of Extensionality, Null Set, Pairing, Union, Infinity, Power Set, Bounded Separation, Foundation, and Choice. **MAC** is naturally related to systems derived from topos-theoretic notions and, moreover, to type theories.

Type theories à la Martin-Löf embodying weak forms of Power Set (such as the calculus of constructions with its impredicative type of propositions) have been studied by Aczel (1986, 2000) and Gambino (1999).

Intuitionistic Zermelo-Fraenkel set theory, **IZF**, is obtained from **CZF**, by adding the full separation axiom scheme and the power set axiom. The strength of **CZF** plus full separation, as has been shown by Lubarsky (2006), is the same as that of second order arithmetic, using a straightforward realizability interpretation in classical second order arithmetic and the fact that second order Heyting arithmetic is already embedded in **CZF** plus full separation. This paper is concerned with the strength of **CZF** augmented by the power set axiom, $\mathbf{CZF}_{\mathcal{P}}$. It will be shown that it is of the same strength as Power Kripke–Platek set theory, $\mathbf{KP}(\mathcal{P})$, as well as a certain system of type theory, $\mathbf{MLV_P}$, which is a version of the calculus of constructions with one universe. It is perhaps worth pointing out that $\mathbf{KP}(\mathcal{P})$ is not the theory **KP** plus power set, **Pow**. An upper bound for the proof-theoretic strength of **KP** + **Pow** is Zermelo's set theory, **Z**, so that it doesn't even prove the existence of $V_{\omega+\omega}$ whereas $\mathbf{KP}(\mathcal{P})$ proves the existence of V_α for any ordinal α.

The reduction of $\mathbf{CZF}_{\mathcal{P}}$ to $\mathbf{KP}(\mathcal{P})$ uses a realizability interpretation wherein a realizer for an existential statement provides a set of witnesses for the existential quantifier rather than a single witness. Tharp (1971) also used realizability to give an interpretation of a semi-intuitionistic set theory closely related to Pozsgay's system. Tharp's realizers are codes for $\Sigma_1^{\mathcal{P}}$ definable partial functions, i.e., functions whose graphs are Σ_1 in the powerset operation $\mathcal{P}(x)$, which is taken as a primitive. For the realizability interpretation he needs a $\Sigma_1^{\mathcal{P}}$-definable search operation on the set-theoretic universe and in point of fact assumes $V = L$. As it turns out, this realizability interpretation could be formalized in $\mathbf{KP}(\mathcal{P}) + V = L$. However, the assumption $V = L$ is not harmless in this context since $\mathbf{KP}(\mathcal{P}) + V = L$ is a much stronger theory than $\mathbf{KP}(\mathcal{P})$ (cf. Mathias 2001; Rathjen 2012), and therefore one would like to remove this hypothesis. This paper shows that this can be achieved by using a notion of realizability with sets of witnesses in the existential quantifier case, and thereby yields a realizability interpretation of a theory in a theory of equal proof-theoretic strength.

The reduction of $\mathbf{KP}(\mathcal{P})$ to $\mathbf{CZF}_{\mathcal{P}}$ is based on results from Rathjen (2012) whose proofs are obtained via techniques from ordinal analysis. They can be used to show that $\mathbf{KP}(\mathcal{P})$ is reducible to **CZF** with the *Negative Power Set Axiom*. As **CZF** plus the negative powerset can be interpreted in $\mathbf{MLV_P}$, utilizing work from Aczel (2000) and Gambino (1999), and the latter type theory has a types-as-classes interpretation in $\mathbf{CZF}_{\mathcal{P}}$, the circle will be completed. We also get a characterization of a subtheory of Tharp's set theory Tharp (1971). The theory in Tharp (1971) has the following axioms (cf. Sect. 15.2.1): Extensionality, Empty Set, Pairing, Union,

Powerset, Infinity, Set Induction, Strong Collection,[1] Excluded Middle for power bounded formulae[2] and an axiom **Ord-Im** which asserts that every set is the image of an ordinal, i.e., for every set x there exists an ordinal α and a surjective function $f : \alpha \twoheadrightarrow x$.

In the Theorem below we use several acronyms. **RDC** stands for the relativized dependent choices axiom. Given a family of sets $(B_a)_{a \in A}$ over a set A we define the dependent product $\prod_{a \in A} B_a$ and the dependent sum $\sum_{a \in A} B_a$ as follows:

$$\prod_{a \in A} B_a := \{f \mid \mathrm{Fun}(f) \wedge \mathrm{dom}(f) = A \wedge \forall z \in A\ f(z) \in B_a\}$$

$$\sum_{a \in A} B_a := \{\langle a, u\rangle \mid a \in A \wedge u \in B_a\}$$

where $\mathrm{Fun}(f)$ signifies that f is a function and $\mathrm{dom}(f)$ stands for its domain.

Let X be the smallest class of sets containing ω and all elements of ω which is closed under dependent products and sums. $\mathbf{\Pi\Sigma-AC}$ asserts that every set A in X is a base, i.e., if $(B_a)_{a \in A}$ is family of sets over A such that B_a is inhabited for every $a \in A$ then there exists a function f with domain A such that $\forall a \in A\ f(a) \in A_a$ (for more information on this axiom see Aczel 1982, Rathjen 2006a, Rathjen and Tupailo 2006).

The negative power set axiom, $\mathbf{Pow}^{\neg\neg}$ for short, asserts that for every set a there exists a set c containing all the subsets x of a for which $\forall u \in a(\neg\neg u \in x \rightarrow u \in x)$ holds.

The intuitionistic version of $\mathbf{KP}(\mathcal{P})$ will be denoted by $\mathbf{IKP}(\mathcal{P})$. Both $\mathbf{KP}(\mathcal{P})$ and $\mathbf{IKP}(\mathcal{P})$ can be subjected to ordinal analysis which reduces them to theories $\mathbf{Z} + \{\text{`}V_\tau \text{ exists'}\}_{\tau \in \mathrm{BH}}$ and $\mathbf{IZ} + \{\text{`}V_\tau \text{ exists'}\}_{\tau \in \mathrm{BH}}$, respectively. Here $\mathbf{Z}$ stands for classical Zermelo set theory and $\mathbf{IZ}$ for its intuitionistic version. BH refers to an ordinal representation system for the Bachmann-Howard ordinal (cf. Rathjen and Weiermann 1993). For $\tau \in \mathrm{BH}$ the statement 'V_τ *exists*' expresses that the powerset operation can be iterated τ times.

Theorem 15.1. *The following theories are of the same proof-theoretic strength.*

(i) $\mathbf{CZF}_{\mathcal{P}}$
(ii) $\mathbf{CZF}_{\mathcal{P}} + \mathbf{RDC} + \mathbf{\Pi\Sigma-AC}$
(iii) $\mathbf{KP}(\mathcal{P})$
(iv) $\mathbf{IKP}(\mathcal{P})$
(v) *Tharp's (1971) quasi-intuitionistic set theory but without* **Ord-Im**.
(vi) $\mathbf{MLV_P}$
(vii) $\mathbf{CZF} + \mathbf{Pow}^{\neg\neg}$
(viii) $\mathbf{Z} + \{\text{`}V_\tau \textit{ exists'}\}_{\tau \in \mathrm{BH}}$
(ix) $\mathbf{IZ} + \{\text{`}V_\tau \textit{ exists'}\}_{\tau \in \mathrm{BH}}$

Presenting a proof of Theorem 15.1 is the main goal of this article.

---

[1] Curiously, Tharp calls this scheme Replacement.

[2] The $\Delta_0^{\mathcal{P}}$-formulae of Definition 15.1.

15.2 The Theories CZF and KP($\mathcal{P}$)

15.2.1 CZF

We briefly summarize the language and axioms of **CZF**, a variant of Myhill's CST (see Myhill 1975). The language of **CZF** is based on the same first order language as that of classical Zermelo-Fraenkel Set Theory, whose only non-logical symbol is $\in$. The logic of **CZF** is intuitionistic first order logic with equality. Among its non-logical axioms are *Extensionality*, *Pairing* and *Union* in their usual forms. **CZF** has additionally axiom schemata which we will now proceed to summarize.

Infinity: $\exists x \forall u \big[u \in x \leftrightarrow \big(\emptyset = u \vee, \exists \exists v \in x\ u = v + 1 \big) \big]$ where $v + 1 = v \cup \{v\}$.

Set Induction: $\forall x[\forall y \in x A(y) \rightarrow A(x)] \rightarrow \forall x A(x)$

Bounded Separation: $\forall a \exists b \forall x[x \in b \leftrightarrow x \in a \wedge A(x)]$

for all *bounded* formulae A. A set-theoretic formula is *bounded* or *restricted* if it is constructed from prime formulae using $\neg, \wedge, \vee, \exists, \rightarrow$, $\forall x \in y$ and $\exists x \in y$ only.

Strong Collection: For all formulae A,

$$\forall a \big[\forall x \in a \exists y A(x, y) \ \rightarrow\ \exists b\, [\forall x \in a\, \exists y \in b\, A(x, y) \wedge \forall y \in b\, \exists x \in a\, A(x, y)] \big].$$

Subset Collection: For all formulae B,

$$\forall a \forall b \exists c \forall u \big[\forall x \in a\, \exists y \in b\ B(x, y, u) \rightarrow$$
$$\exists d \in c\, [\forall x \in a\, \exists y \in d\ B(x, y, u) \wedge \forall y \in d\, \exists x \in a\, B(x, y, u)] \big].$$

The Powerset Axiom, **Pow**, is the following:

$$\forall x\, \exists y\, \forall z\, (z \subseteq x \rightarrow z \in y).$$

Remark 15.1. Subset Collection plays no role when we study $\mathbf{CZF}_{\mathcal{P}}$ since it is a consequence of **Pow** and the other axioms of **CZF**.

To save us work when proving realizability of the axioms of **CZF** it is useful to know that the axiom scheme of bounded separation can be deduced from a single instance (in the presence of strong collection).

Lemma 15.1. *Let* **Binary Intersection** *be the statement* $\forall x \forall y \exists z\, x \cap y = z$. *If* $\mathbf{CZF}_0$ *denotes* **CZF** *without bounded separation and subset collection, then every instance of bounded separation is provable in* $\mathbf{CZF}_0$ + Binary Intersection.

Proof. Aczel and Rathjen (2001, Proposition 4.8) is a forerunner of this result. It is proved in the above form in Aczel and Rathjen (2010, Corollary 9.5.7). □

15.2.2 Kripke–Platek Set Theory

A particularly interesting (classical) subtheory of **ZF** is Kripke–Platek set theory, **KP**. Its standard models are called *admissible sets*. One of the reasons that this is an important theory is that a great deal of set theory requires only the axioms of **KP**. An even more important reason is that admissible sets have been a major source of interaction between model theory, recursion theory and set theory (cf. Barwise 1975). **KP** arises from **ZF** by completely omitting the power set axiom and restricting separation and collection to bounded formulae. These alterations are suggested by the informal notion of 'predicative'. To be more precise, the axioms of **KP** consist of *Extensionality, Pair, Union, Infinity, Bounded Separation*

$$\exists x\, \forall u\, [u \in x \leftrightarrow (u \in a \, \wedge \, A(u))]$$

for all bounded formulae $A(u)$, *Bounded Collection*

$$\forall x \in a\, \exists y\, B(x, y) \rightarrow \exists z\, \forall x \in a\, \exists y \in z\, B(x, y)$$

for all bounded formulae $B(x, y)$, and *Set Induction*

$$\forall x\ [(\forall y \in x\, C(y)) \rightarrow C(x)] \rightarrow \forall x\, C(x)$$

for all formulae $C(x)$.

A transitive set A such that $(A, \in)$ is a model of **KP** is called an *admissible set*. Of particular interest are the models of **KP** formed by segments of Gödel's *constructible hierarchy* L. The constructible hierarchy is obtained by iterating the definable powerset operation through the ordinals

$$\begin{aligned} \mathrm{L}_0 &= \emptyset, \\ \mathrm{L}_\lambda &= \bigcup\{\mathrm{L}_\beta : \beta < \lambda\}\ \lambda \text{ limit} \\ \mathrm{L}_{\beta+1} &= \big\{X : \ X \subseteq \mathrm{L}_\beta;\ X \text{ definable over } \langle \mathrm{L}_\beta, \in\rangle\big\}. \end{aligned}$$

So any element of L of level α is definable from elements of L with levels $< \alpha$ and the parameter L_α. An ordinal α is *admissible* if the structure $(\mathrm{L}_\alpha, \in)$ is a model of **KP**.

Remark 15.2. Our system **KP** is not quite the same as the theory **KP** in Mathias' paper (Mathias 2001, p. 111). There **KP** does not have an axiom of Infinity and set induction only holds for Σ_1 formulae, or what amounts to the same, Π_1 foundation ($A \neq \emptyset \rightarrow \exists x \in A\ x \cap A = \emptyset$ for Π_1 classes A).

15.2.3 Power Kripke–Platek Set Theory

We use subset bounded quantifiers $\exists x \subseteq y \ \ldots$ and $\forall x \subseteq y \ \ldots$ as abbreviations for $\exists x(x \subseteq y \wedge \ldots)$ and $\forall x(x \subseteq y \to \ldots)$, respectively.

We call a formula of $\mathcal{L}_\in$ $\Delta_0^{\mathcal{P}}$ if all its quantifiers are of the form $Q\, x \subseteq y$ or $Q\, x{\in}y$ where Q is $\forall$ or $\exists$ and x and y are distinct variables.

Definition 15.1. The $\Delta_0^{\mathcal{P}}$ formulae are the smallest class of formulae containing the atomic formulae closed under $\wedge, \vee, \exists, \to, \neg$ and the quantifiers

$$\forall x \in a,\ \exists x \in a,\ \forall x \subseteq a,\ \exists x \subseteq a.$$

Definition 15.2. **KP**($\mathcal{P}$) has the same language as **ZF**. Its axioms are the following: Extensionality, Pairing, Union, Infinity, Powerset, $\Delta_0^{\mathcal{P}}$-Separation and $\Delta_0^{\mathcal{P}}$-Collection.

The transitive models of **KP**($\mathcal{P}$) have been termed **power admissible** sets in Friedman (1973b).

Remark 15.3. Alternatively, **KP**($\mathcal{P}$) can be obtained from **KP** by adding a function symbol $\mathcal{P}$ for the powerset function as a primitive symbol to the language and the axiom

$$\forall y\, [y \in \mathcal{P}(x) \leftrightarrow y \subseteq x]$$

and extending the schemes of Δ_0 Separation and Collection to the Δ_0 formulae of this new language.

Lemma 15.2. **KP**($\mathcal{P}$) *is* **not** *the same theory as* **KP** + **Pow**. *Indeed,* **KP** + **Pow** *is a much weaker theory than* **KP**($\mathcal{P}$) *in which one cannot prove the existence of* $V_{\omega+\omega}$.

Proof. Note that in the presence of full Separation and Infinity there is no difference between our system **KP** and Mathias's (2001) **KP**. It follows from Mathias (2001, Theorem 14) that **Z**+**KP**+**AC** is conservative over **Z**+**AC** for stratifiable sentences. **Z** and **Z**+**AC** are of the same proof-theoretic strength as the constructible hierarchy can be simulated in **Z**; a stronger statement is given in (Mathias, 2001, Theorem 16). As a result, **Z** and **Z** + **KP** are of the same strength. As **KP** + **Pow** is a subtheory of **Z** + **KP**, we have that **KP** + **Pow** is not stronger than **Z**. If **KP** + **Pow** could prove the existence of $V_{\omega+\omega}$ it would prove the consistency of **Z**. On the other hand **KP**($\mathcal{P}$) prove the existence of V_α for every ordinal α and hence proves the existence of arbitrarily large transitive models of Z. □

Remark 15.4. Our system **KP**($\mathcal{P}$) is not quite the same as the theory $\mathbf{KP}^{\mathcal{P}}$ in Mathias' paper (Mathias 2001, 6.10). The difference between **KP**($\mathcal{P}$) and $\mathbf{KP}^{\mathcal{P}}$ is that in the latter system set induction only holds for $\Sigma_1^{\mathcal{P}}$ formulae, or what amounts to the same, $\Pi_1^{\mathcal{P}}$ foundation ($A \neq \emptyset \to \exists x \in A\ x \cap A = \emptyset$ for $\Pi_1^{\mathcal{P}}$ classes A).

15.2.4 Extended *E*-Recursive Functions

We would like to have unlimited application of sets to sets, i.e. we would like to assign a meaning to the symbol $[a](x)$ where a and x are sets. In generalized recursion theory this is known as *E-recursion* or *set recursion* (see, e.g., Normann 1978 or Sacks 1990, Chap. X). However, we shall introduce an extended notion of E-computability, christened $E_{\wp}$-computability, rendering the functions $exp(a, b) = {}^{a}b$ and $\mathcal{P}(x) = \{u \mid u \subseteq x\}$ computable as well, (where ${}^{a}b$ denotes the set of all functions from a to b). Moreover, the constant function with value ω is taken as an initial function in $E_{\wp}$-computability. $E_{\wp}$-computability is closely related to power recursion, where the power set operation is regarded to be an initial function. The latter notion has been studied by Moschovakis (1976) and Moss (1995).

There is a lot of leeway in setting up $E_{\wp}$-recursion. The particular schemes we use are especially germane to our situation. Our construction will provide a specific set-theoretic model for the elementary theory of operations and numbers **EON** (see, e.g., Beeson 1985, VI.2, or the theory **APP** as described in Troelstra and van Dalen 1988, Chap. 9, Sect. 3). We utilize encoding of finite sequences of sets by the usual pairing function $\langle\, ,\rangle$ with $\langle x, y\rangle = \{\{x\}, \{x, y\}\}$, letting $\langle x\rangle = x$ and $\langle x_1, \ldots, x_n, x_{n+1}\rangle = \langle\langle x_1, \ldots, x_n\rangle, x_{n+1}\rangle$. We use functions $()_0$ and $()_1$ to retrieve the left and right components, respectively, of an ordered pair $a = \langle x, y\rangle$, i.e., $(a)_0 = x$ and $(a)_1 = y$.

Below we use the notation $[x](y)$ rather than the more traditional $\{x\}(y)$ to avoid any ambiguity with the singleton set $\{x\}$.

Definition 15.3. (**CZF**$_{\mathcal{P}}$, **KP**($\mathcal{P}$)) First, we select distinct non-zero natural numbers $\mathbf{k}$, $\mathbf{s}$, $\mathbf{p}$, $\mathbf{p_0}$, $\mathbf{p_1}$, $\mathbf{s_N}$, $\mathbf{p_N}$, $\mathbf{d_N}$, $\bar{\mathbf{0}}$, $\bar{\boldsymbol{\omega}}$, $\boldsymbol{\gamma}$, $\boldsymbol{\rho}$, $\boldsymbol{v}$, $\boldsymbol{\pi}$, $\boldsymbol{i}_1$, $\boldsymbol{i}_2$, $\boldsymbol{i}_3$, and $\bar{\wp}$ which will provide indices for special $E_{\wp}$-recursive partial (class) functions. Inductively we shall define a class $\mathbb{E}$ of triples $\langle e, x, y\rangle$. Rather than "$\langle e, x, y\rangle \in \mathbb{E}$", we shall write "$[e](x) \simeq y$", and moreover, if $n > 0$, we shall use $[e](x_1, \ldots, x_n) \simeq y$ to convey that

$$[e](x_1) \simeq \langle e, x_1\rangle \wedge [\langle e, x_1\rangle](x_2) \simeq \langle e, x_1, x_2\rangle \wedge \ldots \wedge [\langle e, x_1, \ldots, x_{n-1}\rangle](x_n) \simeq y.$$

We shall say that $[e](x)$ is defined, written $[e](x)\downarrow$, if $[e](x) \simeq y$ for some y. Let $\mathbb{N} := \omega$. $\mathbb{E}$ is defined by the following clauses:

$$
\begin{aligned}
[\mathbf{k}](x, y) &\simeq x \\
[\mathbf{s}](x, y, z) &\simeq [[x](z)]([y](z)) \\
[\mathbf{p}](x, y) &\simeq \langle x, y\rangle \\
[\mathbf{p_0}](x) &\simeq (x)_0 \\
[\mathbf{p_1}](x) &\simeq (x)_1 \\
[\mathbf{s_N}](n) &\simeq n + 1 \text{ if } n \in \mathbb{N}
\end{aligned}
$$

$$
\begin{aligned}
[\mathbf{p_N}](0) &\simeq 0\\
[\mathbf{p_N}](n+1) &\simeq n \text{ if } n \in \mathbb{N}\\
[\mathbf{d_N}](n,m,x,y) &\simeq x \text{ if } n,m \in \mathbb{N} \text{ and } n = m\\
[\mathbf{d_N}](n,m,x,y) &\simeq y \text{ if } n,m \in \mathbb{N} \text{ and } n \neq m\\
[\bar{\mathbf{0}}](x) &\simeq 0\\
[\bar{\boldsymbol{\omega}}](x) &\simeq \omega\\
[\boldsymbol{\pi}](x,y) &\simeq \{x,y\}\\
[\boldsymbol{\upsilon}](x) &\simeq \bigcup x\\
[\boldsymbol{\gamma}](x,y) &\simeq x \cap (\bigcap y)\\
[\boldsymbol{\rho}](x,y) &\simeq \{[x](u) \mid u \in y\} \text{ if } [x](u) \text{ is defined for all } u \in y\\
[\boldsymbol{i_1}](x,y,z) &\simeq \{u \in x \mid y \in z\}\\
[\boldsymbol{i_2}](x,y,z) &\simeq \{u \in x \mid u \in y \to u \in z\}\\
[\boldsymbol{i_3}](x,y,z) &\simeq \{u \in x \mid u \in y \to z \in u\}\\
[\bar{\wp}](x) &\simeq \mathcal{P}(x).
\end{aligned}
$$

Note that $[\mathbf{s}](x,y,z)$ is not defined unless $[x](z)$, $[y](z)$ and $[[x](z)]([y](z))$ are already defined. The clause for $\mathbf{s}$ is thus to be read as a conjunction of the following clauses: $[\mathbf{s}](x) \simeq \langle \mathbf{s}, x\rangle$, $[\langle \mathbf{s}, x\rangle](y) \simeq \langle \mathbf{s}, x, y\rangle$ and, if there exist a, b, c such that $[x](z) \simeq a$, $[y](z) \simeq b$, $[a](b) \simeq c$, then $[\langle \mathbf{s}, x, y\rangle](z) \simeq c$. Similar restrictions apply to $\boldsymbol{\rho}$.

Lemma 15.3. ($\mathbf{CZF}_{\mathcal{P}}$, $\mathbf{IKP}(\mathcal{P})$) *$\mathbb{E}$ is an inductively defined class and $\mathbb{E}$ is functional in that for all e, x, y, y',*

$$\langle e, x, y\rangle \in \mathbb{E} \wedge \langle e, x, y'\rangle \in \mathbb{E} \Rightarrow y = y'.$$

Proof. The inductive definition of $\mathbb{E}$ falls under the heading of Aczel and Rathjen (2001, Theorem 11.4). If $[e](x) \simeq y$ the uniqueness of y follows by induction on the stages (see Aczel and Rathjen 2001, Lemma 5.2) of that inductive definition. □

Definition 15.4. *Application terms* are defined inductively as follows:

(i) The constants The constants $\mathbf{k}$, $\mathbf{s}$, $\mathbf{p}$, $\mathbf{p_0}$, $\mathbf{p_1}$, $\mathbf{s_N}$, $\mathbf{p_N}$, $\mathbf{d_N}$, $\bar{\mathbf{0}}$, $\bar{\boldsymbol{\omega}}$, $\boldsymbol{\gamma}$, $\boldsymbol{\rho}$, $\boldsymbol{\upsilon}$, $\boldsymbol{\pi}$, $\boldsymbol{i_1}$, $\boldsymbol{i_2}$ $\boldsymbol{i_3}$, and $\bar{\wp}$ singled out in Definition 15.3 are *application terms*;
(ii) Variables are *application terms*;
(iii) If s and t are *application terms* then (st) is an *application term*.

Definition 15.5. Application terms are easily formalized in $\mathbf{CZF}_{\mathcal{P}}$. However, rather than translating application terms into the set – theoretic language of $\mathbf{CZF}_{\mathcal{P}}$, we define the translation of expressions of the form $t \simeq u$, where t is an application term and u is a variable. The translation proceeds along the way that t was built up:

$$[c \simeq u]^\wedge \quad \text{is} \quad c = u \text{ if } c \text{ is a constant or a variable;}$$
$$[(st) \simeq u]^\wedge \quad \text{is} \quad \exists x \exists y([s \simeq x]^\wedge \wedge [t \simeq y]^\wedge \wedge \langle x, y, u\rangle \in \mathbb{E}).$$

Abbreviations. For application terms s, t, $t_1, \ldots, t_n$ we will use:

$s(t_1, \ldots, t_n)$ as a shortcut for $((\ldots(st_1)\ldots)t_n)$; *(parentheses associated to the left)*;
$st_1 \ldots t_n$ as a shortcut for $s(t_1, \ldots, t_n)$;
$t\downarrow$ as a shortcut for $\exists x(t \simeq x)^\wedge$; *($t$ is defined)*
$(s \simeq t)^\wedge$ as a shortcut for $(s\downarrow \vee t\downarrow) \rightarrow \exists x((s \simeq x)^\wedge \wedge (t \simeq x)^\wedge)$.

A *closed* application term is an application term that does not contain variables. If t is a closed application term and $a_1, \ldots, a_n, b$ are sets we use the abbreviation

$$t(a_1, \ldots, a_n) \simeq b \quad \text{for } \exists x_1 \ldots x_n \exists y \big(x_1 = a_1 \wedge \ldots \wedge x_n = a_n \wedge y = b \wedge [t(x_1, \ldots, x_n) \simeq y]^\wedge\big).$$

Definition 15.6. Every closed application term gives rise to a partial class function. A partial n-place (class) function Υ is said to be an *$E_\wp$-recursive partial function* if there exists a closed application term t_Υ such that

$$\mathrm{dom}(\Upsilon) = \{(a_1, \ldots, a_n) \mid t_\Upsilon(a_1, \ldots, a_n) \downarrow\}$$

and for all for all sets $(a_1, \ldots, a_n) \in \mathrm{dom}(\Upsilon)$,

$$t_\Upsilon(a_1, \ldots, a_n) \simeq \Upsilon(a_1, \ldots, a_n).$$

In the latter case, t_Υ is said to be an *index* for Υ.

If Υ_1, Υ_2 are $E_\wp$-recursive partial functions, then $\Upsilon_1(\mathbf{a}) \simeq \Upsilon_2(\mathbf{a})$ iff neither $\Upsilon_1(\mathbf{a})$ nor $\Upsilon_2(\mathbf{a})$ are defined, or $\Upsilon_1(\mathbf{a})$ and $\Upsilon_2(\mathbf{a})$ are defined and equal.

The next two results can be proved in the theory **APP** and thus hold true in any applicative structure. Thence the particular applicative structure considered here satisfies the Abstraction Lemma and Recursion Theorem (see e.g. Feferman 1979 or Beeson 1985).

Lemma 15.4 (Abstraction Lemma, cf. Beeson 1985, VI.2.2). *For every application term $t[x]$ there exists an application term $\lambda x.t[x]$ with* $\mathrm{FV}(\lambda x.t[x]) := \{x_1, \ldots, x_n\} \subseteq \mathrm{FV}(t[x])\backslash\{x\}$ *such that the following holds:*

$$\forall x_1 \ldots \forall x_n(\lambda x.t[x]\downarrow \wedge \forall y\,(\lambda x.t[x])y \simeq t[y]).$$

Proof. (i) $\lambda x.x$ is **skk**;
(ii) $\lambda x.t$ is $\mathbf{k}t$ for t a constant or a variable other than x;
(iii) $\lambda x.uv$ is $\big(\mathbf{s}(\lambda x.u)\big)(\lambda x.v)$. □

Lemma 15.5 (Recursion Theorem, cf. Beeson 1985, VI.2.7). *There exists a closed application term* rec *such that for any* f, x,

$$\mathsf{rec} f \downarrow \;\wedge\; \mathsf{rec} f x \simeq f(\mathsf{rec} f)x.$$

Proof. Take rec to be $\lambda f.tt$, where t is $\lambda y\lambda x.f(yy)x$. □

Corollary 15.1. *For any* $E_{\wp}$*-recursive partial function* Υ *there exists a closed application term* τ_{fix} *such that* $\tau_{fix} \downarrow$ *and for all* $\mathbf{a}$,

$$\Upsilon(\bar{e}, \mathbf{a}) \simeq \tau_{fix}(\mathbf{a}),$$

where $\tau_{fix} \simeq \bar{e}$. *Moreover,* τ_{fix} *can be effectively (e.g. primitive recursively) constructed from an index for* Υ.

15.3 Defining Realizability with Sets of Witnesses for Set Theory

Realizability semantics are a crucial tool in the study of intuitionistic theories (see Troelstra 1998, Rathjen 2006b). We introduce a form of realizability based on general set recursive functions where a realizer for an existential statement provides a set of witnesses for the existential quantifier rather than a single witness. Realizability based on indices of general set recursive functions was introduced in Rathjen (2006c) and employed to prove, inter alia, metamathematical properties for **CZF** augmented by strong forms of the axiom of choice in Rathjen and Tupailo (2006, Theorems 8.3 and 8.4). There are points of contact with a notion of realizability used by Tharp (1971) who employed (indices of) Σ_1 definable partial (class) functions as realizers, though there are important differences, too, as Tharp works in a classical context and assumes a definable search operation on the universe which basically amounts to working under the hypothesis $V = L$. Moreover, there are connections with Lifschitz' realizability (Lifschitz 1979) where a realizer for an existential arithmetical statement provides a finite co-recursive set of witnesses (see van Oosten 1990; Chen and Rathjen 2012 for extensions to analysis and set theory).

We adopt the conventions and notations from the previous section. However, we prefer to write $\jmath_0 e$ and $\jmath_1 e$ rather than $(e)_0$ and $(e)_1$, respectively, and instead of $[a](b) \simeq c$ we shall write $a \bullet b \simeq c$.

Definition 15.7. Bounded quantifiers will be treated as quantifiers in their own right, i.e., bounded and unbounded quantifiers are treated as syntactically different kinds of quantifiers.

We use the expression $a \neq \emptyset$ to convey that the set a is inhabited, that is $\exists x\, x \in a$.

We define a relation $a \Vdash_{\mathfrak{w}} B$ between sets a and formulae of set theory. $a \bullet f \Vdash_{\mathfrak{w}} B$ will be an abbreviation for $\exists x[a \bullet f \simeq x \wedge x \Vdash_{\mathfrak{w}} B]$.

$$
\begin{aligned}
a \Vdash_{\mathfrak{w}} A &\text{ iff } A \text{ holds for atomic formulae } A\\
a \Vdash_{\mathfrak{w}} A \wedge B &\text{ iff } \jmath_0 a \Vdash_{\mathfrak{w}} A \ \wedge \ \jmath_1 a \Vdash_{\mathfrak{w}} B\\
a \Vdash_{\mathfrak{w}} A \vee, \exists B &\text{ iff } a \neq \emptyset \wedge (\forall d \in a)([\jmath_0 d = 0 \wedge \jmath_1 d \Vdash_{\mathfrak{w}} A] \vee, \exists\\
&\qquad [\jmath_0 d = 1 \wedge \jmath_1 d \Vdash_{\mathfrak{w}} B])\\
a \Vdash_{\mathfrak{w}} \neg A &\text{ iff } \forall c \, \neg c \Vdash_{\mathfrak{w}} A\\
a \Vdash_{\mathfrak{w}} A \to B &\text{ iff } \forall c \left[c \Vdash_{\mathfrak{w}} A \ \to \ a \bullet c \Vdash_{\mathfrak{w}} B\right]\\
a \Vdash_{\mathfrak{w}} (\forall x \in b)\, A &\text{ iff } (\forall c \in b)\; a \bullet c \Vdash_{\mathfrak{w}} A[x/c]\\
a \Vdash_{\mathfrak{w}} (\exists x \in b)\, A &\text{ iff } a \neq \emptyset \wedge (\forall d \in a)[\jmath_0 d \in b \wedge \jmath_1 d \Vdash_{\mathfrak{w}} A[x/\jmath_0 d]]\\
a \Vdash_{\mathfrak{w}} \forall x A &\text{ iff } \forall c \; a \bullet c \Vdash_{\mathfrak{w}} A[x/c]\\
a \Vdash_{\mathfrak{w}} \exists x A &\text{ iff } a \neq \emptyset \wedge (\forall d \in a)\; \jmath_1 d \Vdash_{\mathfrak{w}} A[x/\jmath_0 d]\\
\Vdash_{\mathfrak{w}} B &\text{ iff } \exists a\, a \Vdash_{\mathfrak{w}} B.
\end{aligned}
$$

In the course of proving that certain formulae are realized, e.g.

$$(A \vee, \exists B) \to [(A \to C) \to ((B \to C) \to C)]$$

or the rule for introducing an existential quantifier in the antecendent of an implication, we will be faced with the problem that we have a non-empty set of realizers where a single realizer is required. The next Lemma shows that we can effectively pass from a set of realizers to a single realizer.

Lemma 15.6. *Let* $\mathbf{x} = x_1, \ldots, x_r$ *and* $\mathbf{a} = a_1, \ldots, a_r$. *To each formula* $A(\mathbf{x})$ *of* **CZF** *(with all free variables among* $\mathbf{x}$*) we can effectively assign (a code of) an* $E_{\wp}$*-recursive partial function* χ_A *such that*

$$\mathbf{IKP}(\mathcal{P}) \vdash \forall \mathbf{a} \forall c \neq \emptyset\, [(\forall d \in c)\, d \Vdash_{\mathfrak{w}} A(\mathbf{a})) \to \chi_A(\mathbf{a}, c) \Vdash_{\mathfrak{w}} A(\mathbf{a})].$$

Proof. We use induction on the buildup of A.

If A is atomic, let $\chi_A(\mathbf{a}, c) := 0$.

Let $A(\mathbf{x})$ be $B(\mathbf{x}) \wedge C(\mathbf{x})$ and χ_B and χ_C be already defined. Then

$$\chi_A(\mathbf{a}, c) := \jmath(\chi_B(\mathbf{a}, \{\jmath_0 x \mid x \in c\}), \chi_C(\mathbf{a}, \{\jmath_1 x \mid x \in c\}))$$

will do the job.

Let $A(\mathbf{x})$ be $B(\mathbf{x}) \to C(\mathbf{x})$ and suppose χ_B and χ_C have already been defined. Assume that $c \neq \emptyset$ and $(\forall d \in c)\, d \Vdash_{\mathfrak{w}} [B(\mathbf{a}) \to C(\mathbf{a})]$. Suppose $e \Vdash_{\mathfrak{w}} B(\mathbf{a})$. Define the $E_{\wp}$-recursive partial function ϑ by

$$\vartheta(c, e) \simeq \{d \bullet e \mid d \in c\}.$$

Then $\vartheta(c, e) \neq \emptyset$ and hence, by the inductive assumption, $\chi_C(\mathbf{a}, \vartheta(c, e)) \Vdash_{\mathfrak{w}} C(\mathbf{a})$, so that

$$\lambda e.\chi_C(\mathbf{a}, \vartheta(c, e)) \Vdash_{\mathfrak{w}} A(\mathbf{a}).$$

Now let $A(\mathbf{x})$ be of the form $\forall y\, B(\mathbf{x}, y)$. Suppose that $c \neq \emptyset$ and $(\forall d \in c)\, d \Vdash_{\mathfrak{w}} A(\mathbf{a})$. Fixing b, we then have $(\forall d \in c)\, d \bullet b \Vdash_{\mathfrak{w}} B(\mathbf{a}, b)$, thus, $\forall d' \in \vartheta(c, b)\; d' \Vdash_{\mathfrak{w}} B(\mathbf{a}, b)$, and therefore, by the inductive assumption, $\chi_B(\mathbf{a}, \vartheta(c, b)) \Vdash_{\mathfrak{w}} B(\mathbf{a}, b)$. As a result

$$\lambda b.\chi_B(\mathbf{a}, \vartheta(c, b)) \Vdash_{\mathfrak{w}} A(\mathbf{a}).$$

The case of $A(\mathbf{x})$ starting with a bounded universal quantifier is similar to the previous case.

In all the remaining cases, $\chi_A(\mathbf{a}, c) := \bigcup c$ will work owing to the definition of realizability in these cases. □

Lemma 15.7 (IKP($\mathcal{P}$)). *Realizers for equality laws:*

(i) $0 \Vdash_{\mathfrak{w}} x = x$.
(ii) $\lambda u.u \Vdash_{\mathfrak{w}} x = y \to y = x$.
(iii) $\lambda u.u \Vdash_{\mathfrak{w}} (x = y \wedge y = z) \to x = z$.
(iv) $\lambda u.u \Vdash_{\mathfrak{w}} (x = y \wedge y \in z) \to x \in z$.
(v) $\lambda u.u \Vdash_{\mathfrak{w}} (x = y \wedge z \in x) \to z \in y$.
(vi) $\lambda u.\jmath_1 u \Vdash_{\mathfrak{w}} (x = y \wedge A(x)) \to A(y)$ *for any formula* A.

Proof. (i)–(v) are obvious. (vi) follows by a trivial induction on the buildup of A. □

Lemma 15.8 (IKP($\mathcal{P}$)). *Realizers for logical axioms: Below we use the $E_{\wp}$-recursive function* $\mathfrak{sg}(a) := \{a\}$.

(IPL1) $\mathbf{k} \Vdash_{\mathfrak{w}} A \to (B \to A)$.
(IPL2) $\mathbf{s} \Vdash_{\mathfrak{w}} [A \to (B \to C)] \to [(A \to B) \to (A \to C)]$.
(IPL3) $\lambda e.\lambda d.\jmath(e, d)\rangle \Vdash_{\mathfrak{w}} A \to (B \to A \wedge B)$.
(IPL4) $\lambda e.\jmath_0 \Vdash_{\mathfrak{w}} A \wedge B \to A$.
(IPL5) $\lambda e.\jmath_1 e \Vdash_{\mathfrak{w}} A \wedge B \to B$.
(IPL6) $\lambda e.\mathfrak{sg}(\jmath(0, e)) \Vdash_{\mathfrak{w}} A \to A\vee, \exists B$.
(IPL7) $\lambda e.\mathfrak{sg}(\jmath(1, e))\rangle \Vdash_{\mathfrak{w}} B \to A\vee, \exists B$.
(IPL8) $\mathfrak{k}(\mathbf{a}) \Vdash_{\mathfrak{w}} (A\vee, \exists B) \to [(A \to C) \to ((B \to C) \to C)]$, *for some $E_{\wp}$-recursive partial function* $\mathfrak{k}$, *where* $\mathbf{a}$ *comprises all parameters appearing in the formula.*

(IPL9) $\lambda e.\lambda d.0 \Vdash_{\mathfrak{w}} (A \to B) \to ((A \to \neg B) \to \neg A)$.
(IPL10) $\lambda e.0 \Vdash_{\mathfrak{w}} A \to (\neg A \to B)$.
(IPL11) $\lambda e.e \bullet b \Vdash_{\mathfrak{w}} \forall x A(x) \to A(b)$.
(IPL12) $\lambda e.\mathfrak{sg}(e) \Vdash_{\mathfrak{w}} A(a) \to \exists x A(x)$.

Proof. As for IPL1 and IPL2, this justifies the combinators **s** and **k**. Combinatory completeness of these two combinators is equivalent to the fact that these two laws together with modus ponens generate the full set of theorems of propositional implicational intuitionistic logic.

Except for IPL8, one easily checks that the proposed realizers indeed realize the pertaining formulae.

So let's check IPL8. $A\vee,\exists B \to ((A \to C) \to ((B \to C) \to C))$. Suppose $e \Vdash_{\mathfrak{w}} A\vee,\exists B$. Then $e \neq \emptyset$. Let $d \in e$. Then $\jmath_0 d = 0 \wedge \jmath_1 d \Vdash_{\mathfrak{w}} A$ or $\jmath_0 d = 1 \wedge \jmath_1 d \Vdash_{\mathfrak{w}} B$. Suppose $f \Vdash_{\mathfrak{w}} A \to C$ and $g \Vdash_{\mathfrak{w}} B \to C$. Define an $E_{\wp}$-recursive partial function $\mathfrak{f}$ by

$$\mathfrak{f}(d', f', g') = [\mathbf{d_N}](\jmath_0 d', 0, f' \bullet (\jmath_1 d'), g' \bullet (\jmath_1 d')).$$

Then

$$\mathfrak{f}(d', f', g') = \begin{cases} f' \bullet (\jmath_1 d') & \text{if } \jmath_0 d' = 0 \\ g' \bullet (\jmath_1 d') & \text{if } \jmath_0 d' = 1 \end{cases}$$

As a result, $\mathfrak{f}(d, f, g) \Vdash_{\mathfrak{w}} C$ and hence $\lambda f.\lambda g.\mathfrak{f}(d, f, g) \Vdash_{\mathfrak{w}} (A \to C) \to ((B \to C) \to C)$. Thus, $\Phi(e, \lambda d.\lambda f.\lambda g.\mathfrak{f}(d, f, g)) \neq \emptyset$ and for all $p \in \Phi(e, \lambda d.\lambda f.\lambda g.\mathfrak{f}(d, f, g))$ we have

$$p \Vdash_{\mathfrak{w}} (A \to C) \to ((B \to C) \to C).$$

Let $E(\mathbf{a}) := (A \to C) \to ((B \to C) \to C)$, where $\mathbf{a}$ comprises all parameters appearing in the formula on the right hand side. The upshot is that by Lemma 15.6 we can conclude

$$\chi_E(\mathbf{a}, \Phi(e, \lambda d.\lambda f.\lambda g.\mathfrak{f}(d, f, g))) \Vdash_{\mathfrak{w}} E(\mathbf{a}).$$

And consequently we have

$$\mathfrak{k}(\mathbf{a}) := \lambda e.\chi_E(\mathbf{a}, \Phi(e, \lambda d.\lambda f.\lambda g.\mathfrak{f}(d, f, g))) \Vdash_{\mathfrak{w}} A\vee,\exists B \to E(\mathbf{a}).$$ □

Theorem 15.2. *Let $D(u_1, \ldots, u_r)$ be a formula of $\mathcal{L}_\in$ all of whose free variables are among $u_1, \ldots, u_r$. If*

$$\mathbf{CZF} + \mathbf{Pow} \vdash D(u_1, \ldots, u_r),$$

then one can effectively construct an index of an $E_{\wp}$-recursive function g such that

$$\mathbf{IKP}(\mathcal{P}) \vdash \forall a_1, \ldots, a_r\; g(a_1, \ldots, a_r) \Vdash_{\mathfrak{w}} D(a_1, \ldots, a_r).$$

Proof. We use a standard Hilbert-type systems for intuitionistic predicate logic. The proof proceeds by induction on the derivation. For the logical axioms and the equality axioms we have already produced appropriate $E_\wp$-recursive functions in Lemmata 15.7 and 15.8. It remains to deal with logical inferences and set-theoretic axioms. We start with the rules.

The only rule from propositional logic is modus ponens. Suppose that we have $E_\wp$-recursive functions g_0 and g_1 such that for all $\mathbf{a}$, $g_0(\mathbf{a}) \Vdash_{\mathfrak{w}} A(\mathbf{a}) \to B(\mathbf{a})$ and $g_1(\mathbf{a}) \Vdash_{\mathfrak{w}} A(\mathbf{a})$. Then $g(\mathbf{a}) \Vdash_{\mathfrak{w}} B(\mathbf{a})$ holds with the $E_\wp$-recursive function $g(\mathbf{a}) := g_0(\mathbf{a}) \bullet g_1(\mathbf{a})$.

For the $\forall$ quantifier we have the rule: from $B(\mathbf{u}) \to A(x, \mathbf{u})$ infer $B(\mathbf{u}) \to \forall x A(x, \mathbf{u})$ if x is not free in $B(\mathbf{u})$. Inductively we have an $E_\wp$-recursive function $\mathfrak{h}$ such that for all $b, \mathbf{a}$,

$$\mathfrak{h}(b, \mathbf{a}) \Vdash_{\mathfrak{w}} B(\mathbf{a}) \to A(b, \mathbf{a}).$$

Suppose $d \Vdash_{\mathfrak{w}} B(\mathbf{a})$. Then $\mathfrak{h}(b, \mathbf{a}) \bullet d \Vdash_{\mathfrak{w}} A(b, \mathbf{a})$ holds for all b, whence $\lambda x.(\mathfrak{h}(x, \mathbf{a}) \bullet d) \Vdash_{\mathfrak{w}} \forall x A(x, \mathbf{a})$. As a result,

$$\lambda d.\lambda x.(\mathfrak{h}(x, \mathbf{a}) \bullet d) \Vdash_{\mathfrak{w}} B(\mathbf{a}) \to \forall x A(x, \mathbf{a}).$$

For the $\exists$ quantifier we have the rule: from $A(x, \mathbf{u}) \to B(\mathbf{u})$ infer $\exists x\, A(x, \mathbf{u}) \to B(\mathbf{u})$ if x is not free in $B(\mathbf{u})$. Inductively we then have an $E_\wp$-recursive function $\mathfrak{g}$ such that for all $b, \mathbf{a}$,

$$\mathfrak{g}(b, \mathbf{a}) \Vdash_{\mathfrak{w}} A(b, \mathbf{a}) \to B(\mathbf{a}).$$

Suppose $e \Vdash_{\mathfrak{w}} \exists x A(x, \mathbf{a})$. Then $e \neq \emptyset$ and for all $d \in e$, $\jmath_1 d \Vdash_{\mathfrak{w}} A(\jmath_0 d, \mathbf{a})$. Consequently, $(\forall d \in e)\, \mathfrak{g}(\jmath_0 d, \mathbf{a}) \bullet \jmath_1 d \Vdash_{\mathfrak{w}} B(\mathbf{a})$. We then have $\Phi(e, \lambda d.\mathfrak{g}(\jmath_0 d, \mathbf{a}) \bullet \jmath_1 d) \neq \emptyset$ and

$$(\forall y \in \Phi(e, \lambda d.\mathfrak{g}(\jmath_0 d, \mathbf{a}) \bullet \jmath_1 d)\; y \Vdash_{\mathfrak{w}} B(\mathbf{a})\,.$$

Using Lemma 15.6 we arrive at $\chi_B(\mathbf{a}, \Phi(e, \lambda d.\mathfrak{g}(\jmath_0 d, \mathbf{a}) \bullet \jmath_1 d)) \Vdash_{\mathfrak{w}} B(\mathbf{a})$; whence

$$\lambda e.\chi_B(\mathbf{a}, \Phi(e, \lambda d.\mathfrak{g}(\jmath_0 d, \mathbf{a}) \bullet \jmath_1 d)) \Vdash_{\mathfrak{w}} \exists x A(x, \mathbf{a}) \to B(\mathbf{a})\,.$$

Next we show that every axiom of **CZF** + **Pow** is realized by an $E_\wp$-recursive function. We treat the axioms one after the other.

(Extensionality): Since $e \Vdash_{\mathfrak{w}} \forall x(x \in a \leftrightarrow x \in b)$ implies $a = b$, and hence $0 \Vdash_{\mathfrak{w}} a = b$, it follows that

$$\lambda u.0 \Vdash_{\mathfrak{w}} [\forall x(x \in a \leftrightarrow x \in b) \to a = b].$$

(Pair): There is an $E_\wp$-recursive function ℓ such that

$$\ell(a,b,c) := \{\jmath(0,a) \mid c = a\} \cup \{\jmath(1,b) \mid c = b\}.$$

We have $\forall u \in \{a,b\}\ \ell(a,b,u) \Vdash_{\mathfrak{w}} (u = a \vee, \exists u = b)$ and hence, letting $c := \{a,b\}$,

$$\lambda u.\ell(a,b,u) \Vdash_{\mathfrak{w}} \forall x \in c(x = u \vee, \exists x = b)\,.$$

We also have $\jmath(0,0) \Vdash_{\mathfrak{w}} (a \in c \wedge b \in c)$, so that

$$\jmath(\lambda u.\ell(a,b,u), \jmath(0,0)) \Vdash_{\mathfrak{w}} \forall x \in c(x = a \vee, \exists x = b) \wedge (a \in c \wedge b \in c).$$

Thus we arrive at

$$\mathfrak{sg}(\jmath(\mathfrak{p}(a,b), \jmath(\lambda u.\ell(a,b,u), \jmath(0,0))) \Vdash_{\mathfrak{w}}$$
$$\exists y\, [\forall x \in y(x = a \vee, \exists x = b) \wedge (a \in y \wedge b \in y)].$$

(Union): Let ℓ_U be the $E_\wp$-recursive function defined by

$$\ell_U(a,u) = \{\jmath(x, \jmath(0,0)) \mid x \in a \wedge u \in x\}.$$

For $u \in \bigcup a$ we then have $\ell_U(a,u) \Vdash_{\mathfrak{w}} \exists x \in a\ u \in x$, and therefore

$$\lambda u.\ell_U(u,a) \Vdash_{\mathfrak{w}} (\forall u \in \bigcup a)(\exists x \in a)\ u \in x.$$

Obviously $\lambda u.\lambda v.0 \Vdash_{\mathfrak{w}} (\forall x \in a)(\forall y \in x)\ y \in \bigcup a$. Therefore we have

$$\mathfrak{sg}(\jmath(\bigcup a, \jmath(\lambda u.\ell_U(u,a), \lambda u.\lambda v.0))) \Vdash_{\mathfrak{w}}$$
$$\exists w\, [(\forall u \in w)(\exists x \in a)\ u \in x \wedge (\forall x \in a)(\forall y \in x)\ y \in w].$$

(Empty Set): Obviously $\mathfrak{sg}(\jmath(\emptyset, \lambda v.0)) \Vdash_{\mathfrak{w}} \exists x\, (\forall u \in x) u \neq u$.

(Binary Intersection): Let $c := a \cap b$. As

$$\lambda v.\jmath(0,0) \Vdash_{\mathfrak{w}} \forall x \in c\, (x \in a \wedge x \in b)$$

and $\lambda u.0 \Vdash_{\mathfrak{w}} \forall x\, (x \in a \wedge x \in b \to x \in c)$ hold, we conclude that

$$\mathfrak{sg}(\jmath(a \cap b, \jmath(\lambda v.\jmath(0,0), \lambda u.0))) \Vdash_{\mathfrak{w}} \exists y\, [\forall x \in y\, (x \in a \wedge x \in b) \wedge$$
$$\forall x\, (x \in a \wedge x \in b \to x \in y)].$$

(Powerset): It suffices to find a realizer for the formula

$$\exists y\, \forall x(x \subseteq a \to x \in y)]$$

since realizability of the power set axiom follows then with the help of Δ_0 Separation. One easily verifies that $e \Vdash_{\mathfrak{w}} \forall u(u \in b \to u \in a)$ implies $b \subseteq a$ and consequently $b \in \mathcal{P}(a)$. Therefore we have

$$\lambda u.\lambda v.0 \Vdash_{\mathfrak{w}} \forall x[x \subseteq a \to x \in \mathcal{P}(a)],$$

thus $\mathfrak{sg}(\jmath(\mathcal{P}(a), \lambda u.\lambda v.0)) \Vdash_{\mathfrak{w}} \exists y\, \forall x[x \subseteq a \to x \in y]$.

(Set Induction): Suppose $e \Vdash_{\mathfrak{w}} \forall x[\forall y(y \in x \to A(y)) \to A(x)]$. Then, for all a,

$$e \bullet a \Vdash_{\mathfrak{w}} [\forall y(y \in a \to A(y)) \to A(a)]\,.$$

Suppose we have an index e^* such that for all $b \in a$, $e^* \bullet b \Vdash_{\mathfrak{w}} A(b)$. As $v \Vdash_{\mathfrak{w}} b \in a$ entails $b \in a$, we get

$$\lambda u.\lambda v.e^* \bullet u \Vdash_{\mathfrak{w}} \forall y(y \in a \to A(y)),$$

and hence

$$(e \bullet a) \bullet (\lambda u.\lambda v.e^* \bullet u) \Vdash_{\mathfrak{w}} A(a). \tag{15.1}$$

By the recursion theorem we can effectively cook up an index $\mathfrak{q}$ such that

$$(\mathfrak{q} \bullet e) \bullet a \simeq (e \bullet a) \bullet (\lambda u.\lambda v.(\mathfrak{q} \bullet e) \bullet u).$$

In view of the above it follows by set induction that for all a, $(\mathfrak{q} \bullet e) \bullet a \downarrow$ and $(\mathfrak{q} \bullet e) \bullet a \Vdash_{\mathfrak{w}} A(a)$. As a result we have $\lambda w.(\mathfrak{q} \bullet e) \bullet w \Vdash_{\mathfrak{w}} \forall x\, A(x)$, yielding

$$\lambda e \lambda w.(\mathfrak{q} \bullet e) \bullet w \Vdash_{\mathfrak{w}} \forall x[\forall y(y \in x \to A(y)) \to A(x)] \to \forall x\, A(x).$$

(Strong Collection): Suppose

$$e \Vdash_{\mathfrak{w}} \forall u(u \in a \to \exists y\, B(u, y)). \tag{15.2}$$

Then we have, for all $b \in a$, $(e \bullet b) \bullet 0 \Vdash_{\mathfrak{w}} \exists y\, B(b, y)$, and so $(e \bullet b) \bullet 0 \neq \emptyset$ and

$$(\forall d \in (e \bullet b) \bullet 0) \;\; \jmath_1 d \Vdash_{\mathfrak{w}} B(b, \jmath_0 d). \tag{15.3}$$

Let

$$C^* := \{\jmath_0 d \mid (\exists x \in a)[d \in (e \bullet x) \bullet 0]\}.$$

C^* is a set in our background theory, using Replacement or Strong Collection.

Now assume $e' \Vdash_{\mathfrak{w}} b \in a$. Then $b \in a$ and hence, by the above, $(e \bullet b) \bullet 0 \neq \emptyset$ and

$$(\forall d \in (e \bullet b) \bullet 0) \;\; \jmath(0, \jmath_1 d) \Vdash_{\mathfrak{w}} [\jmath_0 d \in C^* \wedge B(b, \jmath_0 d)]. \tag{15.4}$$

There is an $E_\wp$-recursive function ℓ_2 defined by

$$\ell_2(e, b) \simeq \{\jmath(\jmath_0 d, \jmath(0, \jmath_1 d)) \mid d \in (e \bullet b) \bullet 0\}.$$

From (15.4) we can infer that $\ell_2(e, b) \Vdash_{\mathfrak{w}} \exists y\, [y \in C^* \wedge B(b, y)]$ and hence

$$\lambda u.\lambda v.\ell_2(e, u) \Vdash_{\mathfrak{w}} \forall x(x \in a \to \exists y\, [y \in C^* \wedge B(x, y)]). \tag{15.5}$$

Now assume $c \in C^*$. Then there exists $b \in a$ and $d \in (e \bullet b) \bullet 0$ such that $c = \jmath_0 d$. Moreover, by (15.3), whenever $b \in a$, $d \in (e \bullet b) \bullet 0$ and $\jmath_0 d = c$, then $\jmath_1 d \Vdash_{\mathfrak{w}} B(b, c)$. Letting ℓ_3 be the $E_\wp$-recursive function defined by

$$\ell_3(a, c, e) \simeq \{\jmath(b, \jmath(0, \jmath_1 d)) \mid b \in a \wedge \exists d \in (e \bullet b) \bullet 0 \; \jmath_0 d = c\},$$

we then have

$$\ell_3(a, c, e) \Vdash_{\mathfrak{w}} \exists x(x \in a \wedge B(x, c)), \tag{15.6}$$

thus

$$\lambda u.\lambda v.\ell_3(a, u, e) \Vdash_{\mathfrak{w}} \forall y[y \in C^* \to \exists x(x \in a \wedge B(x, y))]. \tag{15.7}$$

Finally observe that there is an $E_\wp$-recursive function $\mathfrak{l}$ such that

$$\mathfrak{l}(a, e) := \{\jmath_0 d \mid d \in \bigcup_{x \in a}((e \bullet x) \bullet 0)\} = \{\jmath_0 d \mid (\exists x \in a)[d \in (e \bullet x) \bullet 0]\} = C^*.$$

Thus in view of (15.5) and (15.7) we arrive at

$$\begin{aligned} &\mathfrak{sg}(\jmath(\mathfrak{l}(a, e), \jmath(\lambda u.\lambda v.\ell_2(e, u), \lambda u.\lambda v.\ell_3(a, u, e)))) \\ &\qquad \Vdash_{\mathfrak{w}} \exists z\, [\forall x(x \in a \to \exists y\, [y \in z \wedge B(x, y)]) \\ &\qquad \wedge \forall y[y \in z \to \exists x(x \in a \wedge B(x, y))]]. \end{aligned}$$

As a result, $\lambda w.\lambda q.\mathfrak{sg}(\jmath(\mathfrak{l}(w, q), \jmath(\lambda u.\lambda v.\ell_2(q, u), \lambda u.\lambda v.\ell_3(w, u, q)))))$ is a realizer for each instance of Strong Collection.

(Infinity): By Aczel and Rathjen (2010, Lemma 9.2.2) it suffices to find a realizer for the formula

$$\exists z \forall x(x \in z \leftrightarrow [x = \emptyset \vee, \exists \exists y \in z\, x = y \cup \{y\}]).$$

Here $x = \emptyset$ is an abbreviation for $\forall y(y \in x \to y \neq y)$ and $(\exists y \in z)\, x = y \cup \{y\}$ is an abbreviation for

$$\exists y(y \in z \wedge [\forall w(w \in x \to [w \in y \vee, \exists w = y]) \wedge [\forall w(w \in y \to w \in x) \wedge y \in x]]).$$

We have

$$\lambda u'.\lambda v'.0 \Vdash_{\mathfrak{w}} \forall y(y \in \emptyset \to y \neq y). \tag{15.8}$$

For $n+1 \in \omega$ we have

$$\ell_4(n+1) \Vdash_{\mathfrak{w}} \forall w(w \in n+1 \to (w \in n \vee, \exists w = n)) \tag{15.9}$$

for the $E_{\wp}$-recursive function

$$\ell_4(u) := \lambda w.\lambda v'.\{\jmath(0,0) \mid w \in [\mathbf{p_N}](u)\} \cup \{\jmath(1,0) \mid w = [\mathbf{p_N}](u)\}.$$

We also have $\jmath(\lambda w'.\lambda v'.0, 0) \Vdash_{\mathfrak{w}} \forall w(w \in n \to w \in n+1) \wedge n \in n+1$. Thus

$$\begin{aligned} \ell_5(n+1) \Vdash_{\mathfrak{w}} n \in \omega \wedge [\forall w(w \in n+1 \to (w \in n \vee, \exists w = n)) \\ \wedge [\forall w(w \in n \to w \in n+1) \wedge n \in n+1]]. \end{aligned} \tag{15.10}$$

with $\ell_5(n+1) := \jmath(0, \jmath(\ell_4(n+1), \jmath(\lambda w'.\lambda v'.0, 0)))$. From (15.10) we conclude that

$$\ell_6(n+1) \Vdash_{\mathfrak{w}} (\exists y \in \omega)(n+1 = y \cup \{y\}), \tag{15.11}$$

where $\ell_6(m) := \mathfrak{sg}(\jmath([\mathbf{p_N}](m), \ell_5(m)))$. Now from (15.8) and (15.11) we conclude that for every $m \in \omega$:

$$\mathfrak{sg}([\mathbf{d_N}](0, m, \jmath(0, \lambda u'.\lambda v'.0), \jmath(1, \ell_6(m)))) \Vdash_{\mathfrak{w}} m = \emptyset \vee, \exists \exists y \in \omega\, m = y \cup \{y\}.$$

If $e \Vdash_{\mathfrak{w}} a \in \omega$ then $a \in \omega$, and hence with $\ell_7(\omega) := \lambda u.\mathfrak{sg}([\mathbf{d_N}]$ $(0, u, \jmath(0, \lambda u'.\lambda v'.0), \jmath(1, \ell_6(u))))$,

$$\ell_7(\omega) \Vdash_{\mathfrak{w}} (\forall x \in \omega)[x = \emptyset \vee, \exists \exists y \in \omega\, x = y \cup \{y\}]. \tag{15.12}$$

Conversely, if $e \Vdash_{\mathfrak{w}} \forall y(y \in a \to y \neq y)$, then really $\forall y \in a\, y \neq y$, and hence $a = \emptyset$, so that $a \in \omega$. Also, if $e' \Vdash_{\mathfrak{w}} \exists y \in \omega\; a = y \cup \{y\}$ then by unraveling this definition it turns out that $a \in \omega$ holds. As a result, if $d \Vdash_{\mathfrak{w}} [a = \emptyset \vee, \exists \exists y \in \omega\, a = y \cup \{y\}]$ then there exists $f \in d$ such that $\jmath_0 f = 0$ and $\jmath_1 f \Vdash_{\mathfrak{w}} a = \emptyset$ or $\jmath_0 f = 1$ and $\jmath_1 f \Vdash_{\mathfrak{w}} \exists y \in \omega\, a = y \cup \{y\}$. In either case we have $a \in \omega$, and so

$$\lambda x.\lambda e.0 \Vdash_{\mathfrak{w}} \forall x([x = \emptyset \vee, \exists \exists y \in \omega\, x = y \cup \{y\}] \to x \in \omega). \tag{15.13}$$

Combining (15.12) and (15.13), we have

$$\mathfrak{sg}(\jmath(\omega, \lambda v.\jmath(\lambda d.(\ell_7(\omega) \bullet v), \lambda e.0))) \Vdash_{\mathfrak{w}} \exists z \forall x(x \in z \leftrightarrow$$

$$[x = \emptyset \vee \exists y \in z\, x = y \cup \{y\}]). \tag{15.14}$$

□

We would like to show that **KP**($\mathcal{P}$) also realizes every theorem of Tharp's quasi-intuitionistic set theory without **Ord-Im**. This requires a special Lemma about realizability of bounded formulae.

Definition 15.8. To each $\Delta_0^{\mathcal{P}}$ formula $D(x_1, \ldots, x_r)$ of $\mathcal{L}_\in$ all of whose free variables are among $\mathbf{x} = x_1, \ldots, x_r$, we assign a total $E_\wp$-recursive function $\mathfrak{k}_D$ of arity r as follows:

1. $\mathfrak{k}_D(\mathbf{x}) = \{0\}$ if $D(\mathbf{x})$ is atomic.
2. $\mathfrak{k}_D(\mathbf{x}) = \{\{\langle 0, z\rangle\} \mid z \in \mathfrak{k}_A(\mathbf{x}) \wedge A(\mathbf{x})\} \cup \{\{\langle 1, z\rangle\} \mid z \in \mathfrak{k}_B(\mathbf{x}) \wedge B(\mathbf{x})\}$ if $D(\mathbf{x})$ is of the form $A(\mathbf{x}) \vee, \exists\, B(\mathbf{x})$.
3. $\mathfrak{k}_D(\mathbf{x}) = \{\langle z, w\rangle \mid z \in \mathfrak{k}_A(\mathbf{x}) \wedge w \in \mathfrak{k}_B(\mathbf{x})\}$ if $D(\mathbf{x})$ is of the form $A(\mathbf{x}) \wedge B(\mathbf{x})$.
4. $\mathfrak{k}_D(\mathbf{x}) = \{\lambda v.\chi_B(\mathbf{x}, \mathfrak{k}_B(\mathbf{x}))\}$ if $D(\mathbf{x})$ is of the form $A(\mathbf{x}) \rightarrow B(\mathbf{x})$.
5. $\mathfrak{k}_D(\mathbf{x}) = \{\{\langle z, v\rangle\} \mid z \in x_i \wedge v \in \mathfrak{k}_A(\mathbf{x}, z) \wedge A(\mathbf{x}, z)\}$ if $D(\mathbf{x})$ is of the form $\exists z \in x_i\; A(\mathbf{x}, z)$.
6. $\mathfrak{k}_D(\mathbf{x}) = \{\lambda z.\chi_A(\mathbf{x}, z, \mathfrak{k}_A(\mathbf{x}, z))\}$ if $D(\mathbf{z})$ is of the form $\forall z \in x_i\; A(\mathbf{x}, z)$.
7. $\mathfrak{k}_D(\mathbf{x}) = \{\{\langle z, \langle \lambda y.0, v\rangle\rangle\} \mid z \in \mathcal{P}(x_i) \wedge v \in \mathfrak{k}_A(\mathbf{x}, z) \wedge A(\mathbf{x}, z)\}$ if $D(\mathbf{x})$ is of the form $\exists z \subseteq x_i\; A(\mathbf{x}, z)$.
8. $\mathfrak{k}_D(\mathbf{x}) = \{\lambda y.\lambda z.\chi_A(\mathbf{x}, z, \mathfrak{k}_A(\mathbf{x}, z))\}$ if $D(\mathbf{z})$ is of the form $\forall z \subseteq x_i\; A(\mathbf{x}, z)$.

In the above, we tacitly used the fact that for every $\Delta_0^{\mathcal{P}}$ formula $A(\mathbf{x}, u)$ there is an $E_\wp$-recursive function $\mathfrak{f}_A$ such that $\mathfrak{f}_A(\mathbf{x}, a) = \{u \in a \mid A(\mathbf{x}, u)\}$. This is proved in Rathjen (2012, Lemma 2.20).

For Δ_0-formulae realizability and truth coincide as the following Proposition shows.

Proposition 15.1. *Let $D(\mathbf{x})$ be a $\Delta_0^{\mathcal{P}}$ formula whose free variables are among $\mathbf{x} = x_1, \ldots, x_r$. Then the following are provable in* **IKP**($\mathcal{P}$):

(i) $D(\mathbf{x}) \rightarrow \mathfrak{k}_D(\mathbf{x}) \neq \emptyset \wedge \forall u \in \mathfrak{k}_D(\mathbf{x})\; u \Vdash_{\mathfrak{w}} D(\mathbf{x})$.
(ii) $(\exists e\; e \Vdash_{\mathfrak{w}} D(\mathbf{x})) \rightarrow D(\mathbf{x})$.

Proof. We show (i) and (ii) simultaneously by induction on the complexity of D.

1. For atomic D this is obvious.
2. Let $D(\mathbf{x})$ be of the form $A(\mathbf{x}) \vee, \exists\, B(\mathbf{x})$. First suppose that $D(\mathbf{x})$ holds. Then the induction hypothesis entails that $A(\mathbf{x})$ and $\mathfrak{k}_A(\mathbf{x}) \neq \emptyset$ or $B(\mathbf{x})$ and $\mathfrak{k}_B(\mathbf{x}) \neq \emptyset$. In every case we have $\mathfrak{k}_D(\mathbf{x}) \neq \emptyset$.

   If $u \in \mathfrak{k}_D(\mathbf{x})$ then either $u = \{\langle 0, z\rangle\}$ and $A(\mathbf{x})$ for some $z \in \mathfrak{k}_A(\mathbf{x})$ or $u = \{\langle 1, z\rangle\}$ and $B(\mathbf{x})$ for some $z \in \mathfrak{k}_B(\mathbf{x})$. In the first case the inductive assumption yields $z \Vdash_{\mathfrak{w}} A(\mathbf{x})$ and hence $u \Vdash_{\mathfrak{w}} D(\mathbf{x})$. In the second case the inductive assumption yields $z \Vdash_{\mathfrak{w}} B(\mathbf{x})$ and hence also $u \Vdash_{\mathfrak{w}} D(\mathbf{x})$. This shows (i).

   As to (ii), suppose that $e \Vdash_{\mathfrak{w}} D(\mathbf{x})$. Then there exists $u \in e$ such that $u = \langle 0, d\rangle \wedge d \Vdash_{\mathfrak{w}} A(\mathbf{x})$ or $u = \langle 1, d\rangle \wedge d \Vdash_{\mathfrak{w}} B(\mathbf{x})$ for some d. The induction hypothesis yields $A(\mathbf{x})$ or $B(\mathbf{x})$, thus $D(\mathbf{x})$.
3. Let $D(\mathbf{x})$ be of the form $A(\mathbf{x}) \wedge B(\mathbf{x})$. Then (i) and (ii) are immediate by the induction hypothesis.
4. Let $D(\mathbf{x})$ be of the form $A(\mathbf{x}) \to B(\mathbf{x})$. By definition, $\mathfrak{k}_D(\mathbf{x}) = \{\lambda v.\chi_B(\mathbf{x}, \mathfrak{k}_B(\mathbf{x}))\} \neq \emptyset$. As to (i), assume that $D(\mathbf{x})$ holds and $e \Vdash_{\mathfrak{w}} A(\mathbf{x})$. Then the induction hypothesis (ii) applied to $A(\mathbf{x})$ yields that $A(\mathbf{x})$ holds, which implies that $B(\mathbf{x})$ holds. The induction hypothesis (i) for the latter formula yields that $\mathfrak{k}_B(\mathbf{x}) \neq \emptyset$ and $\forall u \in \mathfrak{k}_B(\mathbf{x})\; u \Vdash_{\mathfrak{w}} B(\mathbf{x})$. An application of Lemma 15.6 thus yields $\chi_B(\mathbf{x}, \mathfrak{k}_B(\mathbf{x})) \Vdash_{\mathfrak{w}} B(\mathbf{x})$. As a result, $\lambda v.\chi_B(\mathbf{x}, \mathfrak{k}_B(\mathbf{x})) \Vdash_{\mathfrak{w}} D(\mathbf{x})$ confirming (i).

   For (ii), suppose $e \Vdash_{\mathfrak{w}} (A(\mathbf{x}) \to B(\mathbf{x}))$ and $A(\mathbf{x})$ holds. By the induction hypothesis (i) for the latter formula, $\mathfrak{k}_A(\mathbf{x}) \neq \emptyset$ and $\forall u \in \mathfrak{k}_A(\mathbf{x})\; u \Vdash_{\mathfrak{w}} A(\mathbf{x})$. Thus, picking $u_0 \in \mathfrak{k}_A(\mathbf{x})$ we have $e \bullet u_0 \Vdash_{\mathfrak{w}} B(\mathbf{x})$, and hence the induction hypothesis (ii) for the latter formula yields that $B(\mathbf{x})$ holds.
5. Let $D(\mathbf{x})$ be of the form $\exists z \in x_i\; A(\mathbf{x}, z)$. To verify (i), suppose $\exists z \in x_i\; A(\mathbf{x}, z)$ holds. Then there is $z \in x_i$ such that $A(\mathbf{x}, z)$. The induction hypothesis (i) for the latter formula yields that $\mathfrak{k}_A(\mathbf{x}, z) \neq \emptyset$, and hence $\mathfrak{k}_D(\mathbf{x}) \neq \emptyset$. Now suppose $u \in \mathfrak{k}_D(\mathbf{x})$. Then $u = \{\langle z, v\rangle\}$ for some $z \in x_i$ and $v \in \mathfrak{k}_A(\mathbf{x}, z)$. As $A(\mathbf{x}, z)$ holds the induction hypothesis (i) yields that $v \Vdash_{\mathfrak{w}} A(\mathbf{x}, z)$, whence $u \Vdash_{\mathfrak{w}} \exists z \in x_i\; A(\mathbf{x}, z)$.

   For (ii), assume $e \Vdash_{\mathfrak{w}} \exists z \in x_i\; A(\mathbf{x}, z)$. Then $e \neq \emptyset$. Picking $d \in e$ we have $\jmath_0 d \in x_i$ and $\jmath_1 d \Vdash_{\mathfrak{w}} A(\mathbf{x}, \jmath_0 d)$, thus $A(\mathbf{x}, \jmath_0 d)$ by the induction hypothesis (ii), thence $\exists z \in x_i\; A(\mathbf{x}, z)$ holds.
6. Let $D(\mathbf{x})$ be of the form $\forall z \in x_i\; A(\mathbf{x}, z)$. To verify (i), suppose $\forall z \in x_i\; A(\mathbf{x}, z)$ is true. By definition, $\mathfrak{k}_D(\mathbf{x}) = \{\lambda z.\chi_A(\mathbf{x}, z, \mathfrak{k}_A(\mathbf{x}, z))\} \neq \emptyset$. If $z_0 \in x_i$ we have $A(\mathbf{x}, z_0)$, so that inductively $\mathfrak{k}_A(\mathbf{x}, z_0) \neq \emptyset$ and $\forall d \in \mathfrak{k}_A(\mathbf{x}, z_0)\; d \Vdash_{\mathfrak{w}} A(\mathbf{x}, z_0)$. Whence, by Lemma 15.6, $\chi_A(\mathbf{x}, z_0, \mathfrak{k}_A(\mathbf{x}, z_0)) \Vdash_{\mathfrak{w}} A(\mathbf{z}, z_0)$. As a result, $\lambda z.\chi_A(\mathbf{x}, z, \mathfrak{k}_A(\mathbf{x}, z)) \Vdash_{\mathfrak{w}} D(\mathbf{x})$.

   As for (ii), suppose $e \Vdash_{\mathfrak{w}} \forall z \in x_i\; A(\mathbf{x}, z)$. Thus $e \bullet z \Vdash_{\mathfrak{w}} A(\mathbf{x}, z)$ for all $z \in x_i$, so that inductively $\forall z \in x_i\; A(\mathbf{x}, z)$ holds.
7. Let $D(\mathbf{x})$ be of the form $\exists z \subseteq x_i\; A(\mathbf{x}, z)$. To verify (i), suppose $\exists z \subseteq x_i\; A(\mathbf{x}, z)$ holds. Then there is $z \in \mathcal{P}(x_i)$ such that $A(\mathbf{x}, z)$. The induction hypothesis (i) for the latter formula yields that $\mathfrak{k}_A(\mathbf{x}, z) \neq \emptyset$, and hence $\mathfrak{k}_D(\mathbf{x}) \neq \emptyset$. Now suppose

$u \in \mathfrak{k}_D(\mathbf{x})$. Then $u = \{\langle z, \langle \lambda y.0, v\rangle\rangle\}$ for some $z \subseteq x_i$ and $v \in \mathfrak{k}_A(\mathbf{x}, z)$. As $A(\mathbf{x}, z)$ holds the induction hypothesis (i) yields that $v \Vdash_{\mathfrak{w}} A(\mathbf{x}, z)$. Also $\lambda y.0 \Vdash_{\mathfrak{w}} z \subseteq x_i$. Whence $u \Vdash_{\mathfrak{w}} \exists z\,(z \subseteq x_i \;\wedge\; A(\mathbf{x}, z))$.

For (ii), assume $e \Vdash_{\mathfrak{w}} \exists z\,[z \subseteq x_i \wedge A(\mathbf{x}, z)]$. Then $e \neq \emptyset$. Picking $d \in e$ we have $\jmath_1 d \Vdash_{\mathfrak{w}} [\jmath_0 d \subseteq x_i \wedge A(\mathbf{x}, \jmath_0 d)]$. This entails $\jmath_0 d \subseteq x_i$ and $\jmath_1(\jmath_1 d) \Vdash_{\mathfrak{w}} A(\mathbf{x}, \jmath_0 d)$. Thus $A(\mathbf{x}, \jmath_0 d)$ by the induction hypothesis (ii), thence $\exists z \subseteq x_i\; A(\mathbf{x}, z)$ holds.

8. Let $D(\mathbf{x})$ be of the form $\forall z \in x_i\; A(\mathbf{x}, z)$. To verify (i), suppose $\forall z \in x_i\; A(\mathbf{x}, z)$ is true. By definition, $\mathfrak{k}_D(\mathbf{x}) = \{\lambda y.\lambda z.\chi_A(\mathbf{x}, z, \mathfrak{k}_A(\mathbf{x}, z))\} \neq \emptyset$. If $y \Vdash_{\mathfrak{w}} z_0 \subseteq x_i$, then $z_0 \subseteq x_i$ holds and we have $A(\mathbf{x}, z_0)$, so that inductively $\mathfrak{k}_A(\mathbf{x}, z_0) \neq \emptyset$ and $\forall d \in \mathfrak{k}_A(\mathbf{x}, z_0)\; d \Vdash_{\mathfrak{w}} A(\mathbf{x}, z_0)$. Whence, by Lemma 15.6, $\chi_A(\mathbf{x}, z_0, \mathfrak{k}_A(\mathbf{x}, z_0)) \Vdash_{\mathfrak{w}} A(\mathbf{z}, z_0)$. As a result, $\lambda y.\lambda z.\chi_A(\mathbf{x}, z, \mathfrak{k}_A(\mathbf{x}, z)) \Vdash_{\mathfrak{w}} D(\mathbf{x})$.

   As for (ii), suppose $e \Vdash_{\mathfrak{w}} \forall z \subseteq x_i\; A(\mathbf{x}, z)$. Thus $e \bullet z \Vdash_{\mathfrak{w}} [z \subseteq x_i \rightarrow A(\mathbf{x}, z)]$ for all z. If $z \subseteq x_i$, then $\lambda y.0 \Vdash_{\mathfrak{w}} z \subseteq x_i$, so that $(e \bullet z) \bullet (\lambda y.0) \Vdash_{\mathfrak{w}} A(\mathbf{x}, z)$, and therefore, by the inductive assumption, $A(\mathbf{x}, z)$ holds. As a result, $\forall z \in x_i\; A(\mathbf{x}, z)$ holds. □

Theorem 15.3. *Let $\mathcal{T}^-$ denote Tharp's (1971) quasi-intuitionistic set theory without* **Ord-Im**. *Let $D(u_1, \ldots, u_r)$ be a formula of $\mathcal{L}_\in$ all of whose free variables are among $u_1, \ldots, u_r$. If*

$$\mathcal{T}^- \vdash D(u_1, \ldots, u_r),$$

then one can effectively construct an index of an $E_\wp$-recursive function g such that

$$\mathbf{KP}(\mathcal{P}) \vdash \forall a_1, \ldots, a_r\; g(a_1, \ldots, a_r) \Vdash_{\mathfrak{w}} D(a_1, \ldots, a_r)\,.$$

Proof. Note that with the exception of excluded middle for power bounded formulae, the axioms of $\mathcal{T}^-$ are axioms of $\mathbf{CZF}_\mathcal{P}$, too. Let $D(\mathbf{u})$ be $\Delta_0^\mathcal{P}$. Define

$$\mathfrak{d}_D(\mathbf{a}) := \{\langle 0, u\rangle \mid u \in \mathfrak{k}_D(\mathbf{a})\} \;\cup\; \{\langle 1, u\rangle \mid u \in \mathfrak{k}_{\neg D}(\mathbf{a})\},$$

with $\mathfrak{k}_D, \mathfrak{k}_{\neg D}$ defined as in Definition 15.8. Note that $\mathfrak{d}_D$ is E-recursive. By Proposition 15.1(i) and classical logic we have that $\mathfrak{d}_D(\mathbf{a}) \neq \emptyset$. Moreover, if $\langle i, u\rangle \in \mathfrak{d}_D(\mathbf{a})$ then either $i = 0$ and $u \Vdash_{\mathfrak{w}} D(\mathbf{a})$ or $i = 1$ and $u \Vdash_{\mathfrak{w}} \neg D(\mathbf{a})$. Thus $\mathfrak{d}_D(\mathbf{a}) \Vdash_{\mathfrak{w}} D(\mathbf{a}) \vee, \exists \neg D(\mathbf{a})$.

In view of the previous Theorem 15.2 we thus found realizers for all theorems of $\mathcal{T}^-$. □

Lemma 15.4. $\mathbf{CZF}_\mathcal{P}$ *is a subtheory of $\mathcal{T}^-$.*

Proof. The only axioms of $\mathbf{CZF}_\mathcal{P}$ that do not already belong to $\mathcal{T}^-$ are the instances of Bounded Separation. Let $A(u)$ be bounded. We shall reason in $\mathcal{T}^-$. Using excluded middle for bounded formulae, Pairing and Emptyset, we have

$$\forall u \in a\, \exists z\,[(A(u) \wedge z = \{u\}) \vee, \exists\, (\neg A(u) \wedge z = 0)].$$

Thus, by Strong Collection, there exists a set b such that

$$\begin{aligned}&\forall u \in a\, \exists z \in b\, [(A(u) \wedge z = \{u\}) \vee, \exists\, (\neg A(u) \wedge z = 0)]\\ &\wedge\ \forall z \in b\, \exists u \in a\, (A(u) \wedge z = \{u\}) \vee, \exists\, (\neg A(u) \wedge z = 0)].\end{aligned} \tag{15.15}$$

By Union, $\bigcup b$ is a set, and by (15.15), $\bigcup b = \{u \in a \mid A(u)\}$. □

15.4 A Type Theory Pertaining to $\mathbf{CZF}_{\mathcal{P}}$

Let $\mathbf{ML}_1$ be Martin-Löf's type theory with a single universe $\mathbf{U}$ but without any W-types (cf. Martin-Löf 1984). The type $\mathbf{U}$ of small types reflects the basic forms of type. These are $\mathbf{N}_0$ (empty type), $\mathbf{N}$ (type of naturals), $(\Pi x : A)F(x)$, $(\Sigma x : A)F(x)$, $A + B$ and $I(A, b, c)$ where A and B are types, F is a family of types over A and $b, c : A$.

$\mathbf{ML_1V}$ is the extension of $\mathbf{ML}_1$ with Aczel's type of iterative sets $\mathbf{V}$ (cf. Aczel 1978). $\mathbf{V}$ is inductively specified by the rule

$$\frac{A : \mathbf{U} \quad x : A \Rightarrow F : \mathbf{V}}{\mathbf{sup}(x : A)F : \mathbf{V}}$$

It is this type $\mathbf{V}$ with the above introduction rule and a corresponding elimination rule (or rule of transfinite recursion on $\mathbf{V}$) that has been used in Aczel (1978) to give an interpretation of constructive set theory (for more details see Aczel 1982; Rathjen 1994).

Remark 15.5. $\mathbf{V}$ can be viewed as a single W-type on top of $\mathbf{U}$. $\mathbf{V}$ should certainly not be construed as an additional universe on top of $\mathbf{U}$. As it turns out, adding $\mathbf{V}$ amounts to the same as adding an elimination rule to $\mathbf{U}$ which renders $\mathbf{U}$ an inductively defined type. $\mathbf{V}$ can then be explicitly defined from $\mathbf{U}$ in extensional $\mathbf{ML}_1$ augmented by the principle of transfinite recursion on $\mathbf{U}$ as has been shown by Palmgren in (1993).

We extend the syntax of $\mathbf{ML_1V}$ with a type constant $\mathbf{P}$ and several other constants pertaining to it. The rules for $\mathbf{P}$ render it an impredicatively Π-closed type universe inside $\mathbf{U}$. The rules governing $\mathbf{P}$ are given by the schemes

$$0_{\mathbf{P}} : \mathbf{P} \qquad \mathbf{P} : \mathbf{U} \qquad \frac{a : \mathbf{P}}{\mathbf{T_P}(a) : \mathbf{U}} \qquad \frac{a : \mathbf{P} \quad b_1 : \mathbf{T_P}(a) \quad b_2 : \mathbf{T_P}(a)}{b_1 = b_2 : \mathbf{T_P}(a)}$$

$$\frac{A : \mathbf{U} \quad x : A \Rightarrow B : \mathbf{P}}{(\pi x : A)B : \mathbf{P}} \qquad \frac{A : \mathbf{U} \quad x : A \Rightarrow B_1 = B_2 : \mathbf{P}}{(\pi x : A)B_1 = (\pi x : A)B_2 : \mathbf{P}}$$

$$\mathbf{T_P}(0_{\mathbf{P}}) = N_0 \qquad \frac{A : \mathbf{U} \quad x : A \Rightarrow B : \mathbf{P}}{s_{A,B} : \mathbf{T_P}((\pi x : A)B) \leftrightarrow (\Pi x : A)\mathbf{T_P}(B)} \quad (\star).$$

The formulation of the rules for the type **P**, embodies the principle that elements of **P** are only codes for types, hence the need for a decoding function $\mathbf{T_P}$ and the π-binder. $0_\mathbf{P}$ represents the false proposition and thus $\mathbf{T_P}(0_\mathbf{P})$ should be the empty type.

With these rules the type **P** behaves like the impredicative type of propositions of the calculus of constructions, with the additional property that all propositions in **P** are proof-irrelevant. The equivalence in the rule ($\star$) was already introduced in Coquand (1990). This type theory will be denoted by $\mathbf{MLV_P}$.

15.5 Reducing $\mathbf{MLV_P}$ to $\mathbf{CZF}_\mathcal{P}$

Here we build on the types-as-classes interpretation from Rathjen (2006c) and Rathjen and Tupailo (2006, Definition 6.7) that uses classes of indices of generalized set recursive functions to interpret large Π-types. Rathjen and Tupailo (2006, Theorem 6.8) shows that this provides a translation of $\mathbf{ML_1V}$ into **CZF**. In this interpretation the type **U** is emulated by the inductively defined class $\mathbf{Y}^*$ introduced in Rathjen and Tupailo (2006, Definition 2.8). A larger class $\mathbf{Y}^{**}$ is obtained by adding a fifth clause to the definition of $\mathbf{Y}^*$ which just says that the powerset of $\{0\}$ and every set $x \subseteq \{0\}$ is in $\mathbf{Y}^{**}$. To deal with $\mathbf{MLV_P}$, **U** will be interpreted as $\mathbf{Y}^{**}$ and the type **V** will then be interpreted as the class $\mathbf{V}(\mathbf{Y}^{**})$ which is defined in the same vein as $\mathbf{V}(\mathbf{Y}^*)$ in Rathjen and Tupailo (2006, Definition 3.1). The type **P** will be interpreted by $\mathcal{P}(\{0\})$, the powerset of $\{0\}$. For sets A and a function $F : A \to \mathcal{P}(\{0\})$ let $\pi(A, F) := \{y \in \{0\} \mid \forall x \in A\, F(x) = \{0\}\}$. This is the way we interpret the π-binder. $\mathbf{T_P}$ will be interpreted as the identity function while $s_{A,B}$ is the unique 1–1 correspondence between the sets $\pi(A, F)$ and $\Pi_{x\in A} F(x)$.

Theorem 15.4. *The types-as-classes translation provides an interpretation of* $\mathbf{MLV_P}$ *in* $\mathbf{CZF}_\mathcal{P}$.

Proof. For details see Rathjen and Tupailo (2006, Theorem 6.8) and Rathjen (1994, Theorem 4.11). □

15.6 Reducing $\mathbf{CZF} + \mathbf{Pow}^{\neg\neg}$ to $\mathbf{MLV_P}$

Recall that the negative power set axiom, $\mathbf{Pow}^{\neg\neg}$, asserts that for every set a there exists a set c containing all the subsets x of a for which $\forall u \in a(\neg\neg u \in x \to u \in x)$ holds. The latter set will be denoted by $\mathcal{P}^{\neg\neg}(a)$.

Lemma 15.10. *The theory obtained from* **CZF** *by adding the axiom '*$\mathcal{P}^{\neg\neg}(\{0\})$ is a set*' is equivalent to* $\mathbf{CZF} + \mathbf{Pow}^{\neg\neg}$.

Proof. Gambino (1999, Lemma 4.3.2). □

Theorem 15.5. *The theory* **CZF+Pow**$^{\neg\neg}$ *can be justified in the type theory* $\mathbf{MLV_P}$.

Proof. For the axioms of **CZF** this is due to Aczel (1978). The validity of the negative power set axiom in a type theory with **P** was shown by Gambino (1999, Lemma 4.3.7). □

15.7 Completing the Circle: The Proof of Theorem 15.1

The main thing we know so far is that $\mathbf{CZF}_{\mathcal{P}}$ is proof-theoretically no stronger than $\mathbf{KP}(\mathcal{P})$ (Theorem 15.2). As for the proof-theoretic equivalence of ((i) and (ii) in Theorem 15.1, we need to show that $\mathbf{CZF}_{\mathcal{P}} + \mathbf{RDC} + \mathbf{\Pi\,\Sigma{-}AC}$ is no stronger than $\mathbf{CZF}_{\mathcal{P}}$. We shall draw on the formulae-as-classes interpretation of Rathjen (2006c) to achieve this.

Theorem 15.6. $\mathbf{CZF}_{\mathcal{P}} + \mathbf{RDC} + \mathbf{\Pi\,\Sigma{-}AC}$ *has a formulae-as-classes interpretation in* $\mathbf{CZF}_{\mathcal{P}}$.

Proof. The interpretation of $\mathbf{CZF} + \mathbf{RDC} + \mathbf{\Pi\,\Sigma{-}AC}$ into **CZF** of Rathjen (2006c, Theorem 4.13) can be lifted to the theories with **Pow** added on both sides if one uses the stronger notion of computability introduced in Definition 15.3. One just needs to show that the power set axiom is validated in this interpretation if one has it in the background theory and uses the stronger notion of computability. This is not very difficult. □

To get back from $\mathbf{KP}(\mathcal{P})$ to $\mathbf{CZF}_{\mathcal{P}}$ we shall rely on Rathjen (2012). Let $\mathcal{OT}(\vartheta) = (\mathrm{BH}, \prec)$ be the primitive recursive ordinal representation system for the Bachmann-Howard ordinal given in Rathjen and Weiermann (1993, Lemma 1.3); here $\mathrm{OT}(\vartheta)$ is a primitive recursive set of naturals equipped with a primitive recursive well-ordering $\prec$ and

$$\mathrm{BH} := \{\alpha \in \mathrm{OT}(\vartheta) \mid \alpha \prec \Omega\}.$$

For $\tau \in \mathrm{BH}$ let

$$V_\tau := \bigcup_{\nu \prec \tau} \mathcal{P}(V_\nu) \tag{15.16}$$

$$V_\tau^{\neg\neg} := \bigcup_{\nu \prec \tau} \mathcal{P}^{\neg\neg}(V_\nu^{\neg\neg}). \tag{15.17}$$

Let 'V_τ *exists*' be the statement

$$\exists F\,[F \text{ function} \wedge \mathrm{dom}(f) = \{\nu \in \mathrm{BH} \mid \nu \prec \tau\} \wedge \forall \nu \prec \tau\ F(\nu) = \bigcup_{\xi \prec \nu} \mathcal{P}(F(\xi))].$$

Lemma 15.11. *For every (meta) $\tau \in$ BH,* **CZF** *proves the scheme of transfinite induction up to τ, i.e.,*

$$\forall \nu \prec \tau[(\forall \mu \prec \nu\, \varphi(\mu)) \rightarrow \varphi(\nu)] \rightarrow \forall \nu \prec \tau\, \varphi(\nu)$$

for all formulae $\varphi(\nu)$.

Proof. This is a consequence of Rathjen (2005, Lemma 4.3, Theorem 4.13). □

Lemma 15.12. *Let $\tau \in$ BH. The following are provable in* **CZF** + **Pow**$^{\neg\neg}$ *for all $\beta \preceq \alpha \preceq \tau$:*

(i) '$V_\alpha^{\neg\neg}$ *exists'.*
(ii) $V_0^{\neg\neg} = \emptyset$.
(iii) *If α is a limit, then* $V_\alpha^{\neg\neg} = \bigcup_{\xi \prec \alpha} V_\xi^{\neg\neg}$.
(iv) $V_{\alpha+1}^{\neg\neg} = V_\alpha^{\neg\neg} \cup \mathcal{P}^{\neg\neg}(V_\alpha^{\neg\neg})$.
(v) $V_\beta^{\neg\neg} \subseteq V_\alpha^{\neg\neg}$.
(vi) $V_\alpha^{\neg\neg}$ *is transitive.*
(vii) $u \in x \in V_\alpha^{\neg\neg} \rightarrow \exists \xi \prec \alpha\; u \in V_\xi^{\neg\neg}$.

Proof. (i) Follows by transfinite recursion on α using Lemma 15.11 and Replacement.

(ii) Holds because $V_0^{\neg\neg} = \bigcup_{\xi \prec 0} V_\xi^{\neg\neg} = \emptyset$.

(iii) : $V_\alpha^{\neg\neg} = \bigcup_{\xi \prec \alpha} \mathcal{P}^{\neg\neg}(V_\xi^{\neg\neg}) = \bigcup_{\xi \prec \alpha} \bigcup_{\zeta \prec \xi} \mathcal{P}^{\neg\neg}(V_\xi^{\neg\neg}) = \bigcup_{\xi \prec \alpha} V_\xi^{\neg\neg}$ when α is a limit.

(iv) :

$$\begin{aligned} V_{\alpha+1}^{\neg\neg} &= \bigcup_{\xi \prec \alpha+1} \mathcal{P}^{\neg\neg}(V_\xi^{\neg\neg}) \\ &= \mathcal{P}^{\neg\neg}(V_\alpha^{\neg\neg}) \cup \bigcup_{\xi \prec \alpha} \mathcal{P}^{\neg\neg}(V_\xi^{\neg\neg}) \\ &= \mathcal{P}^{\neg\neg}(V_\alpha^{\neg\neg}) \cup V_\alpha^{\neg\neg}. \end{aligned}$$

(v) : Suppose $\beta \prec \alpha$. It suffices to show that $V_\beta^{\neg\neg} \in \mathcal{P}^{\neg\neg}(V_\beta^{\neg\neg})$. But this is clearly the case since $V_\beta^{\neg\neg} \subseteq V_\beta^{\neg\neg}$ and (trivially)

$$\forall y \in V_\beta^{\neg\neg}(\neg\neg y \in V_\beta^{\neg\neg} \rightarrow y \in V_\beta^{\neg\neg}).$$

(vi) and (vii): Let $u \in x \in V_\alpha^{\neg\neg}$. Then $u \in x \in \mathcal{P}^{\neg\neg}(V_\xi^{\neg\neg})$ for some $\xi \prec \alpha$. Hence $u \in V_\xi^{\neg\neg}$ for some $\xi \prec \alpha$, so that $u \in V_\alpha^{\neg\neg}$ by (v).

□

Theorem 15.7. (i) *The following theories are proof-theoretically equivalent:*

1. $\mathbf{KP}(\mathcal{P})$
2. $\mathbf{Z} + \{\text{'}V_\tau \textit{ exists'}\}_{\tau \in \mathrm{BH}}$.

(ii) *The following theories are proof-theoretically equivalent:*

1. $\mathbf{IKP}(\mathcal{P})$
2. $\mathbf{IZ} + \{\text{'}V_\tau \textit{ exists'}\}_{\tau \in \mathrm{BH}}$.

Proof. This is shown in Rathjen (2012). □

15.7.1 Reducing $\mathbf{Z} + \{\text{'}V_\tau \textit{ Exists'}\}_{\tau \in \mathbf{BH}}$ to $\mathbf{CZF} + \mathbf{Pow}^{\neg\neg}$

The next step is to employ a double negation interpretation to reduce $\mathbf{Z} + \{\text{'}V_\tau \text{ exists'}\}_{\tau \in \mathrm{BH}}$ to an intuitionistic theory. Here we don't follow Friedman's approach in Friedman (1973c). Instead we use two new relations $=_\infty$ and $\in_\infty$ to interpret $=$ and $\in$, respectively. Moreover, these relations are designed to be stable under double negation. This Ansatz was inspired by a double negation interpretation of Zermelo set theory in $V^{\neg\neg}_{\omega+\omega}$ due to Gambino (see Gambino 1999, Proposition 2.3.21). In it he uses Aczel's a-relations, which combine the idea of bisumulation with stability of doubly negated formulae, to interpret set-theoretic equality (for details see Gambino 1999, Definition 2.2.14). Our interpretation, however, does not employ a-relations since our background theory has only Bounded Separation. Instead it uses an equivalence relation defined by transfinite recursion on the ordinal representations of BH.

Theorem 15.8. *For every $\rho \in$ BH, the theory $\mathbf{Z}$ + '$V^{\neg\neg}_\rho$ exists' has an interpretation in $\mathbf{CZF} + \mathbf{Pow}^{\neg\neg}$.*

The proof of 15.8 will occupy the remainder of this subsection. Given $\rho \in$ BH one can effectively find $\rho^* \in$ BH such that $\rho \prec \rho^*$ and ρ^* is a limit ordinal bigger than ω. In view of Theorem 15.7 we also know that $\mathbf{CZF} + \mathbf{Pow}^{\neg\neg}$ proves '$V^{\neg\neg}_{\rho^*}$ *exists*'. We would like to use the set $V^{\neg\neg}_{\rho^*}$ to provide a model for the theory $\mathbf{Z}$ + 'V_ρ *exists*'. The idea is, of course, to use some kind of double negation interpretation. But as is well known, the extensionality axiom creates a problem when one uses the usual Gödel-Gentzen translation. To overcome this problem we define an equivalence relation $=_\infty$ on $V^{\neg\neg}_{\rho^*}$ which will be used to interpret set-theoretic equality and thereby also membership.

Definition 15.9. Let $x, y \in V^{\neg\neg}_{\rho^*}$. By transfinite recursion on $\alpha \prec \rho^*$ define

$$\begin{aligned} x =_\alpha y \text{ iff } x, y \in V^{\neg\neg}_\alpha \ & \wedge \forall u \in x \neg\neg\exists v \in y\, \exists \beta \prec \alpha\ u =_\beta v \\ & \wedge \forall v \in y \neg\neg\exists u \in x\, \exists \beta \prec \alpha\ u =_\beta v \end{aligned}$$

$$x =_\infty y \text{ iff } \neg\neg\exists\alpha \prec \rho^* \; x =_\alpha y$$

$$x \in_\infty y \text{ iff } \neg\neg\exists\alpha \prec \rho^* \; \exists u \in y \; x =_\alpha u\,.$$

Lemma 15.13. *Let* $x, y \in V_{\rho^*}^{\neg\neg}$ *and* $\alpha \prec \beta \prec \rho^*$. *Then we have*

(i) $x =_\alpha y \to x =_\beta y$.
(ii) $x, y \in V_\alpha^{\neg\neg} \wedge x =_\beta y \to x =_\alpha y$.
(iii) $=_\alpha$ *is a symmetric and transitive relation.* $=_\alpha$ *is a reflexive relation on* $V_\alpha^{\neg\neg}$.

Proof. (i) Suppose $x =_\alpha y$. Then $x, y \in V_\alpha^{\neg\neg}$, thus $x, y \in V_\beta^{\neg\neg}$ by Lemma 15.12(v). Clearly we have $\exists v \in y\, \exists\xi \prec \alpha\; u =_\xi v \to \exists v \in y\, \exists\xi \prec \beta\; u =_\xi v$, thus

$$\neg\neg\exists v \in y\, \exists\xi \prec \alpha\; u =_\xi v \to \neg\neg\exists v \in y\, \exists\xi \prec \beta\; u =_\xi v\,,$$

and hence

$$\forall u \in x\, \neg\neg\exists v \in y\, \exists\xi \prec \alpha\; u =_\xi v \to \forall u \in x\, \neg\neg\exists v \in y\, \exists\xi \prec \beta\; u =_\xi v\,.$$

Likewise, $\forall v \in y\, \neg\neg\exists u \in x\, \exists\xi \prec \alpha\; u =_\xi v \to \forall v \in y\, \neg\neg\exists u \in x\, \exists\xi \prec \beta\; u =_\xi v$. As a result, $x =_\beta y$.

(ii) : We use induction on α. Suppose that $x, y \in V_\alpha^{\neg\neg}$ and $x =_\beta y$. If $u \in x$ and $v \in y$, then $u \in x \in \mathcal{P}^{\neg\neg}(V_{\xi_0}^{\neg\neg})$ and $v \in y \in \mathcal{P}^{\neg\neg}(V_{\xi_1}^{\neg\neg})$ for some $\xi_0, \xi_1 \prec \alpha$. Hence $u \in V_{\xi_0}^{\neg\neg}$ and $v \in V_{\xi_1}^{\neg\neg}$. Due to the linearity of $\prec$ and in view of Lemma 15.12(v), there exists $\alpha_0 \prec \alpha$ such that $u, v \in V_{\alpha_0}^{\neg\neg}$. As a result, if $u \in x$ and $\exists v \in y\, \exists\zeta \prec \beta\; u =_\zeta v$, then the induction hypothesis yields $\exists v \in y\, \exists\zeta \prec \alpha\; u =_\zeta v$. Thus $u \in x$ and $\neg\neg\exists v \in y\, \exists\zeta \prec \beta\; u =_\zeta v$ imply $\neg\neg\exists v \in y\, \exists\zeta \prec \alpha\; u =_\zeta v$. Consequently,

$$\begin{aligned} &\forall u \in x\, \neg\neg\exists v \in y\, \exists\zeta \prec \beta\; u =_\zeta v \\ &\quad\to \forall u \in x\, \neg\neg\exists v \in y\, \exists\zeta \prec \alpha\; u =_\zeta v. \end{aligned} \tag{15.18}$$

Likewise one proves

$$\begin{aligned} &\forall v \in y\, \neg\neg\exists x \in u\, \exists\zeta \prec \beta\; u =_\zeta v \\ &\quad\to \forall v \in y\, \neg\neg\exists u \in x\, \exists\zeta \prec \alpha\; u =_\zeta v. \end{aligned} \tag{15.19}$$

Hence, since we assumed that $x =_\beta y$ we get $x =_\alpha y$ from (15.18) and (15.19).

(iii) Follows by induction on α. As for transitivity, suppose $x, y, z \in V_\alpha^{\neg\neg}$, $x =_\alpha y$, and $y =_\alpha z$. Assume that $u \in x$, $v \in y$, $w \in z$ and $u =_{\xi_0} v$ and $v =_{\xi_1} w$ hold for some $\xi_0, \xi_1 \prec \alpha$. Then, using (i) and the linearity of $\prec$, we find $\xi \prec \alpha$ such that $u =_\xi v$ and $v =_\xi w$, so that, by the induction hypothesis, we get $u =_\xi w$. As a result, letting A be $u \in x \wedge v \in y \wedge \exists\xi_0 \prec \alpha\; u =_{\xi_0} v$,

$$A \to (\exists w \in z\, \exists \xi_1 \prec \alpha\, v =_{\xi_0} w \to \exists w \in z\, \exists \xi \prec \alpha\, u =_{\xi} w)$$
$$A \to (\neg\neg\exists w \in z\, \exists \xi_1 \prec \alpha\, v =_{\xi_0} w \to \neg\neg\exists w \in z\, \exists \xi \prec \alpha\, u =_{\xi} w)$$
$$A \to \neg\neg\exists w \in z\, \exists \xi \prec \alpha\, u =_{\xi} w \quad (\text{since } y =_\alpha z)$$
$$u \in x \to (\exists v \in y\, \exists \xi_0 \prec \alpha\, u =_{\xi_0} v \to \neg\neg\exists w \in z\, \exists \xi \prec \alpha\, u =_{\xi} w)$$
$$u \in x \to (\neg\neg\exists v \in y\, \exists \xi_0 \prec \alpha\, u =_{\xi_0} v \to \neg\neg\exists w \in z\, \exists \xi \prec \alpha\, u =_{\xi} w)$$
$$u \in x \to \neg\neg\exists w \in z\, \exists \xi \prec \alpha\, u =_{\xi} w \quad (\text{since } x =_\alpha y)$$

and hence $\forall u \in x\, \neg\neg\exists w \in z\, \exists \xi \prec \alpha\, u =_\xi w$. Likewise one shows that $\forall w \in z\, \neg\neg\exists u \in x\, \exists \xi \prec \alpha\, u =_\xi w$. Thus $x =_\alpha z$.

Symmetry and reflexivity are established similarly. □

Corollary 15.2. *Let $x, y \in V^{\neg\neg}_{\rho^*}$. Then:*

$$x =_\infty y \leftrightarrow \forall u \in V^{\neg\neg}_{\rho^*}\, (u \in_\infty x \leftrightarrow u \in_\infty y)\,.$$

Proof. "$\to$": Suppose $\alpha, \beta \prec \rho^*$, $v \in x$, $u =_\alpha v$, $w \in y$, and $v =_\beta w$. Letting $\gamma := \max(\alpha, \beta)$, we obtain $u =_\gamma v$ and $v =_\gamma w$ by Lemma 15.13(i), and hence $u =_\gamma w$ by Lemma 15.13(iii). Thus, letting B stand for the conjunction of $\alpha \prec \rho^*$, $v \in x$, and $u =_\alpha v$, we have the following implications:

$$B \wedge \exists \beta' \prec \rho^*\, \exists w' \in y\, v =_{\beta'} w' \to \exists \gamma' \prec \rho^*\, \exists w' \in y\, u =_{\gamma'} w'$$
$$B \wedge \neg\neg\exists \beta' \prec \rho^*\, \exists w' \in y\, v =_{\beta'} w' \to \neg\neg\exists \gamma' \prec \rho^*\, \exists w' \in y\, u =_{\gamma'} w'$$
$$B \wedge \exists \eta \prec \rho^*\, x =_\eta y \to u \in_\infty y$$
$$B \wedge \neg\neg\exists \eta \prec \rho^*\, x =_\eta y \to u \in_\infty y$$
$$B \wedge x =_\infty y \to u \in_\infty y$$
$$\exists \alpha \prec \rho^*\, \exists v \in x\, u =_\alpha v \wedge x =_\infty y \to u \in_\infty y$$
$$\neg\neg\exists \alpha \prec \rho^*\, \exists v \in x\, u =_\alpha v \wedge x =_\infty y \to u \in_\infty y$$
$$u \in_\infty x \wedge x =_\infty y \to u \in_\infty y\,.$$

In the above, we used several times that $C \to \neg\neg C$ and

$$(A \to \neg\neg C) \to (\neg\neg A \to \neg\neg C)$$

are intuitionistically valid propositions.

"$\leftarrow$": Assume that $\forall u \in V^{\neg\neg}_{\rho^*}\, (u \in_\infty x \leftrightarrow u \in_\infty y)$. Choose $\alpha \prec \rho^*$ such that $x, y \in V^{\neg\neg}_\alpha$. Let $u \in x$. Then $u \in V^{\neg\neg}_\xi$ for some $\xi \prec \alpha$ by Lemma 15.11(vii). By Lemma 15.13(iii), we have $u =_\xi u$, which implies $u \in_\infty x$, and hence $u \in_\infty y$ by our standing assumption. We also have

$$\exists \eta \prec \rho^* \, \exists w \in y \; u =_\eta w \;\rightarrow\; \exists \eta \prec \alpha \, \exists w \in y \; u =_\eta w \,,$$

and hence

$$\neg\neg\exists \eta \prec \rho^* \, \exists w \in y \; u =_\eta w \;\rightarrow\; \neg\neg\exists \eta \prec \alpha \, \exists w \in y \; u =_\eta w \,,$$

using Lemma 15.13(ii). Thence, as $u \in_\infty y$, we can conclude that $\neg\neg\exists \eta \prec \alpha \, \exists w \in y \; u =_\eta w$. As a result,

$$\forall u \in x \, \neg\neg\exists \eta \prec \alpha \, \exists w \in y \; u =_\eta w$$

Likewise, we can conclude that $\forall v \in y \, \neg\neg\exists \eta \prec \alpha \, \exists u \in x \; v =_\eta u$, so that $x =_\alpha y$, and consequently $x =_\infty y$. □

Corollary 15.3. *Let* $x, y, z \in V_{\rho^*}^{\neg\neg}$. *Then:*

$$x =_\infty y \wedge x \in_\infty z \rightarrow y \in_\infty z \,.$$

Proof. Suppose $x =_\alpha y$, $x =_\beta u$, and $u \in z$ for some $\alpha, \beta \prec \rho^*$. Pick $\delta \prec \rho^*$ such that $x, y, z \in V_\delta^{\neg\neg}$. By Lemma 15.13 we have $x =_\delta y$, $x =_\delta u$, and thus $y =_\delta u$, which entails $y \in_\infty z$. As a result of the foregoing we have

$$\exists \alpha \prec \rho^* \; x =_\alpha y \;\wedge\; \exists \delta' \prec \rho^* \, \exists u \in z \; x = u \rightarrow y \in_\infty z,$$

$$x =_\infty y \;\wedge\; x \in_\infty z \rightarrow y \in_\infty z \,,$$

exploiting (again) that $y \in_\infty z$ is a twice negated formula. □

Next we will show in $\mathbf{CZF} + \mathbf{Pow}^{\neg\neg}$ that the structure $(V_{\rho^*}^{\neg\neg}, \in_\infty, =_\infty)$ models the double negation translation of all the axioms of $\mathbf{Z}$ + 'V_ρ *exists*' when the elementhood and equality symbols are interpreted as $\in_\infty$ and $=_\infty$, respectively.

Definition 15.10 (N-translation). Let the map $(.)^N$ from the language of set theory into itself be defined as follows:

$$\begin{aligned}
(x \in y)^N &:= x \in_\infty y \\
(x = y)^N &:= x =_\infty y \\
(A \wedge B)^N &:= A^N \wedge B^N \\
(A \vee, \exists B)^N &:= \neg(\neg A^N \wedge \neg B^N) \\
(A \rightarrow B)^N &:= A^N \rightarrow B^N \\
(\neg A)^N &:= \neg\, A^N \\
(\forall x A)^N &:= \forall x \, A^N \\
(\exists x A)^N &:= \neg \forall x \, \neg A^N \,.
\end{aligned}$$

Note that the formulae $x \in_\infty y$ and $x =_\infty y$ are already doubly negated, so that there is no need to put double negations in front of them.

Lemma 15.14. **CZF + Pow$^{\neg\neg}$** $\vdash (V^{\neg\neg}_{\rho^*}, \in_\infty, =_\infty) \models$ *(Extensionality)*^{N}.

Proof. Observe that (Extensionality)^N is

$$\forall x, \forall y\, [x =_\infty y \leftrightarrow \forall u\, (u \in_\infty x \leftrightarrow u \in_\infty y)]\,.$$

So the claimed assertion is a consequence of Corollary 15.2. □

Corollary 15.4. **CZF + Pow$^{\neg\neg}$** $\vdash (V^{\neg\neg}_{\rho^*}, \in_\infty, =_\infty) \models \forall x \forall y\, [A(x) \wedge x = y \to A(y)]$.

Proof. This follows from Lemma 15.14 and Corollary 15.3 by formula induction on $A(x)$. □

Lemma 15.15. **CZF + Pow$^{\neg\neg}$** $\vdash (V^{\neg\neg}_{\rho^*}, \in_\infty, =_\infty) \models$ *(Pairing)*^{N}.

Proof. Let $a, b \in V^{\neg\neg}_{\rho^*}$. Pick $\alpha \prec \rho^*$ such that $a, b \in V^{\neg\neg}_\alpha$ and let

$$c := \{x \in V^{\neg\neg}_\alpha \mid \neg\neg(x =_\infty a \vee, \exists\, x =_\infty b)\}\,.$$

Note that $c \subseteq V^{\neg\neg}_\alpha$. If $u \in V^{\neg\neg}_\alpha$ and $\neg\neg u \in c$, then $\neg\neg(\neg\neg(u =_\infty a \vee, \exists\, u =_\infty b))$, hence $\neg\neg(u =_\infty a \vee, \exists\, u =_\infty b)$, so that $u \in c$. This shows that $c \in \mathcal{P}^{\neg\neg}(V^{\neg\neg}_\alpha)$, thus $c \in V^{\neg\neg}_{\alpha+1}$.

Now suppose $z =_\infty x$ and $x \in c$. Then $\neg\neg(x =_\infty a \vee, \exists\, x =_\infty b)$, and thus, by Corollary 15.4, $\neg\neg(z =_\infty a \vee, \exists\, z =_\infty b)$. Hence, as $z =_\beta x$ implies $z =_\infty x$,

$$\begin{aligned}
\beta \prec \rho^* \wedge x \in c \wedge z =_\beta x &\to \neg\neg(z =_\infty a \vee, \exists\, z =_\infty b)\\
\exists \beta \prec \rho^*\, \exists x \in c\; z =_\beta x &\to \neg\neg(z =_\infty a \vee, \exists\, z =_\infty b)\\
\neg\neg\exists \beta \prec \rho^*\, \exists x \in c\; z =_\beta x &\to \neg\neg(z =_\infty a \vee, \exists\, z =_\infty b)\\
z \in_\infty c &\to \neg\neg(z =_\infty a \vee, \exists\, z =_\infty b)\,.
\end{aligned}$$

Conversely, $z =_\infty a \vee, \exists\, z =_\infty b$ implies $z \in_\infty c$ by Corollary 15.4 since $a \in_\infty c$ and $b \in_\infty c$. Thus $\neg\neg(z =_\infty a \vee, \exists\, z =_\infty b)$ implies $z \in_\infty c$ since the latter formula starts with a negation. □

Lemma 15.16. **CZF + Pow$^{\neg\neg}$** $\vdash (V^{\neg\neg}_{\rho^*}, \in_\infty, =_\infty) \models$ *(Union)*^{N}.

Proof. Let $a \in V^{\neg\neg}_{\rho^*}$. Pick $\alpha \prec \rho^*$ such that $a \in V^{\neg\neg}_\alpha$ and let

$$c := \{v \in V^{\neg\neg}_\alpha \mid \neg\neg\exists z \in a\; v \in_\infty z\}\,.$$

Note that $c \subseteq V^{\neg\neg}_\alpha$. If $v \in V^{\neg\neg}_\alpha$ and $\neg\neg v \in c$, then $\neg\neg(\neg\neg\exists z \in a\; v \in_\infty z)$, hence $\neg\neg\exists z \in a\; v \in_\infty z$, so that $v \in c$. This shows that $c \in \mathcal{P}^{\neg\neg}(V^{\neg\neg}_\alpha)$, thus $c \in V^{\neg\neg}_{\alpha+1}$.

For $x \in V^{\neg\neg}_{\rho^*}$ we have:

$$\beta \prec \rho^* \;\wedge\; v \in c \;\wedge\; x =_\beta v \rightarrow x =_\infty v \;\wedge\; \neg\neg\exists z \in a\; v \in_\infty z$$

$$\beta \prec \rho^* \;\wedge\; v \in c \;\wedge\; x =_\beta v \rightarrow x =_\infty v \;\wedge\; \neg\neg\exists z \in V^{\neg\neg}_{\rho^*}\, (z \in_\infty a \;\wedge\; v \in_\infty z)$$

$$\beta \prec \rho^* \;\wedge\; v \in c \;\wedge\; x =_\beta v \rightarrow \neg\neg\exists y \in V^{\neg\neg}_{\rho^*}\, (y \in_\infty a \;\wedge\; x \in_\infty y) \quad \text{(by Corollary 15.4)}$$

$$\neg\neg\exists\gamma \prec \rho^*\, \exists u \in c\; x =_\beta u \rightarrow \neg\neg\exists y \in V^{\neg\neg}_{\rho^*}\, (y \in_\infty a \;\wedge\; x \in_\infty y)$$

$$x \in_\infty c \rightarrow \neg\neg\exists y \in V^{\neg\neg}_{\rho^*}\, (y \in_\infty a \;\wedge\; x \in_\infty y)\,.$$

Conversely, let $x, y, z \in V^{\neg\neg}_{\rho^*}$ and $\beta, \delta \prec \rho^*$. Then:

$$y \in a \;\wedge\; u \in y \rightarrow u \in V^{\neg\neg}_\alpha \;\wedge\; \exists z \in a\; u \in_\infty z$$

$$y \in a \;\wedge\; u \in y \rightarrow u \in V^{\neg\neg}_\alpha \;\wedge\; \neg\neg\exists z \in a\; u \in_\infty z$$

$$y \in a \;\wedge\; u \in y \rightarrow u \in_\infty c$$

$$y \in a \;\wedge\; u \in y \;\wedge\; x =_\beta u \rightarrow x \in_\infty c \quad \text{(by Corollary 15.4)}$$

$$y \in a \;\wedge\; \neg\neg\exists\beta' \prec \rho^*\, \exists u' \in y\; x =_{\beta'} u' \rightarrow x \in_\infty c$$

$$y \in a \;\wedge\; x \in_\infty y \rightarrow x \in_\infty c$$

$$y \in a \;\wedge\; z =_\delta y \;\wedge\; x \in_\infty z \rightarrow x \in_\infty c \quad \text{(by Corollary 15.4)}$$

$$\exists\delta' \prec \rho^*\, \exists y' \in a\; z =_\delta y' \;\wedge\; x \in_\infty z \rightarrow x \in_\infty c$$

$$\neg\neg\exists\delta' \prec \rho^*\, \exists y' \in a\; z =_\delta y' \;\wedge\; x \in_\infty z \rightarrow x \in_\infty c$$

$$z \in_\infty a \;\wedge\; x \in_\infty z \rightarrow x \in_\infty c$$

$$\neg\neg\exists z' \in V^{\neg\neg}_{\rho^*}(z' \in_\infty a \;\wedge\; x \in_\infty z') \rightarrow x \in_\infty c\,.$$

From the above we conclude that $(V^{\neg\neg}_{\rho^*}, \in_\infty, =_\infty) \models (\text{Union})^N$. □

Lemma 15.17. **CZF + Pow**$^{\neg\neg} \vdash (V^{\neg\neg}_{\rho^*}, \in_\infty, =_\infty) \models$ *(full Separation)*^{N}.

Proof. Let $a \in V^{\neg\neg}_{\rho^*}$ and let $A(v)$ be a formula with parameters from $V^{\neg\neg}_{\rho^*}$ and at most the free variable v. Let $A^*(v)$ arise from $A(v)$ by first applying the N-translation and subsequently restricting all unbounded quantifiers to $V^{\neg\neg}_{\rho^*}$. Pick $\alpha \prec \rho^*$ such that $a \in V^{\neg\neg}_\alpha$ and let

$$c := \{x \in V^{\neg\neg}_\alpha \mid x \in_\infty a \;\wedge\; (V^{\neg\neg}_{\rho^*}, \in_\infty, =_\infty) \models A^*(x)\}\,.$$

c is a set by bounded Separation in our background theory. Obviously $c \subseteq V^{\neg\neg}_\alpha$. Suppose $u \in V^{\neg\neg}_\alpha$ and $\neg\neg u \in c$. Then $\neg\neg u \in_\infty a$ and $\neg\neg A^*(u)$, thus $u \in_\infty a$ and $A^*(u)$ since both formulae are negative. As a result, $c \in V^{\neg\neg}_{\alpha+1}$.

Now let $x \in V^{\neg\neg}_{\rho^*}$ and $\beta \prec \rho^*$. Then:

$$
\begin{aligned}
u \in c &\to u \in_\infty a \wedge A^*(u) \\
u \in c \wedge x =_\beta u &\to x \in_\infty a \wedge A^*(x) \quad \text{(by Corollary 15.4)} \\
\exists \beta' \prec \rho^* \, \exists u \in c \; x =_\beta u &\to x \in_\infty a \wedge A^*(x) \\
\neg\neg\exists \beta' \prec \rho^* \, \exists u \in c \; x =_\beta u &\to x \in_\infty a \wedge A^*(x) \quad \text{(succedent is negative)} \\
x \in_\infty c &\to x \in_\infty a \wedge A^*(x).
\end{aligned}
$$

$$
\begin{aligned}
u \in a \wedge A^*(u) &\to u \in c \\
u \in a \wedge A^*(u) &\to u \in_\infty c \\
u \in a \wedge x =_\beta u \wedge A^*(x) &\to x \in_\infty c \quad \text{(by Corollary 15.4)} \\
\exists \beta' \prec \rho^* \, \exists u \in a \; x =_\beta u \wedge A^*(x) &\to x \in_\infty c \\
x \in_\infty a \wedge A^*(x) &\to x \in_\infty c\,.
\end{aligned}
$$

As a result of the above we have

$$(V^{\neg\neg}_{\rho^*}, \in_\infty, =_\infty) \models \exists z \, \forall x [x \in_\infty z \leftrightarrow (x \in_\infty a \wedge A^N(x))]. \qquad \square$$

Lemma 15.18. **CZF + Pow**$^{\neg\neg}$ $\vdash (V^{\neg\neg}_{\rho^*}, \in_\infty, =_\infty) \models (\textit{Set Induction})^N$.

Proof. Let $A(v)$ be a formula with parameters from $V^{\neg\neg}_{\rho^*}$ and at most the free variable v. Let $A^*(v)$ arise from $A(v)$ by first applying the N-translation and subsequently restricting all unbounded quantifiers to $V^{\neg\neg}_{\rho^*}$. Assume that

$$\forall z \in V^{\neg\neg}_{\rho^*} [\forall y \in V^{\neg\neg}_{\rho^*} (y \in_\infty z \to A^*(y)) \to A^*(z)]. \tag{15.20}$$

Let $a \in V^{\neg\neg}_{\alpha}$ where $\alpha \prec \rho^*$. The aim is to show that $A^*(a)$ holds. To this end we proceed by induction on α. If $u \in a$ then $u \in V^{\neg\neg}_{\xi}$ for some $\xi \prec \alpha$ by Lemma 15.12(vii), thus $A^*(u)$ holds by the inductive assumption. For $x \in V^{\neg\neg}_{\rho^*}$ we thus have

$$
\begin{aligned}
x =_\beta u \wedge u \in a &\to A^*(x) \quad \text{(by Corollary 15.4)} \\
\exists \beta \prec \rho^* \, \exists u \in a \; x =_\beta u &\to A^*(x) \\
x \in_\infty a &\to A^*(x) \quad (A^*(x) \text{ being negative})\,.
\end{aligned}
$$

In view of our assumption (15.20) we thus have $A^*(a)$. $\square$

Lemma 15.19. **CZF + Pow**$^{\neg\neg} \vdash (V_{\rho^*}^{\neg\neg}, \in_\infty, =_\infty) \models (\textit{Power Set})^N$.

Proof. For $x, y \in V_{\rho^*}^{\neg\neg}$ define $x \subseteq_\infty y$ as $\forall u \in V_{\rho^*}^{\neg\neg}\,(u \in_\infty x \to u \in_\infty y)$.

Let $a \in V_{\alpha+1}^{\neg\neg}$ for some $\alpha \prec \rho^*$. Let

$$c := \{x \in V_{\alpha+1}^{\neg\neg} \mid x \subseteq_\infty a\}.$$

Then $c \subseteq V_{\alpha+1}^{\neg\neg}$ and we have

$$\begin{aligned}
w \in V_{\alpha+1}^{\neg\neg} \wedge \neg\neg w \in c &\to \neg\neg w \subseteq_\infty a\\
&\to \neg\neg \forall u \in V_{\rho^*}^{\neg\neg}\,(u \in_\infty w \to u \in_\infty a)\\
&\to \forall u \in V_{\rho^*}^{\neg\neg}\,\neg\neg(u \in_\infty w \to u \in_\infty a)\\
&\to \forall u \in V_{\rho^*}^{\neg\neg}\,(\neg\neg u \in_\infty w \to \neg\neg u \in_\infty a)\\
&\to \forall u \in V_{\rho^*}^{\neg\neg}\,(u \in_\infty w \to u \in_\infty a)\\
&\to w \subseteq_\infty a\\
&\to w \in c\,.
\end{aligned}$$

This shows that $c \in V_{\alpha+2}^{\neg\neg}$.

Now suppose $y \subseteq_\infty a$. Let

$$y^* := \{v \in V_\alpha^{\neg\neg} \mid v \in_\infty y\}.$$

Then $y^* \subseteq_\infty y$. Let $u \in_\infty y$. Then $u \in a$ and hence

$$\begin{aligned}
\beta \prec \rho^* \;\wedge\; u =_\beta v \;\wedge\; v \in a &\to v \in_\infty y\\
\beta \prec \rho^* \;\wedge\; u =_\beta v \;\wedge\; v \in a &\to v \in_\infty y^*\\
\beta \prec \rho^* \;\wedge\; u =_\beta v \;\wedge\; v \in a &\to u \in_\infty y^*\\
\exists \beta \prec \rho^*\, \exists v \in a\; u =_\beta v &\to u \in_\infty y^*\\
\neg\neg\,\exists \beta \prec \rho^*\, \exists v \in a\; u =_\beta v &\to u \in_\infty y^*\\
u \in_\infty y &\to u \in_\infty y^*\,.
\end{aligned}$$

So $y \subseteq_\infty y^*$, which together with $y^* \subseteq_\infty y$ yields $y =_\infty y^*$, and hence $y \in_\infty c$. As a result,

$$y \subseteq_\infty a \to y \in_\infty c\,. \tag{15.21}$$

Conversely, suppose $y \in_\infty c$. Then we have

$$
\begin{aligned}
\beta \prec \rho^* \;\wedge\; y =_\beta z \;\wedge\; z \in c &\rightarrow z \subseteq_\infty a \\
\beta \prec \rho^* \;\wedge\; y =_\beta z \;\wedge\; z \in c &\rightarrow y \subseteq_\infty a \\
\exists\beta \prec \rho^* \,\exists z \in c\; y =_\beta z &\rightarrow y \subseteq_\infty a \\
\neg\neg\,\exists\beta \prec \rho^* \,\exists z \in c\; y =_\beta z &\rightarrow y \subseteq_\infty a \\
y \in_\infty c &\rightarrow y \subseteq_\infty a\,. \qquad (15.22)
\end{aligned}
$$

Equations 15.21 and 15.22 imply that the Powerset axiom holds in $(V_{\rho^*}^{\neg\neg}, \in_\infty, =_\infty)$.
□

Lemma 15.20. **CZF + Pow**$^{\neg\neg} \vdash (V_{\rho^*}^{\neg\neg}, \in_\infty, =_\infty) \models$ *(Infinity)*^{N}.

Proof. By recursion on $n \in \omega$ define

$$
\begin{aligned}
0^* &:= \emptyset \\
(n+1)^* &:= \{u \in V_{n+1}^{\neg\neg} \mid \neg\neg(u \in_\infty n^* \vee,\exists\, u =_\infty n^*)\} \\
\omega^* &:= \{u \in V_\omega^{\neg\neg} \mid \neg\neg\,\exists n \in \omega\; u =_\infty n^*\}\,.
\end{aligned}
$$

By induction on n one readily verifies that $n^* \in V_{n+1}^{\neg\neg}$. Also $\omega^* \in V_{\omega+1}^{\neg\neg}$. Moreover, it is by now routine (though tedious) to verify that the following statement holds in $(V_{\rho^*}^{\neg\neg}, \in_\infty, =_\infty)$:

$$
\begin{aligned}
&\forall x\,[x \in \omega^* \;\leftrightarrow\; \neg\neg\,(\neg\neg\,\exists u\, u \in x \vee,\exists\,\neg\neg\,\exists y\,[y \in \omega^* \;\wedge\; \forall v \\
&\quad (v \in_\infty x \;\leftrightarrow\; \neg\neg(v \in y \vee,\exists\, v = y))])]\,. \qquad (15.23)
\end{aligned}
$$

It is a consequence of (15.23) that the N-translation of the Infinity axiom holds in $(V_{\rho^*}^{\neg\neg}, \in_\infty, =_\infty)$.
□

Lemma 15.21. **CZF + Pow**$^{\neg\neg} \vdash (V_{\rho^*}^{\neg\neg}, \in_\infty, =_\infty) \models$ *(V_ρ exists)*^{N}.

Proof. The elements of the ordinal representation system $\mathcal{OT}(\vartheta) = (\mathrm{OT}(\vartheta) \cap \Omega, \prec)$ are elements of ω. In the proof of Lemma 15.20 we defined the internalization $n^* \in V_{n+1}^{\neg\neg}$ of $n \in \omega$ in the structure $(V_{\rho^*}^{\neg\neg}, \in_\infty, =_\infty)$. We will now define the internalization $H(\alpha)$ of the ordered pair $\langle \alpha^*, V_\alpha^{\neg\neg} \rangle$ for each $\alpha \prec \rho$. Recall that we chose ρ to be of the form ω^{ρ_0} for some $\rho_0 \succ 1$, so that for all $\alpha \prec \rho$, $\alpha + \omega \prec \rho$. For $x \in V_{\rho^*}^{\neg\neg}$ we use $x \in_\infty OP(\alpha^*, V_\alpha^{\neg\neg})$ to abbreviate the following formula:

$$
\begin{aligned}
&\neg\neg\,[\forall v \in V_{\rho^*}^{\neg\neg}(v \in_\infty x \;\leftrightarrow\; v =_\infty \alpha^*) \vee,\exists\, \forall v \in V_{\rho^*}^{\neg\neg} \\
&\quad (v \in_\infty x \;\leftrightarrow\; \neg\neg\,(v =_\infty \alpha^* \vee,\exists\, v =_\infty V_\alpha^{\neg\neg}))]\,.
\end{aligned}
$$

For $\alpha \prec \rho$ define

$$H(\alpha) := \{x \in V^{\neg\neg}_{\omega+\alpha+2} \mid x \in_\infty OP(\alpha^*, V^{\neg\neg}_\alpha)\}$$
$$V^*_\rho := \{z \in V^{\neg\neg}_\rho \mid \neg\neg\exists\alpha \prec \rho\, z =_\infty H(\alpha)\}.$$

One readily checks that $H(\alpha) \in V^{\neg\neg}_{\omega+\alpha+3}$ and $V^*_\rho \in V^{\neg\neg}_{\rho+1}$. It remains to show that V^*_ρ is the set witnessing that $(V^{\neg\neg}_{\rho^*}, \in_\infty, =_\infty) \models (V_\rho \text{ exists})^N$ holds. This is so by design of V^*_ρ but it is rather tedious to check in detail. □

15.7.2 Proof of Theorem 15.1

We use $\leq$ and $\equiv$ for the relations of being proof-theoretically reducible and proof-theoretically equivalent, respectively. We have $\mathbf{CZF}_\mathcal{P} + \mathbf{RDC} + \mathbf{\Pi\Sigma-AC} \leq \mathbf{CZF}_\mathcal{P} \leq \mathbf{IKP}(\mathcal{P})$ using Theorem 15.6 and Theorem 15.2. By Lemma 15.4 and Theorem 15.3 we get $\mathbf{CZF}_\mathcal{P} \leq \mathcal{T}^- \leq \mathbf{KP}(\mathcal{P})$. $\mathbf{CZF} + \mathbf{Pow}^{\neg\neg} \leq \mathbf{MLV_P} \leq \mathbf{CZF}_\mathcal{P}$ holds by Theorems 15.5 and 15.4. Theorems 15.8 and 15.7 yield $\mathbf{KP}(\mathcal{P}) \leq \mathbf{Z} + \{`V_\tau \text{ exists'}\}_{\tau\in\mathrm{BH}} \leq \mathbf{CZF} + \mathbf{Pow}^{\neg\neg}$. Moreover, $\mathbf{IKP}(\mathcal{P}) \equiv \mathbf{IZ} + \{`V_\tau \text{ exists'}\}_{\tau\in\mathrm{BH}}$ holds by Theorem 15.7. The upshot of these results is thus that all theories of Theorem 15.1 are proof-theoretically equivalent. □

References

Aczel, P. 1978. The type theoretic interpretation of constructive set theory. In *Logic Colloquium '77*, ed. A. MacIntyre, L. Pacholski, and J. Paris. Amsterdam: North–Holland.

Aczel, P. 1982. The type theoretic interpretation of constructive set theory: Choice prinicples. In *The L.E.J. brouwer centenary symposium*, ed. A.S. Troelstra and D. van Dalen. Amsterdam: North–Holland.

Aczel, P. 1986. The type theoretic interpretation of constructive set theory: Inductive definitions. In *Logic, methodology and philosophy science VII*, ed. R.B. Marcus et al., 17–49. Amsterdam: North–Holland.

Aczel, P. 2000. On relating type theories and set theories. In *Types '98*, Lecture notes in computer science, vol. 1257, ed. T. Altenkirch, W. Naraschewski, and B. Reus. Berlin: Springer.

Aczel, P., and M. Rathjen. 2001. Em notes on constructive set theory, Technical report 40, Institut Mittag-Leffler. Stockholm: The Royal Swedish Academy of Sciences. http://www.ml.kva.se/preprints/archive2000-2001.php

Aczel, P., and M. Rathjen. 2010. *Constructive set theory*, book draft.

Barwise, J. 1975. *Admissible sets and structures*. Berlin/Heidelberg/New York: Springer.

Beeson, M. 1985. *Foundations of constructive mathematics*. Berlin: Springer.

Chen, R.-M., and M. Rathjen. 2012. *Lifschitz realizability for intuitionistic Zermelo-Fraenkel set theory*, to appear in: Archive for Mathematical Logic.

Coquand, T. 1990. Metamathematical investigations of a calculus of constructions. In *Logic and Computer science*, ed. P. Oddifreddi, 91–122. London: Academic.

Friedman, H. 1973a. Some applications of Kleene's method for intuitionistic systems. In *Cambridge summer school in mathematical logic*, Lectures notes in mathematics, vol. 337, ed. A. Mathias and H. Rogers, 113–170. Berlin: Springer.

Friedman, H. 1973b. Countable models of set theories. In *Cambridge summer school in mathematical logic*, Lectures Notes in Mathematics, vol. 337, ed. A. Mathias and H. Rogers, 539–573. Berlin: Springer.

Friedman, H. 1973c. The consistency of classical set theory relative to a set theory with intuitionistic logic. *Journal of Symbolic Logic* 38: 315–319.

Feferman, S. 1979. Constructive theories of functions and classes. In *Logic colloquium '78*, ed. M. Boffa, D. van Dalen, and K. McAloon, 1–52. Amsterdam: North-Holland.

Gambino, N. 1999. Types and sets: A study on the jump to full impredicativity, Laurea dissertation, Department of Pure and Applied Mathematics, University of Padua.

Lifschitz, V. 1979. CT0 is stronger than CT0! *Proceedings of the American Mathematical Society* 73: 101–106.

Lubarsky, R.S. 2006. CZF and second order arithmetic. *Annals of Pure and Applied Logic* 141: 29–34.

Mac Lane, S. 1992. *Form and function*. Berlin: Springer.

Martin-Löf, P. 1984. *Intuitionistic type theory*. Naples: Bibliopolis.

Mathias, A.R.D. 2001. The strength of Mac Lane set theory. *Annals of Pure and Applied Logic* 110: 107–234.

Moschovakis, Y.N. 1976. Recursion in the universe of sets, mimeographed note.

Moss, L. 1995. Power set recursion. *Annals of Pure and Applied Logic* 71: 247–306.

Myhill, J. 1975. Constructive set theory. *Journal of Symbolic Logic* 40: 347–382.

Normann, D. 1978. Set recursion. In *Generalized recursion theory II*, 303–320. Amsterdam: North-Holland.

Palmgren, E. 1993. Type-theoretic interpretations of iterated, strictly positive inductive definitions. *Archive for Mathematical Logic* 32: 75–99.

Pozsgay, L. 1971. Liberal intuitionism as a basis for set theory, in Axiomatic set theory. *Proceedings Symposium Pure Mathematics* 12(1): 321–330.

Pozsgay, L. 1972. Semi-intuitionistic set theory. *Notre Dame Journal of Formal Logic* 13: 546–550.

Rathjen, M. 1994. The strength of some Martin-Löf type theories. *Archive for Mathematical Logic* 33: 347–385.

Rathjen, M. 2005. Replacement versus collection in constructive Zermelo-Fraenkel set theory. *Annals of Pure and Applied Logic* 136: 156–174.

Rathjen, M. 2006a. Choice principles in constructive and classical set theories. In *Logic colloquium 2002*, Lecture notes in logic, vol. 27, ed. Z. Chatzidakis, P. Koepke, and W. Pohlers, 299–326. Wellesley: A.K. Peters.

Rathjen, M. 2006b. Realizability for constructive Zermelo-Fraenkel set theory. In *Logic Colloquium 2003*, Lecture notes in logic, vol. 24, ed. J. Väänänen and V. Stoltenberg-Hansen, 282–314. Wellesley: A.K. Peters.

Rathjen, M. 2006c. The formulae-as-classes interpretation of constructive set theory. In *Proof technology and computation*, Proceedings of the international summer school marktoberdorf 2003, ed. H. Schwichtenberg and K. Spies, 279–322. Amsterdam: IOS Press.

Rathjen, M., and S. Tupailo. 2006. Characterizing the interpretation of set theory in Martin-Löf type theory. *Annals of Pure and Applied Logic* 141: 442–471.

Rathjen, M. 2012. An ordinal analysis of Power Kripke–Platek set theory, Infinity Conference, Centre de Recerca Matemàtica Barcelona preprint series.

Rathjen, M. 2012. From the weak to the strong existence property. *Annals of Pure and Applied Logic*. doi:10.1016/j.apal.2012.01.012

Rathjen, M., and A. Weiermann. 1993. Proof-theoretic investigations on Kruskal's theorem. *Annals of Pure and Applied Logic* 60: 49–88.

Sacks, G.E. 1990. *Higher recursion theory*. Berlin: Springer.

Tharp, L. 1971. A quasi-intuitionistic set theory. *Journal of Symbolic Logic* 36: 456–460.

Thiele, E.J. 1968. Über endlich axiomatisierbare Teilsysteme der Zermelo-Fraenkel'schen Mengenlehre. *Zeitschrift für Mathematische Logik und Grundlagen der Mathematik* 14: 39–58.
Troelstra, A.S. 1998. Realizability. In *Handbook of proof theory*, ed. S.R. Buss, 407–473. Amsterdam: Elsevier.
Troelstra, A.S., and D. van Dalen. 1988. *Constructivism in mathematics, volumes I, II*. Amsterdam: North Holland.
van Oosten, J. 1990. Lifschitz's realizability. *The Journal of Symbolic Logic* 55: 805–821.
Wolf, R.S. 1974. Formally intuitionistic set theories with bounded predicates decidable. Ph.D. thesis, Stanford University.

Chapter 16
Coalgebras as Types Determined by Their Elimination Rules

Anton Setzer

Dedicated to Per Martin-Löf on the occasion of his retirement.

16.1 Introduction

Most programs in computing are interactive programs. This means that they are not batch programs, which, once started, are guaranteed to terminate after a certain amount of time and deliver their result. They are programs which keep running and interacting with user input, until they are terminated by the user. Such programs correspond to non-well-founded trees: Nodes are labelled by commands and the branching degree of a node labelled by a command is the set of responses to this command. A computation which goes on for ever corresponds to an infinite path in this tree. More details of this can be found in a series of articles by the author and Hancock and Setzer (1999, 2000a,b, 2004, 2005). Colists are simple trees with branching degrees 0 or 1, and for ease of presentation, we restrict ourselves in this article to colists.

Martin-Löf type theory is supposed to be a language in which programs can be written and in which we can prove correctness properties of such programs. In order to be able to write interactive programs and reason about them, we need to represent non-well-founded structures. Coalgebras originating from category theory provide a theory of non-well-founded structures. They allow to represent the elements of such structures in a finitary way. Elements are not per se infinitary – in fact we will

*Supported by EPSRC grant EP/G033374/1

A. Setzer (✉)
Department of Computer Science, Swansea University, Singleton Park, Swansea, SA2 8PP, UK
e-mail: a.g.setzer@swansea.ac.uk

P. Dybjer et al. (eds.), *Epistemology versus Ontology*, Logic, Epistemology, and the Unity of Science 27, DOI 10.1007/978-94-007-4435-6_16,

represent them in type theory as finitary objects. As part of a coalgebra we have a case distinction operation. In case of colists, the result of applying it to a colist is the information whether the element represents the empty list or a list formed from a given head and a given tail. By iteratively applying case distinction, a colist then unfolds to a potentially infinite list.

The goal of this article is to introduce a notion of coalgebras into type theory and provide meaning explanations for them. We want coalgebras to be first class citizens, i.e. they are not encoded in terms of other data types. This seems to be the general way of moving forward in type theory. In most other mathematical theories the goal is to define a minimal closed theory, which allows to encode all structures needed in mathematics. In type theory it is usual practice to continuously extend the theory in such a way that new structures needed are represented directly.

In this article we develop the theory of coalgebras in type theory, while closely following the categorical notions. One main focus is to develop meaning explanations for coalgebras, in order to fully integrate them into the theoretical setting of type theory. Whereas coalgebras only extend the expressiveness, not the proof theoretic strength, of type theory, we hope that this project will help to develop the basis for future proof theoretic strong extensions of type theory.

We start by exploring the notion of inductive data types, which correspond to initial algebras. We will as well review meaning explanations for them. Then we develop the notion of a final coalgebra. We will see that a simple form of guarded recursion is nothing but the introduction rule of final coalgebras, which represent the existence of morphisms in the defining diagrams for coalgebras. We will develop a slight extension of guarded recursion as well. We then explore limitations of coalgebras needed in order to maintain decidable equality. For this reason we will switch to weakly final coalgebras with an extended version of guarded recursion. We will see that in a decidable type theory we cannot assume that every element is introduced by a coconstructor. This is the underlying reason for the failure of subject reduction in implementations of type theory and problems with dependent case distinction. Next, we develop type theoretic rules for coalgebras based on extended guarded recursion.

In the last part, we will develop meaning explanations for coalgebras. We will need to change the setting of meaning explanations in order to be able to explain coalgebras. As in the original meaning explanations by Martin-Löf, inductive data types are given by explaining how to introduce its elements and when two elements introduced are equal. So the elements are determined by their introduction rules. The elimination rules are justified by verifying that they operate correctly for every element introduced. Meaning explanations of coalgebras are given differently. Elements of coalgebras are given by defining how to compute other elements from them. Elements are equal if the computed results are equal. Therefore elements are given by their elimination rules. The introduction rules are justified by verifying that they introduce elements which allow to apply the elimination principle.

Related work. The use of coalgebras in non-dependent functional programming was to the author's knowledge first introduced 1987 in the PhD thesis of Hagino

(1987) (see as well Hagino 1989). He used the terminology codatatype for coalgebras defined by their elimination rules. Aczel introduced 1988 in his book Aczel (1988) non-well-founded set theory. Non-well-founded sets are necessarily infinite objects, which can be introduced by the anti-foundation axiom, a form of guarded recursion. Based on Hagino's work, Cockett, Fukushima and Spencer developed 1992 the non-dependent functional programming language Charity with a very clean categorical syntax. Leclerc and Paulin-Mohring (1994) used the impredicative types in Coq in order to represent streams and define the sieve of Eratosthenes. Coquand 1994 introduced in Coquand (1994) the concept of guarded recursion. Giménez (1994, 1996) developed 1994 an extension of the calculus of constructions by inductive and coinductive types. He showed how to reduce general forms of guarded recursion to coalgebras. Already in his PhD thesis Giménez (1996), he discovered problems with subject reduction, which will discussed later in this paper. Paulson implemented 1994 axioms for coinduction in Isabelle Paulson (1994). Telford and Turner (1997) and Turner (2004, 1995) starting 1995 promoted the use of codata as truly infinite data types introduced by their introduction rules, and implemented them in the functional programming language Miranda. The author has together with Hancock since 1999 developed in Hancock and Setzer (1999, 2000a,b, 2004, 2005) interactive programs in dependent type theory. This included in Hancock and Setzer (2004, 2005) a definition of the rules for guarded recursion and weakly final coalgebras in Martin-Löf Type Theory (2004). Coalgebras have been introduced in the interactive theorem prover Coq. The "Coq-book" (Bertot and Castéran 2004) by Bertot and Castéran contains an extensive Chap. 14 on the development of coinductive data types and proofs of their properties. See as well the note (Bertot 2006) by Bertot. Coinductive data types have as well been implemented in Agda Norell (2007) by Norell, Danielsson, Abel and other members of the Agda development team – see intense discussions on the Agda email list (Agda 2011). The latest version, which is currently implemented in Agda using a notion for coalgebraic arguments, was presented in Altenkirch and Danielsson (2010). McBride has written a short paper (McBride 2009) on the problem of subject reduction in coalgebras, and how to develop coalgebras in observational type theory. We will discuss this paper later in detail.

General setting and notations. This paper is heavily based on Martin-Löf Type Theory (1984), mainly on the version presented in the second part of Nordström et al. (1990), with the restriction to the small logical framework outlined below. As usual we have the basic judgements A : Set, $A = B$: Set, $a : A$ and $a = b : A$. Hypothetical judgements will be written as $\Gamma \Rightarrow \theta$, where θ is a basic judgement and Γ a context. Contexts Γ have the form $x : A_1, \ldots, x_n : A_n$, where $x_1 : A_1, \ldots, x_{i-1} : A_{i-1} \Rightarrow A_i$: Set. If $\emptyset$ is the empty context, we write instead of $\emptyset \Rightarrow \theta$ simply θ.

We will develop type theory based on the small logical framework, see for instance Setzer (2008). If A : Set and $x : A \Rightarrow B$: Set, we can form the dependent function set $(x : A) \to B$: Set. (This type is often written as $\Pi x : A.B$. However, in Martin-Löf Type Theory $\Pi x : A.B$ is reserved for the inductive data type having constructor $\lambda : ((x : A) \to B) \to \Pi x : A.B)$.

The canonical elements of $(x : A) \to B$ are terms $(x)t$ where $x : A \Rightarrow t : B$, which is sometimes written as $\lambda x.t$. Following the conventions in Martin-Löf Type Theory, we reserve λ for the constructor of $\Pi x : A.B$. Application is written in functional style in the form $(s\ t)$. We use usual abbreviations such as writing $(r\ s\ t)$ for $((r\ s)\ t)$ (the outermost brackets are only for better readability). Furthermore $(x : A, y : B, z : C) \to D$ denotes $(x : A) \to ((y : B) \to ((z : C) \to D))$.

Note that large types such as $(x : A) \to \mathrm{Set}$ are only allowed in the full logical framework. The reason for restricting ourselves to the small logical framework is that we have a satisfactory understanding of how to develop meaning explanations for it. One central part of this article is the discussion of meaning explanation for coinductive types.

Because of the restriction to the small logical framework, arguments referring to elements of type Set are presented as premises in rules. For practical applications, the use of the full logical framework, as it is implemented for instance in Agda, is preferred. Then these arguments can easily be abstracted.

Apart from the standard structural rules and the rules for the dependent function sets, we add rules for the intensional equality type $a ==_A b$ (where $A : \mathrm{Set}$, $a : A$ and $b : A$), the one element set $\mathbf{1}$ with only element $* : \mathbf{1}$, the binary product $(A \times B)$ (where $A, B : \mathrm{Set}$), the disjoint union $(A + B)$ (again $A, B : \mathrm{Set}$), and the set of natural numbers $\mathbb{N}$. The use of $\mathbb{N}$ is not crucial for the development of type theory in this article, we just use it as a convenient example set.

We will use expressions such as $C(x)$, $\mathrm{step}_{\mathrm{cons}}(n, l)$ for terms depending on free variables x or n, l. After using $C(x)$, the expression $C(t)$ is the result of substituting the term t for x (where we identify α-equivalent terms and resolve substitution problems as usual). After a premise of a rule $x : A \Rightarrow C(x) : \mathrm{Set}$ we write simply C rather than $(x)C(x)$ for the argument C. The same applies to similar expressions as well.

16.2 Initial Algebras Defined by Their Introduction Rules

The set of lists in Martin-Löf Type Theory. In Martin-Löf type theory, types are usually introduced by their introduction rules. Let us consider the type of lists of natural numbers. It has formation rule

$$\mathrm{List}_{\mathbb{N}} : \mathrm{Set}$$

and introduction rules

$$\mathrm{nil} : \mathrm{List}_{\mathbb{N}} \qquad \mathrm{cons} : \mathbb{N} \to \mathrm{List}_{\mathbb{N}} \to \mathrm{List}_{\mathbb{N}}$$

The elimination rules express that $\mathrm{List}_{\mathbb{N}}$ is the least set closed under these operations, as expressed by the principle of higher type primitive recursion over lists:

$$\frac{x : \mathrm{List}_{\mathbb{N}} \Rightarrow C(x) : \mathrm{Set}}{\begin{aligned}\mathrm{Rec}^{\mathrm{List}}_{C} : {} & (\mathrm{step}_{\mathrm{nil}} : C(\mathrm{nil})) \\ & \to (\mathrm{step}_{\mathrm{cons}} : (n : \mathbb{N}, l : \mathrm{List}_{\mathbb{N}}) \to C(l) \to C(\mathrm{cons}\ n\ l)) \\ & \to (l : \mathrm{List}_{\mathbb{N}}) \\ & \to C(l)\end{aligned}}$$

The equality rules, where we omit the obvious assumptions on types of the parameters, are as follows:

$$\begin{aligned}\mathrm{Rec}^{\mathrm{List}}_{C}\ \mathrm{step}_{\mathrm{nil}}\ \mathrm{step}_{\mathrm{cons}}\ \mathrm{nil} &= \mathrm{step}_{\mathrm{nil}} \\ \mathrm{Rec}^{\mathrm{List}}_{C}\ \mathrm{step}_{\mathrm{nil}}\ \mathrm{step}_{\mathrm{cons}}\ (\mathrm{cons}\ n\ l) &= \mathrm{step}_{\mathrm{cons}}\ n\ l\ (\mathrm{Rec}^{\mathrm{List}}_{C}\ \mathrm{step}_{\mathrm{nil}}\ \mathrm{step}_{\mathrm{cons}}\ l)\end{aligned}$$

By the type theoretic rules for $\mathrm{List}_{\mathbb{N}}$ we mean the rules above.

Meaning explanations were introduced by Martin-Löf (1984, 1987, 1996, 1998). They are part of a program to develop a theory in such a way that we have a direct insight that everything proved in it is correct. By Gödel's incompleteness theorem we know that there is no proof of the consistency of any reasonable mathematical theory by weaker methods. Therefore, there is no mathematical argument which guarantees that the mathematical theories used for proving theorems are actually consistent, and which wouldn't be prone to the danger of using an inconsistency of the theory in question. So any justification for the consistency of a reasonable mathematical theory needs ultimately be based on a philosophical argument. Such an argument can never be fully formal – otherwise we would obtain a mathematical proof of the consistency of the theory in question. What meaning explanations by Martin-Löf provide is the to the author's knowledge at this time best possible way of getting a direct insight into the validity of the judgements derivable in Martin-Löf type theory. They are a way of making as precise as possible the reasons why all judgements derivable in this theory are valid.

In meaning explanations one gives a meaning to each judgement and investigates for each rule that we obtain valid judgements in the conclusion from valid judgements in the premise. The meaning of a set is given by explaining what the elements are and when two elements are equal. Two sets are equal if an element of one set is an element of the other, and if two elements are equal in one set they are so in the other.

One should note that meaning explanations, as the author understand them, justify extensional equality. For colists, as defined later, they will even justify bisimilarity as equality (which will be introduced below). We do not see any inherent problem in it. The reason for having intensional equality is that we want to decide for every proposition whether a term p is a proof of this proposition. Hence we need decidable type checking.

Meaning explanations for $\mathrm{List}_{\mathbb{N}}$. In these explanations, elements are determined by their introduction rules. $\mathrm{List}_{\mathbb{N}}$ is a set. We have nil is a canonical element of $\mathrm{List}_{\mathbb{N}}$, and for n a natural number, and l an element of $\mathrm{List}_{\mathbb{N}}$ we have that $(\mathrm{cons}\ n\ l)$

is a canonical element of $\text{List}_{\mathbb{N}}$. Non-canonical elements of $\text{List}_{\mathbb{N}}$ are programs which evaluate to canonical elements of $\text{List}_{\mathbb{N}}$. The element nil is equal to itself. The elements (cons n l) and (cons n', l') are equal, if n and n' are equal elements of $\mathbb{N}$ and l and l' are equal elements of $\text{List}_{\mathbb{N}}$. The elements nil and (cons n l) are not equal. Non-canonical elements are equal, if the results of evaluating them to canonical elements are equal.

The elimination and equality rules are explained by showing how to compute from elements of $\text{List}_{\mathbb{N}}$ elements of other sets. Their explanation uses that we have determined what the canonical elements of $\text{List}_{\mathbb{N}}$ are, so it makes use of the introduction rules for $\text{List}_{\mathbb{N}}$. The explanation of $\text{Rec}_C^{\text{List}}$ is as follows: Assume $C(x)$ is a set, depending on an element x of $\text{List}_{\mathbb{N}}$. So for every element l of $\text{List}_{\mathbb{N}}$ we have that $C(l)$ is a set. Assume step_{nil} is an element of $C(\text{nil})$ and $\text{step}_{\text{cons}}$ is a function, which maps elements n of $\mathbb{N}$, l of $\text{List}_{\mathbb{N}}$ and p of $C(l)$ to elements of $C(\text{cons}\ n\ l)$. Assume l is an element of $\text{List}_{\mathbb{N}}$. Then ($\text{Rec}_C^{\text{List}}$ step_{nil} $\text{step}_{\text{cons}}$ l) is a program which computes an element of $C(l)$. This element is computed as follows: First l is computed which evaluates to a canonical element of $\text{List}_{\mathbb{N}}$. If this element is nil, then ($\text{Rec}_C^{\text{List}}$ step_{nil} $\text{step}_{\text{cons}}$ l) evaluates to the result of computing step_{nil} which is an element of $C(\text{nil})$ and therefore as well of $C(l)$. Otherwise l evaluates to (cons n l'), where n is an element of $\mathbb{N}$ and l' is an element of $\text{List}_{\mathbb{N}}$. Before we introduce l we have introduced l' and therefore $c' := \text{Rec}_C^{\text{List}}\ \text{step}_{\text{nil}}\ \text{step}_{\text{cons}}\ l'$ is an element of $C(l')$. Now ($\text{Rec}_C^{\text{List}}$ step_{nil} $\text{step}_{\text{cons}}$ l) is evaluated by computing ($\text{step}_{\text{cons}}$ n l' c') which has as result an element of $C(\text{cons}\ n\ l')$ and therefore of $C(l)$. The equality rules follow since the left hand side is evaluated by evaluating the right hand side.

$\text{List}_{\mathbb{N}}$ as an initial algebra. Assume a category having finite products (including an initial object $\mathbf{1}$ which is the empty product), and a binary coproduct $(A + B)$ for objects A, B. Assume as well a natural numbers object $\mathbb{N}$ (we will not need any specific properties about it). Elements a of objects A are arrows $a : \mathbf{1} \to A$, and we write $a : A$ for such elements. Let F_{List} be the functor with object part $\text{F}_{\text{List}}\ X = \overline{\text{nil}} + \overline{\text{cons}}(\mathbb{N}, X)$. Here $\overline{\text{nil}} + \overline{\text{cons}}(\mathbb{N}, X)$ is a notation for $\mathbf{1} + (\mathbb{N} \times X)$, where we write $\overline{\text{nil}} := \text{inl} *$ for the element $* : \mathbf{1}$ (corresponding to $\text{id} : \mathbf{1} \to \mathbf{1}$) and ($\overline{\text{cons}}$ n x) for the element (inr $\langle n, x\rangle$) where $n : \mathbb{N}$ and $x : X$. The name $\overline{\text{nil}}$ signifies a nil-shape and $\overline{\text{cons}}$ a cons-shape. For $f : A \to B$ we obtain an obvious morphism part $\text{F}_{\text{List}}\ f : \text{F}_{\text{List}}\ A \to \text{F}_{\text{List}}\ B$. An F_{List}-algebra is a pair (A, f) where A is an object and $f : \text{F}_{\text{List}}\ A \to A$. A morphism between F_{List}-algebras (A, f) and (B, g) is a function $h : A \to B$ s.t. the following diagram commutes:

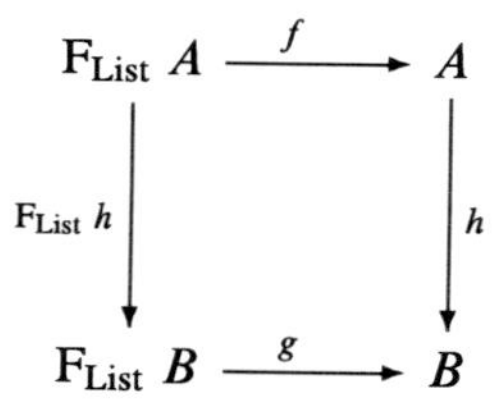

An initial $\mathrm{F}_{\mathrm{List}}$-algebra $(\mathrm{List}_{\mathbb{N}}, \mathrm{intro})$ is an initial object in the category with objects $\mathrm{F}_{\mathrm{List}}$-algebras and morphism being $\mathrm{F}_{\mathrm{List}}$-morphisms. So we have a morphism intro : $\mathrm{F}_{\mathrm{List}}(\mathrm{List}_{\mathbb{N}}) \to \mathrm{List}_{\mathbb{N}}$, and if we have any other $\mathrm{F}_{\mathrm{List}}$-algebra (A, f), i.e. if we have $f : \mathrm{F}_{\mathrm{List}}\ A \to A$, then there exists a unique $g : \mathrm{List}_{\mathbb{N}} \to A$ s.t. the following diagram commutes:

$$\begin{array}{ccc} \mathrm{F}_{\mathrm{List}}(\mathrm{List}_{\mathbb{N}}) & \xrightarrow{\mathrm{intro}} & \mathrm{List}_{\mathbb{N}} \\ \Big\downarrow {\scriptstyle \mathrm{F}_{\mathrm{List}}\ g} & & \Big\downarrow {\scriptstyle \exists! g} \\ \mathrm{F}_{\mathrm{List}}\ A & \xrightarrow{f} & A \end{array}$$

Consider now the specific category, in which objects are elements of Set (where definitionally equal sets are identified) derivable in Martin-Löf type theory. Let morphism $f : A \to B$ be functions of this type derivable in type theory. Let $f, f' : A \to B$. Consider f equal to f' as morphisms in category theoretic diagrams, if and only if f, f' are equal extensionally, i.e. $\forall a : A.f\ a ==_B f'\ a$, where $==_B$ is the intensional equality type. Assume the type theoretic rules for $\mathrm{List}_{\mathbb{N}}$. Let intro : $\mathrm{F}_{\mathrm{List}}(\mathrm{List}_{\mathbb{N}}) \to \mathrm{List}_{\mathbb{N}}$, intro $\overline{\mathrm{nil}} = \mathrm{nil}$ and intro $(\overline{\mathrm{cons}}\ n\ l) = \mathrm{cons}\ n\ l$. Then $(\mathrm{List}_{\mathbb{N}}, \mathrm{intro})$ is an initial $\mathrm{F}_{\mathrm{List}}$-algebra: It is obviously an $\mathrm{F}_{\mathrm{List}}$-algebra. Furthermore, assume (A, f) is another $\mathrm{F}_{\mathrm{List}}$-algebra. Then we can define using the elimination rule for $\mathrm{F}_{\mathrm{List}}$ a function $g : \mathrm{List}_{\mathbb{N}} \to A$ such that $g\ \mathrm{nil} = f\ \overline{\mathrm{nil}}$, $g\ (\mathrm{cons}\ n\ l) = f\ (\overline{\mathrm{cons}}\ n\ (g\ l))$. It follows in type theory that g is the unique $\mathrm{F}_{\mathrm{List}}$-algebra morphism $g : (\mathrm{List}_{\mathbb{N}}, \mathrm{intro}) \to (A, f)$: That it is a $\mathrm{F}_{\mathrm{List}}$-algebra morphism is obvious. Further, if there is any other $\mathrm{F}_{\mathrm{List}}$-algebra morphism $g' : (\mathrm{List}_{\mathbb{N}}, \mathrm{intro}) \to (A, f)$, one can show by induction on $l : \mathrm{List}_{\mathbb{N}}$ (which corresponds to the elimination rule for $\mathrm{List}_{\mathbb{N}}$) $\forall l : \mathrm{List}_{\mathbb{N}}.g(l) ==_{\mathrm{List}_{\mathbb{N}}} g'(l)$, so g and g' are equal morphisms.

Therefore the rules of type theory for $\mathrm{List}_{\mathbb{N}}$ imply the principle that $\mathrm{List}_{\mathbb{N}}$ is an initial algebra. One can show as well that the principle of $(\mathrm{List}_{\mathbb{N}}, \mathrm{intro})$ being an initial algebra implies the type theoretic rules for $\mathrm{List}_{\mathbb{N}}$. However, this direction requires extensional equality. This result is in fact a special case of Dybjer and Setzer (2003). The type theoretic rules for $\mathrm{List}_{\mathbb{N}}$ and the principle of $(\mathrm{List}_{\mathbb{N}}, \mathrm{intro})$ being an initial algebra are therefore extensionally equivalent, but are intensionally different (although we have no formal proof for this). In this sense we can regard the type theoretic rules without extensional equality as one possible representation of the rules of an initial algebra.

16.3 Weakly Final Coalgebras

Colist. We will introduce the type of colists, which are elements which can be unfolded to potentially infinite lists of natural numbers. Colists will be defined as weakly final coalgebras. Coalgebras are the dual of algebras, and are obtained by inverting the direction of the arrows in the category theoretic formulation of algebras. An $\mathrm{F_{List}}$-coalgebra is a pair (A, f) where $f : A \to \mathrm{F_{List}}\ A$, and as for algebras we sometimes omit f when it is obvious from the context. An $\mathrm{F_{List}}$-coalgebra morphism between coalgebras (A, f) and (B, g) is a function $h : A \to B$ s.t. the following diagram commutes:

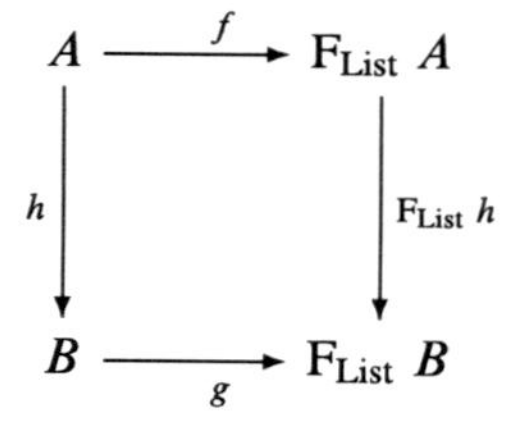

A final $\mathrm{F_{List}}$-coalgebra (coList, case) is a terminal object in the category of $\mathrm{F_{List}}$-coalgebras. Therefore, it is an $\mathrm{F_{List}}$-coalgebra. Furthermore, for any other coalgebra (A, f), i.e. $f : A \to \mathrm{F_{List}}\ A$ there exists a unique coalgebra morphism $g : (A, f) \to (\mathrm{coList}, \mathrm{case})$, i.e. a unique $g : A \to \mathrm{coList}$ s.t. the following diagram commutes:

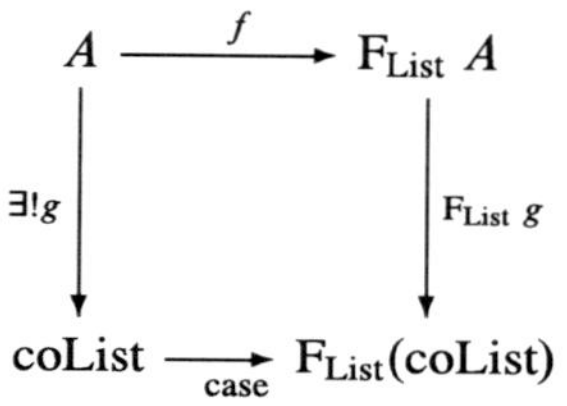

Weakly final $\mathrm{F_{List}}$-coalgebras are weakly terminal objects in the category of $\mathrm{F_{List}}$-coalgebras, which means that we omit the condition that g as above is unique. Assume in the following (coList, case) is a weakly final $\mathrm{F_{List}}$-coalgebra.

The function $\mathrm{case} : \mathrm{coList} \to (\overline{\mathrm{nil}} + \overline{\mathrm{cons}}(\mathbb{N}, \mathrm{coList}))$ determines for an element of coList whether it is of the form $\overline{\mathrm{nil}}$ or $(\overline{\mathrm{cons}}\ n\ l)$. Note that we can apply case to l again. So an element of coList is an element which can, by iteratively applying case to it, be unfolded to a potentially infinite list. For instance an element $a : \mathrm{coList}$ s.t. $\mathrm{case}\ a = \overline{\mathrm{cons}}\ 0\ a$ represents what would in a framework of infinite terms be the infinite list (cons 0 (cons 0 (cons 0 $\cdots$))).

Codata types and guarded recursion. In functional programming, codata types (Turner 2004) are often considered as variants of algebraic data types which allow the formation of infinitely many applications of constructors. For instance one could define the codata type of colists, which has constructors nil and cons. Then it is possible to have infinite nesting of cons and define a colist (cons 0 (cons 0 (cons 0 $\cdots$))) directly. One sees immediately that this destroys normalisation. We will see below that decidable type checking is not possible, if we assume that each element of a coalgebra is introduced by a constructor. Coalgebras are a version of codata types, where elements are not per se infinitary, but unfold to infinite objects.

Relationship to guarded recursion. Guarded recursion was introduced by Coquand (1994) in a setting of infinitary terms. Bertot and Castéran use in Chap. 13 of the "Coq-book" Bertot and Castéran (2004) guarded recursion and codata types extensively for the development of infinite objects and proofs for these objects. Guarded recursion allows to define elements of codata types recursively, by allowing full recursion, as long as recursive calls are guarded by at least one (possibly more) constructors of the codata type in question, and no other functions are applied to the result of a recursive call. A simple form of guarded recursion is where we always have one recursive call guarded by exactly one constructor.

We can see now that in the coalgebraic setting the existence of the $\mathrm{F_{List}}$-coalgebra morphism $g : A \to \mathrm{coList}$ for any $\mathrm{F_{List}}$-coalgebra (A, f) corresponds to this simple form of guarded recursion: We have

$$\mathrm{case}\ (g\ a) = \begin{cases} \overline{\mathrm{nil}} & \text{if } f\ a = \overline{\mathrm{nil}}, \\ \overline{\mathrm{cons}}\, n\ (g\ a') & \text{if } f\ a = \overline{\mathrm{cons}}\, n\ a'. \end{cases}$$

By choosing suitable f we can therefore define $g : A \to \mathrm{coList}$ by guarded recursion, s.t. for $a : A$ we have $\mathrm{case}\ (g\ a) = \overline{\mathrm{nil}}$ or $\mathrm{case}\ (g\ a) = \overline{\mathrm{cons}}\ n\ (g\ a')$. Which of the two cases holds and the choice of n and a' can be decided depending on a. Note that there are no conditions on a' to be smaller than a. This principle is the simple form of the principle of guarded recursion. The difference to the setting using codata types is that $(g\ a)$ is not equal to nil or $(\mathrm{cons}\ n\ (g\ a'))$, but unfolds when applying case to it to an element having the shape $\overline{\mathrm{nil}}$ or $(\overline{\mathrm{cons}}\ n\ (g\ a'))$.

An example of guarded recursion is that we can define a function $g : \mathbb{N} \to \mathrm{coList}$ s.t. $\mathrm{case}\ (g\ n) = \overline{\mathrm{cons}}\ n\ (g\ (n+1))$. Then $(g\ 0)$ represents the infinite list $(\mathrm{cons}\ 0\ (\mathrm{cons}\ 1\ (\mathrm{cons}\ 2\ \cdots)))$.

Extended guarded recursion. Let $(\overline{\mathrm{nil}'} + \overline{\mathrm{cons}^{\mathrm{r}}}(\mathbb{N}, A) + \overline{\mathrm{cons}^{\mathrm{n}}}(\mathbb{N}, \mathrm{coList}))$ be the set having elements $\overline{\mathrm{nil}'}$, $(\overline{\mathrm{cons}^{\mathrm{r}}}\ n\ a)$ for $n : \mathbb{N}, a : A$ and $(\overline{\mathrm{cons}^{\mathrm{n}}}\ n\ l)$ for $n : \mathbb{N}, l : \mathrm{coList}$. We are going to show that, if $g : A \to (\overline{\mathrm{nil}'} + \overline{\mathrm{cons}^{\mathrm{r}}}(\mathbb{N}, A) + \overline{\mathrm{cons}^{\mathrm{n}}}(\mathbb{N}, \mathrm{coList}))$, then we can define a function $f : A \to \mathrm{coList}$ s.t.

$$\mathrm{case}\ (f\ a) = \begin{cases} \overline{\mathrm{nil}} & \text{if } g\ a = \overline{\mathrm{nil}'}, \\ \overline{\mathrm{cons}}\, n\ (g\ a') & \text{if } g\ a = \overline{\mathrm{cons}^{\mathrm{r}}}\, n\ a', \\ \overline{\mathrm{cons}}\, n\ l & \text{if } g\ a = \overline{\mathrm{cons}^{\mathrm{n}}}\, n\ l \end{cases}$$

So $(g\ a)$ decides whether $(f\ a)$ is of nil-shape (constructor $\overline{\mathrm{nil}'}$); of cons-shape with a recursive call to $(g\ a')$ (therefore the name $\overline{\mathrm{cons}^{\mathrm{r}}}$); or of non-recursive cons-shape (therefore the name $\overline{\mathrm{cons}^{\mathrm{n}}}$). This principle adds to the principle of guarded recursion the possibility of defining $(\mathrm{case}\ (f\ a))$ by a non-recursive cons shape.

We show the existence of f just given, provided that coList is a final coalgebra.

Here $(\overline{\mathrm{nil}'} + \overline{\mathrm{cons}^{\mathrm{r}}}(\mathbb{N}, A) + \overline{\mathrm{cons}^{\mathrm{n}}}(\mathbb{N}, \mathrm{coList}))$ will be a notation for the disjoint union $(\mathbf{1} + ((\mathbb{N} \times A) + (\mathbb{N} \times \mathrm{coList})))$ where $\overline{\mathrm{nil}'} := \mathrm{inl}\ *$, $\overline{\mathrm{cons}^{\mathrm{r}}}\ n\ a := \mathrm{inr}\ (\mathrm{inl}\ \langle n, a\rangle)$ and $\overline{\mathrm{cons}^{\mathrm{n}}}\ n\ l := \mathrm{inr}\ (\mathrm{inr}\ \langle n, l\rangle)$.

Assume g as just given. Define $A' := A + \text{coList}$, and $g' : A' \to (\overline{\text{nil}} + \overline{\text{cons}}(\mathbb{N}, A'))$,

$$g'\ (\text{inl}\ a) = \begin{cases} \overline{\text{nil}} & \text{if } f\ a = \overline{\text{nil}'}, \\ \overline{\text{cons}}\ n\ (\text{inl}\ a') & \text{if } f\ a = \overline{\text{cons}^{\text{r}}}\ n\ a', \\ \overline{\text{cons}}\ n\ (\text{inr}\ l) & \text{if } f\ a = \overline{\text{cons}^{\text{n}}}\ n\ l. \end{cases}$$

$$g'\ (\text{inr}\ l) = \begin{cases} \overline{\text{nil}} & \text{if case } l = \overline{\text{nil}}, \\ \overline{\text{cons}}\ n\ (\text{inr}\ l') & \text{if case } l = \overline{\text{cons}}\ n\ l'. \end{cases}$$

Let $f' : A' \to \text{coList}$ be the coalgebra morphism such that the following diagram commutes:

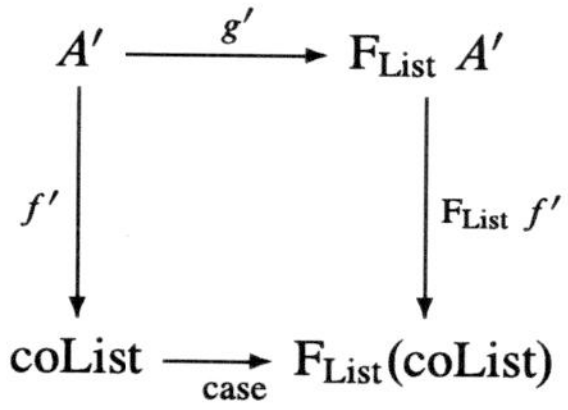

If coList is a final coalgebra, then one can see that $(f'\ (\text{inr}\ l))$ is equal to l. The reason for defining $(f'\ (\text{inr}\ l))$ was that it allows to replace the non-recursive call to l in f by a recursive call to $(f'\ (\text{inr}\ l))$. Let $f := f' \circ \text{inl} : A \to \text{coList}$. We obtain that f indeed fulfils the desired equations.

We call the principle that, for every $g : A \to (\overline{\text{nil}'} + \overline{\text{cons}^{\text{r}}}(\mathbb{N}, A) + \overline{\text{cons}^{\text{n}}}(\mathbb{N}, \text{coList}))$ we can define $f : A \to \text{coList}$ such that the equations for $(\text{case}\ (f\ a))$ just given hold the principle of extended guarded recursion. Full details will be found in Setzer (2012). Note that we chose in the third case not to escape directly to an element $l : \text{coList}$, but only to an element l such that $\text{case}\ l = \overline{\text{cons}}\ n\ l'$ for given n, l'. The reason for this is that this allows to define cons as given before.

Giménez shows in (1994) how to derive more general forms of guarded recursion for coalgebras.

The coconstructors nil, cons. In case of final coalgebras it follows (e.g. Jacobs 2005, Lemma 2.3.3) that $\text{case} : \text{coList} \to \text{F}_{\text{List}}(\text{coList})$ is an isomorphism. Let case^{-1} be its inverse and define $\text{nil} := \text{case}^{-1}\ \overline{\text{nil}}$, $\text{cons}\ n\ l := \text{case}^{-1}\ (\overline{\text{cons}}\ n\ l)$. Then we have that $\text{case nil} = \overline{\text{nil}}$ and $\text{case}\ (\text{cons}\ n\ l) = \overline{\text{cons}}\ n\ l$. cons^{-1} is surjective, so every $l : \text{coList}$ is equal to nil or $(\text{cons}\ n\ l')$ for some n, l'. Especially, $\text{case}\ l = \overline{\text{nil}}$ if and only if $l = \text{nil}$, and $\text{case}\ l = \overline{\text{cons}}\ n\ l'$ if and only if $l = \text{cons}\ n\ l'$. By iterating it we obtain that if $l : \text{coList}$, then for every k we have that $l = \text{cons}\ n_1\ (\text{cons}\ n_1\ \cdots\ (\text{cons}\ n_i\ \text{nil}) \cdots)$ for some $i < k$ and $n_1, \ldots, n_i : \mathbb{N}$ or l is equal to $(\text{cons}\ n_1\ (\text{cons}\ n_1\ \cdots\ (\text{cons}\ n_k\ l') \cdots))$ for some $n_1, \ldots, n_k : \mathbb{N}$ and $l' : \text{coList}$. Roughly speaking, an element of coList is a potentially infinite list of natural numbers. Furthermore, the principle of extended guarded recursion can be rewritten as follows: We can define $g : A \to \text{coList}$ s.t. depending on a we can choose $g\ a = \text{nil}$, $g\ a = \text{cons}\ n\ (g\ a'))$ for some n, a' or $g\ a = \text{cons}\ n\ l$ for some n, l.

Bisimilarity as equality. A weakly final F_{List}-coalgebra (coList, case) is final if and only if equality on coList is bisimilarity. Here bisimilarity on colists is the largest relation $\sim$ s.t., if $l \sim l'$, then $(\text{case}\ l) = \overline{\text{nil}} = (\text{case}\ l')$ or $(\text{case}\ l) = (\overline{\text{cons}}\ n\ l_0)$ and $(\text{case}\ l') = (\overline{\text{cons}}\ n\ l_0')$ for some $l_0 \sim l_0'$. Bisimilarity can be introduced as an indexed coalgebra (as will e.g. be shown in the current context in Setzer 2012). Bisimilarity is equality on final F_{List}-coalgebras (see e.g. Jacobs 2005, Theorem 3.4.1) and one can easily show as well that weakly final coalgebras are final coalgebras, if bisimilar elements are equal. Full details will be presented in Setzer (2012).

Coconstructors in case of weakly final coalgebras. In case of weakly final coalgebras we can define nil by case nil $= \overline{\text{nil}}$. We can define $(\text{cons}\ n\ l)$ s.t. $\text{case}\ (\text{cons}\ n\ l) = \overline{\text{cons}}\ n\ l'$ for some l which is bisimilar to l'. This can be done by defining $A = (\mathbb{N} \times \text{coList}) + \text{coList}$, $f : A \to (\overline{\text{nil}} + \overline{\text{cons}}(\mathbb{N}, A))$, $f\ (\text{inl}\ \langle n, l\rangle) = \overline{\text{cons}}\ n\ (\text{inr}\ l)$, $f\ (\text{inr}\ l) = \overline{\text{nil}}$ if case $l = \overline{\text{nil}}$, $f\ (\text{inr}\ l) = \overline{\text{cons}}\ n\ (\text{inr}\ l')$ if case $l = \overline{\text{cons}}\ n\ l'$. Then one can easily see that $f\ (\text{inr}\ l) \sim l$, and define therefore $\text{cons}\ n\ l = f\ (\text{inl}\ \langle n, l\rangle)$. We obtain $\text{case}\ (\text{cons}\ n\ l) = \overline{\text{cons}}\ n\ l'$ for some $l' \sim l$.

Combining the above we obtain a version of case^{-1} as well. The function case^{-1} is not surjective. In case of $\overline{\text{cons}}$, the equality holds only up to bisimilarity. If we add the principle of extended guarded recursion to weakly final coalgebras, we can define case^{-1} in such a way that the equality holds definitionally (however case^{-1} will not be surjective): Define $\text{case}^{-1} : \text{F}_{\text{List}}(\text{coList}) \to \text{coList}$, $\text{case}\ (\text{case}^{-1}\ \overline{\text{nil}}) = \overline{\text{nil}}$, $\text{case}\ (\text{case}^{-1}\ (\overline{\text{cons}}\ n\ l)) = \overline{\text{cons}}\ n\ l$. In order to allow this definition we defined the non-recursive case in case of extended guarded recursion the way we did it.

Undecidability results. Bisimilarity on F_{List}-coalgebras is undecidable: Define $\text{toColist} : (\mathbb{N} \to \mathbb{N}) \to \mathbb{N} \to \text{coList}$, $\text{case}\ (\text{toColist}\ f\ n) = \overline{\text{cons}}\ (f\ n)\ (\text{toColist}\ f\ (n+1))$. Therefore, in case of final coalgebras we have $\text{toColist}\ f\ n = \text{cons}\ (f\ n)\ (\text{cons}\ (f\ (n+1))\ (\text{cons}\ (f\ (n+2))\ \cdots))$. Now it follows immediately that f, g are extensionally equal if and only if $(\text{toColist}\ f\ 0) \sim (\text{toColist}\ g\ 0)$. Since extensional equality on $\mathbb{N} \to \mathbb{N}$ is undecidable, bisimilarity is undecidable as well. Therefore, if we want decidable definitional equality, we cannot define final coalgebras, only weakly final coalgebras.

In Setzer (2012) we will show that the assumption that case^{-1} is surjective results in an undecidable equality as well. So, if we want decidable equality on a weakly final coalgebra, we cannot assume that every element of it is of the form nil or $(\text{cons}\ n\ l)$ for some n, l. This implies that pattern matching on coalgebras in the setting of decidable type checking is misleading, since it suggests that every element of a coalgebra is introduced by a coconstructor, and therefore contains the hidden assumption that case^{-1} is surjective.

Problem of Subject Reduction. The problems of pattern matching have been discussed intensively on the Agda email list. Giménez (1996, Sect. 3.4) discovered that dependent case distinction results in a problem with subject reduction. Later Nicolas Oury found a very short program in a previous version of Agda, which exposes this problem, and which he orally communicated to N. Danielsson, who then posted it in Danielsson (2008). Oury then converted it to Coq and posted it in Oury (2008). A detailed analysis can be found in McBride (2009). There were as well intensive discussions on the Agda and Coq club mailing lists, to which the author contributed. Some changes have been made to Agda which avoid this problem, see Altenkirch and Danielsson (2010). The author would prefer a more aesthetically clear solution, based on what is presented in his article. The goal would be to have a solution which presents algebras and coalgebras in a symmetric way. In Coq the problem of subject reduction seems to persist.

Type theoretic rules for weakly final coalgebras. Because of the undecidability of equality in final coalgebras, we can only introduce rules for weakly final coalgebras, if we want to preserve decidable type checking. For weakly final coalgebras we can still derive the principle of extended guarded recursion, but the equations we want to satisfy will only hold up to bisimilarity as equality. For initial algebras we observed that the fact that the type theoretic rules for $\mathrm{List}_{\mathbb{N}}$ are extensionally equal but intensionally stronger than the rules for $\mathrm{List}_{\mathbb{N}}$ being an initial algebra. In the same way we are defining rules for coList which are up to bisimilarity equivalent but without bisimilarity as equality stronger than the rules for coList being a weakly final coalgebra. The principle of a weakly final coalgebra plus the principle that bisimilarity is equality is equivalent to the principle of a final coalgebra. If we take the rules for coiteration derived from the diagram, we get type theoretic rules which are up to bisimilarity equivalent to the rules of a weakly final coalgebra. If we extend the principle of guarded recursion to extended guarded recursion, we get a principle which is up to bisimilarity derivable, but without it stronger than the principle of simple guarded recursion. Therefore extended guarded recursion plus the principle of (coList, case) being a coalgebra is without bisimilarity as equality stronger, with it equivalent to that of a weakly final coalgebra. As in case of $\mathrm{List}_{\mathbb{N}}$ we use the rules of (coList, case) being a coalgebra augmented by the principle of extended guarded recursion as one possible type theoretic formulation of the rules for (coList, case) being a weakly final coalgebra. It is not the only possible one. In general one can think of adding rules which imply further definitional equalities, which are provable up to bisimilarity, as long as the rules behave well (we have decidable type checking, subject reduction and other good properties). One reason for including extended guarded recursion is that it allows us to define the coconstructor cons by defining cons $n\ l$: coList, s.t. case (cons $n\ l$) $= \overline{\mathrm{cons}}\ n\ l$).

For completeness, we introduce rules for dealing with $(\overline{\mathrm{nil}} + \overline{\mathrm{cons}}(X, Y))$ and $(\overline{\mathrm{nil}'} + \overline{\mathrm{cons}^{\mathrm{r}}}(X, Y) + \overline{\mathrm{cons}^{\mathrm{n}}}(Z, Z'))$. (Note that if as above we treat these definitions as abbreviations, these rules can be derived from the rules for $\mathbf{1}$, $+$ and $\times$). We borrow notations for case distinction from Cockett and Fukushima (1992):

Assume in the following rules X, Y, Z, Z' : Set.
Formation rule: $(\overline{\text{nil}'} + \overline{\text{cons}^{\text{r}}}(X, Y) + \overline{\text{cons}^{\text{n}}}(Z, Z'))$: Set.

Introduction rules:
$$\begin{array}{lll} \overline{\text{nil}'} & : & (\overline{\text{nil}'} + \overline{\text{cons}^{\text{r}}}(X, Y) + \overline{\text{cons}^{\text{n}}}(Z, Z')), \\ \overline{\text{cons}^{\text{r}}} & : X \to Y \to & (\overline{\text{nil}'} + \overline{\text{cons}^{\text{r}}}(X, Y) + \overline{\text{cons}^{\text{n}}}(Z, Z')), \\ \overline{\text{cons}^{\text{n}}} & : Z \to Z' \to & (\overline{\text{nil}'} + \overline{\text{cons}^{\text{r}}}(X, Y) + \overline{\text{cons}^{\text{n}}}(Z, Z')). \end{array}$$

Elimination rule:
$$\frac{\begin{array}{c} x : (\overline{\text{nil}'} + \overline{\text{cons}^{\text{r}}}(X, Y) + \overline{\text{cons}^{\text{n}}}(Z, Z')) \Rightarrow C(x) : \text{Set} \\ \text{step}_{\overline{\text{nil}'}} : C(\overline{\text{nil}'}) \\ x : X, y : Y \Rightarrow \text{step}_{\overline{\text{cons}^{\text{r}}}}(x, y) : C(\overline{\text{cons}^{\text{r}}}\ x\ y) \\ z : Z, z' : Z' \Rightarrow \text{step}_{\overline{\text{cons}^{\text{n}}}}(z, z') : C(\overline{\text{cons}^{\text{n}}}\ z\ z') \end{array}}{\left\{\begin{array}{ll} \overline{\text{nil}'} & \mapsto \text{step}_{\overline{\text{nil}'}} \\ \overline{\text{cons}^{\text{r}}}\ x\ y & \mapsto \text{step}_{\overline{\text{cons}^{\text{r}}}}(x, y) \\ \overline{\text{cons}^{\text{n}}}\ z\ z' & \mapsto \text{step}_{\overline{\text{cons}^{\text{n}}}}(z, z') \end{array}\right\} : (x : (\overline{\text{nil}'} + \overline{\text{cons}^{\text{r}}}(X, Y) + \overline{\text{cons}^{\text{n}}}(Z, Z'))) \to C(x)}$$

Equality rules:
$$\begin{array}{l} \{\cdots\}\ \overline{\text{nil}'} = \text{step}_{\overline{\text{nil}'}} \\ \{\cdots\}\ (\overline{\text{cons}^{\text{r}}}\ x\ y) = \text{step}_{\overline{\text{cons}^{\text{r}}}}(x, y) \\ \{\cdots\}\ (\overline{\text{cons}^{\text{n}}}\ z\ z') = \text{step}_{\overline{\text{cons}^{\text{n}}}}(z, z') \end{array}$$
where $\{\cdots\}$ is the expression introduced in the elimination rule

Now we can define the rules for colist:

Formation rule: coList : Set
Elimination rule: case : coList $\to (\overline{\text{nil}} + \overline{\text{cons}}(\mathbb{N}, \text{coList}))$

Introduction rule:
$$\frac{A : \text{Set}}{\begin{array}{l} \text{intro}_A : (A \to (\overline{\text{nil}'} + \overline{\text{cons}^{\text{r}}}(\mathbb{N}, A) + \overline{\text{cons}^{\text{n}}}(\mathbb{N}, \text{coList})) \\ \qquad \to A \to \text{coList} \end{array}}$$

Equality rule:
$$\text{case}\ (\text{intro}_A\ f\ a) = \left\{\begin{array}{ll} \overline{\text{nil}'} & \mapsto \overline{\text{nil}} \\ \overline{\text{cons}^{\text{r}}}\ n\ a' & \mapsto \overline{\text{cons}}\ n\ (\text{intro}_A\ f\ a') \\ \overline{\text{cons}^{\text{n}}}\ n\ l & \mapsto \overline{\text{cons}}\ n\ l \end{array}\right\} (f\ a)$$

Note that the introduction rule is complex because a generic form of guarded recursion in the same way that the elimination rule for algebraic data types is complicated, because it is generic. Specific instances can be described more easily. For instance we can define

$$\begin{array}{l} \text{toColist} : (\mathbb{N} \to \mathbb{N}) \to \mathbb{N} \to \text{coList} \\ \text{case}\ (\text{toColist}\ f\ n) = \overline{\text{cons}}\ (f\ n)\ (\text{toColist}\ f\ (n+1)) \end{array}$$

The coconstructors nil and cons can be defined by

$$\begin{array}{ll} \text{nil} : \text{coList} & \text{cons} : \mathbb{N} \to \text{coList} \to \text{coList} \\ \text{case nil} = \overline{\text{nil}} & \text{case } (\text{cons } n\ l) = \overline{\text{cons}}\ n\ l \end{array}$$

We observe that the elimination rules are simple whereas the introduction rules seem to be complicated and refer to all sets. This is dual to the setting for initial algebras where the introduction rules are simple and the elimination rules refer to all sets. So a weakly final coalgebra is given by its elimination rules, which essentially expresses: elements of coList are programs, to which we can apply case and obtain $\overline{\text{nil}}$ or $(\overline{\text{cons}}\ n\ l)$ for some other colist l.

Problems with dependent case distinction. McBride (2009) discussed dependent case distinction, as it occurs in the PhD thesis by Giménez (1996) and is implemented in Coq. In our notation it reads

$$\frac{x : \text{coList} \Rightarrow B(x) : \text{Set}}{\begin{array}{l} \text{depcase}_B : (\text{step}_{\text{nil}} : B(\text{nil})) \\ \qquad \to (\text{step}_{\text{cons}} : (n : \mathbb{N}, l : \text{coList}_{\mathbb{N}}) \to B(\text{cons } n\ l)) \\ \qquad \to (l : \text{coList}) \\ \qquad \to B(l) \end{array}}$$

$$\begin{array}{ll} \text{depcase}_B\ \text{step}_{\text{nil}}\ \text{step}_{\text{cons}}\ \text{nil} & = \text{step}_{\text{nil}} \\ \text{depcase}_B\ \text{step}_{\text{nil}}\ \text{step}_{\text{cons}}\ (\text{cons } n\ l) & = \text{step}_{\text{cons}}\ n\ l \end{array}$$

There is an equality rule missing, namely for an element introduced by intro. Such a rule should be (the case for $\overline{\text{cons}^{\text{n}}}$ was added by the author to stay in accordance with the rest of the current article):

$$\begin{array}{l} \text{depcase}_B\ \text{step}_{\text{nil}}\ \text{step}_{\text{cons}}\ (\text{intro}_A\ f\ a) \\ = \left\{ \begin{array}{ll} \overline{\text{nil}'} & \mapsto \text{step}_{\text{nil}} \\ \overline{\text{cons}^{\text{r}}}\ n\ a' & \mapsto \text{step}_{\text{cons}}\ n\ (\text{intro}_A\ f\ a') \\ \overline{\text{cons}^{\text{n}}}\ n\ l & \mapsto \text{step}_{\text{cons}}\ n\ l \end{array} \right\}\ (f\ a) \end{array}$$

McBride states that this is the source of the problem discovered/communicated by Giménez (1996), Oury (2008), and Danielsson (2008). As McBride observes, it does not even type check: in case of $f\ a = \overline{\text{nil}'}$, the two sides of the equations have types $B(\text{intro}_A\ f\ a)$ and $B(\text{nil})$, but $\text{intro}_A\ f\ a \neq \text{nil}$.

As observed by McBride dependent case distinction results, if we omit the last rule, in non-canonical terms for the intensional equality type. In fact the situation is even worse: We get non-canonical elements of $\mathbb{N}$ in normal form: Let $f = \text{depcase}_{(x)\mathbb{N}}\ 0\ ((n, l)0) : \text{coList} \to \mathbb{N}$. Let $\text{zeroStream} = \text{intro}_1\ ((x)(\overline{\text{cons}^{\text{r}}}\ 0\ *))\ * :$ coList. We have that $(f\ \text{ zeroStream})$ is a non-canonical closed element of $\mathbb{N}$ in normal form. The reason is of course that we do not have an equality rule for depcase applied to an element introduced by intro.

The underlying problem is that dependent case distinction expresses that every element of coList is of the form nil or (cons n l), i.e. that case^{-1} is surjective. In order to repair this problem, McBride suggests to switch to observational type theory. This means essentially to define for all types a propositional equality together with some additional axioms. In case of coList, this equality would be bisimilarity. Since, if we add to weakly final coalgebras bisimilarity as equality, we obtain final coalgebras, the problem vanishes. However, it does not solve the problem, what the correct rules regarding definitional equalities in intensional type theory are.

16.4 Meaning Explanations for Coalgebraic Types as Determined by Their Elimination Rules

We give now meaning explanations for coList based on the principle that elements of coalgebras are determined by their elimination rules. coList is a set. Elements of coList are programs l, which, if we apply case to them, evaluate to $\overline{\mathrm{nil}}$ or ($\overline{\mathrm{cons}}$ n l') for some n in $\mathbb{N}$, and some other element l' of coList. Note that we do not demand that l' is defined before l. Several elements of coList might be introduced simultaneously. Two elements l, l' of coList are equal if after applying case to it, both evaluate to $\overline{\mathrm{nil}}$ or they evaluate to ($\overline{\mathrm{cons}}$ n l_0) and ($\overline{\mathrm{cons}}$ n' l_0') where n, n' are equal elements of $\mathbb{N}$ and l_0, l_0' are equal elements of coList. Again we do not demand that the equality of l_0, l_0' is established before the equality of l, l' is established. Assume A is a set and f a function mapping an element of A to an element of $(\overline{\mathrm{nil}'} + \overline{\mathrm{cons}^{\mathrm{r}}}(\mathbb{N}, A) + \overline{\mathrm{cons}^{\mathrm{n}}}(\mathbb{N}, \mathrm{coList}))$. Then for every $a : A$, (intro_A f a) is an element of coList. For this we determine (case (intro_A f a)): Compute (f a). If (f a) evaluates to $\overline{\mathrm{nil}'}$ then (case (intro_A f a)) evaluates to $\overline{\mathrm{nil}}$. If (f a) evaluates to ($\overline{\mathrm{cons}^{\mathrm{r}}}$ n a'), then (case (intro_A f a)) evaluates to ($\overline{\mathrm{cons}}$ n (intro_A f a')). If (f a) evaluates to ($\overline{\mathrm{cons}^{\mathrm{n}}}$ n l), then (case (intro_A f a)) evaluates to ($\overline{\mathrm{cons}}$ n l).
Assume A, A' are equal sets, f, f' map elements of A to equal elements of $(\overline{\mathrm{nil}'} + \overline{\mathrm{cons}^{\mathrm{r}}}(\mathbb{N}, A) + \overline{\mathrm{cons}^{\mathrm{n}}}(\mathbb{N}, \mathrm{coList}))$. For all a, a' equal elements of A we have that (intro_A f a) and ($\mathrm{intro}_{A'}$ f' a') are equal elements of coList: Assume a and a' are equal elements of A.

Assume (f a) evaluates to $\overline{\mathrm{nil}'}$. Then, since f is equal to f' and a is equal to a', f' a' evaluates to $\overline{\mathrm{nil}'}$ as well. Then (case (intro_A f a)) and (case ($\mathrm{intro}_{A'}$ f' a')) both evaluate to the same element $\overline{\mathrm{nil}}$.

Assume (f a) evaluates to ($\overline{\mathrm{cons}^{\mathrm{r}}}$ n a_0). Then (f' a') evaluates to ($\overline{\mathrm{cons}^{\mathrm{r}}}$ n' a_0') for some n' equal to n and a_0' equal to a_0. Then (case (intro_A f a)) evaluates to (cons n (intro_A f a_0)) and (case ($\mathrm{intro}_{A'}$ f' a')) evaluates to ($\overline{\mathrm{cons}}$ n' ($\mathrm{intro}_{A'}$ f' a_0')). n and n' are equal elements of $\mathbb{N}$, and (intro_A f a_0) and ($\mathrm{intro}_{A'}$ f' a_0') are equal elements of coList. Therefore (case (intro_A f a)) and (case ($\mathrm{intro}_{A'}$ f' a')) evaluate to equal elements.

Assume $(f\ a)$ evaluates to $(\overline{\text{cons}^{\text{r}}}\ n\ l)$. Then $(f\ a')$ evaluates to $(\overline{\text{cons}^{\text{r}}}\ n'\ l')$ for some n equal to n' and l equal to l'. Therefore (case l) and (case l') and therefore as well (case $(\text{intro}_A\ f\ a)$ and (case $(\text{intro}_{A'}\ f'\ a')$ evaluate to equal elements. Therefore $(\text{intro}_A\ f\ a)$ and $(\text{intro}_{A'}\ f'\ a')$ are equal.

Function sets as determined by their elimination rules. We can see now that the elements of the function type of the logical framework are as well introduced by their elimination rules: Assume A is a set and $B(x)$ is a set depending on elements x of A. Then $(x : A) \to B(x)$ is a set. An element of $(x : A) \to B(x)$ is a program t which, when applied to an element a of A evaluates to an element of $B(a)$. Two elements t, t' of $(x : A) \to B(x)$ are equal, if, when applied to an element a of A, they evaluate to equal elements of $B(a)$. Assume that for every x of A we have that t is an element of $B(x)$. Then $(x)t$ is the following element of $(x : A) \to B(x)$: If applied to $a : A$ it first substitutes in t the variable x by a. Let the result be s. Then s is evaluated, which is the result returned. Since for x of A, t is an element of $B(x)$, s is an element of $B(a)$. So $(x)t$ is an element of $(x : A) \to B$. Assume that t, t' are equal elements of $B(x)$, depending on x of type A. Then if $(x)t$ and $(x)t'$ are applied to an element a of type A, we obtain s, s' which are equal elements of $B(a)$. So $(x)t$ and $(x)t'$ are equal elements of $(x : A) \to B(x)$.

More advanced examples of coalgebras. coList is only the simplest example of a coalgebra. More advanced examples are the definition of bisimilarity on colists or on other transition systems. In Hancock and Setzer (1999, 2000a,b, 2004, 2005) we discussed how to define state-dependent interactive programs in Martin-Löf type theory, and in Hancock and Setzer (2004) we showed how to define them as an indexed coalgebra. More examples can be found for instance in Chap. 13 of Bertot and Castéran (2004).

16.5 Conclusion

We have seen that coalgebras can be introduced in Martin-Löf type theory using formation, elimination, introduction and equality rules. Meaning explanations can be given by defining as elements of coalgebras those which allow elimination rules. One can then explain that the introduction rules indeed introduce elements of the coalgebra. So elements of coalgebras are given by their elimination rules, the introduction rules can be considered as being derived. This is similar to algebraic data types, for which the elements are given by their introduction rules, and the elimination rules are derived. We have seen as well that the elements of the function types from the logical framework are as well determined by their elimination rules. One can as well develop models of coalgebras, in which coalgebras are interpreted as the set of those terms which allow to apply the elimination principle.

Acknowledgements We want to thank the anonymous referee for valuable comments on earlier version of this articles. We want to thank as well our PhD student Fredrik Nordvall Forsberg for diligent proof reading and valuable remarks. Part of this work was carried out while the author was a visiting fellow of the Isaac Newton Institute for Mathematical Sciences, Cambridge, UK, in 2012.

References

Aczel, P. 1988. *Non-wellfounded set theory, CSLI lecture notes*, vol. 14, xx+137. Stanford: Stanford University, Center for the Study of Language and Information.

Agda. 2011 Email list archive. Available at https://lists.chalmers.se/pipermail/agda/.

Altenkirch, T. 2004. Codata. Talk given at the TYPES workshop in Jouy-en-Josas. Available from http://www.cs.nott.ac.uk/~txa/talks/types04.pdf, Dec 2004.

Altenkirch, T., and N.A. Danielsson. 2010. Termination checking nested inductive and coinductive types. In *Proceedings of workshop on partiality and recursion in interactive theorem provers, PAR'10*, Edinburgh.

Bertot, Y. 2006. CoInduction in Coq. Available from http://arxiv.org/abs/cs/0603119, Mar 2006.

Bertot, Y., and P. Castéran. 2004. *Interactive theorem proving and program development. Coq'Art: The calculus of inductive constructions*. Berlin/New York: Springer.

Cockett, J., and D. Spencer. 1992. Strong categorical datatypes I. In *Category theory 1991: proceedings of an international summer category theory meeting, held June 23–30, 1991*, vol. 13, ed. R.A.G. Seely, 141. Providence: American Mathematical Society.

Cockett, J.R.B., and D. Spencer. 1995. Strong categorical datatypes II: A term logic for categorical programming. *Theoretical Computer Science* 139(1–2): 69–113.

Cockett, R., and T. Fukushima. 1992. About charity. Technical report, Department of Computer Science, The University of Calgary, June 1992. Yellow series report no. 92/480/18.

Coquand, T. 1994. Infinite objects in type theory. In *Types for proofs and programs*, Lecture notes in computer science, vol. 806, ed. H. Barendregt and T. Nipkow, 62–78. Berlin: Springer

Danielsson, N.A. 2008. Codata oddity. Email posted on the Agda email list. Available from http://thread.gmane.org/gmane.comp.lang.agda/226.

Díez, G.F. 2000. Five observations concerning the intended meaning of the intuitionistic logical constants. *Journal of Philosophical Logic* 29(4): 409–424.

Díez, G.F. 2002. The logic of constructivism. *Disputatio* 12: 37–41.

Díez, G.F. 2003. Is the language of intuitionistic mathematics adequate for intuitionistic purposes? *L&PS - Logic and Philosophy of Science* 1(1): 1–9. Available at http://www2.units.it/episteme/L&PS_Vol1No1/contents_L&PS_Vol1No1.htm.

Dummett, M. 2000. *Elements of intuitionism. Oxford logic guides*, 2nd ed. New York: Oxford University Press.

Dybjer, P., and A. Setzer. 2003. Induction–recursion and initial algebras. *Annals of Pure and Applied Logic* 124: 1–47.

Dybjer, P., and A. Setzer. 2006. Indexed induction–recursion. *Journal of Logic and Algebraic Programming* 66: 1–49, 2006.

Giménez, C.E. 1996. *Un calcul de constructions infinies et son application à la vérification de systèmes communicants. (English: A calculus of infinite constructions and its application to the verification of communicating systems)*. Ph.D. thesis, Ecole normale supérieure de Lyon, Lyon, France.

Giménez, E. 1994. Codifying guarded definitions with recursive schemes. In *Proceedings of the 1994 workshop on types for proofs and programs*, Bastad, LNCS No. 996, 39–59. Springer.

Hagino, T. 1987. *A categorical programming language*. Ph.D. thesis, Laboratory for Foundations of Computer Science, University of Edinburgh. Available from http://www.tom.sfc.keio.ac.jp/~hagino/thesis.pdf.

Hagino, T. 1989. Codatatypes in ml. *Journal of Symbolic Computation* 8(6): 629–650.

Hancock, P., and A. Setzer. 1999. The IO monad in dependent type theory. In *Electronic proceedings of the workshop on dependent types in programming, Göteborg, 27–28, Mar 1999*. Available at http://www.md.chalmers.se/Cs/Research/Semantics/APPSEM/dtp99.html.

Hancock, P., and A. Setzer. 2000a. Interactive programs in dependent type theory. In *Computer science logic: 14th international workshop, CSL 2000. Lecture notes in computer science*, Vol. 1862, ed. P. Clote and H. Schwichtenberg, 317–331. Berlin: Springer.

Hancock, P., and A. Setzer. 2000b. Specifying interactions with dependent types. In *Workshop on subtyping and dependent types in programming, Portugal, 7 July 2000*. Electronic proceedings, available at http://www-sop.inria.fr/oasis/DTP00/Proceedings/proceedings.html.

Hancock, P., and A. Setzer. 2004. Interactive programs and weakly final coalgebras (extended version). In *Dependently typed programming*, ed. T. Altenkirch, M. Hofmann, and J. Hughes, number 04381 in Dagstuhl Seminar Proceedings. Internationales Begegnungs- und Forschungszentrum (IBFI), Schloss Dagstuhl, Germany. Available at http://drops.dagstuhl.de/opus/.

Hancock, P., and A. Setzer. 2005. Guarded induction and weakly final coalgebras in dependent type theory. In *From sets and types to topology and analysis. Towards practicable foundations for constructive mathematics*, ed. L. Crosilla and P. Schuster, 115–134, Oxford: Clarendon Press.

Jacobs, B. 1995. Objects and classes, co-algebraically. In *Object orientation with parallelism and persistence*, ed. B. Freitag, C.B. Jones, C. Lengauer, and H.-J. Schek, 83–103. Boston: Kluwer.

Jacobs, B. 1998. Coalgebraic reasoning about classes in object-oriented languages. *Electronical Notes in Computer Science* 11: 231–242. Special issue on the workshop coalgebraic methods in computer science (CMCS 1998).

Jacobs, B. 1999. *Categorical logic and type theory*, Number 141 in studies in logic and the foundations of mathematics. Amsterdam: North Holland.

Jacobs, B. 2005. Introduction to coalgebra. Towards mathematics of states and oberservations. Available at http://www.cs.ru.nl/B.Jacobs/CLG/JacobsCoalgebraIntro.pdf.

Jacobs, B., and J. Rutten. 1997. A tutorial on (co)algebras and (co)induction. *EATCS Bulletin* 62: 62–222.

Leclerc, F., and C. Paulin-Mohring. Programming with streams in Coq. A case study: The sieve of eratosthenes. In *Types for proofs and programs*, Lecture notes in computer science, vol. 806, ed. H. Barendregt and T. Nipkow, 191–212. Berlin/Heidelberg: Springer.

Martin-Löf, P. 1972. Infinite terms and a system of natural deduction. *Compositio Mathematica* 24(1): 93–103.

Martin-Löf, P. 1984. *Intuitionistic type theory*. Naples: Bibliopolis.

Martin-Löf, P. 1987. Truth of a proposition, evidence of a judgement, validity of a proof. *Synthese* 73: 407–420.

Martin-Löf, P. 1996. On the meaning of the logical constants and the justification of the logical laws. *Nordic Journal of Philosophical Logic* 1(1): 11–60.

Martin-Löf, P. 1998. An intuitionistic theory of types. In *Twenty-five years of constructive type theory*, ed. G. Sambin and J. Smith, 127–172. Oxford: Oxford University Press. Reprinted version of an unpublished report from 1972.

McBride, C. 2009. Let's see how things unfold: Reconciling the infinite with the intensional. In *Proceedings of the 3rd international conference on algebra and coalgebra in computer science, CALCO'09*, Lecture notes in computer science, ed. A. Kurz, M. Lenisa, and A. Tarlecki, 113–126. Berlin/Heidelberg: Springer.

Nanevski, A., F. Pfenning, and B. Pientka. 2008. Contextual modal type theory. *ACM Transactions on Computational Logic* 9(3):23:1–23:49.

Nordström, B., K. Petersson, and J.M. Smith. 1990. *Programming in Martin-Löf's type theory: An introduction*. New York: Oxford University Press.

Norell, U. 2007. *Towards a practical programming language based on dependent type theory*. Ph.D. thesis, Department of Computer Science and Engineering, Chalmers University of Technology, SE-412 96 Göteborg.

Oury, N. 2008. Coinductive types and type preservation. Email posted 6 June 2008 at science.mathematics.logic.coq.club. Available at https://sympa-roc.inria.fr/wws/arc/coq-club/2008-06/msg00022.html?checked_cas=2.

Paulson, L. 1994. A fixedpoint approach to implementing (Co) inductive definitions. In *Automated deduction CADE-12*, Lecture notes in computer science, vol. 814, ed. A. Bundy, 148–161. Berlin/Heidelberg: Springer.

Setzer, A. 2008. Universes in type theory part I – Inaccessibles and Mahlo. In *Logic colloquium '04*, Lecture notes in logic 29, ed. A. Andretta, K. Kearnes, and D. Zambella, Association of Symbolic Logic, 123–156. Cambridge: Cambridge University Press.

Setzer, A. 2012. Why codata should be replaced by coalgebras. In preparation.

Sundholm, G. 1983. Constructions, proofs and the meaning of logical constants. *Journal of Philosophical Logic* 12: 151–172. 10.1007/BF00247187.

Telford, A. and D. Turner. 1997. Ensuring streams flow. In *Algebraic methodology and software technology: 6th international conference, AMAST'96 Sydney, December 13 – 17, 1997*, Lecture notes in computer science, ed. M. Johnson, 509–523, 1349. Berlin: Springer.

Turner, D. 1995. Elementary strong functional programming. In *Elementary strong functional programming*, Lecture notes in computer science, vol. 1022, eds. P. Hartel and R. Plasmeijer, 1–13.

Turner, D.A. 2004. Total functional programming. *Journal of Universal Computer Science* 10(7): 751–768.

Chapter 17
Second Order Logic, Set Theory and Foundations of Mathematics

Jouko Väänänen*

17.1 Introduction

The distinction between first and second order logic did not arise as a serious matter before model theory was developed in the early part of the twentieth century. The current view, supported by model theory, is that first and second order logic are about as far from each other as it is possible to imagine. On the other hand, in a proof theoretic account first and second order logic behave very similarly, even though the latter is in general somewhat stronger than the former. Thus it seems necessary that in any discussion on second order logic and first order set theory one must be very clear whether the framework is model theoretic or proof theoretic. However, in a foundational discussion we should be able to make judgements that are free from frameworks. A framework is just a tool.

When second order logic is thought of as a foundation of mathematics, it is nowadays taken in the model theoretic sense with reference to its power to characterize classical mathematical structures up to isomorphism and thereby capture our intuition in an exact way. This is often contrasted to the situation with the model theory of set theory, where the first order Zermelo–Fraenkel axioms have countable models and models with non-standard integers, contrary to our intuition about the set theoretic universe. So the huge distance between the model theory of second order logic and the model theory of first order logic manifests itself in

*Research partially supported by grant 40734 of the Academy of Finland and by the EUROCORES LogICCC LINT programme.

J. Väänänen (✉)
Department of Mathematics and Statistics, University of Helsinki, Helsinki, Finland

Institute for Logic, Language and Computation, University of Amsterdam, Amsterdam, The Netherlands
e-mail: jouko.vaananen@helsinki.fi

P. Dybjer et al. (eds.), *Epistemology versus Ontology*, Logic, Epistemology, and the Unity of Science 27, DOI 10.1007/978-94-007-4435-6_17,

discussions about foundations of mathematics, with second order logic appearing to emerge as the logic which more accurately captures the intended meaning of mathematical concepts.

I will argue below that despite their great apparent differences, second order logic and first order set theory turn out to be virtually indistinguishable as far as capturing mathematical concepts is concerned.

Another difference between second order logic and set theory is that the latter builds up a transfinite power-set hierarchy while the former settles with one layer of power-sets. This is a genuine difference which, unlike claims of categoricity, cannot be explained away. However, if second order logic is extended to third and higher order logics and eventually to type theory, this difference to set theory becomes respectively smaller. Still type theory maintains an explicit typing of objects while set theory is type-free. On the other hand, every set has a rank, an ordinal, which more or less works like a type. So the difference between type theory and set theory is really only in that set theory has an internal mechanism for generating higher and higher types while in type theory this is part of the set-up of the language. From the point of view of foundational questions this difference seems like a minute one.

17.2 Second Order Logic

The approach of second order logic to the foundations of mathematics is that mathematical propositions have the form

$$\mathfrak{A} \models \phi, \tag{17.1}$$

where $\mathfrak{A}$ is a structure, typically one of the classical structures such as integers or reals, and ϕ is a mathematical statement written in second order logic. This seems at least at first sight like an excellent approach since almost any statement in mathematics can be succinctly written as a second order property of one of the classical structures. An example is provided by Fermat's Last theorem (where exponentiation is a defined symbol):

$$\mathfrak{A} \models \forall x > 0 \forall y > 0 \forall z > 0 \forall n > 2 \neg(x^n + y^n = z^n), \tag{17.2}$$

where $\mathfrak{A} = (\mathbb{N}, +, \cdot, <)$.

If $\mathfrak{A}$ is one of the structures, such as $(\mathbb{N}, +, \cdot, <)$ or $(\mathbb{R}, +, \cdot, <, \mathbb{N})$, for which there is a second order sentence $\theta_{\mathfrak{A}}$ such that $\forall \mathfrak{B}(\mathfrak{B} \models \theta_{\mathfrak{A}} \leftrightarrow \mathfrak{B} \cong \mathfrak{A})$, then (17.1) can be expressed as a second order semantic logical truth

$$\models \theta_{\mathfrak{A}} \to \phi. \tag{17.3}$$

As was known already in the thirties, the second order semantic logical consequence relation

$$\models \psi \tag{17.4}$$

is not axiomatizable (i.e. not r.e.). The Levy-hierarchy Lévy (1965) of set theory is one possible way to measure how far (17.4) is from being axiomatizable. As it turns out, (17.4) is Π_2-complete—and thus a fortiori not Σ_2—in set theory (Väänänen 2001). In other words, the only way to become convinced of (17.4) for a given ψ requires, in the unavoidable worst case, going through the entire set theoretical universe in search of evidence. To see what this means, suppose (17.4) is written as $\forall x \exists y \Phi(x, y)$, where $\Phi(x, y)$ is Σ_0. Then to be convinced of the truth of (17.4) one has to go through every set x and then look for y with $\Phi(x, y)$.

The powerful Levy Reflection Theorem (Lévy 1965) says that if $\Psi(x, y)$ is Σ_0, $\kappa > 0$ and $a_1, \ldots, a_n \in H(\kappa)$[1] such that for *some* b we have $\Psi(b, a_1, \ldots, a_n)$, then there is $b \in H(\kappa)$ such that $\Phi(b, a_1, \ldots, a_n)$. That means that if we are given $a_1, \ldots, a_n \in H(\kappa)$ and we want to find b such that $\Phi(b, a_1, \ldots, a_n)$ we need only look for such a b in $H(\kappa)$, rather than in the vast universe V of set theory.

So going back to (17.4) and its Π_2-representation $\forall x \exists y \Phi(x, y)$ we first take an arbitrary x in the universe of sets and then search for y. The message of Levy's Reflection Principle is that we only need to look for y in the "neighborhood" of x (in the sense that if $x \in H(\kappa)$, $\kappa > \omega$, then we only need to look for y in $H(\kappa)$). So the good news is that we only need to hang around x, but the bad news is that x may be anywhere in the universe V of sets. For example, there is no a priori bound on the hereditary cardinality of x.

So the difference to $\models \phi$ where ϕ is first order is great for in this case we only need to look for a natural number that codes a proof of ϕ. The difference is also great to $\models \phi$ for $\phi \in L_{\omega_1 \omega}$, where we only need to look for a real number that codes a proof of ϕ. Likewise with $\models \phi$ for $\phi \in L_{\omega_1 \omega_1}$, where we only need[2] to look for sets of reals that code possible models of ϕ. No matter how badly behaving $L_{\omega_1 \omega_1}$ otherwise is,[3] at least validity in $L_{\omega_1 \omega_1}$ can be checked without going further afield than sets of reals. Not so for second order logic.

It is true, to check $\models \phi$ for second order ϕ it is sufficient to check sets of hereditary cardinality less than the first supercompact cardinal (and nothing less will do) (Magidor 1971). This is however a far cry from going through all natural numbers, going through all reals, or going through all sets of reals.

What does "going through" mean, for surely we cannot really "go through" all natural numbers, let alone all reals, or all sets of reals. It is all a question of how to deal with abstract objects, something where logic is supposed to help us. The great thing about natural numbers is that we can "look" at them. We can take natural numbers one by one and check what they are like. We can even imagine ourselves looking at a real number and wondering whether it codes a proof of

[1]The set of sets that have hereditary cardinality $< \kappa$ i.e. are included in a transitive set of cardinality $< \kappa$.

[2]Thanks to the Löwenheim–Skolem Theorem of $L_{\omega_1 \omega_1}$.

[3]For example, its Hanf number can be bigger than the first measurable cardinal (Kunen 1970).

something or not. But when we come to sets of real numbers there is phase transition to something infinitely more complex. We leave the world of concrete or almost concrete objects and enter the world of abstract objects.

The situation with (17.3) is a little simpler than with (17.4). In (17.3) one only has to search for evidence in the power-set of $\mathfrak{A}$. If $\mathfrak{A}$ is continuum-sized, then we have to search through sets of reals. What was said above about going through all sets of reals applies. It is abstract thinking.

The main result of Gödel's doctoral thesis was the Completeness Theorem for first order logic which tells us that the semantic logical consequence relation of first order logic is, in contrast to second order logic, axiomatizable in the way suggested by Hilbert and Ackermann. Thus the combination of Gödel's Completeness Theorem and his Incompleteness Theorem established the sharp difference between the semantics of first and second order logics. In first order logic truth in all models can be reduced to the existence of a finite proof. In second order logic truth in all models cannot be reduced to the existence of a finite proof in any sensible way.

If (17.1) is the general form of a mathematical proposition, then what is the general form of a proof of (17.1)? Logicians have a formal concept of a proof, but we can ask more generally what is the basic form or nature of something we can assert that would make asserting (17.1) legitimate? An obvious answer would seem to be that if we have already grounds to assert

$$\mathfrak{A} \models \psi, \tag{17.5}$$

and we moreover know that every model of ψ is a model of ϕ, i.e.

$$\models \psi \to \phi, \tag{17.6}$$

then we can assert by the logical rule of Modus Ponens that (17.1) holds. But we just observed that (17.6) is an even more complex notion than (17.1). In view of our discussion above there are no rules that completely explain (17.6) that would be easier to present and use than (17.1). It is rather the opposite: we can prove in set theory that one cannot be convinced of (17.6) without going through the entire universe of sets (or at least up to a supercompact cardinal), while one can be convinced of (17.1) by "just" going through all subsets of $\mathfrak{A}$.

Of course there are *some* rules that govern (17.6), like

$$\models \phi(P) \to \exists X \phi(X)$$

and are used all the time even if it is known that they cannot produce all cases of (17.6).

There are two stronger versions of (17.3), namely

$$ZFC \vdash \forall \mathfrak{B}(\mathfrak{B} \models \theta_{\mathfrak{A}} \to \mathfrak{B} \models \phi) \tag{17.7}$$

and

$$CA \vdash \theta_{\mathfrak{A}} \rightarrow \phi, \tag{17.8}$$

where *CA* is the usual axiomatization of second order logic i.e. the comprehension schema and the axiom of choice. These two conditions are Σ^0_1-properties of ϕ. Even if one wants to assert the weaker (17.3) it seems reasonable to give (17.7) or (17.8) as the justification. Indeed, it is the habit of mathematicians to aim at the strongest statements, whenever they seem within reach. Also, one can justify (17.7) or (17.8) by giving a proof, which is a finite object, typically even surveyable, while justifying (17.3) without recourse to (17.7) or (17.8) involves dealing directly with infinite objects.

One may ask, whether indeed (17.3) is ever asserted with certainty without the certainty arising from knowing (17.7) or (17.8). Be this as it may, there is nothing wrong in believing (17.3) on the basis of (17.7) or (17.8), and the recourse to (17.7) or (17.8) does not in any way render (17.3) meaningless.

If we compare (17.7) and (17.8), we may observe that while (17.8) may be harder to prove, (17.7) certainly gives a shorter proof. In fact, there is no recursive function h such that the following holds: If ϕ is a second order sentence and there is a proof in ZFC of $\forall M(M \models \phi)$ with n lines, then there is a proof of ϕ of $\leq h(n)$ lines from *CA*.[4]

I have called (17.7) and (17.8) *stronger* forms of (17.3) because I take it for granted that *ZFC* and *CA* are *true* axioms. It is not the main topic of this paper to investigate how much *ZFC* and *CA* can be weakened in this or that special instance of (17.7) and (17.8), as such considerations do not differentiate second order logic and set theory from each other in any essential way.

It is not easy to give an example of a ϕ such that (17.1) would not hold because of (17.7) or (17.8), especially since we may simply replace *ZFC* and *CA* with stronger theories if needed, e.g. if ϕ is *Con*(*ZFC*) or the Paris-Harrington sentence. When one moves to topology and measure theory more examples start to emerge. Let us consider the statement ϕ_0 stating that the Lebesgue measure has an extension to a total σ-additive measure on the reals. This is a second order statement about the structure $\mathfrak{A}_0 = (\mathcal{P}(\mathbb{R}), \in, \mathbb{R}, +.\cdot, <, \mathbb{N})$ (or alternatively, a third order statement about $(\mathbb{R}, +.\cdot, <, \mathbb{N})$). No answer is known for the question $\mathfrak{A}_0 \models \phi_0$ nor to the question $\mathfrak{A}_0 \models \neg\phi_0$, although trivially

$$\mathfrak{A}_0 \models \phi_0 \text{ or } \mathfrak{A}_0 \models \neg\phi_0. \tag{17.9}$$

This is an example of the non-classical nature of the logic of correct judgements.

No generally accepted version CA^* of axiomatization of second order logic is known which would give

$$CA^* \vdash \theta_{\mathfrak{A}_0} \rightarrow \phi, \tag{17.10}$$

when $\phi = \phi_0$ or $\phi = \neg\phi_0$. Still the respective version of (17.3) is true either for $\phi = \phi_0$ or $\phi = \neg\phi_0$. So the weaker claim (17.3) holds one way or the other, but at

[4] Joint work with Moshe Vardi.

the moment we cannot strengthen this observation to (17.10) one way or the other. All we know is that CA^* cannot be the usual CA, since ϕ_0 violates CH but is true in a model of CA constructed from a measurable cardinal. To make progress on the question of the truth of $\mathfrak{A}_0 \models \phi_0$ one would have to study *true* extensions CA^* of CA. We have the implications

$$CA \vdash \theta_{\mathfrak{A}_0} \to \phi \quad \Rightarrow \quad CA^* \vdash \theta_{\mathfrak{A}_0} \to \phi \quad \Rightarrow \quad \models \theta_{\mathfrak{A}_0} \to \phi. \tag{17.11}$$

The current situation is roughly speaking that for trivial reasons the rightmost implicant holds for either ϕ_0 or its negation, and for totally non-trivial reasons the leftmost implicant holds for neither. A lot of effort is put into trying to establish the middle implicant for either ϕ_0 or its negation, with the right choice of CA^*. The fact that the rightmost implicant holds for either ϕ_0 or its negation is, of course, of little use unless we know which case holds.

An obvious complaint about (17.8) in comparison to the weaker (17.1) is that (17.1) merely asserts that ϕ should hold in the standard model, while (17.8) seems to assert that ϕ holds in the whole pack of "non-standard" models in addition to the "standard" model $\mathfrak{A}$. While every student of logic knows how to prove this, there is a more subtle sense in which this is not really so. To see this, let us consider $\mathfrak{A} = (\mathbb{N}, +, \cdot)$. Let us consider two versions of $\phi_{\mathfrak{A}}$, one, let us call it $\phi^1_{\mathfrak{A}}$, in the vocabulary $\{+_1, \cdot_1\}$ and the other, let us call it $\phi^2_{\mathfrak{A}}$, in the vocabulary $\{+_2, \cdot_2\}$. If CA denotes the axiomatization of second order logic in a vocabulary that includes both $\{+_1, \cdot_1\}$ and $\{+_2, \cdot_2\}$, then

$$CA \vdash (\phi^1_{\mathfrak{A}} \wedge \phi^2_{\mathfrak{A}}) \to \mathrm{Isom}_{1,2}, \tag{17.12}$$

where $\mathrm{Isom}_{1,2}$ denotes the statement of second order logic stating that there is a bijection f such that for all x, y: $f(x +_1 y) = f(x) +_2 f(y)$ and $f(x \cdot_1 y) = f(x) \cdot_2 f(y)$. So in this subtle sense (17.8) really asserts the truth of ϕ in one and only one model, namely the standard model (Feferman and Hellman 1995; Parsons 2008; Väänänen 2012).

There are good reasons to believe that the situation described above is not characteristic of arithmetic but applies equally to other structures that can be categorically characterized in second order logic. Curiously, (17.2) is an example where we do not know at the moment for sure what the right CA^* would be, although it is generally believed that CA will suffice.

Naturally, CA itself has non-standard models but they should not be the concern in connection with (17.8) because we are not studying CA but the structure $\mathfrak{A}$. In fact the whole concept of a model of CA is out of place here as CA is used as a medium of evidence for (17.3). We can convince ourselves of the correctness of the evidence by simply looking at the proof given in CA very carefully. There is no infinitistic element in this.

17.3 Set Theory

The approach of set theory to the foundations of mathematics is that mathematical propositions have the form

$$\Phi(a), \tag{17.13}$$

where $\Phi(x)$ is a first order formula with variables ranging over the universe of sets, and a is a set. If we compare (17.1) and (17.13), we observe that the former is restricted to one presumably rather limited structure $\mathfrak{A}$ while (17.13) refers to the entire universe. This is one often quoted difference between second order logic and set theory. Second order logic takes one structure at a time and asserts second order properties about that structure, while set theory tries to govern the whole universe at a time. This observation requires two qualifications.

First, while it is true that (17.13) refers to the entire universe, typical mathematical propositions are really statements about some V_α such that $a \in V_\alpha$. It requires some effort to find a mathematical theorem outside the realm of set theory which could not be written as a set theoretical fact about $V_{\omega+\omega}$. Borel Determinacy is one such (Friedman 1971). But what about V_{ω_1}? In fact, (17.1) can be easily written in set theory as a first order property of $V_{\alpha+1}$ as soon as $\mathfrak{A} \in V_\alpha$. So we can reformulate the set theoretical approach to mathematical propositions as follows: they are of the form

$$V_\alpha \models \Phi, \tag{17.14}$$

where Φ is a first order sentence and α is some large enough ordinal; in most areas of mathematics we can take $\alpha \leq \omega + \omega$. This demonstrates that it is not essential in (17.13) that the quantifiers range over the entire universe, and there is no essential difference to (17.1). It is a different matter if we study set theory itself. Then it is essential that (17.13) is not limited to any portion of the universe. Still, if we restrict (17.13) to formulas $\Phi(x)$ of quantifier-rank $\leq n$ for a fixed n then there is a closed unbounded class of ordinals α such that (17.13) is equivalent to $V_\alpha \models \Phi(a)$.

Let us now turn to the question when can we assert justifiably (17.13)? The complexity of (17.13) is of course beyond description, as (17.13) is undefinable in set theory. Even the stronger

$$\forall\alpha(V_\alpha \models \Phi) \tag{17.15}$$

is a Π_2-complete property of (the Gödel number of) Φ. Just as with (17.1) there is a different stronger formulation of (17.13):

$$ZFC \vdash \Phi(a), \tag{17.16}$$

where we assume that a is a definable term. As in our above discussion on truth and provability in second order logic we can view (17.16) as potentially surveyable evidence for (17.13) without drawing the conclusion that (17.16) is the meaning of (17.13). The most famous example of a difference between (17.13) and (17.16) is the Continuum Hypothesis *CH*. One of the intriguing problems of set theory is to find a *true* extension ZFC^* of *ZFC* which decides *CH*. For any extension ZFC^* of *ZFC* we have the analogue of (17.11):

$$ZFC \vdash \Phi \quad \Rightarrow \quad ZFC^* \vdash \Phi \quad \Rightarrow \quad \Phi. \tag{17.17}$$

Here Φ can be any mathematical statement. For both *CH* and $\neg CH$ we know that the leftmost implicant is false. The rightmost implicant holds for *CH* or $\neg CH$ but we do not know for which. Kreisel seems to be the first to pay attention to this curious situation.

Discussing implications like (17.11) and (17.17) does not mean that we are anywhere near finding CA^* or ZFC^*, or even that it is unproblematic to suggest that CA^* or ZFC^* can be found. The point is that the situation is entirely *similar* in second order logic and in set theory. In both cases we are equally far or equally close to a solution.

All the usual mathematical structures can be characterized up to isomorphism in set theory by appeal to their second order characterization but letting the second order variables range over sets that are subsets of the structure to be characterized.

The only difference to the approach of second order logic is that in set theory these structures are indeed explicitly defined while in second order logic they are merely described. In this respect second order logic is closer to the standard mathematical practice of not paying attention to what the "objects" e.g. complex numbers really are, as long as they obey the right rules. However, it is important also in second order logic to prove the *existence* of the structure to be characterized. After the existence has been proved, the object can be forgotten. In set theory the existence is proved by defining the object and showing that the definition is legitimate. When we move on to more counter-intuitive structures this difference disappears. Take for example the structure

$$(\mathcal{P}(\omega), <) \tag{17.18}$$

where $<$ is a well-order of the order-type of the cardinal $2^{\aleph_0}$. There is a second order sentence θ such that (17.18) is the only model of θ, up to isomorphism.[5] Neither second order logic nor set theory can define such a well-ordering.

If set theory is formalized with two $\in$-relations, say $\in_1$ and $\in_2$, and the ZFC axioms are adopted in the common vocabulary $\{\in_1, \in_2\}$, then the equation

$$F(x) = \{_2 F(y) : y \in_1 x\}_2 \tag{17.19}$$

[5]Take the sentence which says that $<$ is a well-order of the universe such that every initial segment is of smaller cardinality than the whole universe. Add a conjunct stating the existence of an element a and a binary relation E such that a is the least limit element of $<$ but not the least element of $<$, $\forall x \forall y(\forall z(zEy \leftrightarrow zEy) \rightarrow x = y)$ and $\forall X \exists y \forall x(N(x) \rightarrow (X(x) \leftrightarrow xEy))$.

defines a class function F which is an isomorphism between the $\in_1$-sets and the $\in_2$-sets.[6] In this sense set theory, like second order logic, has *internal categoricity*.

17.4 Foundations of Mathematics

Which is the right way to do mathematics: second order logic or set theory? Let us leave aside the question whether the higher ordinals that exist in set theory are really needed. The point is that set theory is just a "taller" version of second order logic, and if one does not need (or like) the tallness, then one can replace set theory by second (or higher) order logic. However, this does not yield more categoricity, for both second order logic and set theory are equally "internally categorical". If we look at second order logic and set theory from the outside we enter metamathematics. Then we can build formalizations of the semantics of either second order logic or set theory and prove their categoricity in "full" models as well as their non-categoricity in "Henkin" models.

But is there an "outside" position for the language of mathematics? If there is a framework where one can position oneself outside mathematics, what is that framework? If we are going to prove metamathematical results in that framework, the framework has to involve a lot of mathematics itself. It would seem natural to identify that supposedly outside framework with mathematics itself and construe the metamathematics in some other way. An alternative approach to metamathematics is that in our language of mathematics we formalize the various languages that we are interested in, including their semantics. Then we use our mathematics to prove illuminating (completeness, incompleteness, categoricity, non-categoricity, etc) results about those formal languages and about their semantics. These results tell something about our "real" language of mathematics to the extent that our formalizations reflect this "real" language. However, properties of this reflection can only be observed, never proved.

The "real" language is not a mathematical concept, even less its semantics. Only the formal languages and their semantics are mathematically defined and can be subjected to mathematical proofs.

There is no outside point of view in foundations of mathematics. Formalization does not take us outside but rather inside. Formalization does not give a more general view but a more restricted view. Therefore foundational conclusions made from mathematical results concerning formal systems have always an element of doubt (See also Kennedy (to appear).). Still formalization endowed with conceptual analysis is a correct way to deepen our understanding of foundations of mathematics, and has been highly successful in explaining why some mathematical questions have turned out so difficult to solve and why some cannot be solved with current methods.

---

[6]We use the symbols $\{_2$ and $\}_2$ to denote the usual set formation symbols $\{$ and $\}$ in the sense of the epsilon symbol $\in_2$.

References

Feferman, S., and G. Hellman. 1995. Predicative foundations of arithmetic. *Journal of Philosophical Logic* 24(1): 1–17.
Friedman, H.M. 1970/1971. Higher set theory and mathematical practice. *Annals of Mathematical Logic* 2(3): 325–357.
Kennedy, J. (to appear). On formalism freeness.
Kunen, K. 1970. Some applications of iterated ultrapowers in set theory. *Annals of Mathematical Logic* 1: 179–227.
Lévy, A. 1965. A hierarchy of formulas in set theory. *Memoirs of the American Mathematical Society* 57: 76.
Magidor, M. 1971. On the role of supercompact and extendible cardinals in logic. *Israel Journal of Mathematics* 10: 147–157.
Parsons, C. 2008. *Mathematical thought and its objects*. Cambridge: Cambridge University Press.
Väänänen, J. 2001. Second-order logic and foundations of mathematics. *Bulletin of Symbolic Logic* 7(4): 504–520.
Väänänen, J. 2012. Second order logic or set theory? *Bulletin of Symbolic Logic* 18(1): 91–121.

Index

P. Dybjer et al. (eds.), *Epistemology versus Ontology*, Logic, Epistemology, and the Unity of Science 27, DOI 10.1007/978-94-007-4435-6,